T0401724

The Art and Science of Surround and Stereo Recording

Edwin Pfanzagl-Cardone

The Art and Science of Surround and Stereo Recording

Including 3D Audio Techniques

Edwin Pfanzagl-Cardone
Acoustics Department
Salzburg Festival
Salzburg, Austria

ISBN 978-3-7091-4889-1 ISBN 978-3-7091-4891-4 (eBook)
https://doi.org/10.1007/978-3-7091-4891-4

This Springer imprint is published by the registered company Springer-Verlag GmbH, AT part of Springer Nature.
The registered company address is: Prinz-Eugen-Str. 8-10, 1040 Wien, Austria

'If one takes only a brief look at the matter,
one would not believe
that it will be possible to change anything ...'
from the book 'Opening the Eye of New Awareness,' Wisdom Publications,
1984 (Dalai Lama)

The author would like to dedicate this book to
his first teacher of sound-engineering
Prof. Ing. Hellmut Gottwald
(1938–2004)
as well as
his father, the mathematician
o. Univ. Prof. Dr. Dr. h.c. Johann Pfanzagl
(1928–2019)
and all other people who live their lives
in the constant search for insight

Preface

This book is written by a sound-engineer for sound-engineers and all people interested in stereophonic (two-channel) and multichannel sound recording and reproduction, as well as the new field of '3D' or 'immersive' audio and the basics of spatial hearing.

Sound-engineering is based on two realms: science and art. To obtain good sounding recordings, it is necessary—or at least helpful—to have a solid understanding of the laws of acoustics and psychoacoustics, but also to be willing to occasionally experiment and deviate from 'well-established' recording techniques, put their validity under discussion and—most important—let your ears have the final decision.

One big advantage of surround and 3D audio recordings in comparison with two-channel stereo recordings is their potential to be able to convey an enhanced sense of spaciousness. This is due to the fact that in a multichannel loudspeaker setup—e.g. 5.1 surround—the direct sound and diffuse sound (reverb) contained in a recording can reach the listener's ears from different directions (as is the case in the concert hall), while in a traditional stereo setup both of them are reproduced from the front.

Various research has shown that listener preference is strongly connected to the perceived spaciousness, as well as 'naturalness' of an audio recording; both factors are undoubtedly linked to the overall product quality and therefore also important in respect of a potential buying decision of such a recording.

In the past years, a significant number of papers has been published on the matter of surround and stereo microphone techniques, usually trying to evaluate and compare their quality through subjective listening tests or judge their 'technical correctness' (mostly in respect of localisation accuracy) via mathematical calculations. Unfortunately, sometimes the underlying models—based on the psychoacoustics of human hearing—for the sake of feasibility have had to be simplified to an extent which puts their validity under question.

There are hardly any studies which try to tackle both sides: Compare and rate several microphone techniques via subjective listening tests and also analyze on an acoustical/technical level how their physically measurable parameters are related to their sonic 'character':

In an Audio Engineering Society Convention Preprint of 2002 (Pfanzagl-Cardone 2002), I have published the first results on measurements of the most common stereo microphone techniques, making use of a—back then—new method, which analyzes signal correlation over frequency (FCC, i.e. 'frequency-dependent cross-correlation coefficient'). This somewhat unusual approach provides information about the stereophonic width (apparent source width or ASW), the depth and—mainly important—the spaciousness contained in a recording.

One of the—probably surprising—outcomes of these measurements is the fact that the majority of stereo microphone techniques does not produce a stereophonic output over the entire frequency range: With lowering frequency, many of them become more and more monophonic, which also significantly reduces their ability to convey the impression of spaciousness, which is related mainly to signal de-correlation below 500 Hz.

What may have been very helpful in the course of this research which has been conducted over almost two decades is the fact that in my position as head of sound at the Salzburg Festival I am not only obliged to do various kinds of opera, orchestral, ensemble and soloist recordings, but also a lot of sound reinforcement, which has helped to enrich my experience in a different direction: A microphone technique which may sound well over loudspeakers spaced apart by only 3–4 m (as is the case in a domestic environment) may not sound convincing at all, if the loudspeakers are at a distance to each other of 20 m (60 ft) and more, a situation commonly found in the world of theater. From these very 'hands-on' (or should one rather say: 'ears on') experiences, it became clear quite soon that only microphone techniques providing high signal de-correlation also at low frequencies are suited to provide a satisfactory degree of 'naturalness' and 'spaciousness' when replayed on a large sound system. An all too high degree of correlation at low frequencies may quickly lead to a 'boxy' sound, which is perceived as quite unnatural by most listeners.

As spaciousness is not necessarily one of the strongest points of two-channel stereo recordings, this aspect has never received as much attention, as might have been appropriate; for surround and 3D recordings on the other hand, it is a *must* to use appropriate recording techniques, able to capture a sound event in a manner to ensure optimum spatial reproduction (otherwise you miss out on one of the main advantages of surround and 3D over stereo recordings).

Unfortunately, many of the current surround microphone techniques are derived from stereo counterparts which do not provide enough signal de-correlation at low frequencies. In terms of localisation accuracy, their output may be convincing—or at least satisfactory—to the listener due to the high-frequency content involved, but on the low-frequency side they fail to capture the sound source in an adequate manner. This is true not only for many of the most well-known stereo microphone techniques—such as 'small AB', for example—but even for some of the most commonly used surround microphone systems (e.g. OCT surround—'Optimum Cardioid Triangle', KFM—Schoeps 'Kugelflächenmikrofon').

In my AES Convention Preprints from 2002 and 2008, I have proposed microphone techniques (AB-PC 'AB-Polycardioid Centerfill' and BPT 'Blumlein-Pfanzagl-Triple') for large (e.g. symphony orchestra), as well as small (e.g. soloist or chamber ensemble) sound-source recording. Both techniques are based on the concept of capturing the sound-source in a 'de-correlated' manner, and they function well in a surround as well as stereo context. The latter technique—which is now also being manufactured in the form of a compact 3-capsule microphone, the NEVATON BPT—is covered by patents in a few countries.

This book does not only give a timely overview of surround and stereo microphone techniques, but it also—for the first time—offers an answer to the question of why RCA's 'Living Stereo' series of legacy recordings is appreciated by music lovers worldwide despite their use of a seemingly 'wrong' recording technique in psychoacoustic terms.

In doing so, this book draws upon knowledge gained through the studying of the acoustics of concert halls, which—being ideal locations for musical performances—should remain the basic reference points on 'how recordings should sound,' at least according to my opinion (a point of view which is often discussed among 'Tonmeisters').

The five surround microphone techniques studied in much detail in this book have been analyzed through subjective listening tests with 50 participants, as well as through measuring signal correlation over frequency. The outcome of the subjective listening test is compared with the objective physical factors, which have been measured by means of the correlation function. The insight obtained through this research project is—in principle—of high interest also to the classical segment of the recording industry, as it manages to provide a clear indication on how to achieve more naturally sounding recordings.

In terms of practical recording, the application of the 'de-correlated microphone technique' principle has led me to create a new sound aesthetics, for which I have coined the term 'natural perspective' recording, which has its foundation in the sound which can be found in a 'best seat' location of a concert hall (with an appropriate direct/diffuse sound ratio) and not in a seeming reality that is mainly made by a mix of close-up spot microphones set up at arm's length from the various instruments, which—at least in my ears—will never sound convincing (rem.: obviously, the use of a minimum amount of appropriately placed spot-microphones in the mix will usually be necessary to arrive at a 'correct' direct/diffuse sound ratio).

As an example: If you try to capture a large orchestra with choir and vocal soloists (e.g. 'St. Luke's Passion' by Penderecky) with a plenitude of more than 90 microphones—as was put in practice by a colleague—you may be able to zoom in on the details of the various instrumental groups, but in the process of trying to tame the 'masses of sound' resulting from such an abundance of signals, one is likely to lose the 'overall picture' that should be created in terms of sound.

Along the way, a number of people have influenced or contributed in a direct or indirect way to the making of this book. I would like to thank the most important of these here (in alphabetical order): Victor Baranow (Nevaton), Tony Faulkner (Green Room Productions), Peter Freedman (RODE Mic), Prof. Ing. Hellmut Gottwald (MdW), Dr. David Griesinger (Lexicon), o. Univ. Prof. Mag. art. DI Dr.techn. Robert Höldrich (IEM @ KUG), Mag. Martin Klebahn (4Tune), Prof. DI(FH) Karl-Heinz Müller (Müller-BBM), Dr.rer.nat. DI Markus Noisternig (IRCAM—CNRS—Sorbonne University), Eberhard Sengpiel (UdK Berlin), Heinrich Schläfer, Dr. Alois Sontacchi (IEM @ KUG), Dr. Floyd Toole (Harman), Mag. Markus Waldner, Dr. Helmut Wittek (Schoeps), DI Jörg Wuttke (Schoeps) and the British company SoundField.

Another reason which has led me to the point where I am now (and, therefore, also to the writing of this book) is a willingness 'to go that extra mile' in order to experiment and maybe find out something new or the answer to some unresolved question.

Almost 20 years ago, in my first paper on the matter, which was based on previous research by David Griesinger, presented at a convention of the VDT (Verband Deutscher Tonmeister) in 2002, with the title 'Über die Wichtigkeit ausreichender Dekorrelation bei 5.1 Surround-Mikrofonsignalen zur Erzielung besserer Räumlichkeit,' I tried to point out the importance of sufficient signal de-correlation in 5.1 surround recordings in order to achieve a better spatial impression. At that time, I think there was hardly any awareness among colleagues on the importance of how not only direct sound, but also diffuse sound gets picked up by a specific microphone technique in use. Furthermore, certain authorities in the field actually reinforced an opposite opinion of a minimum signal coherence necessary for a ' … natural reproduction of space and envelopment …' (for more detail, see Chap. 2, Sect. 2.2). Fortunately, during the years to follow more papers were published by various authors which were in the same vain as mine, and about ten years later an important expert in the field concluded that ' … the diffuse sound field plays an enormously important role for spatial perception, as well as for sound colour. Therefore it must be ensured that it is reproduced with low correlation. Many good or bad properties of a stereophonic [microphone] arrangement do not have anything to do with their localization properties, but only with their ability to reproduce a nice, open-sounding spaciousness. XY cardioids are a good example: good localisation properties, but bad spatial reproduction. In the future, much more importance should be paid to diffuse field correlation. Very often only the effective recording angle is considered when choosing a specific arrangement. Diffuse field correlation is at least as important' (from Wittek H (2012) Mikrofontechniken für Atmoaufnahmen in 2.0 und 5.1 und deren Eigenschaften. In: Proceedings to the 27. Tonmeistertagung des VDT, Cologne, 2012, pg. 804).

At the same convention—as part of another paper presentation—the authors noted that ' … Diffuse sound (i.e. reverb or background noise) needs to be reproduced diffusely. This can be achieved using Auro-3D if appropriate signals are fed to the extra speakers. Diffuse signals must be sufficiently different on each speaker, that is, they need to be decorrelated over the entire frequency range.

A sufficient degree of independence is necessary, in particular, in the low-frequency range as it is the basis of envelopment perception. However, increasing the number of channels that need to be independent makes recording more complex. It is a tough job to generate decorrelated signals using first-order microphones—for example, a coincident array such as a double-MS array or a Soundfield microphone allows for generating a maximum of four channels providing a sufficient degree of independence. Therefore, the microphone array needs to be enlarged to ensure decorrelation. It is worth noting here that measuring diffuse-field correlation is not trivial. There are two reasons for this: First, measuring the correlation requires the diffuse sound level to be much higher than direct and reflection levels, so the distance from the source needs to be sufficiently long. Secondly, considering the degree of correlation is not sufficient; this does not account for the fact that low-frequency (de)correlation is particularly important …' (from Theile G, Wittek H (2012) 3D Audio Natural Recording. In: Proceedings to the 27. Tonmeistertagung des VDT, Cologne, Nov 2012, pp. 731).

Fortunately, in the meantime also measurement technology has 'caught up' and is providing the necessary means for Tonmeisters to have an informed, frequency-dependent look on the cross-correlation of the signals involved in their mix: While for the measurements documented in this book I had to use a self-programmed non-real-time MATLAB computation of previously selected appropriate sections of sound to arrive at the display of the FCC (or FIACC—in the case of measuring the frequency-dependent inter-aural cross-correlation coefficient), since 2019 a marvelous real-time VST plugin is available by company MAAT digital, the '2BC multiCORR' correlation meter (see appendix for more details).

During the last 10–15 years, several papers have been published which are analyzing, mostly on a theoretical level, the inter-channel correlation of 2-channel stereo microphone techniques; their results coincide very well with my findings from almost 20 years ago. However, it seems there is still only scarce material published on the inter-channel correlation of surround microphone systems. I believe to be the first one to have undertaken research in this particular area of interest, which is also very relevant when it comes to understanding (and practically working on) 3D audio.

As part of the research I am presenting in this book—in order to make the FCC (frequency-dependent cross-correlation coefficient) more meaningful—an artificial human head has been introduced (Neumann KU81i, as well as a software plugin, using the HRTFs of the KEMAR dummy head) to serve as a 'human reference' when measuring the FIACC.

By measuring the signal cross-correlation over frequency between channels for the various microphone techniques (or the binaural equivalent FIACC of the final result of replaying the recordings via such microphone arrays over a standard 2-channel stereo or 5.1 surround loudspeaker setup), it has been shown that every stereo and surround microphone technique is characterized by an individual 'sonic fingerprint.'

The specific pattern of signal correlation/de-correlation in the various frequency ranges is quite meaningful and can serve as a first indicator for the sonic properties of the respective microphone technique under test.

As a consequence of my findings, I have also proposed the BQI_{rep} ('Binaural Quality Index for reproduced music,' derived from the BQI as defined by Keet in 1968), which is intended to serve as a measure for the (spatial) quality of reproduced music.

Especially with 3D audio—for which the number of channels interacting with each other has been augmented significantly in relation to 2D surround sound—it does make very much sense to make use of the abovementioned 'human reference' (binaural dummy head recording) in the form of measuring the FIACC to arrive at a qualitative evaluation of a respective microphone technique: no matter how many reproduction loudspeaker channels are involved as part of a 3D replay setup—finally, the sound vibrations have to enter the two ear canals of the human head, so the resulting FIACC, which is effective between the eardrums, is what counts in the end.

Salzburg, Austria
2020

Edwin Pfanzagl-Cardone

Contents

About the Author

Edwin Pfanzagl-Cardone is Head of sound at the Acoustics Department of the 'Salzburg Festival' of classical music in Austria. After completing his degree in electronics engineering and information technology at TGM, Vienna, in 1988, he graduated from the 'University of Music and Performing Arts' in Vienna in 1991 and received a Tonmeister (Sound Master) degree. In 1994–1999, he was Lecturer on the theory of sound-engineering at the 'Institute of Electro-Acoustics' at the same university. In 2000, he completed his M.A. in audio production at the University of Westminster, London, and in 2011 received his Ph.D. in musical acoustics and psychoacoustics from KUG—University of Music and Performing Arts, Graz, Austria. Since the early 1990s, he has been working as Sound-engineer for music recording and live sound reinforcement, and for film and TV, mainly in Europe, but also in Japan and the USA. As an arranger and composer, he has released recordings with BMG and Sony in the field of pop music and has provided content for international library-music labels, and for radio and TV commercials. The author of AES and VDT Convention Preprints, he has published more than 60 articles in magazines for sound-engineers, such as Pro Sound News Europe, Studio Sound, Media Biz and Prospect. He has been Member of the AES (Audio Engineering Society) since 1991 and Member of the 'Austrian Sound-Engineer's and Music Producer's Association' ÖTMV. In the area of classical music, he has worked with well-known conductors such as Abbado, Barenboim, Boulez, Dudamel, Gatti, Gergiev and Harnoncourt. He also worked with Jansons, Maazel, Mehta, Metzmacher, Minkowski, Muti, Sir Simon Rattle, Orozco-Estrada, Salonen, Savall and many others. He recorded the Vienna and Berlin Philharmonic Orchestra, The Mozarteum Orchestra Salzburg and a large number of other European and foreign orchestras. For the Salzburg Summer Festival 2018, he was responsible for the 3D audio sound effects for Mozart's 'Magic Flute' conducted by Constantinos Carydis. Since March 2010, he has been teaching Sound Reinforcement Technology at the Faculty of Design, Media and Arts at the University of Applied Sciences in Salzburg. As a composer, he has released four international CDs. In addition to several hundred archival recordings for the Salzburg Festival, his discography as a sound-engineer consists of about thirty CDs and three LPs with music labels such as Deutsche Grammophon and Orfeo. He is the inventor of three microphone techniques: the AB-Polycardioid Centerfill (AB-PC), the ORTF-Triple (ORTF-T) and the Blumlein-Pfanzagl-Triple (BPT), and holds a patent in surround microphone technology.

Abbreviations

AB-PC	AB-Polycardioid Centerfill
AES	Audio Engineering Society
AITD	Average Interaural Time Delay
ALT	Acoustic Lens Technology
ANOVA	ANalysis Of VAriance
ASI	Acoustic Spatial Impression
ASW	Apparent Source Width
BACCH™	Band-Assembled Crosstalk Cancellation Hierarchy
BPT	Blumlein-Pfanzagl-Triple
BQI	Binaural Quality Index
BQI_{rep}	Binaural Quality Index of reproduced music
C	Clarity
CD	Critical Distance (also: d_{crit})
CD	Compact Disk
CHAB	Center Hemicardioid AB
D	Deutlichkeit
DFC	Diffuse-Field Correlation
DFI	Diffuse-Field Image
DI	Directivity Index
DHAB	Delayed Hemicardioid AB
DMS	Double-MS
DPE	Direct Point Evaluation
DTF	Diffuse-Field Transfer Function
EBU	European Broadcasting Union
EQ	Equalizer
ER	Early Reflection
ERB	Equivalent Rectangular Critical Band
ERD	Equivalent Rectangular Duration
ESMA	Equal Segment Mic Array
f_{crit}	Critical Frequency
f_{opm}	Out-of-Phase Maximum frequency

FCC	Frequency-Dependent Cross-correlation Coefficient
FEC	Free-Air Equivalent Coupling
FIACC	Frequency-dependent Inter-Aural Cross-correlation Coefficient
FOA	First-Order Ambisonics
FSCC	Frequency-dependent Spatial Cross-Correlation (same as FCC)
G	Strength Factor
HOA	Higher-Order Ambisonics
HRTF	Head-Related Transfer Function
IACC	Inter Aural Cross-Correlation
ICCC	Interaural Cross-Correlation Coefficient
ILD	Interaural Level Difference
INA	Ideale Nieren-Anordnung
INDISCAL	Individual Difference Scaling
IPD	Interaural Phase Difference
IR	Impulse Response
IRT	Institut für Rundfunktechnik
ITD	Interaural Time Difference
ITU	International Telecommunications Union
K	Wave Vector
KFM	Kugelflächenmikrofon
LEDT	Lateral Early Decay Time
LEV	Listener EnVelopment
LF	Lateral Fraction
LFC	Lateral Fraction Coefficient
LG	Lateral Gain
LR	Late Reflection
MEDUSA	Multichannel Enhancement of Domestic User Stereo Applications
MS	Mid-Side (microphone technique)
MUSHRA	MUltiple Stimuli with Hidden Reference and Anchor
NOS	Nederlandse Omroep Stichting
OCT	Optimum Cardioid Triangle
ORTF	Office de Radiodiffusion Télévision Française
OSIS	Optimal Sound Image Space
OSS	Optimal Stereo Signal
PDR	Pressure Division Ratio
PICC	Perceptual Interaural Cross-correlation Coefficient
Q	'Quality' (i.e. for loudspeakers, microphones: directivity factor)
RCA	Radio Corporation of America
RT_{60}	Reverb Time
SACD	Super Audio Compact Disk
SC	Spectral Centroid
SHAB	Side Hemicardioid AB
SMPTE	Society of Motion Picture and Television Engineers
SOA	Second-Order Ambisonics
SPL	Sound Pressure Level

SSF	Surround Sound Forum
THD	Total Harmonic Distortion
TUT	Technique Under Test
VR	Virtual Reality

Chapter 1
Spatial Hearing

Abstract The fundamentals of spatial perception in human hearing are outlined. We are taking a look at this phenomenon mainly from the perspectives of physical acoustics and psychoacoustics as these two disciplines are majorly relevant in respect to the research and fields of interest which will be presented in this book. For the sake of compactness, we are refraining from giving an outline of the historic development of research into human hearing, but are trying to give a short summary of the current state of knowledge: HRTFs, cone of confusion, ILDs and ITDs, ERB, clarity, 'Deutlichkeit', various acoustic measures related to 'spatial impression'—to name a few relevant topics. Special focus is put on frequency-dependent aspects of human hearing concerning localization in the horizontal and vertical plane, distance perception and spatial impression both for 'live hearing,' as well as in relation to loudspeaker production. In doing so, this chapter draws on findings by many specialists in the field of psychoacoustics, as well as scientific research in the realm of audio engineering, as well as concert hall acoustics.

Keywords Spatial hearing · 3D sound · Immersive audio · Spaciousness · HRTF · Localization perception

Sound impinging on the human head (Fig. 1.1) is altered mainly by the 'pinnae' (outer ear) as well as the shoulders, chest and—of course—the head, before it reaches the entrance to the ear canal (cavum conchae). After that, the frequency response (and other parameters) of sound is affected by the ear canal (meatus), which leads to the middle ear. The middle ear consists of the tympanic membrane, hammer, anvil and stirrup. The mechanical energy which arrives to the middle ear by means of the sound wave is submitted to the inner ear via the 'oval window' and leads to a change of pressure of the liquid inside the inner ear (cochlea). This pressure leads to frequency-dependent excitation patterns on the Basilar membrane, which causes the hair cells to respond and fire nerve impulses, which in turn triggers electrical action potentials in the neurons of the auditory system. These nerve impulses are dealt with

---

Content of this chapter is based on—and partly cites—the following papers: Merimaa and Pulkki (2005), with some additions from Rumsey (2001) and Toole (2008), as well as translated text-citations from Fellner and Höldrich (1998a, b).

E. Pfanzagl-Cardone, *The Art and Science of Surround and Stereo Recording*,
https://doi.org/10.1007/978-3-7091-4891-4_1

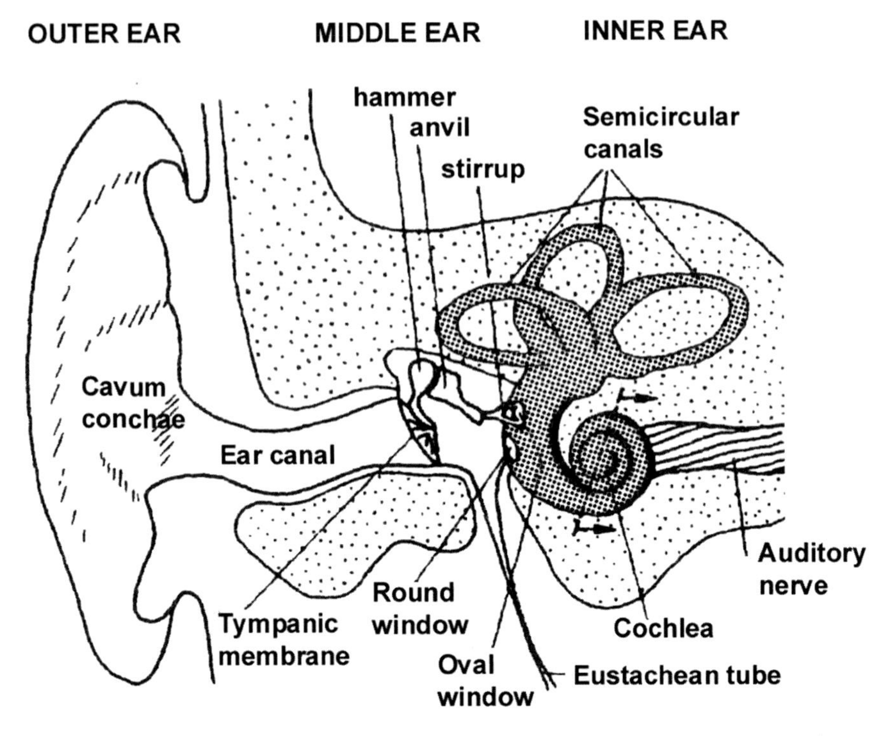

Fig. 1.1 Anatomy of the human inner, middle and outer ear (after Zollner and Zwicker 2003)

and combined with the information which arrives from the other ear on a higher level of the brain (Fig. 1.2).

Perception of a sound source in space is determined by the following acoustical parameters:

localization in the horizontal (azimuth) and vertical plane (elevation), perception of distance, perception of spatial impression (diffuse sound/impulse response of the room).

The following sections try to give a short explanation of the underlying psychoacoustic principles related to these parameters.

1.1 Mechanisms of Localization

The localization of sound sources is mainly based on the following five frequency-dependent information contained in a 'live' sound event (meaning a sound event, which happens in real space, as opposed to an electronically created sound):

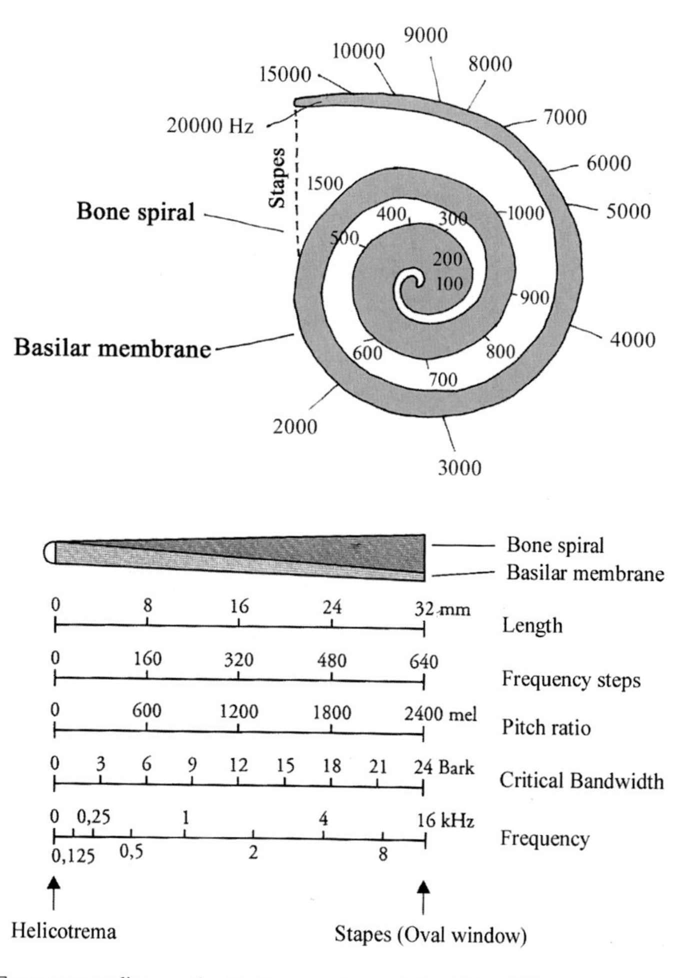

Fig. 1.2 Frequency coding on the Basilar membrane (after Eska 1997, Fig. 34, pg. 123) top: Areas which are most sensitive to indicated frequencies (after K.-H. Plattig); bottom: Schematic of stretched Cochlea (from Zwicker and Fastl 1990)

1. Interaural Time Difference—ITD and
2. Interaural Level Difference—ILD are the two most important information, based on which human hearing decides in which 'cone of confusion' a sound source is situated (Figs. 1.3 and 1.4).

Lord Rayleigh was among the first to understand that—based on his experiments concerning human hearing toward the end of the nineteenth century—sound with a wavelength smaller than the diameter of the human head will effectively be shaded off for the ear on the other side, which results in an Interaural Level Difference (ILD) between the two ears. In addition to that, sound takes a different amount of time to arrive at both ears, which results in Interaural Time Differences (ITDs) (Fig. 1.3).

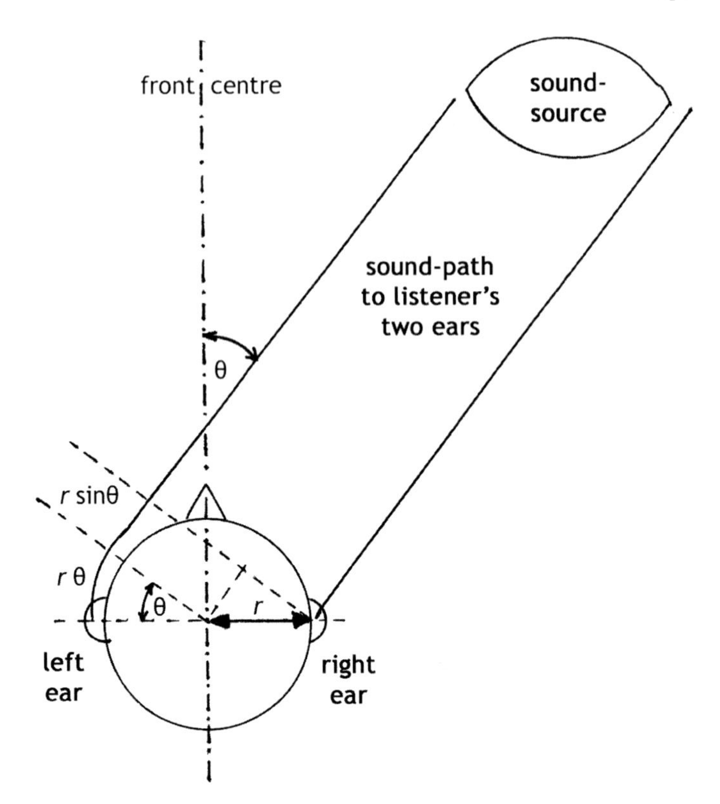

Fig. 1.3 Effective ITD for an individual listener depends on the angle of incidence, as this determines the additional pathway, which the sound needs to travel in order to arrive at the other ear. In this model, ITD can be calculated as follows: $\Delta t = r(\theta + \sin\theta)/c$ (with speed of sound $c = 340$ m/s and θ in radians); (after Rumsey 2001, Fig. 2.1, pg. 22)

The maximum ITD of natural sound sources is reached for sound impinging from the side of the head at 90°. This amounts to 650 μs (=0.65 ms) for a mean ear spacing of 21 cm (see (Rumsey 2001), p. 22 and (Blauert 1997), p. 143). Other research (Yanagawa et al. 1976; Suzuki and Tohyama 1981; Tohyama and Suzuki 1989) showed that the acoustically effective distance between the ears is—in reality—much larger with about 33 cm (see (Blauert 1997), p. 146–149). Based on the speed-of-sound with 340 m/s, this would result in an ITD of 0.96 ms.

Lord Rayleigh was able to show that ITDs are especially important at low frequencies, at which no ILDs at relevant levels occur (Rayleigh 1907). His conclusion was that at low frequencies localization is based on ITDs, while at high frequencies it is determined by ILDs. The crossover frequency range between these two psychoacoustic mechanisms lies approximately at 1.5 kHz.

If one does not take into account the filter effect caused by the pinnae, for an abstract sphere-shaped head, symmetrical lateral sound-source positions slightly in front (a) or behind (b) the dummy head in Fig. 1.6 will have the same ITDs and

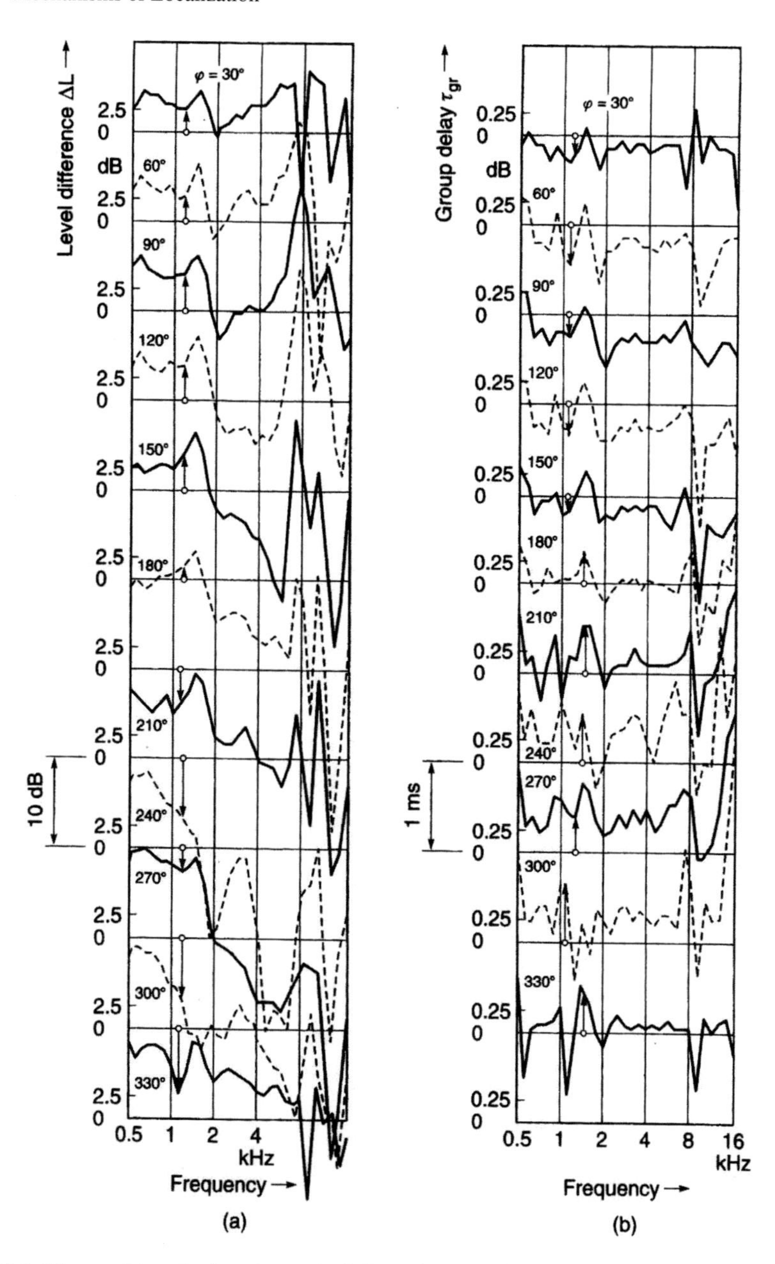

Fig. 1.4 Monaural transfer function at the left ear for several directions in the horizontal plane, (front: $\varphi = 0°$, acoustically damped room. Loudspeaker-distance 2 m (6–7 ft), 25 listeners, complex mean**a** level difference,**b** time difference [Fig. 2.2 from (Rumsey 2001), pg. 24, after (Blauert 1997)]

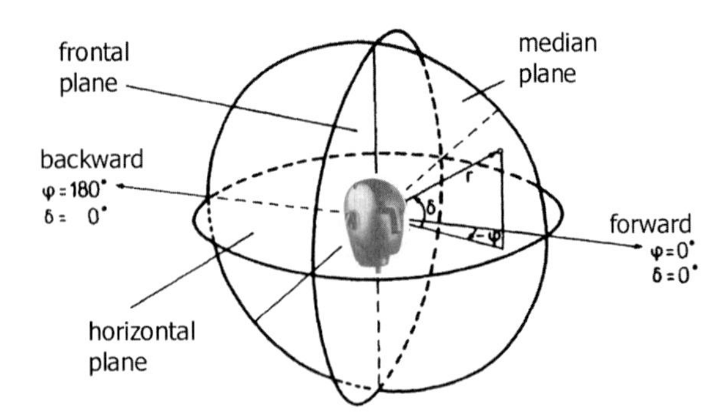

Fig. 1.5 Head-related coordinate system, as being used for listening experiments; definition of median, frontal and horizontal plane; r is the distance to the sound-source, φ is the azimuth, and δ is elevation (after Blauert 1997)

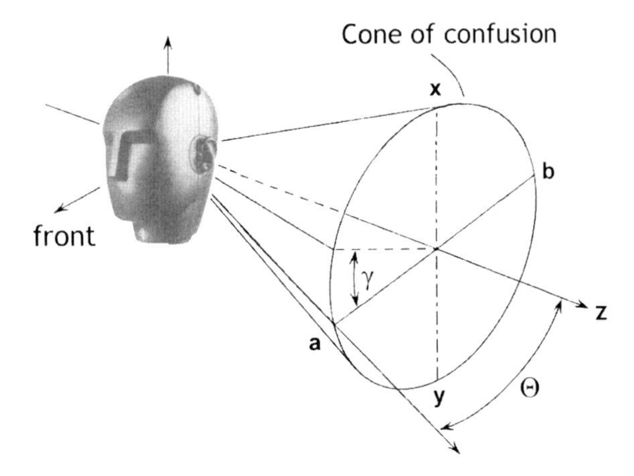

Fig. 1.6 'Cone of confusion': a mere lateral localization of a sound source needs only determination of angle θ, while for a precise localization also γ needs to be detected by the human hearing mechanism [graphic after (Hall 1980), pg. 342, German edition; after (von Hornbostel and Wertheimer 1920)]

ILDs. In analogy to this front-back ambiguity, there exists also a so-called elevation-ambiguity for sound source positions x (up) and y (down).

With natural hearing (as opposed to listening via headphones) two additional, very helpful mechanisms come into play:

3. The frequency response of monaural signals ('pinnae frequency response') usually helps to achieve higher localization accuracy within the 'cone of confusion' (see Figs. 1.4, 1.5 and 1.6)

Already Mach was convinced that the pinnae had a certain 'directivity' and therefore should be important also for localization. In the 1960s, D.W. Batteau published his research on the effects of the pinnae-reflections in the time domain (see Batteau 1967). In his view, the pinnae were serving as directional reflectors, which generated characteristic echo patterns depending on the angle of incidence (azimuth, elevation) of the sound. In his measurements, he was able to detect ITDs in the range from 10 to 300 μs. According to Batteau, it would then be the task of the inner ear to analyze the echo patterns and derive the corresponding angle of incidence. However, in order for this mechanism to function, the human ear would need to have a much higher accuracy in the time domain, than it actually has.

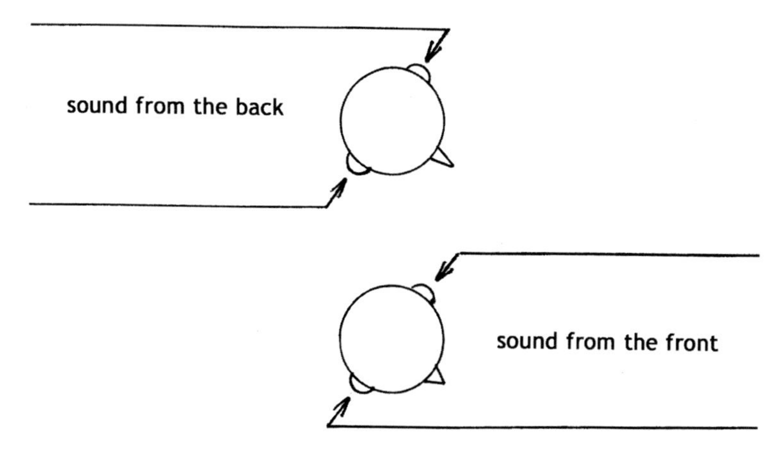

Fig. 1.7 Change of ITDs—caused by a head turn—is of opposite sense for sound from the front or from the back (after Blauert 1997, p. 180)

Hebrank and Wright (1974) found that Batteau's pinnae-reflections result in corresponding spectral 'colorations', which indeed can be decoded by the brain. Later examinations by various researchers came to the conclusion that the resulting spectral filtering due to the pinnae is especially important for localization in the median plane as well as for front-back distinction.

4. In addition small head turns and the changes in the ITDs, which go along with them, help humans to detect the position of a sound source if in doubt (Blauert 1997) (Fig. 1.7).
5. Furthermore, humans are very sensitive to inter-aural signal coherence [see (Boehnke et al 2002) and the references cited therein, as well as (Tohyama and Suzuki 1989)], which has been acknowledged as an important factor for human listening and sound source localization in reverberant rooms and acoustic environments with several simultaneous sound sources (see Faller and Merimaa 2004).

All of the abovementioned features are of rather individual nature and depend strongly on the size and shape of the head, the pinnae, as well as chest and shoulders. These inter-individual differences can be analyzed best by measuring a listener's HRTF ('head-related transfer function').

The resolution of human hearing in the time and frequency domain has been researched extensively for monaural hearing (see, e.g. Moore 1997). For binaural listening, the frequency resolution seems to be quite similar to that of monaural (Moore 1997; van der Hejden and Trahiotis 1998), even though for some of the test signals higher bandwidths have been found (Kollmeier and Holube 1992; Holube et al. 1998). Therefore, it seems valid to use the 'equivalent rectangular critical band' (ERB) resolution, which has been found accurate for monaural listening, also for hearing in general (Glasberg and Moore 1990) (see Fig. 1.8). On this occasion, it should also be mentioned that not all frequencies seem to have the same importance

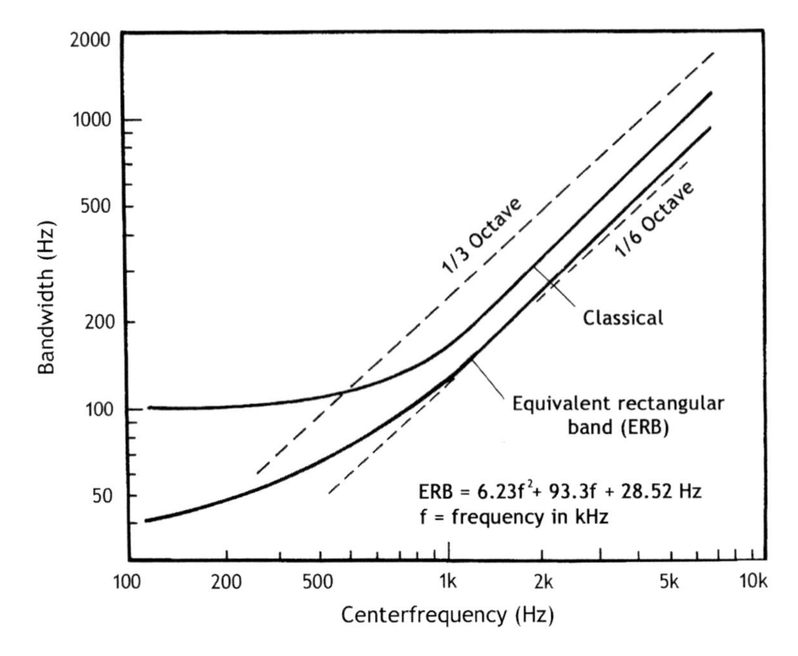

Fig. 1.8 ERB—'equivalent rectangular critical band': shown is a comparison of the bandwidths of 1/3rd and 1/6th octave band, the 'critical bandwidth' of human hearing, as well as the 'equivalent rectangular critical band' (ERB), which has been calculated according to the formula shown in the graphic (after Everest 1994)

in terms of contributing to localization: The 'Weighted-Image Lateralization Model' by Stern et al. (1988) shows a dominant region around 600 Hz, and Wightman and Kistler (1992) have managed to prove the high importance of ITD's of the LF band it in this respect.

To determine the temporal resolution of human hearing is more difficult: It was possible to prove that human listeners are capable of detecting spatial movement of sound sources, which cause sinusoidal fluctuations of ITDs an ILDs, but only up to a fluctuation frequency of about 2.4–3.1 Hz (see Blauert 1972). Despite this fact, Grantham and Wightman (1978) were able to show that their test listeners were able to discern ITD fluctuations up to 500 Hz, but primarily based on the resulting broadening of the sound source's ASW (apparent source width). Several studies on this subject came to the conclusion that binaural perception apparently functions similar to signal processing with a double-sided exponential, rounded exponential or Gaussian time window (see Holube et al. 1998; Grantham and Wightman 1979; Kollmeier and Gilkey 1990; Culling and Summersfield 1999; Akeroyd and Summersfield 1999, as well as Breebart et al. 2002). The window lengths which have been found in these studies are quite varied, but the mean 'Equivalent Rectangular Duration' (ERD), i.e. the mean integration time (or time constant τ) of human hearing seems to lie in the order of 100 ms. However, also much shorter time constants have been found by some researchers (see (Akeroyd and Bernstein 2001) with an ERD of 10 ms, as well as (Bernstein et al. 2001) with an ERD of 13.8 ms). A possible explanation has been given in that these varied time constants may correspond to different parts of the binaural 'hearing mechanism' (Kollmeier and Gilkey 1990; Akeroyd and Bernstein 2001; Bernstein et al. 2001).

The so-called Precedence effect (or 'Haas effect', a.k.a. 'Law of the first wavefront'; see Haas (1951) which also plays a very important role in connection with localization. Of two closely following sound events arriving at the listener from different directions, the first one will determine the apparent localization direction, as long as both sound events are separated only by a few milliseconds (Blauert 1997; Litovsky et al. 1999). Upon replay of a mono test signal via headphones, it seems to shift or 'wander' from the center ('in head localization') to the ear at which the sound arrives first, for ITDs from 0 to 0.6 ms. For the time range of Δt equal 0.6–35 ms, the sound is being localized at the 'advanced' ear; for Δt larger than 35 ms, two distinct sound events will be perceived in the form of an echo.

1.1.1 HRTF Phase-Characteristics

Time-of-arrival differences Δt can also be looked at as phase differences if pure sinusoidal signals are concerned. It is a—quite common—misbelieve (among sound-engineers, but not only) that we (humans) cannot detect phase differences, unless we have the case of 'reversed polarity' of one loudspeaker in a (two channel) stereo replay system, in which the cable was attached 'out of phase' at one of the loudspeaker terminals. Quite to the contrary, there is evidence that the human ear is capable of detecting phase differences at least at low frequencies up to about 700 Hz. Around this frequency, the head diameter becomes equal to half the wavelength of sound and hence the 'phase detection' mechanism of human hearing starts to falter, as a clear detection at which ear the sound arrives first is no longer possible.

In addition, phase differences between signals at both ears can also be misleading if they are caused by room modes (which is more likely when listening in small rooms than in large rooms) or reflections.

Figure 1.9 shows the interaural phase difference IPD for a listener of a sound event impinging from 0° to 150° in steps of 30° in the horizontal plane. The thick lines show the corresponding interaural time delays Δt at 0.25, 0.5 and 1.0 ms.

As mentioned above, localization based on ITDs works very well up to approximately 1.5 kHz.

Above this frequency, the phase relationship starts to be ambiguous to our perception, but the human hearing mechanism is nevertheless capable of analyzing the ITDs of the amplitude envelope of a sound event if it is appropriately structured (i.e. a non-continuous sound).

Apart from the outer ear, which has the largest spectral influence, also head, shoulders and torso interact with sound and cause slight sound colorations. In Fig. 1.10, the direction-dependent and direction-independent contributors to sound coloration

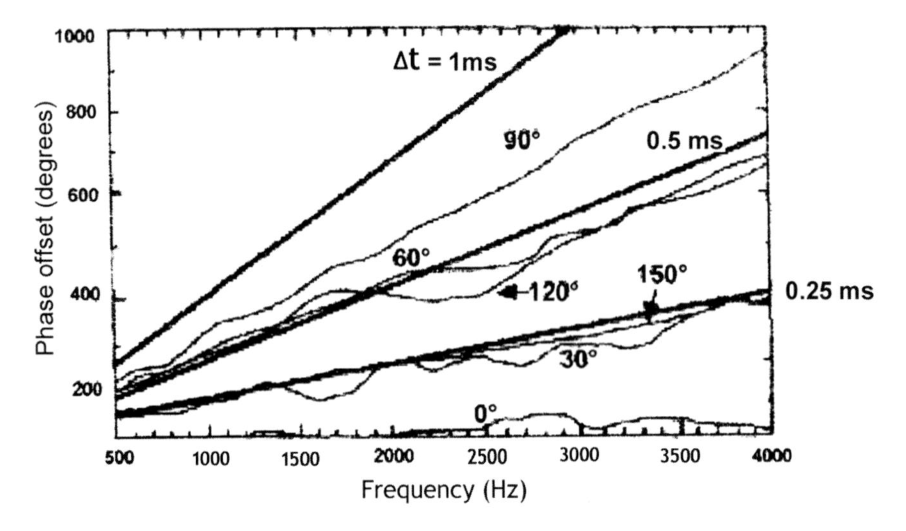

Fig. 1.9 Continuous interaural phase difference IPD (after Begault 1994)

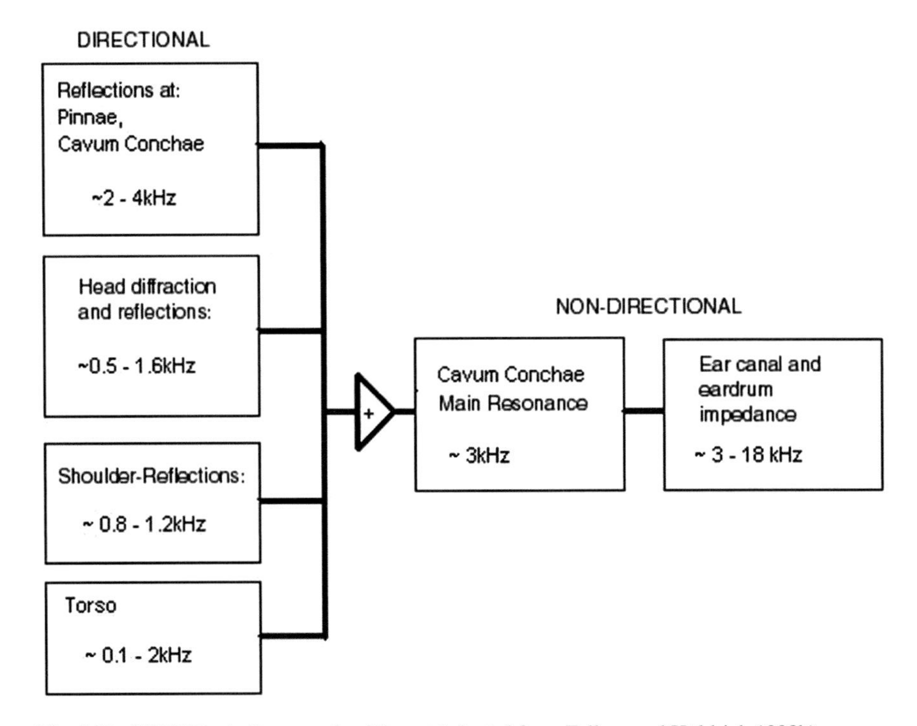

Fig. 1.10 HRTF block diagram after Blauert (adapted from Fellner and Höldrich 1998b)

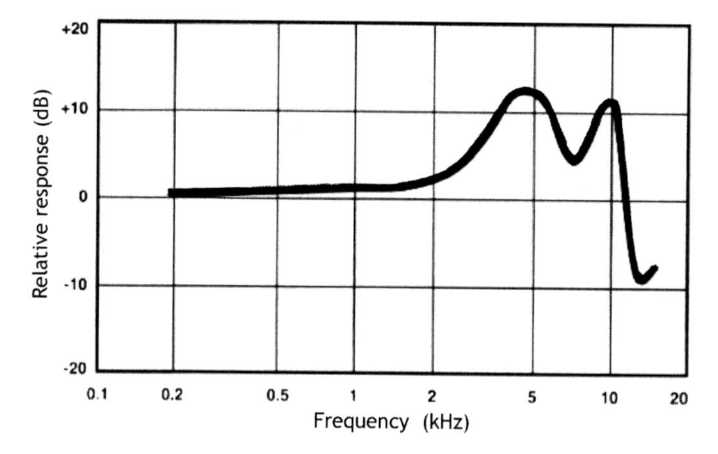

Fig. 1.11 Typical transfer function of the auditory canal—which can be considered to act as a resonant pipe. The transfer function of the ear canal is a fixed entity that is superimposed on each of the highly variable transfer functions of the outer ear. This affects the character of what we hear and the directional cues as well (adapted from Streicher and Everest 2006, after Mehrgardt and Mellert 1977)

within the human hearing system are pictured according to their rank order (top to bottom).

The 'Cavum Conchae' is the central cavity and the biggest area of resonance of the pinna at the entrance to the ear canal. The outer ear canal leads from the cavum conchae to the tympani and is about 2.5 cm long with a diameter of 7–8 mm. Due to these physical dimensions, the ear canal is characterized by a strong acoustic resonance usually around 4 kHz (Fig. 1.11).

According to the current state of science, it is assumed that it is actually the ear canal resonance which is responsible for 'externalization' of sound perception (as opposed to 'in-head-localization'; see Rumsey 2001). Due to this reason, efforts have been made to excite the ear canal from various directions (when sound is replayed via headphones) in order to overlay this direction-specific EQ, which is different and characteristic for each individual listener, before sound arrives to the tympanum, in an attempt to achieve better 'externalization' (Tan and Gan 2000).

The spectral influence of the shoulders is in the order of ±5 dB, while for the torso it is only about ±3 dB.

1.1.2 Localization and HRTFs

One of the most important results of Blauert's research in the 1970s was the discovery of the 'directional bands', responsible for localization in human hearing (Blauert 1997).

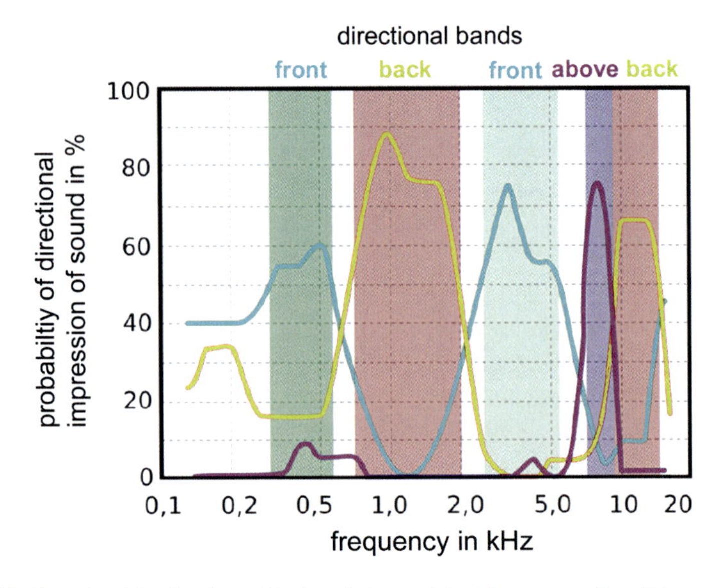

Fig. 1.12 Directional bands: pictured is the relative statistical frequency of test listeners, who are giving one of the answers 'back', 'above', 'front' more often than the other two taken together. Bands drawn at a 90% confidence level; colored background areas indicate most probable answers (after Blauert 1974)

As we have already seen in Fig. 1.4 with the individual HRTF curves, there are characteristic peaks and dips in the frequency response, dependent on the sound source position relative to the human head. This is effectively a kind of 'frequency coding' of sound source positions, which the human brain seems to learn at a young age. Typically, sound perception for sources behind the listener is characterized by less amplitude at high frequencies, which is due to the shadowing effect of the pinna.

As can be seen in Fig. 1.12 ,the frequency range around 8 kHz is mainly responsible for 'above' localization, while 'front' localization is being determined by signal content in the frequency bands from 300 to 600 Hz, as well as 3000 to 6000 Hz. 'Back' localization is related to the frequency bands around 1200 Hz as well as 12 kHz. Similar results were found also by Hebrank and Wright (1974), as well as Begault (1994). It is interesting to note that—looking at Blauert's findings—the relevant frequency bands seem to be at a distance of one decade from each other for both front-, as well as rear localization mechanisms.

1.2 Mechanisms of Distance Perception

The following explanations are excerpts taken from Nielsen (1993), with additions from Fellner and Höldrich (1998a), as well as Rumsey (2001).

1.2.1 Sound Intensity

For humans, the most important 'anchor' for distance perception is the sound intensity, which is transformed into the individual perception of 'loudness' by our hearing mechanism. Loudness is the individually perceived amplitude or level of a sound event, which is also frequency dependent, according to the sensitivity of the human ear (see the Flechter-Munson curves) (Fig. 1.13).

Under free-field conditions sound pressure suffers a reduction in level by 6dB_{SPL}, when doubling the distance to the sound source. A 'halving' of the loudness perceived by the listener—which is marked in 'Sone'—corresponds to a loss in sound pressure level by 10 dB. (As human hearing is 'nonlinear' in its behavior in many respects, the exact amount of attenuation (in dB) neccesary to achieve the impression of 'half the loudness' is largely signal dependent, but the SPL level drop by 10 dB is usually true at least for the frequency range from 400 Hz to 5 kHz and sound pressure levels in-between 40 dB and approx. 100 dB). In this context, it needs to be added that usually the estimation of distance of a sound source in a natural environment almost always happens in connection with a corresponding visual stimulus, which is also why it is easier to judge the distance of a sound-source in an environment that is well-known. Whether the perceived loudness is a meaningful indicator of distance for the listener, or not, depends largely on the question whether he or she knows the 'normal' loudness of the sound source. (see Nielsen 1993).

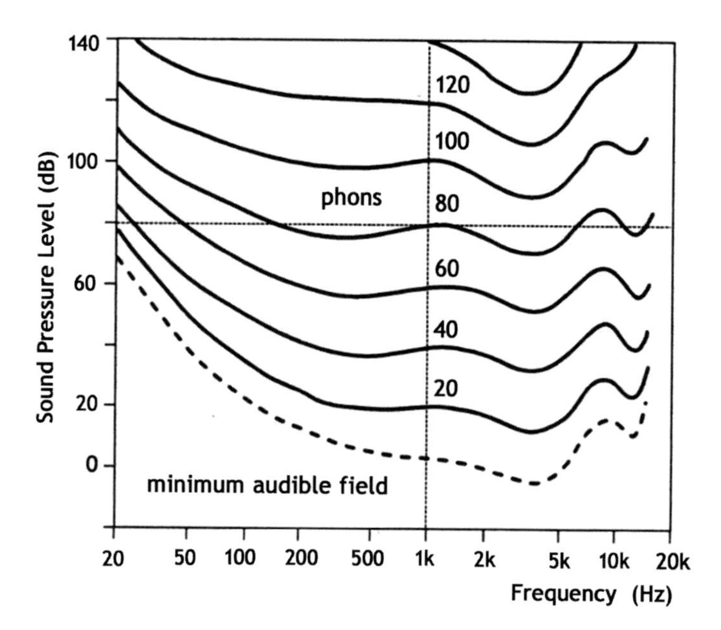

Fig. 1.13 Equal loudness contours of hearing, sound pressure level in dB re 20 $\mu\text{N/m}^2$ (0dB_{SPL} is about the threshold of hearing) versus frequency in Hz, from ISO 226 (after Fletcher–Munson). These curves show that at no level is the sensation of loudness flat with frequency; more energy is required in the bass range to sound equally loud as the mid-range

1.2.2 Diffuse Sound

Due to the energy contained in diffuse sound, the level drop when doubling the distance to the sound source in an average room is not 6 dB, as would be the case in a completely damped room or under free-field conditions, but less. The direct-/diffuse-sound ration seems to be an even stronger indicator for distance perception than the mere sound pressure level at the listener position. According to a hypothesis by Peter Craven, the perceived distance of a sound source is determined by the relative time delay and the relative amplitude of the early reflections in relation to the direct sound signal (see Gerzon 1992).

If the acoustic properties of the listening room are not known to the listener (as can be the case under test conditions), apparently the human hearing mechanism is capable of extracting the acoustic information about the room from what is being heard (Nielsen 1993). In their experiments, Mershon and Bowers (1979) found that it made no difference whether their test listeners were already familiar with the acoustic conditions of the listening room before the test, or not. However, the diffuse sound triggered off by a sound stimulus seems to be an important source of information for the human listener: in a sound-deadened room, test listeners did not manage to establish a relationship between the perceived and the real physical distance to the sound source (Nielsen 1993).

1.2.3 Frequency Response

As the HF part of a sound signal is more heavily damped when propagating through air than LF signals (air absorption effect), for the human listener also the frequency spectrum of an acoustic sound event contains distance information. A prerogative for this is that the listener is familiar with the regular spectral content of the sound event. In a room with diffuse sound components (reverb) the spectral balance of the reverb changes due to the frequency-dependent absorptive behavior of the surrounding surfaces, as well as due to the air absorption effect. The absorption coefficient of air depends on temperature, as well as relative humidity, which can lead to relevant changes in sound especially with live outdoor sound reinforcement systems. Also, local climate factors (wind, etc.) can lead to a change of the frequency response or spectral balance.

1.2.4 Binaural Differences

In connection with sound sources that are close to the listener, there can be relevant changes of the HRTFs of the listener. Especially, the ILDs can become much larger for low and high frequencies (see Huopaniemi 1999).

1.3 Spatial Impression

The resulting 'spatial impression' or 'spaciousness' for a human listener is primarily based on two components: the so-called apparent source width (ASW), i.e. the perceived base width of a sound event, as well as 'listener envelopment' (LEV), i.e. the acoustic (or musical) envelopment of a listener—the feeling of being 'surrounded by sound'. In the time domain, the acoustic properties of a room can be represented by the 'impulse response' (IR) and the 'reflectogram', which can be derived from the IR (Fig. 1.14).

When analyzing the temporal aspects of a diffuse sound field in a room, acousticians distinguish between the 'early reflections' (ER) which occur during the first 80 ms right after the sound event, and the 'late reflections' (LR), which happen afterward. While the first (single) reflections from the walls, ceiling and floor of a venue can help a listener to get an impression of his (or her) position in the room, the later reflections are of much more diffuse nature and also much weaker in level, as they result from multiple reflections at the borders of the room and the objects contained within.

Based on psychoacoustic research, which takes into account and distinguishes between the early reflections (which happen within the first 80 ms) and the late reflections, the acoustic measure of 'clarity' (C) has been defined by Reichardt and Kussev in the early 1970s (see Reichardt and Kussev 1972).

$$C_{80} = 10 \log \frac{\int_0^{80\ \mathrm{ms}} p^2(t)\mathrm{d}t}{\int_{80\ \mathrm{ms}}^{\infty} p^2(t)\mathrm{d}t} (\mathrm{dB}) \tag{1.1}$$

with

p sound pressure.

The abovementioned ASW is primarily influenced by the sound reflections which arrive during the first 80 ms at the listener, a context in which the frequency range

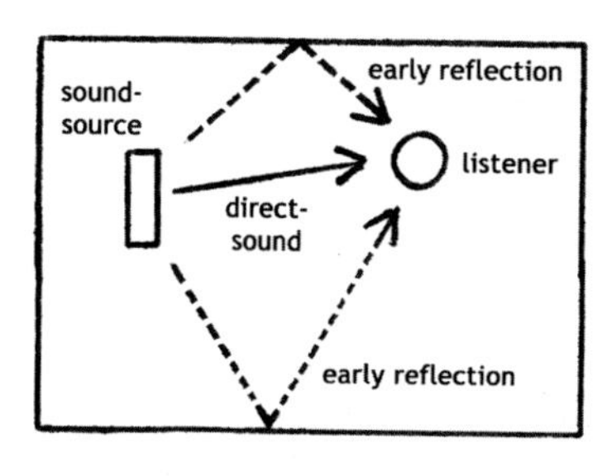

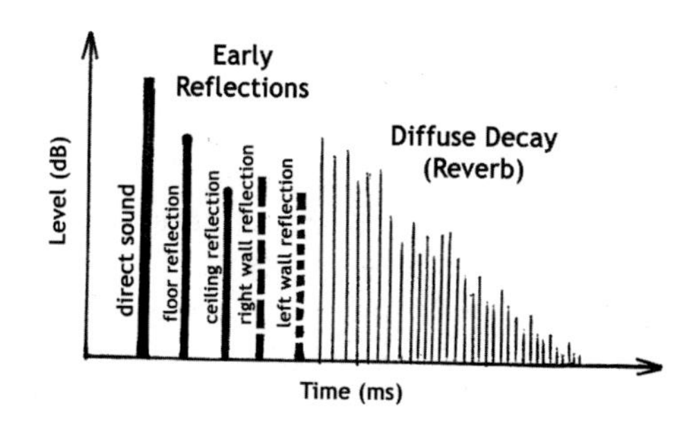

Fig. 1.14 Response of an enclosed space to a single sound impulse: Left: The direct path from source to listener is the shortest, followed by early reflections from the nearest surfaces. Right: The impulse response in the time domain shows the direct sound, followed by some discretely identifiable early reflections, followed by a gradually more dense reverberant tail that decays exponentially

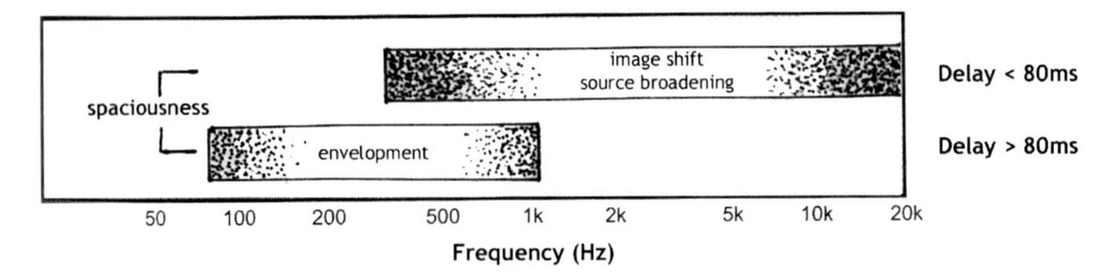

Fig. 1.15 Approximate frequency and delay ranges over which reflected sound contributes to the perception of different spatial aspects (ASW and LEV) (after Toole 2008, Fig. 7.1, pg. 97)

from 1 to 10 kHz is most important (see Fig. 1.15). On the other hand, for LEV, it is the low-frequency signal components (mainly below 200 Hz and up to 500 Hz) arriving after 80 ms and from various directions, which are most important for the listener.

According to research by Hidaka et al. (1997) ASW is mainly influenced by reflections in the octave bands around 500, 1000 and 2000 Hz, which cause low inter-aural cross-correlation (IACC). As defined in Hidaka et al,. the $IACC_{E3}$ is the mean value of the IACCs in these three octave bands measured during the first 80 ms (E = early), which is directly related to the impression of ASW. Ando (1977) had already showed that in respect to the angle of incidence of a single reflection, apparently there is a listener preference for 60°, which finds its correspondence in a reduced IACC around that angle range (see Fig. 1.16). The findings of this study are reinforced through the outcome of later research by Barron and Marshall (1981).

Research by Griesinger has shown that the ideal angle of incidence to evoke the sensation of envelopment in listeners is frequency dependent: while LF signal components under 700 Hz should ideally arrive laterally (at ±90° from the listener), at higher frequencies LEV can also be achieved with sounds much closer to the median plane (see Griesinger 1999).

As can be seen in Fig. 1.17 for a single reflection of a 1000 Hz signal arriving at 45°, there is a high fluctuation of ITD, resulting in a strong sensation of acoustic envelopment (LEV). For frequencies around 2000 Hz arriving in the same angular segment (40°–45°), there is much less fluctuation and hence a much lower sensation of envelopment.

Unfortunately, it seems that there is not really a consensus among scientists on the exact meaning of acoustical terms used especially in the realm of spatial perception. While Toole (2008) essentially declares 'spaciousness' and 'envelopment' to be synonymous (ibid. pg. 99), Rumsey also uses the words 'room impression' and—rightly so—sees a connection between listener envelopment and the 'externalization' of a sound event (see Rumsey 2001, pg. 38).

In Lehnert (1993), 'spaciousness' and 'spatial impression' are defined as follows (after Kuhl 1978):

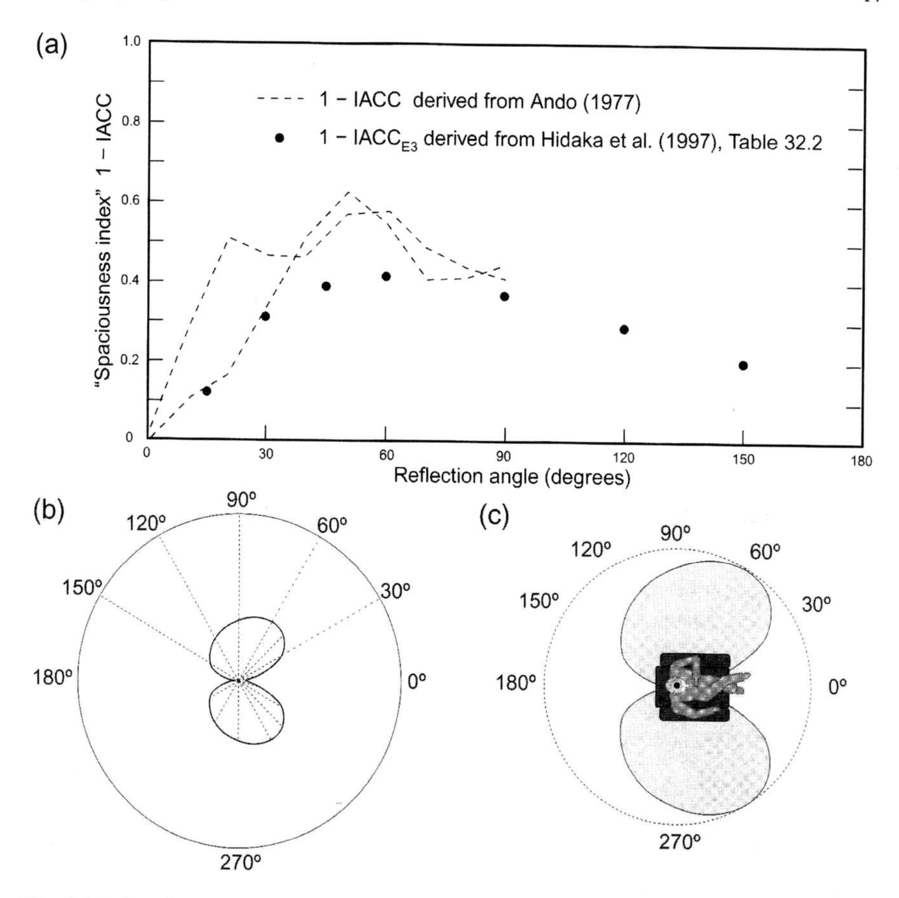

Fig. 1.16 'Spaciousness index' (1-IACC) depending on the angle of incidence of a reflection; (various forms of graphical representation (**a–c**) from (Toole 2008), Fig. 7.5, pg. 106).**a** Both curves from Ando (1977) and Hidaka et al. (1997) reach a maximum around 60°.**b** The 1-IACC$_{E3}$ data from Hidaka et al. (1997) reformatted to a polar plot.**c** The intention is made even more clear with the addition of a listener: a 'spatial effect balloon' surrounding a listener showing estimates of ASW and image broadening generated by early reflections incident from different directions

- Auditory Spatial Impression (German: 'Raumeindruck'): The concept of the type and size of an actual or simulated space to which a listener arrives spontaneously when he/she is exposed to an appropriate sound field. Primary attributes of auditory spatial impression are auditory spaciousness and reverberance.
- Auditory Spaciousness (German: 'Räumlichkeit'): Characteristic spatial spreading of the auditory events, in particular, the apparent enlarged extension of the auditory image compared to that of the visual image. Auditory spaciousness is mainly influenced by early lateral reflections.

Sarroff and Bello have proposed and tested two quantitative models for the estimation of the spatial impression contained in a stereo recording: (a) (for ASW) based on

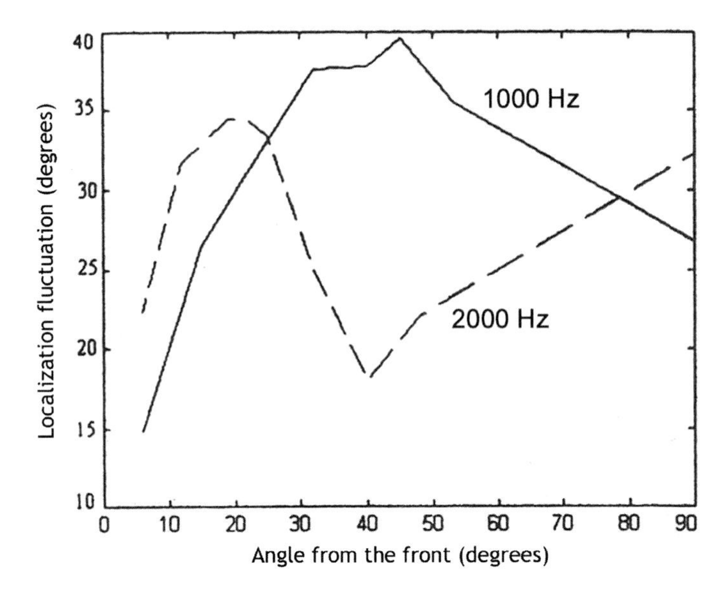

Fig. 1.17 Fluctuations of the ITD for a single reflection depending on the angle of incidence plotted for two frequencies (computer simulation). This shows that the resulting spatial impression is both angle- and frequency-dependent (after Griesinger 1997)

the analysis of stereophonic width (or ASW) of several panned single sound sources, and (b) (for LEV) the analysis of the quantification of reverb as part of the overall signal (see Sarroff and Bello 2008).

In Avni and Rafaely (2009), it is analyzed if and how spatial impression can be explained by the connection between IACC and spatial correlation in diffuse sound fields via 'spherical harmonics'. (see Avni and Rafaely 2009).

1.4 Physical Measures in Relation to Spatial Impression

In the last subchapter, it has already been reported that the individual impression of ASW is best represented through measuring the $IACC_{E3}$ (see Hidaka et al. 1997).

Already back in 1975 *R*, the 'Measure of Spatial Impression' has been defined by Lehmann (1975) (see also Reichardt et al. 1975). For the practical measurement of *R*, a 40° funnel around the 0° axis of main orientation has been defined. In this context, sound signals arriving with more than 80 ms delay within the funnel, as well as those arriving laterally (i.e. outside of the funnel) with a delay of 25–80 ms are regarded to increase spatial impression. Not in favor of increased spatial impression are lateral reflections arriving within the first 25 ms, as well as frontal reflections within the time window of 25–80 ms.

If we denote the sound energy components, which are positive for spatial impression with E_R, and the sound energy components which are negative for spatial impression with E_{NR}, then R can be described as follows:

$$R = \lg \frac{E_{\mathrm{R}}}{E_{\mathrm{NR}}} (\mathrm{dB}) \tag{1.2}$$

The practical implementation of measuring R requires the use of a directional microphone in addition to the pure pressure transducer (omni-pattern), and R can be obtained by performing the following calculation:

$$R = 10 \log \frac{\int_{25\,\mathrm{ms}}^{\infty} p_{\mathrm{omni}}^2(t)\mathrm{d}t - \int_{25\,\mathrm{ms}}^{80\,\mathrm{ms}} p_{\mathrm{shotgun,front}}^2(t)\mathrm{d}t}{\int_{0\,\mathrm{ms}}^{25\,\mathrm{ms}} p_{\mathrm{omni}}^2(t)\mathrm{d}t + \int_{25\,\mathrm{ms}}^{80ms} p_{\mathrm{shotgun,front}}^2(t)\mathrm{d}t} (\mathrm{dB}) \tag{1.3}$$

with 'shotgun, front' denoting the output signal of a highly directional (shotgun) microphone, directed toward the sound source.

For the sake of completeness also R, the 'Reverb-Measure' (German: 'Hallmaß' or ratio of reverberant sound energy to early sound energy), as defined by Beranek and Schultz (1965) should be mentioned:

$$R = 10 \log \frac{\int_{50\,\mathrm{ms}}^{\infty} p^2(t)\mathrm{d}t}{\int_{0\,\mathrm{ms}}^{50\,\mathrm{ms}} p^2(t)\mathrm{d}t} (\mathrm{dB}) \tag{1.4}$$

Concerning listener envelopment LEV, different measures such as 'lateral fraction' (LF), as well as 'lateral gain' (LG_{80}) have been proposed.

Research by Marshall (1968) and Barron (1971) shows the importance of lateral reflections for spatial impression for the listener in the concert hall. The 'lateral energy fraction' (LF) has been used as a measure for ASW in Barron and Marshall (1981) and is defined as follows:

$$LF = 10 \log \frac{\int_{0\,\mathrm{ms}}^{80\,\mathrm{ms}} p_F^2(t)\mathrm{d}t}{\int_{0\,\mathrm{ms}}^{80\,\mathrm{ms}} p_O^2(t)\mathrm{d}t} (\mathrm{dB}) \tag{1.5}$$

with $p_{\mathrm{F}}(t)$ being the sound pressure of the impulse response, which has been measured in the concert hall with a side-oriented Fig. 8 microphone (i.e. 'on-axis' is oriented at 90° from the sound source), while $p_{\mathrm{O}}(t)$ is the sound pressure at the same point in the concert hall, but measured with an omni-directional microphone. Usually, signals of both microphones are integrated over the first 80 ms for the calculation of LF.

The definition of 'lateral gain' can be found in Bradley and Soulodre (1995), which sets the energy of the late-arriving (i.e. after 80 ms) lateral diffuse sound in relation to the energy picked up by an omni-directional microphone, both at the listener position in the hall. In Soulodre et al. (2003) LG_{80} is defined as follows:

$$LG_{80} = 10 \log \frac{\int_{80\,\mathrm{ms}}^{\infty} p_F^2(t)\mathrm{d}t}{\int_{0\,\mathrm{ms}}^{\infty} p_A^2(t)\mathrm{d}t} (\mathrm{dB}) \tag{1.6}$$

with $p_F(t)$ being measured by a side-oriented Fig. 8 microphone and $p_A(t)$ with an omni-directional microphone, both at a distance of 10 m from the sound-source under 'free-field' conditions.

In this paper, it is also analyzed that the restriction to a fixed time interval of 80 ms (independent of the frequency band concerned) is not sufficient if one wants to take into account the frequency-dependent 'inertia' (or time constant) of human hearing. This is why Soulodre et al. propose an integration time of 160 ms up to 500 Hz, above that a shorter integration time (75 ms at 1000 Hz) and only 45 ms above 2000 Hz.

Further they plead for a new objective measure for LEV which they denote with $\mathrm{GS}_{\mathrm{perc}}$

$$GS_{\mathrm{perc}} = 0.5G_{\mathrm{perc}} + S_{\mathrm{perc}}(\mathrm{dB}) \tag{1.7}$$

in which G_{perc} is a gain component and S_{perc} is a component of spatial distribution.

G_{perc} is essentially the acoustic measure of 'strength' from concert hall acoustics, also known as 'strength factor' (G) (see Beranek 2004, p. 617), which has been first defined by Lehmann in 1976 (see Lehmann 1976):

$$G_x = 10 \log \frac{\int_{x\,\mathrm{ms}}^{\infty} p_O^2(t)\mathrm{d}t}{\int_{0\,\mathrm{ms}}^{\infty} p_A^2(t)\mathrm{d}t} (\mathrm{dB}) \tag{1.8}$$

According to the proposal of Soulodre et al., G_x is a measure for the relative sound pressure level of the late energy, while—depending on the octave band which is being measured—the integration times x of Table 1.1 are used. With this new measure for listener envelopment—which can be considered a refinement of LG_{80}—it has been possible to further the correlation of LEV by another 4%, reaching a total of 0.98. In connection with the search for a physical measure for the spatial distribution S_{perc}, Soulodre argues that a measure based on IACC may be considered, as this is generally used to quantify the spatial distribution of a sounds field (see Soulodre et al. 2003, p. 838, 2nd paragraph).

Table 1.1 Frequency-dependent integration times

Octave bands (Hz)	Integration limit x (ms)
*	*
63	*
125	160
250	160
500	160
1000	75
2000	55
4000	45
8000	45

Adapted to human psychoacoustics (after Soulodre et al. 2003)

In this context, it is interesting to note that IACC can be used not only in connection with the first component of spatial impression, namely ASW (as pointed out above), but also a part of LEV (i.e. the component which concerns the spatial distribution S_{perc}) could be represented through a measure related to IACC. These are both strong arguments for taking a further look at IACC as a measure for spatial impression with human listeners.

In relation to the acoustics of small rooms, Griesinger (1996) proposes the 'lateral early decay time' (LEDT) as a measure for spaciousness:

$$LEDT_{350} = \frac{60*350 \text{ ms}}{(S(0) - SD(350))1000 \text{ ms/s}} \tag{1.9}$$

with $S(t)$ being the Schröder integral of the impulse response and $SD(t)$ being the Schröder integral of the interaural level difference (ILD).

A multitude of factors contributes to spatial impression: Research by Griesinger (1997) has shown that it is the fluctuations of ITD and ILD, which contribute significantly to the spaciousness of a sound event. Research by Mason and Rumsey (2002) seems to confirm this: measurements based on the $IACC_E$ showed the highest correlation to the perceived source width as well as environmental width (i.e. the acoustically perceived width of the performance space).

ASW and IACC are very useful in connection with acoustic measurements of concert halls, but not equally suited for the measurement of smaller rooms. This is why Griesinger (1998) has proposed two additional measures: The 'diffuse field transfer function' (DTF) as a measure of envelopment, which is equally suited for small and large rooms, and the 'average interaural time delay' (AITD) as a measure for externalization.

As can be seen from all the information provided above, the individual listening impression in respect to spatial impression is determined by ASW and LEV, which are both caused by reflected sound, which arrives at the listener in various time slots and from different directions. The result for both is a decreased IACC at the ears of the listener (due to lateral reflections), which is important as research has shown that in respect to the perceived acoustic quality of concert halls listener preference increases with lowering IACC (see Hidaka et al. 1995; Beranek 2004). In this context also the 'Binaural Quality Index' BQI has been defined in Keet (1968). According to Beranek (2004), it is 'one of the most effective indicators of acoustic quality of concert halls' (ibidem, pg. 506) and defined as follows:

$$\text{BQI} = (1 - \text{IACC}_{E3}) \tag{1.10}$$

The subindex $_{E3}$ denotes the early sound energy in the time window from 0 to 80 ms in the octave bands with center frequencies at 500 Hz, 1 kHz and 2 kHz. Figure 1.18 clearly shows the correlation between hall ranking and BQI.

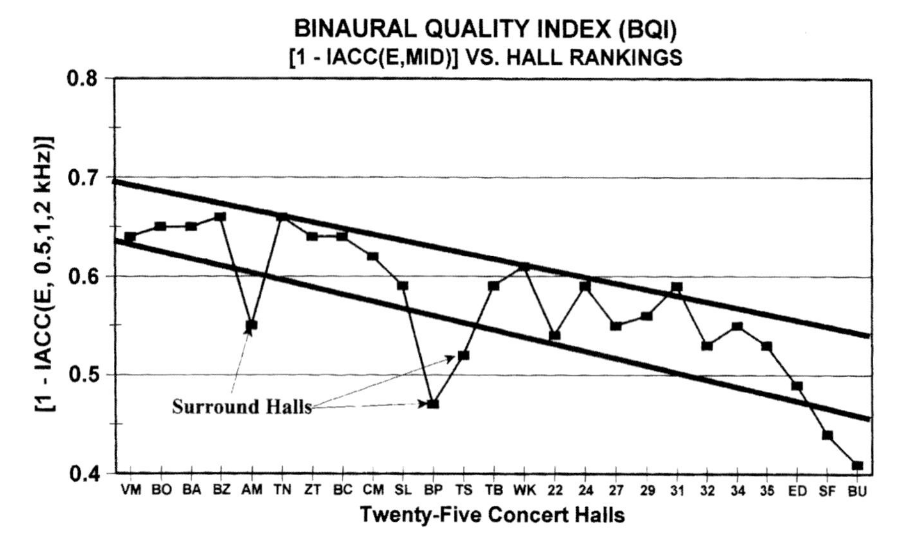

Fig. 1.18 Binaural Quality Index (BQI) for 25 concert halls, measured when unoccupied, plotted versus the subjective rank orderings of acoustical quality; average standard deviation 0.11 s (from Beranek 2004, pg. 509)

1.5 The Influence of Loudspeaker Quality and Listening Room Acoustics on Listener Preference

When undertaking the task of evaluating the quality and various aspects of microphone techniques, it must also be assumed that the acoustical quality of the reproduction loudspeakers as well as the acoustical properties of the listening room will have an influence on the result. Changes in the signal from the source (e.g. musical instrument) to the recipient (ear of human listener) are unavoidable due to technical limitations in the form of qualitative limits of the electronics and electroacoustic transducers involved (microphones, loudspeakers, AD and DA converters, etc.) as well as limitations of the transmission channel or storage medium.

In connection with the selection of appropriate loudspeakers, the question arises whether their directivity index (DI) would have an influence on listener evaluations of test material. Various researchers seem to have divergent answers to this: Kates (1960) favors loudspeakers with higher directionality, as he thinks that it is of advantage to avoid unnecessary room reflections. Also Zacharov (1998) shares this opinion, but encounters strong criticism by important leaders within the audio engineering community (see—among others—Holman 2000a). To underline his arguments, Holman points out the results of his own research (Holman 1991), in which he was able to show that a higher directivity index of loudspeakers is helpful for better localization, but a smaller directivity index is in favor of better 'envelopment' for the listener (especially with a surround-replay setup). Likewise (see Toole 2008, pg. 137)—based on the analysis of his previous listening tests—Toole (1985, 1986) arrives at the conclusion that the majority of listeners apparently prefers loudspeakers with a broader dispersion angle (i.e. smaller directivity index). In his case, this also had to do with the fact that he was using listening rooms with acoustically untreated sidewalls which caused relevant reflections due to the wide dispersion angle of the loudspeakers. In turn, these reflections were responsible for a lowering of the IACC at the listeners which lead to an increase in perceived ASW. This resulted in a higher listener preference for loudspeakers which had a wider dispersion characteristic.

That a lager base width or ASW is preferred by listeners coincides with the findings of Berg and Rumsey (2001), as well as Bech (1998). In the case of Bech's research, the variation in base width of the sound source had been achieved on the reproduction side by means of a varied spacing of the (stereo-) front speakers (see Fig. 1.19). Additional research which arrived to similar results can be found in Moulton et al. (1986), Moulton (1995) and Augspurger (1990).

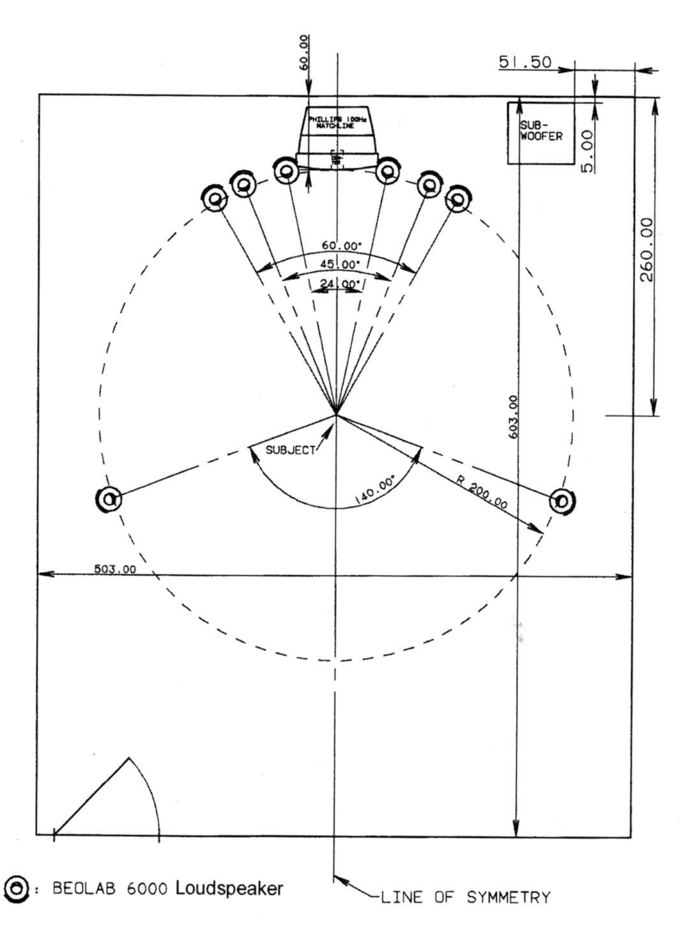

Fig. 1.19 Loudspeaker and TV setup in the listening room for research on the influence of base width on perceived audio quality (from Bech 1998)

1.6 Psychoacoustic Effects Concerning Localization and Spatial Impression with Loudspeaker Reproduction

When playing back sound via a standard stereo loudspeaker setup (i.e. speakers at −30° and +30°), sufficiently large level or time differences (ILDs and ITDs) between the two signals will essentially lead to localizing the stimulus only in one of the two speakers. In order for this to happen, the level difference needs to be about 15 dB or the time difference about 1.1 ms, or a suitable combination of level and time difference values in between (see Fig. 1.20). First research in that direction can already be found in de Boer (1940), and also in later publications such as Theile et al. (1988) and Gernemann (1994). In his research, Sengpiel (1992) arrives at slightly different results with 18 dB and 1.5 ms, respectively.

When comparing the results from various researches, it becomes clear that the deviation between values is much larger for time-of-arrival-based localization than for level-based localization. The big discrepancies for time-of-arrival-based stereophony is also backed by the findings of a study by Wittek and Theile (2002).

As will be shown in a later chapter in Table 3.1, the differences in perceived localization width (regarding the resulting recording angle) can be more than 100% (!) (compare the results of Wittek with those of Sengpiel for a small AB pair with a capsule spacing of 50 cm, or roughly 2.7 ft).

In Knothe and Plenge (1978), various studies are named which make it clear that the effect of frequency dependence of sound source localization has been known

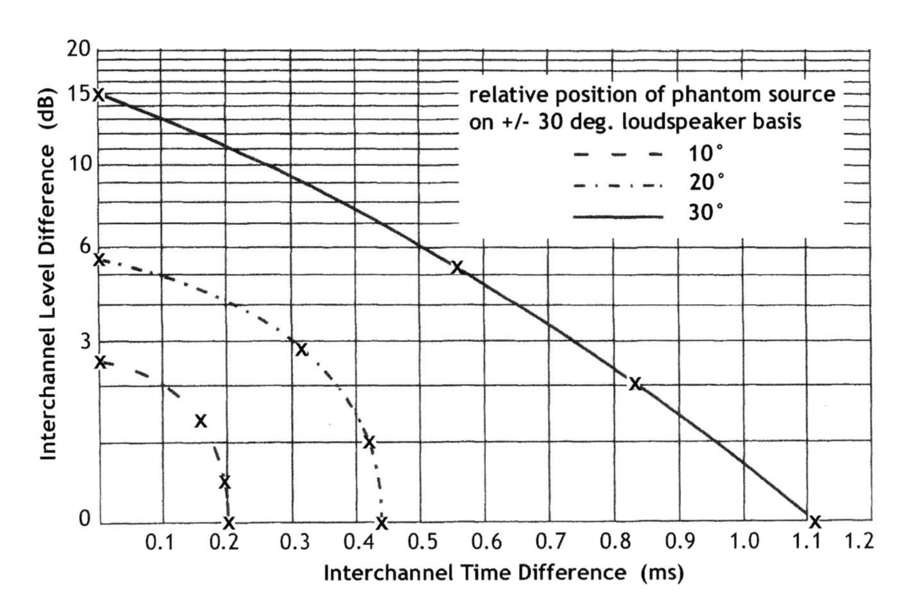

Fig. 1.20 Localization curves (after Williams 1987; Simonsen 1984; Rumsey 2001)

at least since 1934. More recent research on this topic can be found in Griesinger (2002), among others.

Reasons for the differences between results of the studies examined in Wittek and Theile (2002) may be the following:

- Difference in the test signals which have been used for various studies,
- Varying acoustical characteristics of the loudspeakers used for the different studies [concerning this topic see (Gernemann 1998)],
- Large inter-individual differences in acoustic perception among the test listeners, in combination with localization-distortion which happens at mid-frequencies [see (Benjamin and Brown 2007), as well as Fig. 1.17)] (Fig. 1.21).

If one looks at the differences of more than 100%, as pointed out in the study of Wittek and Theile (2002), it seems clear that the results achieved through small AB-based recording techniques are not consistent enough in terms of localization (or stereophonic "base width") so that inter-individual differences between listeners (as well as qualitative differences between loudspeakers) could be neglected. Therefore, small AB cannot be recommended as a reliable microphone technique what concerns the reproduction of localization and spatial distribution of sound sources.

Through experiments regarding the localization of sound sources replayed through a 5.1 loudspeaker setup according to (ITU-R BS.775.1, 1994) it was found out that already a time difference of 0.6 ms was sufficient to cause full localization of the sound source in just one of the two rear speakers. Most likely this has to

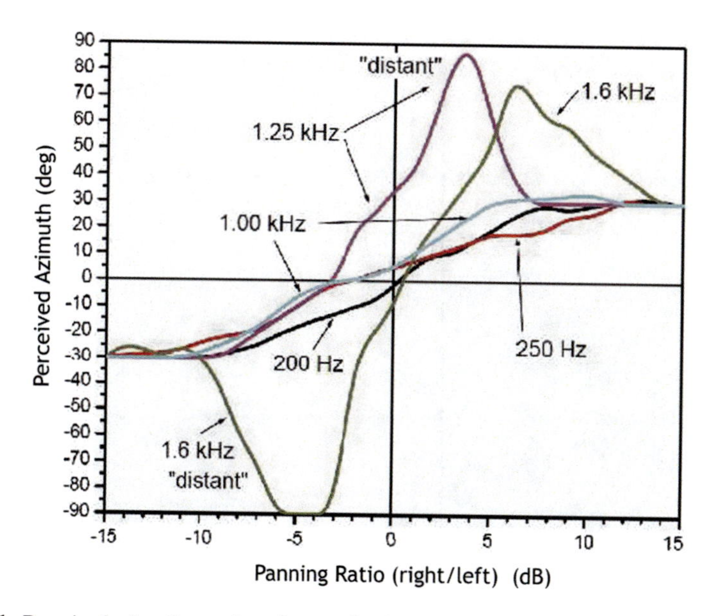

Fig. 1.21 Perceived azimuth at various frequencies for signals (narrowband Gaussian sine-bursts), which have been panned between the stereo-channels by means of level difference (from Benjamin 2006)

do with the much larger distance (or greater opening angle of 120°) between the rear speakers (in contrast to the 60° between the L and R front speaker), relative to the listener (see Rumsey 2001, pg. 32). Another conclusion of this research was that level-based localization (ILD) results in a more stable listener impression than time-of-arrival-based (ITD) localization.

1.6.1 Frequency-Dependent Localization Distortion in the Horizontal Plane

As has been pointed out in Knothe and Plenge (1978), localization of a panned mono sound-source is strongly frequency-dependent (see Fig. 1.22). The level difference ΔS, achieved on the summing bus due to panning, needs to be larger at lower frequencies in order to achieve the same impression of position in terms of localization than with higher frequencies. The more lateral the perception of the sound source is supposed to be on the ±30° stereo base width, the larger ΔS needs to be: for a localization impression at 10° the difference in necessary ΔS between high and low

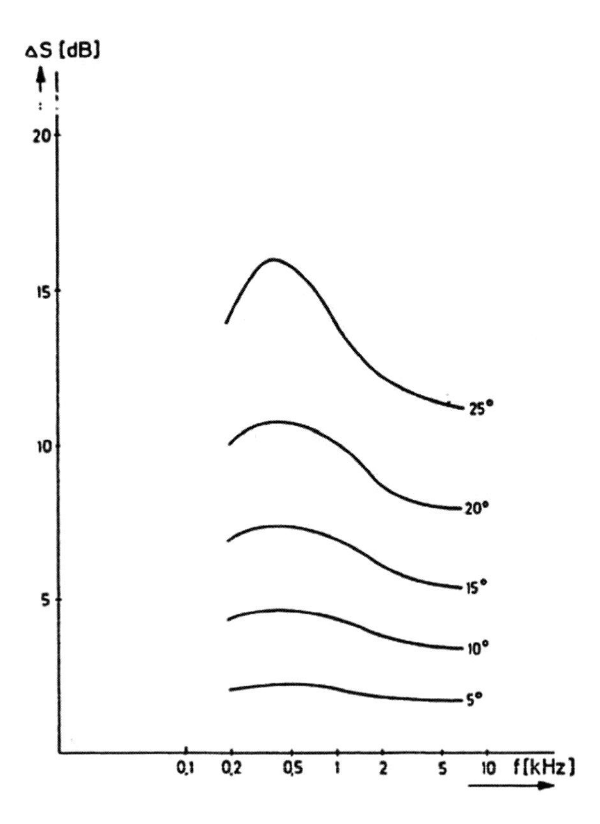

Fig. 1.22 Frequency-dependent localization of a sound-source and ΔS of relative level difference between L and R channel in a stereo loudspeaker system (from Knothe und Plenge 1978)

frequencies is about 1–2 dB, for 25° azimuth the necessary difference is already up to 5 dB (compare ΔS at 400 Hz and 7 kHz in Fig. 1.22).

Also Griesinger (2002) arrives at similar conclusions in respect to frequency-dependent localization both for stereo, as well as surround signals. For spectral components above 600 Hz, localization is heavily 'biased' toward the loudspeaker to which the signal gets panned. According to Griesinger's research, this is due to interference of the direct signal from one loudspeaker with the signal from the other loudspeaker, which diffracts around the listener's head.

1.6.2 Frequency-Dependent Localization in the Vertical Plane

As research by Ferguson and Cabrera (2005) has shown, also in the vertical plane localization distortion between signals of high and low frequencies takes place: while high-frequency sound sources are normally localized at their correct physical position, this is not the case for low-frequency sound sources. These are usually localized below their real physical position. In the case of a broadband sound signal, localization is dominated by the high-frequency signal components (see also Morimoto et al. 2003).

1.6.3 Effects Concerning the Reproduction of Spaciousness

In practical recording situations, the frequency-dependent localization of sound signals can be especially difficult for instruments that are not only broadband in terms of frequency spectrum, but also broad or wide in a physical sense—like a piano for example: for the human listener the source width of the piano seems much smaller for low frequencies, than for high frequencies; at least for microphone techniques which are purely level-based, which is the case for the coincident "Blumlein-Pair", consisting of two crossed Fig. 8 microphones. In this context, methods for "spatial equalization" have been proposed by various researchers [see, e.g. (Gerzon 1986), as well as (Griesinger 1986)].

The change in terms of spaciousness toward low frequencies can be checked by listening selectively to isolated frequency bands to see (or better: hear) which spatial impression a microphone technique is able to provide at different frequency ranges. In this respect, the entire frequency band below approx. 800 Hz is especially important, as the human head is not yet effective as a baffle and sound signals bend around it. Above 800 Hz, the shadowing effect of the human head becomes more and more evident and thus human hearing is based mainly on ILD, while at low frequencies it is mainly based on an analysis of phase and time differences (see Fig. 1.23).

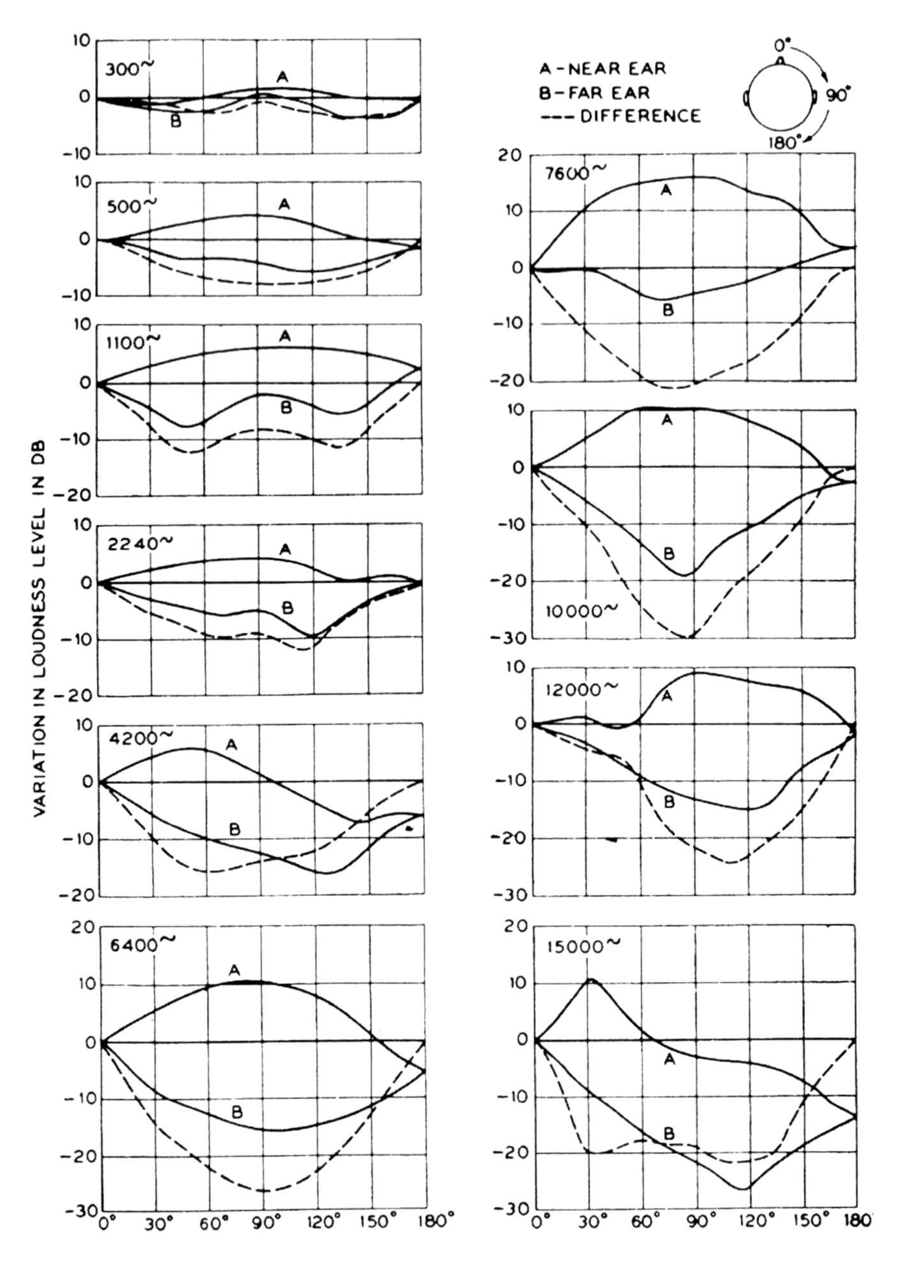

Fig. 1.23 Graphic from Steinberg and Snow (1934)

In this context, research by Yost et al. (1971) should be mentioned, which has shown that the low-frequency components of transient binaural signals are of highest importance for localization: the high-pass filtering of clicks (similar to Dirac impulses) with a cutoff frequency at 1500 Hz leads to a clear deterioration of localization, while low-pass filtering with the same frequency resulted only in a minimal change (deterioration) of localization.

Research by Hirata (1983) deals with the phenomenon of localization distortion of the low-frequency components of a stereo signal upon loudspeaker playback. He proposes PICC, a "Perceptual Interaural Cross-correlation Coefficient":

$$PICC = DR_0 + (1 - D)R_\mathrm{E} \tag{1.11}$$

with D (Deutlichkeit), as it has been defined by R. Thiele (see Thiele 1953)

$$D = \frac{\int_0^{50\ \mathrm{ms}} p^2(t)\mathrm{d}t}{\int_{0\ \mathrm{ms}}^{\infty} p^2(t)\mathrm{d}t} \tag{1.12}$$

with:

R_0 the interaural cross-correlation coefficient of direct sound (which is 1 for sound from 0°),

R_E the interaural cross-correlation coefficient of diffuse sound, which is defined as:

$$R_\mathrm{E} = \frac{\sin k\, r(f)}{k\, r(f)} \tag{1.13}$$

and the wave number k as $k = 2\,\pi\, f/c$

with c, the speed of sound and

$r(f)$ the effective acoustic distance between the human ears, which is 30 cm [see (Yanagawa et al. 1976), as well as (Suzuki and Tohyama 1981)].

In addition, he defines ASI, an "Index of Acoustic Spatial Impression", as follows:

$$ASI = 100(1 - D)(\%) \tag{1.14}$$

Total spatial impression signifies ASI = 100%, while the complete absence of spatial impression is ASI = 0%.

Figure 1.24 shows that in a standard listening room ($RT_{60} = 0.3$ s), ASI is small for frequencies below 800 Hz, but high for frequencies above 800 Hz in comparison with ASI at a seat in the concert hall, for which ASI equals 60%. Research by Griesinger concerning listener envelopment (Griesinger 1999) has shown that for rising frequency the ideal loudspeaker position moves toward the median plane. For frequencies below 700 Hz instead, a speaker layout is ideal which enables maximal lateral separation between the transducers, i.e. at a position left and right of the listener at ±90° (see Fig. 1.25).

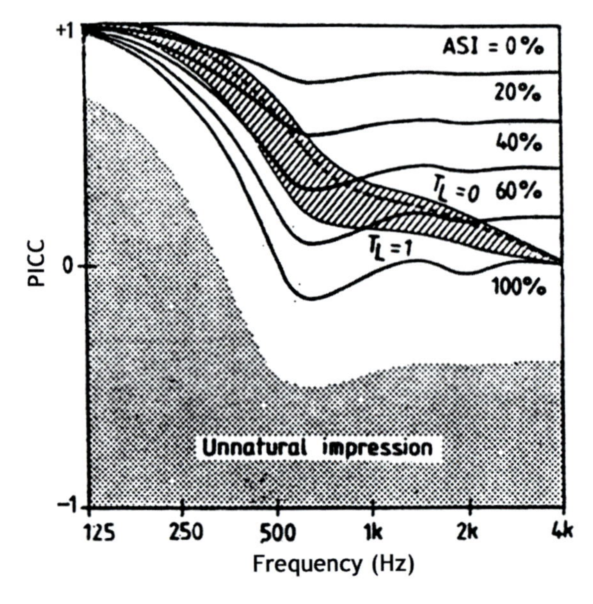

Fig. 1.24 PICC curves for stereo reproduction in a listening room with a reverb time T_L (0–1 s) show small ASI values at low frequencies compared to an ASI = 60% for the seats in the middle section of a concert hall. The dashed curve stands for T_L = 0.3 s (from Hirata 1983)

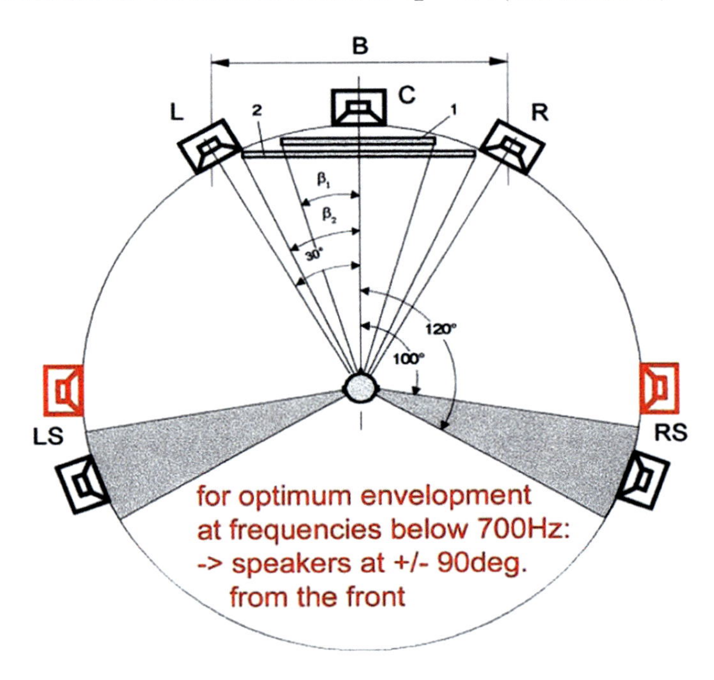

Fig. 1.25 A '5.2' arrangement with subwoofers at the sides at ±90°, optimized for the spatial reproduction of low frequencies, according to the recommendation by Griesinger (1999)

The abovementioned research by Hirata and Griesinger shows that the standard loudspeaker layout for stereo (as well as 5.1 surround, see ITU specification according to (BS.775-1)) with loudspeakers at ±30° are far from ideal for the reproduction of low-frequency signal components.

For this reason, it is very important to choose—already at the stage of the recording process—a microphone technique, which captures the sound signal in a de-correlated manner over the whole frequency range, as a deterioration (in respect to a 'forced' increase in correlation) upon replay has to be expected anyway (see Hirata 1983).

References

Ando Y (1977) Subjective preference in relation to objective parameters of music sound fields with a single echo. J Acoust Soc Am 62:1436–1441

Akeroyd MA, Summerfield AQ (1999) A binaural analog of gap detection. J Acoust Soc Am 105:2807–2820

Akeroyd MA, Bernstein LR (2001) The variation across time of sensitivity to interaural disparities: behavioural measurements and quantitative analyses. J Acoust Soc Am 110:2516–2526

Augspurger GL (1990) Loudspeakers in control rooms and listening rooms. Paper presented at the Audio Eng Soc 8th int conference

Avni A, Rafaely B (2009) Inter-aural cross correlation in a sound field represented by spherical harmonics. J Acoust Soc Am 125(4):2545

Barron M (1971) The subjective effects of first reflections in concert halls—the need for lateral reflections. J Sound Vib 15:475–494

Barron M, Marshall AH (1981) Spatial impression due to early lateral reflections in concert halls: the derivation of a physical measure. J Sound Vib 77:211–232

Batteau DW (1967) The role of the Pinna in human localization. Proc Roy Soc B168(1011):158–180

Bech S (1998) The influence of stereophonic width on the perceived quality of an audiovisual presentation using a multichannel sound system. J Audio Eng Soc 46(4):314–322

Begault D (1994) 3-D sound for virtual reality and multimedia. Academic Press, USA

Benjamin E (2006) An experimental verification of localization in two-channel stereo. Paper 6968 presented at the 121st Audio Eng Soc Convention

Benjamin E, Brown R (2007) The effect of head diffraction on stereo localization in the mid-frequency range. Paper 7018 presented at the 122nd Audio Eng Soc Convention, Vienna, 2007

Beranek L (2004) Concert halls and opera houses: music, acoustics and architecture, 2nd edn. Springer, New York

Beranek LL, Schultz TJ (1965) Some recent experiences in the design and testing of concert halls with suspended panel arrays. Acustica 15:307

Berg J, Rumsey F (2001) Verification and correlation of attributes used for describing the spatial quality of reproduced sound. Paper presented at the Audio Eng Soc 19th int conference

Bernstein LR, Trahoitis C, Akeroyd MA, Hartung K (2001) Sensitivity to brief changes of interaural time and interaural intensity. J Acoust Soc Am 109:1604–1615

Blauert J (1972) On the lag of lateralization caused by interaural time and intensity differences. Audiology 11:265–270

Blauert J (1974) Räumliches hören. S. Hirzel Verlag, Stuttgart

Blauert J (1997) Spatial hearing. The MIT Press

Boehnke SE, Hall SE, Marquadt T (2002) Detection of static and dynamic changes in interaural correlation. J Acoust Soc Am 112:1617–1626

Bradley J, Soulodre G (1995) Objective measures of listener envelopment. J Acoust Soc Am 98:2590–2597

Breebart J, van der Par S, Kohlrausch A (2002) A time-domain binaural signal detection model and its predictions for temporal resolution data. Acta Acustica-Acustica 88:110–112
Culling JF, Summerfield AQ (1999) Measurement of the binaural temporal window using a detection task. J Acoust Soc Am 103:3540–3553
de Boer K (1940) Plastische Klangwiedergabe. Philips Technische Rundschau 5(4)
Eska G (1997) Schall und Klang: wie und was wir hören. Birkhäuser Verlag
Everest FA (1994) The master handbook of acoustics, 3rd edn. TAB Books McGraw-Hill
Faller C, Merimaa J (2004) Source localization in complex listening situations: selection of binaural cues based on interaural coherence. J Acoust Soc Am 116:3075–3089
Fellner M, Höldrich R (1998a) Physiologische und psychoakustische Grundlagen des räumlichen Hörens. IEM-Report 03 KUG: Univ f Musik u darst Kunst, Graz
Fellner M, Höldrich R (1998b) Außenohr-Übertragungsfunktion – Messung und Datensätze. IEM-Report 04, KUG: Univ f Musik u darst Kunst, Graz
Ferguson S, Cabrera D (2005) Vertical localization of sound from multiway loudspeakers. J Audio Eng Soc 53(3):163–173
Gernemann A (1994) Summenlokalisation im Stereodreieck—Überlegungen zu psychoakustischen Untersuchungen mit dynamischem Testsignal und hochpräzisen Schallwandlern. Manus, Düsseldorf
Gernemann A (1998) Rösner T (1998) Die Abhängigkeit der stereophonen Lokalisation von der Qualität der Wiedergabelautsprecher. Proceedings to the 20. Karlsruhe, Tonmeistertagung des VDT, p 828
Gerzon M (1986) Stereo shuffling: new approach, old technique. Studio Sound July 1986:122–130
Gerzon M (1992) Psychoacoustic decoders for multispeaker stereo and surround sound. Paper 3406 presented at 103rd Audio Eng Soc Convention, San Francisco, 1–4 Oct
Glasberg BR, Moore BCJ (1990) Derivation of auditory filter shapes from notched-noise data. Hear Res 47:103–138
Grantham DW, Wightman FL (1978) Detectability of varying interaural temporal differences. J Acoust Soc Am 63:511–523
Grantham DW, Wightman FL (1979) Detectability of a pulsed tone in the presence of a masker with time-varying interaural correlation. J Acoust Soc Am 65:1509–1517
Griesinger D (1986) Spaciousness and localization in listening rooms and their effects on the recording technique. J Audio Eng Soc 34(4):255–268
Griesinger D (1996) Spaciousness and envelopment in musical acoustics. In: Proceedings to the 19. Tonmeistertagung des VDT, pp 375–391
Griesinger D (1997) Spatial impression and envelopment in small rooms. Paper 4638 presented at the 103rd Audio Eng Soc Convention
Griesinger D (1998) General overview of spatial impression, envelopment, localization and externalization. In: Proceedings to the Audio Eng Soc 15th int conference on small rooms
Griesinger D (1999) Objective measures of spaciousness and envelopment. Paper 16-003 presented at the Audio Eng Soc 16th int conference on spatial sound reproduction
Griesinger D (2002) Stereo and surround panning in practice. Paper 5564 presented at the 112th Audio Eng Soc Convention, Munich, May 2002
Haas H (1951) The influence of a single echo on the audibility of speech. (German), Acoustica 1(2)
Hall DE (1980) Musical acoustics. Brooks/Cole Publ Company, California. German edition: Musikalische Akustik – ein Handbuch, Schott-Verlag
Hebrank J, Wright D (1974) Spectral cues in the localization of sound sources on the median plane. J Acoust Soc Am 56(3):1829–1834
Hidaka T, Beranek L, Okano T (1995) Interaural cross-correlation, lateral fraction, and low- and high-frequency sound levels as measures of acoustical quality in concert halls. J Acoust Soc Am 98(2), Aug 1995
Hidaka T, Beranek L, Okano T (1997) Some considerations of interaural cross correlation and lateral fraction as measures of spaciousness in concert halls. In: Ando Y, Noson D (eds) Music and concert hall acoustics. Academic Press, London

Hirata Y (1983) Improving stereo at L.F. Wireless World, Oct 1983, pp 60
Holman T (1991) New factors in sound for cinema and television. J Audio Eng Soc 39:529–539
Holman T (2000a) Comments on the 'Subjective appraisal of loudspeaker directivity for multichannel reproduction'. J Audio Eng Soc 48(4):314–317
Holube I, Kinkel M, Kollmeier B (1998) Binaural and monaural auditory filter bandwidths and time constants in probe tone detection experiments. J Acoust Soc Am 104:2412–2425
Huopaniemi J (1999) Virtual acoustics and 3D sound in multimedia signal processing. Dissertation, Helsinki University of Technology
Kates JM (1960) Optimum loudspeaker directional patterns. J Audio Eng Soc 28:787–794
Keet de WV (1968) The influence of early lateral reflections on spatial impression. In: 6th int congress on acoustics, Tokyo
Knothe J, Plenge G (1978) Panoramaregler mit Berücksichtigung der frequenzabhängigen Pegeldifferenzbewertung durch das Gehör. In: Proceedings to the 11. Tonmeistertagung des VDT, Berlin, 1978
Kollmeier B, Holube I (1992) Auditory filterbandwidths in binaural and monaural listening conditions. J Acoust Soc Am 92:1889–1901
Kollmeier B, Gilkey RH (1990) Binaural forward and backward masking: evidence for sluggishness in binaural detection. J Acoust Soc Am 87:1709–1719
Kuhl W (1978) Räumlichkeit als eine Komponente des Höreindrucks. Acustica 40:167–168
Lehmann U (1975) Untersuchung zur Bestimmung des Raumeindrucks bei Musikdarbietungen und Grundlagen der Optimierung. Dissertation, TU Dresden
Lehmann P (1976) Über die Ermittlung raumakustischer Kriterien und deren Zusammenhang mit subjektiven Beurteilungen der Hörsamkeit. Dissertation, TU Berlin
Lehnert H (1993) Auditory spatial impression. In: Proceedings of the Audio Eng Soc 12th int conf on the perception on reproduced sound, pp 40–46
Litovsky RY, Colburn HS, Yost WA, Guzman SJ (1999) The precedence effect. J Acoust Soc Am 106:1633–1654
Marshall AH (1968) Acoustical determinants for the architectural design of concert halls. Arch Sci Rev Australia 11:81–87
Martin G, Woszczyk W, Corey J, Quesnel R (1999) Sound source localization in a five channel surround sound reproduction system. Paper 4994 presented at the 107th Audio Eng Soc Convention, New York, Sept 1999
Mason R, Rumsey F (2002) A comparison of objective measurements for predicting selected subjective spatial attributes. Paper 5591 presented at the 112th Audio Eng Soc Convention, Munich
Mehrgardt S, Mellert V (1977) Transformation characteristics of the external human ear. J Acoust Soc Amer 61(6):1567–1576
Merimaa J, Pulkki V (2005) Spatial impulse response rendering I: analysis and synthesis. J Audio Eng Soc 53(12)
Mershon DH, Bowers JN (1979) Absolute and relative cues for the auditory perception of egocentric distance. Perception 8:311–322
Moore BCJ (1997) An introduction to the psychology of hearing, 4th edn. Academic Press, London UK
Morimoto M, Yairi M, Iida K, Itoh M (2003) The role of low frequency components in median plane localization. Acoust Sci Technol 24:76–82
Moulton D (1995) The significance of early high-frequency reflections from loudspeakers in listening rooms. Paper 4094 presented at the 99th Audio Eng Soc Convention
Moulton D, Ferralli M, Hebrock S, Pezzo M (1986) The localization of phantom images in an omnidirectional stereophonic loudspeaker system. Paper 2371 presented at the 81st Audio Eng Soc Convention
Nielsen SH (1993) Auditory perception in different rooms. J Audio Eng Soc 41(10)
Rayleigh (1907) On our perception of sound direction. Phil Mag 13

Reichardt W, Kussev A (1972) title unknown. In: Zeitschrift elektr Inform u Energietechnik 3(2):66, Leipzig, (rem.: without title citation (see (Cremer und Müller, 1978), footnote 2, p. 345)
Reichardt W, Abdel Alim O, Schmidt W (1975) Definitionen und Messgrundlage eines objektiven Maßes zur Ermittlung der Grenze zwischen brauchbarer und unbrauchbarer Durchsichtigkeit bei Musikdarbietung. Acustica 32:126
Rumsey F (2001) Spatial audio. Focal Press (Elsevier)
Sarroff A, Bello JP (2008) Measurements of spaciousness for stereophonic music. Paper 7539 presented at the 125th Audio Eng Soc Convention
Sengpiel E (1992) Grundlagen der Hauptmikrophon-Aufnahmetechnik – Skripten zur Vorlesung (Musikübertragung). Hochschule der Künste, Berlin. http://www.sengpielaudio.de. Accessed 2004
Simonsen G (1984) Master's thesis, Techn Univ of Lyngby, Denmark (no title information available)
Soulodre GA, Lavoie MC, Norcross SG (2003) Objective measures of Listener Envelopment in Multichannel Surround Systems. J Audio Eng Soc 51(9)
Steinberg JC, Snow WB (1934) Auditory perspective—physical factors. Electr Eng 53(1):12–15
Stern RM, Zeiberg AS, Trahoitis C (1988) Lateralization of complex binaural stimuli: a weighted image model. J Acoust Soc Am 84:156–165
Streicher R, Everest A (2006) The new Stereo Soundbook, 3rd edn. Audio Eng Associates
Suzuki A, Tohyama M (1981) Interaural cross-correlation coefficient of Kemar head and torso simulator. IECE Japan, Tech Rep EA80–78
Tan CJ, Gan WS (2000) Direct Concha excitation for the introduction of individualized hearing cues. J Aud Eng Soc 48(7/8):642–653
Theile G et al (1988) Raumbezogene Stütztechnik – eine Möglichkeit zur Optimierung der Aufnahmequalität. In: Proceedings to the 15. Tonmeistertagung des VDT
Thiele R (1953) Richtungsverteilung und Zeitfolge der Schallrückwürfe in Sälen. Acustica 3:291–302
Tohyama M, Suzuki A (1989) Interaural cross-correlation coefficients in stereo-reproduced sound fields. J Acoust Soc Am 85(2). Reprinted in: Rumsey F (ed) 2006 An anthology of articles on, spatial sound techniques—part 2: multichannel audio techniques. Audio Eng Soc, New York
Toole FE (1985) Subjective measurements of loudspeaker quality and listener performance. J Audio Eng Soc 33(1/2):2–32
Toole FE (1986) Loudspeaker measurements and their relationship to listener preferences. J Audio Eng Soc 34:227–235
Toole FE (2008) Sound reproduction—loudspeakers and rooms. Focal Press (Elsevier)
Van der Hejden M, Trahiotis C (1998) Binaural detection as a function of interaural correlation and bandwidth of masking noise: implications for estimates of spectral resolution. J Acoust Soc Am 103:1609–1614
von Hornbostel EM, Wertheimer M (1920) Über die Wahrnehmung der Schallrichtung. Report to the Acad of Sciences, Berlin, pp 388–396
Wightman FL, Kistler DJ (1992) The dominant role of low-frequency interaural time differences in sound localization. J Acoust Soc Am 91:1648–1661
Williams M (1987) Unified theory of microphone systems for stereophonic sound recording. Paper 2466 presented at the 82nd Audio Eng Soc Convention, 1987
Wittek H, Theile G (2002) The recording angle—based on localisation curves. Paper 5568 presented at the 112th Audio Eng Soc Convention, Munich
Yanagawa H, Higashi H, Mori S (1976) Interaural correlation coefficients of the dummy head and the feeling of wideness. Acoust Soc Jap Tech Rep H-35–1
Yost WA, Wightman FL, Green DM (1971) Lateralisation of filtered clicks. J Acoust Soc Am 50:1526–1531
Zacharov N (1998) Subjective appraisal of loudspeaker directivity for multichannel reproduction. J Audio Eng Soc 46(4):288–303
Zollner M, Zwicker E (2003) Elektroakustik, 3rd edn. Springer
Zwicker E, Fastl H (1990) Psychoacoustics. Springer, Berlin

Chapter 2
Correlation and Coherence

Abstract First, the fundamentals of signal correlation and signal coherence are outlined. After presenting the mathematical definitions of correlation and coherence, we are looking into the practical aspect of signal correlation in stereo microphone systems from a theoretical point of view (e.g. signal correlation coefficient vs. capsule distance). A few keywords: IACC—interaural Cross-Correlation, FSCC—Frequency-dependent Spatial Cross-Correlation and MSC—Magnitude Squared Coherence. Also, a psychoacoustically very relevant aspect such as 'diffuse sound distribution'—as transmitted through the various stereo microphone techniques—is dealt with. A look at 'optimized signal coherence/correlation' in surround microphone systems is taken, both from the point of view of researchers in the field of sound-engineering, as well as psychoacoustics and concert hall acoustics. The results from the search for an 'optimized interaural correlation coefficient' are presented, along with a short analysis concerning the interaction of loudspeaker directivity and listener envelopment, an aspect which is of major importance for spatial impression and related to listener preference.

Keywords Signal correlation · Signal coherence · Stereo · Surround microphone systems · Diffuse sound · Optimized correlation

A detailed analysis of the differences between signal correlation and signal coherence can be found—among others—in Cremer (1976) and a short introduction in Martin (2005). An in-depth analysis can both be found in Hesselmann (1993), and in a chapter of the Encyclopedia of Acoustics written by Piersol (1997), where the following mathematical definitions and explanations have been taken from:

Given two stationary random signals $x(t)$ and $y(t)$, $0 < t < T$, any linear relationship between these two signals will be extracted by the *cross-correlation function* $R_{xy}(\tau)$, defined by:

$$R_{xy}(\tau) = \lim_{T\to\infty} \frac{1}{T-\tau} \int_{0}^{T-\tau} x(t)y(t+\tau)\mathrm{d}t \qquad (2.1)$$

E. Pfanzagl-Cardone, *The Art and Science of Surround and Stereo Recording*,
https://doi.org/10.1007/978-3-7091-4891-4_2

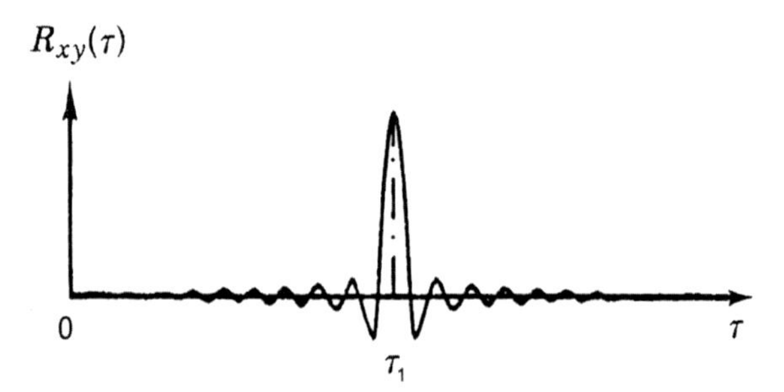

Fig. 2.1 Typical cross-correlation function between wide-bandwidth random signals representing measurements at two points along a propagating acoustic wave (propagation time $\tau = \tau_1$) (after Piersol 1997)

A typical plot of a cross-correlation function between two correlated random signals with wide bandwidths is shown in Fig. 2.1. In this plot, the signals could be interpreted as two measurements along the path of a propagating acoustic wave, where the propagation time is τ seconds. The units of the plot are the product of the magnitudes of $x(t)$ and $y(t)$ (usually Pascals squared for acoustic signals) versus time displacement in seconds.

For some applications, it is more convenient to estimate a normalized cross-correlation function, called the *correlation coefficient function* $\rho_{xy}(\tau)$ given by:

$$\rho_{xy}(\tau) = \frac{R_{xy}(\tau)}{\sigma_x \sigma_y} \tag{2.2}$$

where $x(t)$ and/or $y(t)$ have a zero mean value, and σ_x and σ_y are the standard deviations of $x(t)$ and $y(t)$, respectively. The quantity $\rho^2{}_{xy}(\tau)$ is bounded by zero and unity and defines the fraction of the variance (the square of the standard deviation) of $y(t)$ that is linearly related to $x(t)$; that is, $\rho^2{}_{xy}(\tau) = 0$ means there is no linear relationship and $\rho^2{}_{xy}(\tau) = 1$ means there is a perfect linear relationship between $x(t)$ and $y(t)$ at the time displacement τ. There is also the possibility of $\rho_{xy}(\tau) = -1$ which would indicate a perfect linear relationship between $x(t)$ and $y(t)$, but of opposed (inverted) signal-polarity (or 'out-of-phase' relationship between the two signals, as one would say in sound-engineering terms).

For applications in the realm of digital signal processing, it is important that there is a clearly defined relation between the *cross-correlation function* to its system function. In the frequency domain, this system function is the so-called *cross-spectral density function*:

Given two signals $x(t)$ and $y(t)$, $0 \le t \le T$, any linear relationship between these two signals at various different frequencies will be extracted by the *cross-spectral density function* $G_{xy}(f)$, also called the *cross-spectrum*.

It can be shown that the *cross-spectral density function* is equal to the Fourier transform of the *cross-correlation function*; that is:

$$G_{xy}(f) = 2 \int_{-\infty}^{\infty} R_{xy}(\tau) e^{-j2\pi f \tau} \, d\tau, \quad f \ge 0 \tag{2.3}$$

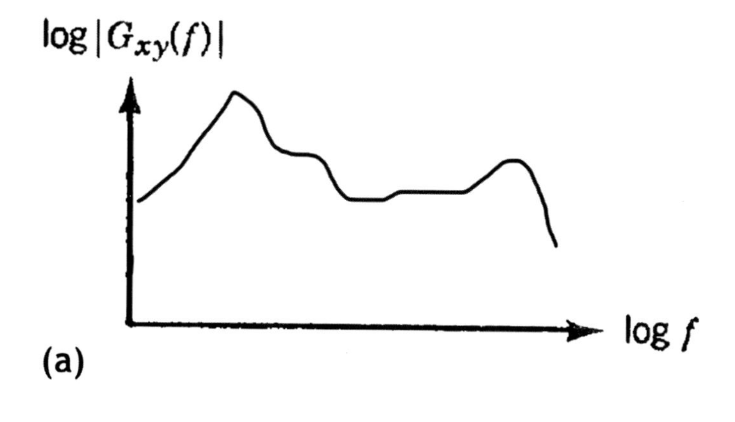

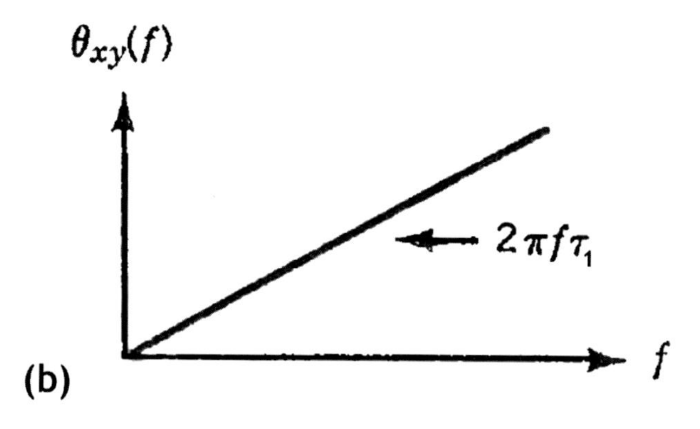

Fig. 2.2 Typical cross-spectral density function between wide-bandwidth random signals representing measurements at two points along a propagating acoustic wave (propagation time $\tau = \tau_1$): **a** magnitude; **b** phase (after Piersol 1997)

where the factor of 2 appears because $G_{xy}(f)$ is a one-sided spectrum. The cross-spectrum can also be expressed in terms of magnitude and phase.

A typical plot of the cross-spectral density function between two correlated random signals with wide bandwidths, having the same origin as discussed in Fig. 2.1 is presented in Fig. 2.2. The units of the plot are the product of the magnitudes of $x(t)$ and $y(t)$ per unit frequency (usually Pascals squared per Hertz for acoustic signals) versus frequency in Hertz. The cross-spectrum may be interpreted as a measure of the linear dependence between two signals as a function of frequency, although the coherence function discussed next is more useful for this application.

It is often convenient to normalize the cross-spectral density magnitude to obtain a quantity called the *coherence function* $\gamma^2{}_{xy}(f)$ (sometimes called 'coherency squared' or 'magnitude squared coherence'), given by:

$$\gamma^2{}_{xy}(f) = \frac{\left|G_{xy}(f)\right|^2}{G_{xy}(f)G_{yy}(f)} \tag{2.4}$$

where $G_{xy}(f)$ is the cross-spectrum and $G_{xx}(f)$ and $G_{yy}(f)$ are the power spectra. The coherence function is bounded by $0 \leq \gamma^2{}_{xy}(f) \leq 1$ at all frequencies and essentially identifies the fractional portion of linear dependence (or correlation) between two signals $x(t)$ and $y(t)$ as a function of frequency. Specifically, $\gamma^2{}_{xy}(f) = 0$ means there is no linear relationship, and $\gamma^2{}_{xy}(f) = 1$ means there is a perfect linear relationship between $x(t)$ and $y(t)$ at frequency f. For values between zero and unity, the coherence can be interpreted as the fractional portion of $G_{yy}(f)$ that can be determined from a knowledge of $G_{xx}(f)$. A plot of $\gamma^2{}_{xy}(f)$ is a dimensionless number versus frequency.

For the measurements undertaken by the author and presented in this book, the 'mscohere'-function in MATLAB has been used, which finds the magnitude squared coherence estimate using Welch's averaged, modified periodogram method.

2.1 Signal Correlation in Stereo Microphone Systems

The signals of both channels of a microphone system for stereo recording result from the summing of direct and diffuse sound at the positions of each of the capsules. Therefore, the content of the stereo signal depends on the distance between sound source and microphone system (direct/diffuse sound ratio), the acoustic properties of the room (geometry and absorption coefficients of the boundary surfaces of the room and of objects contained within), as well as the relative position of sound source and microphones in the room (Fig. 2.3).

An early research concerning the cross-correlation of the signals of two pressure transducers in a diffuse sound field can be found in Cook et al. (1955), which is around the time during which various record companies had started to deal with the challenges of the still very new 'stereophonic' recording techniques.

To explain the constants and variables used in Fig. 2.4:

$$R = (\sin kr)/kr \tag{2.5}$$

with: $k = 2\pi/\lambda$ (k = wave-number; λ = wavelength), therefore

$$kr = 2\pi r/\lambda \tag{2.6}$$

or $k = \omega/c$, since $\lambda = c/f$ (with c … speed of sound, i.e. approx. 340 m/s). The complete deduction can be found in Elko (2001).

In the same publication, the calculation for signal coherence of various microphone arrays is found, which are combinations of two capsules each with varying polar patterns and orientations (0°, 90°, 180°). A few examples:

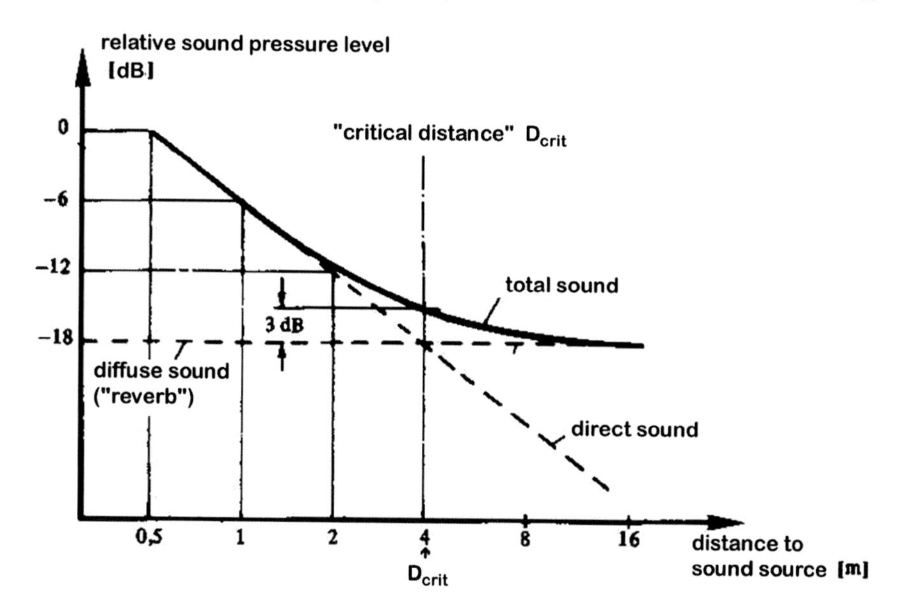

Fig. 2.3 Relation between total sound, direct and diffuse sound

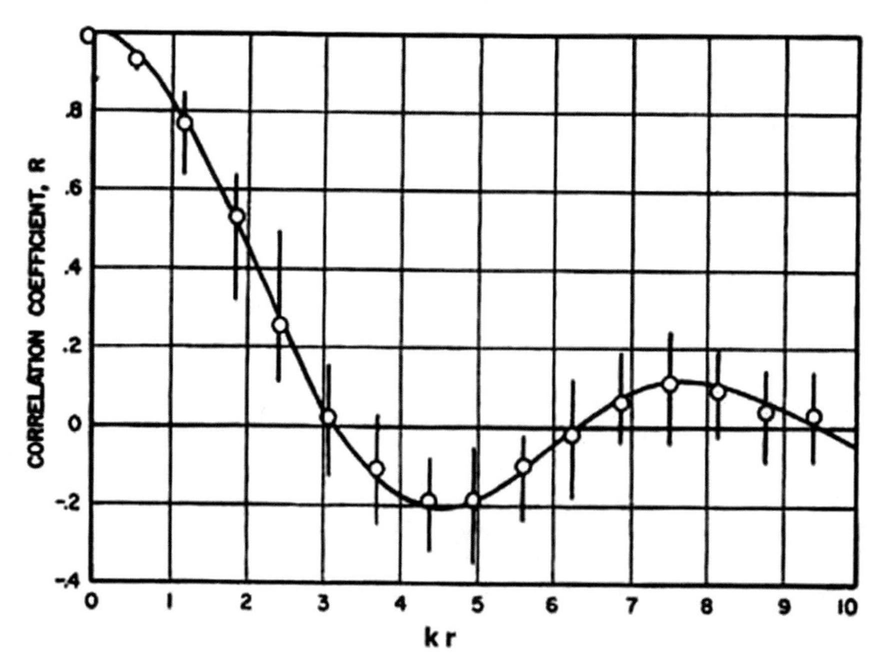

Fig. 2.4 Correlation-coefficient R versus capsule distance of two microphones (solid line: theoretical calculation; o = measured; from Cook et al. 1955)

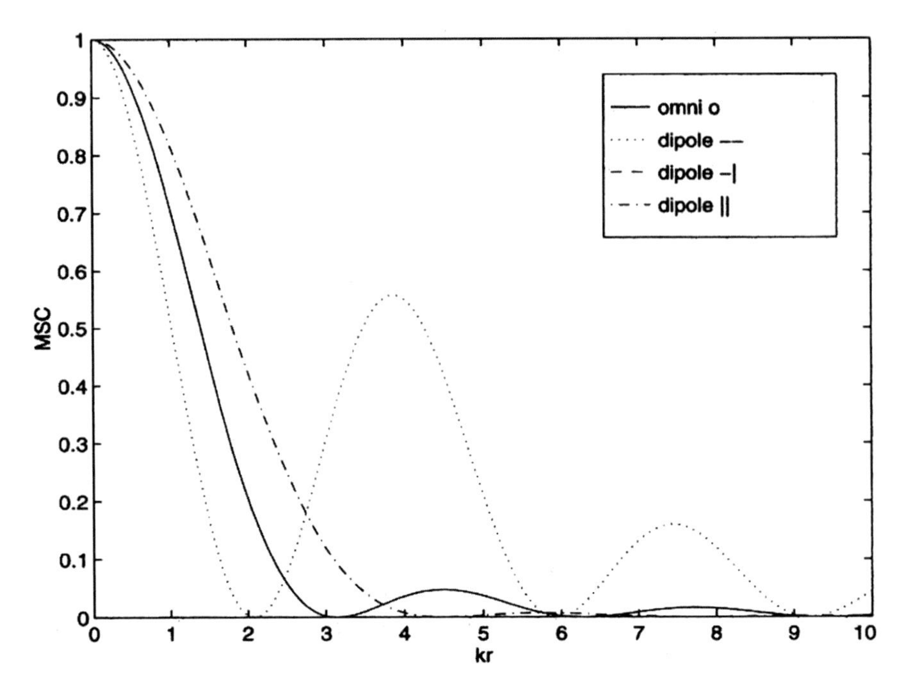

Fig. 2.5 Magnitude-squared coherence (MSC) for omni-directional and bidirectional (i.e. fig-8) microphones in a spherical isotropic sound field. Note that the coherence function for the orthogonal fig-8s coincides with the x-axis (from Elko 2001)

While the two fig-8 microphones (dipoles) which are orthogonal to each other (i.e. at an included angle of 90°) are basically a 'Blumlein-Pair,' the parallel fig-8 microphones can be regarded as a 'Faulkner Phased Array' (dash-dot line). An arrangement in which both fig-8s are turned side-wise (with their null-axis pointing toward the sound source) can be seen as one half of a 'Hamasaki Square' (dotted line in Fig. 2.5).

The combination of omni and fig-8 transducer in Fig. 2.6 is essentially an MS-stereo microphone technique with an omni for the 'mid' microphone. In Pfanzagl-Cardone (2002), it was shown on an empirical level that the use of the same microphone technique in acoustically different rooms will cause the same characteristic patterns in terms of signal cross-correlation over frequency. This indicates that there is a kind of individual 'fingerprint' of correlation over-frequency property for any stereo or surround microphone technique (see also Pfanzagl-Cardone and Höldrich 2008).

In later research by Muraoka et al. (2007) frequency-dependent cross-correlation for signals of a few stereo microphone techniques (AB, Wavefront [i.e. similar to the 'microphone curtain,' consisting of an inner and outer AB pair], ORTF, XY and MS) have been calculated (Fig. 2.7).

The graphs obtained from the measurements of cross-correlation over frequency (FCC) for various microphone systems as carried out by the author (as presented in

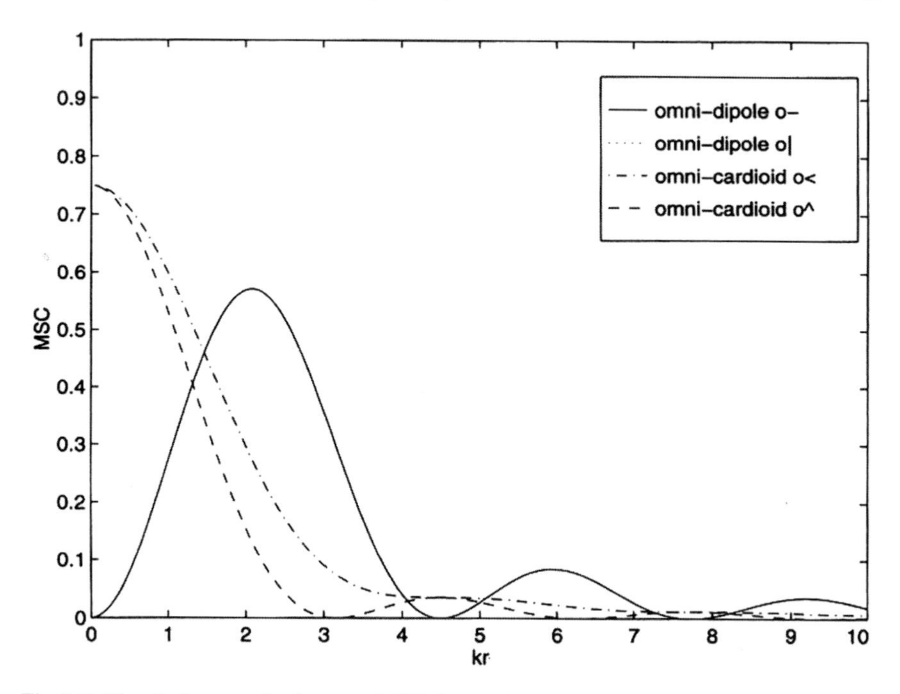

Fig. 2.6 Magnitude squared coherence (MSC) for omni-directional, bidirectional (dipole or fig-8) and cardioid capsules in an isotropic sound field. The curves for the two combinations 'omni-dipole' share the same solid curve (from Elko 2001)

Chaps. 7 and 8) are differing from the theoretical calculations mainly because of the following two reasons:

1. Directional patterns of real microphones are almost never ideal (e.g. cardioid patterns slightly change shape in different frequency bands, with a tendency to become more omni-directional toward low frequencies; likewise, omni-capsules can exhibit an increase in directional pickup-sensitivity toward high frequencies, depending on their type of construction), and—probably even more relevant,
2. The acoustic properties of the room, particularly with respect to the characteristics of room modes and the diffuse sound in that room, which has a great influence on the resulting overall signal correlation, and
3. The graphs display the combination of direct *and* diffuse sound (and not only the diffuse sound component, as in an isotropic sound field).

The calculation of the effective recording angle of a stereo- (or surround-) microphone system usually concerns only direct sound from the source. The question how a microphone technique picks up diffuse sound has usually received much less attention in this respect, even though it is a majorly important aspect in relation as to how a microphone technique of choice is able to capture the sound event as a whole in a performance venue.

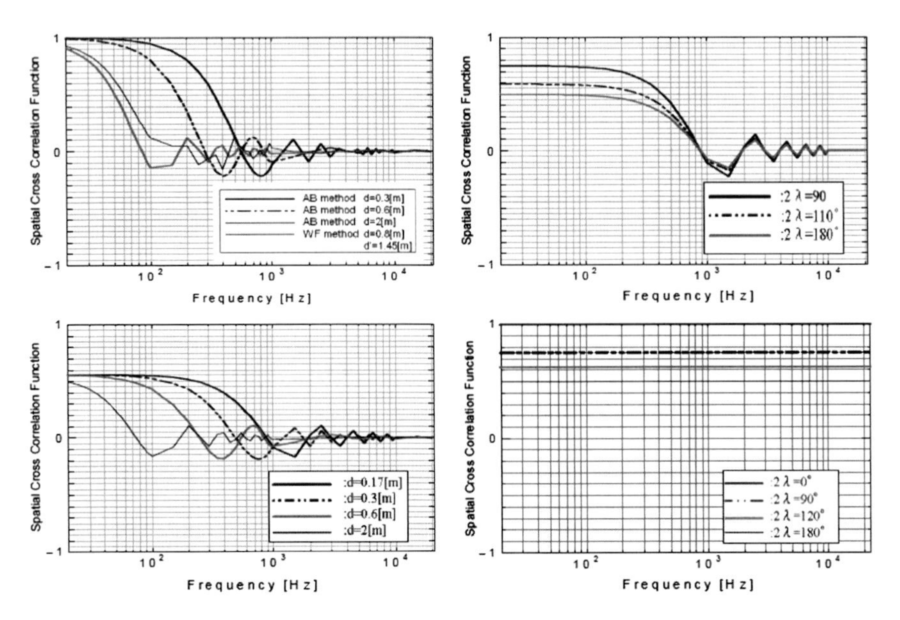

Fig. 2.7 Calculation of the cross-correlation for various stereo microphone techniques; FSCC = Frequency-dependent Spatial Cross-Correlation (from Muraoka et al. 2007). Left up: FSCC-patterns of AB microphones and WF-microphone; Left down: FSCC of ORTF microphones (capsule spacing-dependent); Right up: FSCC of ORTF microphones (directional azimuth-dependent); Right down: FSCC of XY microphone

While it is relatively easy to capture direct sound in a 'de-correlated' manner by use of directional microphones and applying appropriate opening angles in-between, for a 'de-correlated' pickup of diffuse sound it is necessary to employ either microphones with strong directivity (e.g. hyper-cardioids or fig-8 microphones) in combination with appropriate opening angles (i.e. 90° in case of a Blumlein-Pair), or—when pure pressure transducers are used—the capsule spacing must be larger than the reverberation radius.

For 'real-world' microphones (even if they are of high quality) the directivity often deviates from ideal with frequency (e.g. cardioids may become almost omnis toward low frequencies), so that also diffuse sound is no longer picked up in a de-correlated manner. From concert hall acoustics and psychoacoustics we know however, that for good and convincing spatial impression binaural signals at the listener's ears need to be de-correlated especially in the low-frequency band below 500 Hz (and even more so below 200 Hz). For this reason, it can happen that microphone techniques with small capsule spacings may provide a satisfactory 'stereophonic image' (ASW) at a few kHz, but fail to deliver sufficient de-correlation in the low-frequency region to also ensure proper spatial impression, resulting in an almost monophonic sound for the bass region (see Griesinger 1997, p. 7).

In this respect—regarding the question how well a microphone technique is able to capture also diffuse sound—the two abovementioned systems (Blumlein-Pair and

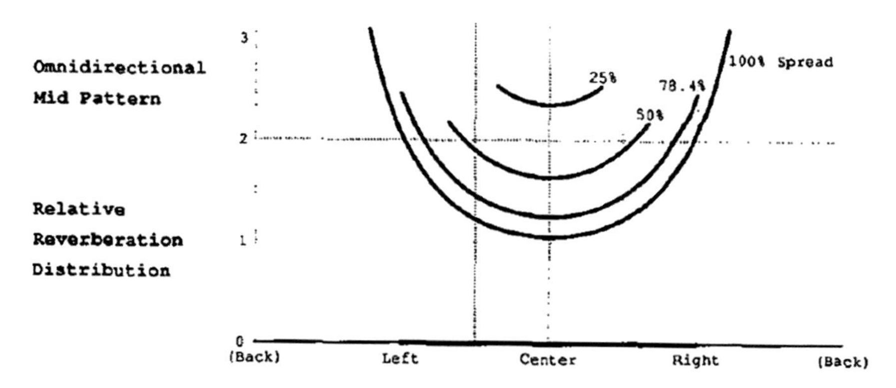

Fig. 2.8 Diffuse sound (reverb) distribution for (MS) stereo-encoding with mid-microphone in omni-mode: level of side-signal versus spread (in [%]) (from Julstrom 1991)

Large-AB with a capsule spacing larger than the reverberation radius) represent two opposed extremes (in terms of microphone pattern, as well as capsule positioning), for which complete de-correlation is ensured (assuming 'ideal' directional patters of the microphones). The vast majority of stereo microphones does not manage to fulfill this criterion in relation to diffuse sound pickup. This may be one of the main reasons, why these two techniques stand out sonically from the rest, according to the opinion of the author, but also measurable in acoustic terms (see FCC and FIACC measurements in later chapters).

In Julstrom (1991), an analysis can be found concerning how various coincident microphone systems (XY and MS) pickup and 'map' (or project) direct and diffuse sound components onto the stereo base of a regular stereo loudspeaker setup. For this purpose, Julstrom had converted XY techniques into the respective equivalent MS technique (with the same recording angle, etc.): the virtual S-microphone (side-microphone) always has fig-8 characteristics, while the virtual mid-microphone assumes various characteristics. The level ratios between side- and mid-microphone are altered according to the desired recording angle; in Figs. 2.8, 2.9, 2.10 and 2.11, a 'spread' of—say—25% (depending on the directional pattern) is equivalent to a relative side-signal level of -14 dB (fig-8) to -11 dB (omni), while a spread of 100% corresponds to a relative level of $+3$ dB (omni) to 0 dB (fig-8). In Figs. 2.8, 2.9, 2.10 and 2.11, the distribution for direct and diffuse sound for 'stereo surround' reproduction is depicted for some of the most common microphone techniques (Rem.: In case of a pure 2-channel stereo reproduction, only the inner part of the horizontal left-center-right scale is relevant and the parts lying outside are being mirrored back to the 'inner scale'; in case of a multichannel surround loudspeaker system and appropriate decoding, the diffuse sound signals could also be replayed on the rear-quadrants).

In Figs. 2.8 and 2.11, the two extremes in terms of distribution of diffuse sound (or reverb) are worth to be pointed out:

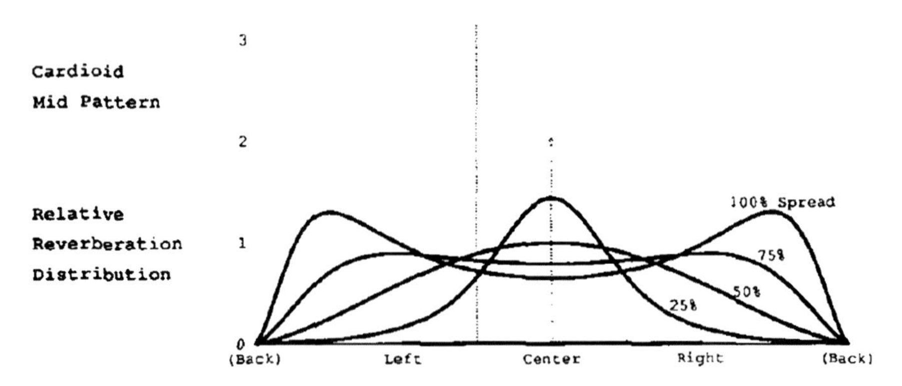

Fig. 2.9 Diffuse sound (reverb) distribution for (MS) stereo-encoding with mid-microphone in cardioid-mode: level of side-signal versus spread (in [%]) (from Julstrom 1991)

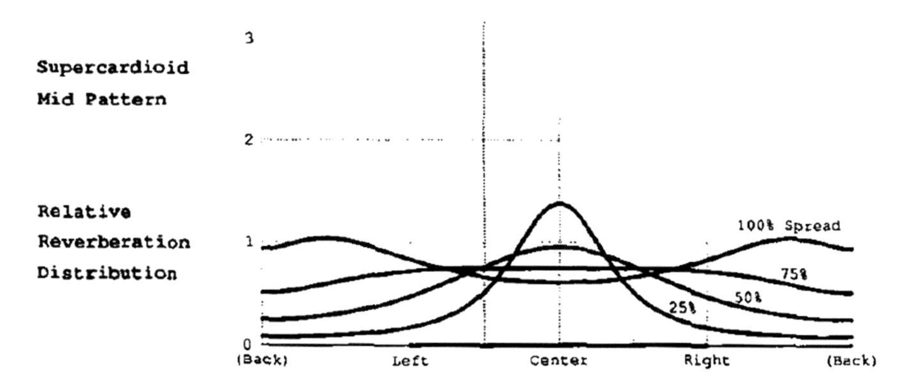

Fig. 2.10 Diffuse sound (reverb) distribution for (MS) stereo-encoding with mid-microphone in supercardioid-mode: level of side-signal versus spread (in [%]) (from Julstrom 1991)

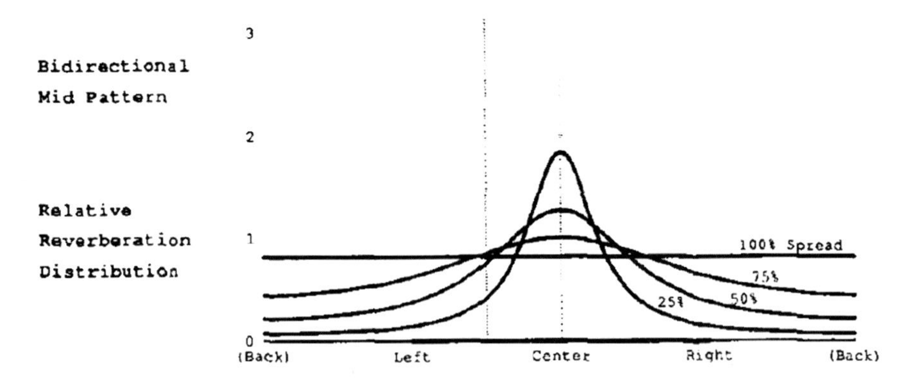

Fig. 2.11 Diffuse sound (reverb) distribution for (MS) stereo-encoding with mid-microphone in figure-of-8 (= bidirectional)-mode: level of side-signal versus spread (in [%]) (from Julstrom 1991)

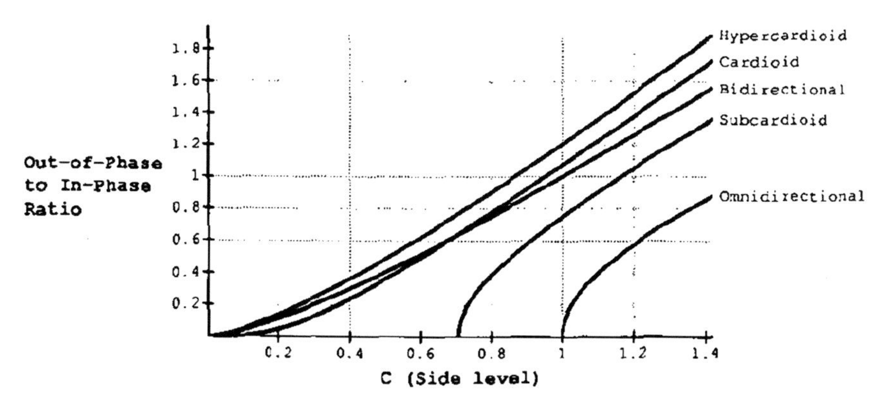

Fig. 2.12 Ratio of out-of-phase signal components versus in-phase signal components (from Julstrom 1991)

1. When using a mid-microphone with omni pattern, both direct and diffuse sound—due to lack of directional characteristic—are represented on the frontal basis *L*, *C*, *R*.
2. When using a fig-8 as mid-microphone, we have the singular event of an equal level distribution between side- and mid-microphone ($C = 0$ dB) over 360°. Therefore, this technique (i.e. Blumlein-Pair) is the only one which—in case of surround-replay—allows for a natural, even distribution of diffuse sound.

The combination of two fig-8 microphones as an MS-pair with a spread of 100% can be converted into the traditional 'Blumlein-Pair' (with the capsules on-axis oriented at −45°/+45°). This represents a special case among the techniques with strong directional patterns (i.e. super- and hypercardioid, fig-8), as with the Blumlein-Pair the ratio between in-phase and out-phase signal components always is in favor of the in-phase components up to equal level ratio of mid- to side-signal (see Fig. 2.12).

According to Julstrom (1991, p. 644) with an analysis of the 'Difference-to-Sum' ratio of an MS-signal, the aspect of stereophonic spaciousness (of the microphone technique under test) can be predicted. With small values of the 'Difference-to-Sum' ratio, the sonic image (stereophonic width) tends toward mono. Griesinger (1987) proposes a ratio ≥ 1 in order to achieve low correlation between the stereo loudspeaker signals. In the case of surround-replay, signal components that occur in the sum-signal will be reproduced frontally, while difference-signal components will be reproduced from behind, so that the 'Diffuse-to-Sum' ratio basically determines the front/back reverb balance. In any case, it seems advisable to avoid ratios which are higher than 1 (Rem.: Due to the technology available at the time of his research, Julstrom refers to an older Dolby-Surround coding standard, which contained only one (mono) rear channel, the information of which was coded into the analogue 'Dolby-Stereo' signal in the form of a difference-signal).

The following considerations concerning the ratio of difference-to-sum signal show that microphones with a directivity index in-between cardioid and fig-8, the recommendation of Griesinger (i.e. 'Difference-to-Sum' ratio ≥ 1) can be fulfilled,

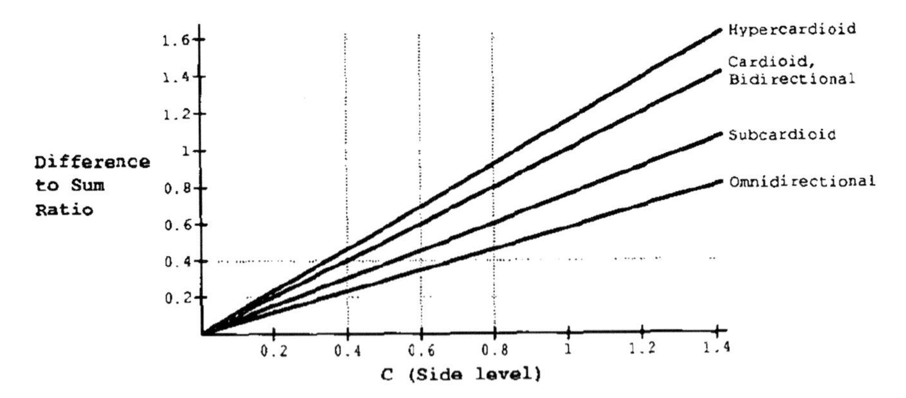

Fig. 2.13 Ratio of diffuse sound (reverb) in the stereo signal (side-signal) to sum-signal (mid-signal, mono) (from Julstrom 1991)

but with one disadvantage though: as we have seen in Fig. 2.12 for all microphone patterns with a directivity higher than cardioid (except fig-8) with $C = 1$ the out-phase signal components prevail, which is not to the benefit of sonic perception at the listener, as out-phase components in the L/R signal are likely to impact the impression of naturalness of a recording in a negative way (Fig. 2.13).

According to the analysis above—based on an interpretation of the results by Julstrom—the 'Blumlein-Pair' technique, consisting of two fig-8 microphones at an included angle of 90°, stands out as a singular case which is characterized by 'ideal' reproduction characteristics (as it favors a natural 360° distribution of sound, especially in respect to diffuse sound).

2.2 Optimized Signal Coherence in Surround Microphone Systems

In various of his publications (see, e.g., Theile 2000), Günther Theile conveys the idea that a minimum signal coherence $\gamma \geq 0.35$ is necessary for a (quadraphonic) surround microphone system in order to achieve '... natural reproduction of space and envelopment ... ' (ibid., p. 419).

He arrives at that conclusion upon his interpretation of the research of Damaske (1967). For the majority of his experiments Damaske had used band-limited noise and music signals in the frequency range from 250 to 2000 Hz with different values of coherence. The purpose of which was to let the test listeners evaluate the perceived degree of 'diffusion,' depending on the spatial distribution (i.e. placement) of the loudspeakers, as well as the value of coherence of the test-signal (Fig. 2.14).

In his paper, Damaske writes (translated) '... to name a coherence factor for music is problematic, as it certainly depends on the momentary bandwidth of the music signal.' Based upon the results of his experiments, he assumes that the coherence of

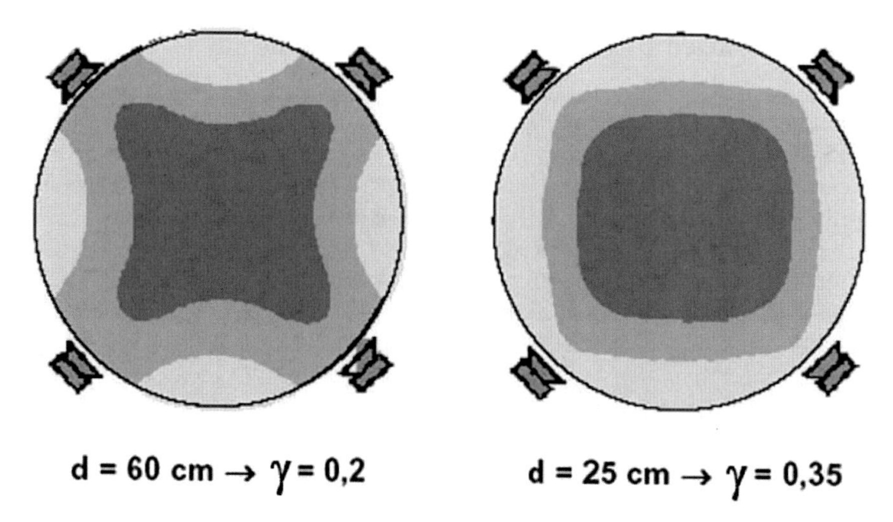

Fig. 2.14 Perceived directions in a diffuse field generated by four loudspeakers radiating noise (0.25–2.5 kHz) with inter-channel coherence factors γ. The noise signal was recorded in a reverberation chamber by means of spaced omni microphones with capsule spacing d (from Theile 2000)

music signals should not deviate much from that of band-limited noise signals (see Damaske 1967, p. 203). In his publication, Damaske also writes that test listeners were describing that—with incoherent sound fields—phantom sources would appear to '... come from different directions,' not necessarily limited to the connecting lines (base widths) between the loudspeakers. In addition, the listeners reported that the resulting spatial impression was independent of head-movement and more pleasant than the impression in a coherent sound field (ibid., p. 204).

Considering the conclusions of Damaske, it seems very daring that Theile simply takes the optimal values of coherence found with band-limited noise by Damaske and claims their validity also in relation to music signals. In fact, as will be pointed out in Chap. 7, which deals with signal correlation and coherence in surround microphone systems, there is strong evidence that the optimal coherence value of $\gamma \geq 0.35$, as proposed by Theile, is not suited to evoke a 'natural spatial impression'; instead the results of various studies indicate that a high degree of de-correlation (or incoherence) in all frequency bands is preferable in order to achieve convincing spatial impression.

In this context, it seems appropriate to raise the question if Theile's biased interpretation of Damaske's research—together with the opinions of other researchers (see Sengpiel 1992 and Wuttke 1998)—may be misleading for many sound-engineers and researchers in the field: to look for surround microphone systems with high correlation (i.e. a minimum coherence of 0.35, but preferably higher), while in reality they should be looking for the lowest correlation possible in all frequency bands instead.

2.3 Optimized Interaural Correlation Coefficient

The abovementioned proposal by Theile concerning optimal coherence with $\gamma \geq 0.35$ also contradicts research by Gottlob (1973): For the performance of classical music, there is an optimal reverb time preferred by the listeners, which—depending on the size of the room—usually lies between 1.8 and 2.1 s [compare Beranek (2004, p. 494) and Pierce (1983, p. 151) and Gottlob (1973). Rem.: for a precise calculation of optimal reverb time vs. room volume see Watson (1923)]. In analogy to that, Gottlob concludes in his research that the optimal interaural cross-correlation IACC (for live music in a concert hall) equals 0.23 (see Cremer und Müller 1978, p. 482).

More recent research by Muraoka et al. (2007) seems to indicate that for the reproduction of recorded music the correlation of the recorded stereo signal should optimally be zero. This evaluation seems to be based primarily on an informal evaluation by the team of the authors. In their paper, 'Ambience sound recording utilizing dual MS (Mid-Side) microphone systems based upon Frequency-dependent Spatial Cross-Correlation (FSCC)' the authors claim that they '…studied the sound pickup characteristics of typically used main microphones to clarify their ambience representation, employing Frequency-dependent Spatial Cross-Correlation (FSCC) as a criterion, and concluded that MS-microphone set with a directionality azimuth at 132° is best. FSCC of the microphone output becomes uniformly zero [Rem.: see also Chap. 8, especially Figs. 8.7 and 8.8 for more details]. This makes [the] listener feel the most natural ambience, and the effect was certified through actual orchestral recording.' The fundamental requirements for the main microphone systems under test—as set out by the authors—were:

1. sufficient ambience description and
2. exact sound source localization.

In this context research by Nakahara (2005) should also be mentioned, which states that a low correlation between the loudspeaker signals (L/R vs. C vs. LS/RS) ensures better compatibility between different reproduction environments as well as to an enlargement of the 'sweet spot' (see also Prokofieva 2007). This is one of the reasons why—preferably—microphone techniques should be used, if possible, which are characterized by low signal correlation over the whole frequency range (Fig. 2.15).

2.4 Evaluations Concerning the Interaction Between Loudspeaker Directivity and Listener Envelopment

Many sound-engineers know the phenomenon that a surround microphone technique may provide convincing 'acoustic envelopment' and spatial impression as long as the recording and reproduction of musical signals is concerned, but as soon as the applause of the audience is heard, it becomes evident to the critical listener that he/she

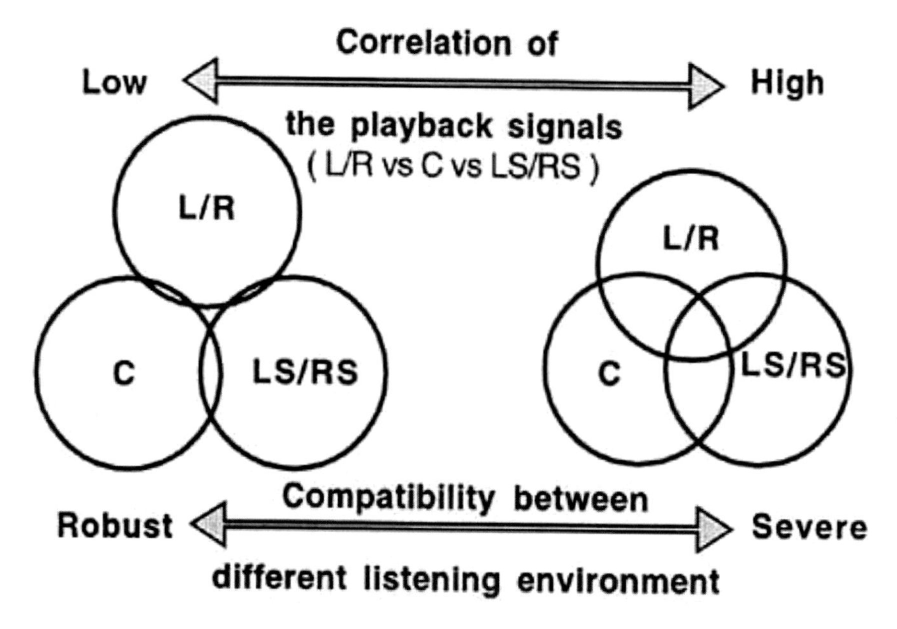

Fig. 2.15 Amount of signal cross-correlation in relation to compatibility between different listening environments (from Nakahara 2005)

is listening to a reproduction via loudspeakers. The main reason for this seems to be the different spectral composition of music vs. noise signals (applause). While acoustic music usually has less energy in the higher frequency bands, applause can be compared to colored noise (similar to 'pink noise,' which is characterized by a spectral power distribution proportional to 1/f).

Loudspeakers usually have a strongly frequency-dependent radiation characteristic and while low frequencies are being radiated in a near omni-directional manner, mid and high frequencies are being radiated in an increasingly directional way (see Fig. 2.16). This fact is most likely responsible for the effect that listeners may localize sound sources 'in the speaker,' instead of feeling a well-balanced distribution of—or even envelopment by—sound, which apparently can happen rather easily when the reproduced sound signal contains a high amount of high-frequency signal.

In this context, the assumption may be expressed that the use of loudspeakers with a well defined, but largely frequency-independent radiation pattern (see, e.g., the loudspeaker models Beolab 5 and Beolab 3 by Bang & Olufsen), which use 'Acoustic Lens Technology' (patented by Sausalito Audio) may have caused different results if they had been used in the experiments of Damaske (Figs. 2.17 and 2.18).

With traditional n-way (with $n \geq 2$) dynamic loudspeaker systems (constructed according to the 'direct radiator' principle, i.e. with direct, directional radiation characteristics), the signal of interest is split into several frequency bands. In case of the abovementioned noise or applause signals, the much smaller high-frequency tweeter

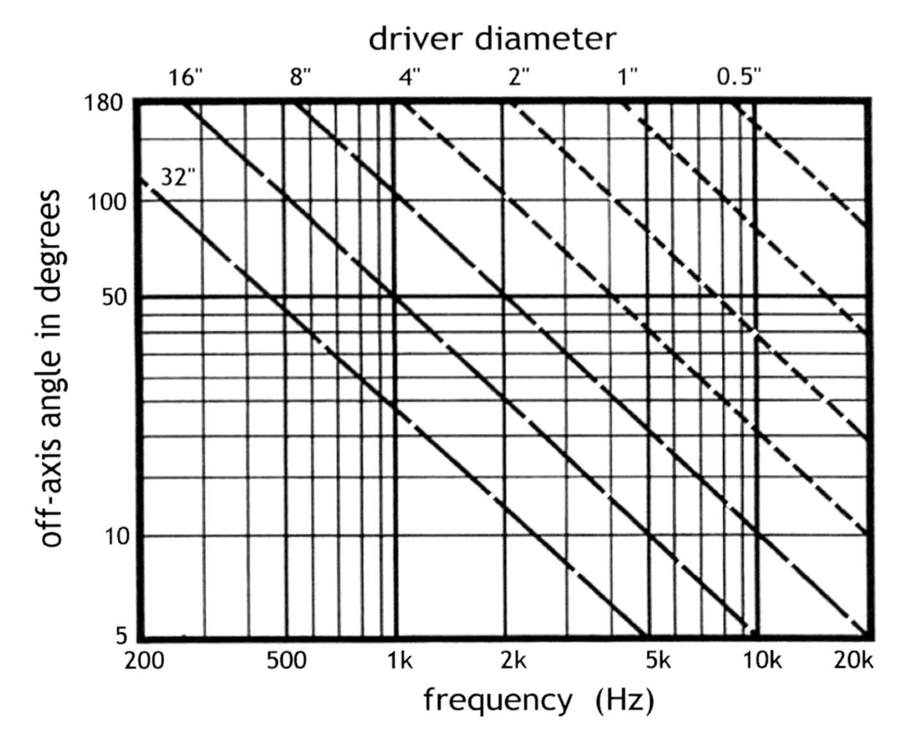

Fig. 2.16 Radiation angle of a loudspeaker membrane (diaphragm or cone driver) versus frequency for selected driver-diameters (after Fig. 0.8 from Dickason 2000)

has to radiate much more energy per square centimeter (or inch) than the mid- or low-frequency driver. This, combined with the stronger directivity of a high-frequency membrane (if they are not being used in connection with a horn or waveguide system) enhances the effect that the listener will easily perceive the HF-driver as a more clearly distinguishable point source than would be the case with an LF-driver (always under the condition of using an appropriate sound signal, which allows for this effect).

Most likely this effect will not appear with loudspeakers with lower directivity (or even frequency-independent, omni-directional radiation characteristic) or at least would appear to a much lesser extent. To this group belong all loudspeakers which use membranes, that are large in relation to the wavelength of the highest frequency they have to radiate, like electro-static and magneto-static loudspeakers (usually with a rectangular membrane of significant surface dimensions), or dynamic speakers using the 'acoustic lens' principle (see Bang & Olufsen 'BeoLab' (Fig. 2.19), as well as Yamaha 'Relit' (Fig. 2.20), among others) or systems, which use a different principle of radiation, like—for example—if 'indirect radiation' in the form of reflection at larger surfaces is employed (see Pfanzagl 2007). [Regarding the interaction between loudspeaker directivity and adjacent reflective surfaces like side-walls, floor, ceiling, furniture please refer to Sect. 1.5 which deals with 'The influence of loudspeakers

Fig. 2.17 'BeoLab'3- way loudspeaker with acoustic lens technology (ALT) (from Bang & Olufsen promotional folder)

and listening room acoustics on listener preference' and the publications by Olive et al. (1994) and Toole (2008)].

2.5 Summary and Conclusion

If Theile strongly favors a coherence of $\gamma \geq 0.35$ for (4-channel) surround signals (see Sects. 2.2 and 2.3, as well as Theile 2000), claiming that only under this condition also for applause listeners will have the impression of being enveloped (instead of perceiving the loudspeakers as isolated sources), he completely ignores the spectral differences between applause and music (with the latter being the real signal of interest) and hence arrives at wrong conclusions regarding the optimum degree of coherence for music signals, which is also in stark contrast with the findings of optimal interaural signal correlation of 0.23 for listeners in a concert hall, as claimed by Gottlob (1973), as well as the findings by Muraoka (see Muraoka et al. 2007), which point in the direction of complete de-correlation.

We will see in later Chaps. 8 and 9, that only a few stereo microphone systems comply to the criterion of de-correlation:

- Large-AB with two omni capsules, spaced apart by more than the reverberation radius of the room,
- MS technique with an effective recording angle of 132° (see Muraoka et al. 2007),

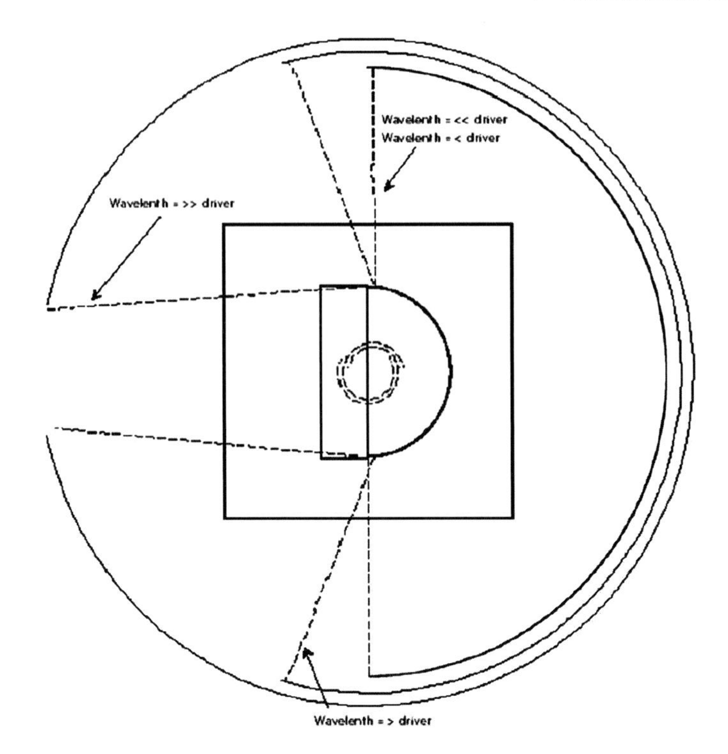

Fig. 2.18 Schematic view of horizontal radiation angles of 'Acoustic Lens Technology' depending on the relation of wavelength to driver-diameter (from Moulton 1998)

- Hypercardioids at an included angle of 133° are largely de-correlated as well (see Griesinger 1987), and finally
- A Blumlein-Pair of two crossed figure-of-eight microphones at 90° is ideally de-correlated over the entire frequency range (see Fig. 2.5 from Elko 2001).

2.6 Further Notice

I would also like to point the interested reader to a visual documentation series of measurements which are related to real-time frequency-dependent cross-correlation measurements (FCCs), which can be found on my Youtube channel 'futuresonic100,' when looking for clips with the term 'mic tech analysis' in the title. These video clips have been produced utilizing MAAT's '2BC multiCORR' VST plugin, which displays the cross-correlation of two audio channels *over frequency* in realtime. Several audio samples, based on recordings with well-documented microphone techniques, are analyzed with respect to their signal correlation, which unveils interesting details concerning their overall sonic character (Rem.: more information on the '2BC multiCORR' plugin can be found in Appendix C).

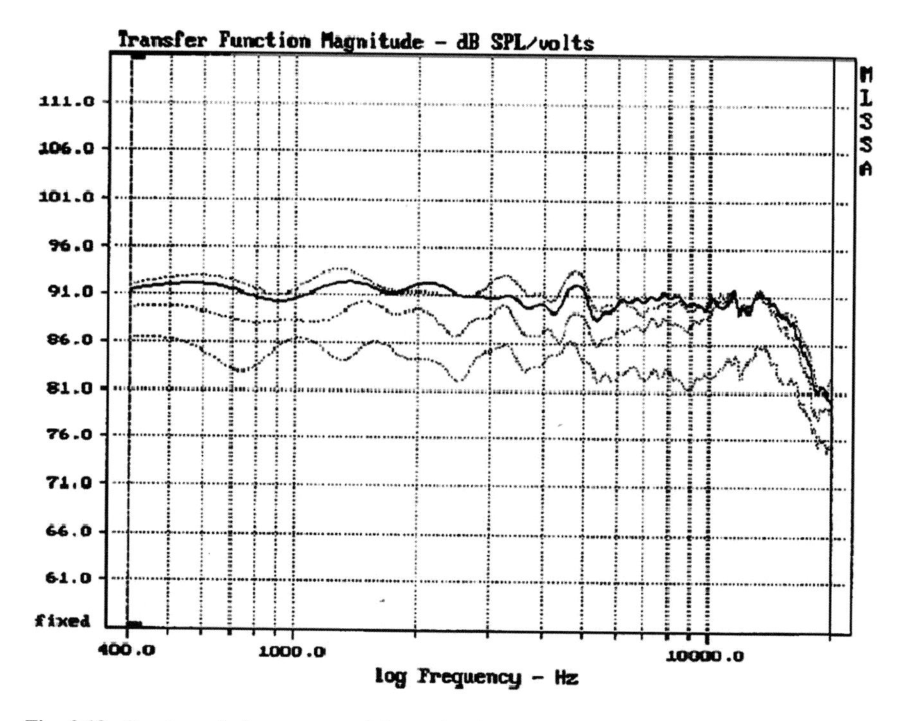

Fig. 2.19 Quasi-anechoic response of 3-way loudspeaker prototype, measured at 0°, 30° (the design axis), 60° and 90°, using acoustic lens technology for midrange and HF-tweeter sections (from Moulton 1998)

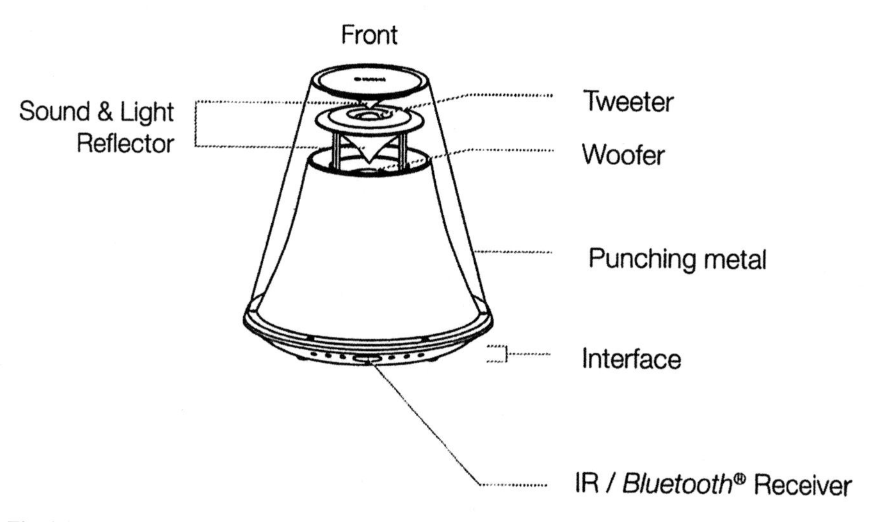

Fig. 2.20 Another example of loudspeaker with 360° radiation angle: the Yamaha LSX-170

References

Beranek L (2004) Concert halls and opera houses: music, acoustics and architecture, 2nd edn. Springer, New York

Cook RK, Waterhouse RV, Berendt RD, Edelman S, Thompson MC (1955) Measurement of correlation coefficients in reverberant sound fields. J Acoust Soc Am 27(6):1072

Cremer L (1976) Zur Verwendung der Worte 'Korrelationsgrad' und 'Kohärenzgrad'. Acustica 35:215–218

Cremer L, Müller HA (1978) Die wissenschaftlichen Grundlagen der Raumakustik. (Band 1). Hierzel-Verlag, Stuttgart

Damaske P (1967/68) Subjective investigation on sound fields. Acustica 19:199–213

Dickason V (2000) The loudspeaker design cookbook. Audio Amateur Publishing

Elko GW (2001) Spatial coherence functions for differential microphones in isotropic noise fields. In: Brandstein M, Ward D (eds) Microphone arrays. Springer, p 61

Griesinger D (1987) New perspectives on coincident and semi-coincident microphone arrays. Paper 2464 presented at the 82nd Audio Eng Soc Convention, May 1987

Griesinger D (1997) Spatial impression and envelopment in small rooms. Paper 4638 presented at the 103rd Audio Eng Soc Convention

Gottlob D (1973) Vergleich objektiver akustischer Parameter mit Ergebnissen subjektiver Untersuchungen an Konzertsälen. Dissertation, Universität Göttingen

Hesselmann N (1993) Digitale Signalverarbeitung – Rechnergestützte Erfassung. Analyse und Weiterverarbeitung analoger Signale, Vogel, Würzburg

Julstrom S (1991) An intuitive view of coincident stereo microphones. J Audio Eng Soc 39(9)

Martin G (2005) A new microphone technique for five-channel recording. Paper 6427 presented at the 118th Audio Eng Soc Convention, Barcelona

Moulton D (1998) The use of an acoustic lens to control the high frequency dispersion of conventional soft dome radiators. Paper presented at the American Loudspeaker Manufacturer's Association Symposium, Las Vegas, Jan 1998

Muraoka T, Miura T, Ifukuba T (2007) Ambience sound recording utilizing dual MS (Mid-Side) microphone systems based upon Frequency dependent Spatial Cross Correlation (FSCC). Paper 6997 to the 122nd Audio Eng Soc Convention, Vienna

Nakahara M (2005) Multichannel monitoring tutorial booklet (M2TB) rev. 3.5.2. Yamaha Corp 2005, SONA Corp 2005, p 41

Olive SE, Schuck PL, Sally SL, Bonneville ME (1994) The effects of loudspeaker placement on listener preference ratings. J Audio Eng Soc 42(9):651–669

Pfanzagl E (2007) Vom Punkt zum Flächenstrahler. PROSPECT – Magazin der OETHG für Bühnen und Veranstaltungstechnik, Dec 2007, p 32

Pfanzagl-Cardone E (2002) In the light of 5.1 surround: why AB-PC is superior for symphony-orchestra recording. Paper 5565 presented at the 112th Audio Eng Soc Convention, Munich

Pfanzagl-Cardone E, Höldrich R (2008) Frequency-dependent signal-correlation in surround- and stereo-microphone systems and the Blumlein-Pfanzagl-Triple (BPT). Paper 7476 presented at the 124th Audio Eng Soc Convention, Amsterdam

Pierce JR (1983) The science of musical sound. Scientific American Books

Piersol AG (1997) Statistical theory of acoustic signals. In: Crocker MJ (ed) Encyclopedia of acoustics. Wiley, New York

Prokofieva E (2007) Relation between correlation characteristics of sound field and width of listening zone. Paper 7089 presented at the 122nd Audio Eng Soc Convention, Vienna

Sengpiel E (1992) Grundlagen der Hauptmikrophon-Aufnahmetechnik – Skripten zur Vorlesung (Musikübertragung). Hochschule der Künste, Berlin. http://www.sengpielaudio.de. Accessed 5 May 2004

Theile G (2000) Mikrofon- und Mischungskonzepte für 5.1 Mehrkanal-Musikaufnahmen. In: Proceedings to the 21. Tonmeistertagung des VDT, Hannover, pp 348

Toole FE (2008) Sound reproduction – loudspeakers and rooms. Focal Press (Elsevier)

Watson FR (1923) Acoustics of auditoriums. New York
Wuttke J (1998) Das Mikrofon zwischen Physik und Emotion. In: Proceedings to the 20. Tonmeistertagung des VDT, Karlsruhe, p 460

Chapter 3
Stereo Microphone Techniques

Abstract In this chapter, almost 20 stereo microphone techniques are explained in detail and partly analyzed. Starting out with coincident techniques (XY) with various patterns, via MS techniques to seemingly odd systems using 'coincident omni-directional' capsules. Systems using three capsules (like BPT or DECCA-triangle) are being covered, the former of which is also analyzed with respect to its localization properties by use of the 'Image Assistant 2.0' Java-App. Pitfalls and advantages of small and large AB techniques are pointed out by means of the newly introduced FCC-analysis (Frequency-dependent correlation coefficient). More exotic techniques such as the 'Faulkner Phased Array' (also in an enhanced version) and the 'Jecklin disk' are presented before we move on to examine several techniques which belong to the group of 'equivalence stereophony' (ORTF, NOS, etc.) Toward the end of this chapter, we deal with 'Tree'-based techniques: DECCA-tree, OCT as well as the newly presented DHAB, SHAB and CHAB techniques. To finish up, there is a section on 'combined microphone techniques' (e.g. AB-Polycardioid Centerfill (AB-PC) and AB-Blumlein Centerfill) which excel in combining diverse positive aspects of the microphone techniques involved in order to obtain a sonic result that is superior to what any single technique could achieve on its own.

Keywords Microphone technique · Stereo · Coincident · AB pair · Equivalence stereophony · DECCA-tree

The following description of the most common stereo microphone techniques is partly based on Streicher and Dooley (1985).

3.1 Coincident Techniques

In coincident arrangements, the microphone capsules are usually aligned on top of each other along a vertical axis, while including a certain 'opening angle' in the horizontal axis. The signal differences between the two channels of the stereo signal

E. Pfanzagl-Cardone, *The Art and Science of Surround and Stereo Recording*,
https://doi.org/10.1007/978-3-7091-4891-4_3

are caused by the volume differences that occur due to the directional characteristics of the capsules. The usual patterns are subcardioid, cardioid, hyper-cardioid and figure-of-eight.

3.1.1 XY with Cardioids and Hyper-cardioids

In this case, the horizontal (physical) opening angle is usually between 60 and 120°, which—in turn—determines the so-called recording angle. Since individual microphones pick up the largest part of a sound-source only 'off-axis,' it is desirable for them to have an overall very flat frequency response in order to keep sound coloration to a minimum. Due to this reason, for stereo recordings so-called matched pairs of microphones get used, which are very similar in terms of frequency response, as well as polar pattern.

When using microphones with cardioid patterns, the most common opening angles are between 90 and 120°, while the axis of the microphones may sometimes be pointing toward the L and R edge of the sound source.

When hyper-cardioids are used, usually a narrower physical opening angle is preferred in order to achieve a more stable image at the center of the sound source. Due to the stronger directional characteristic, hyper-cardioids can be positioned further away from the sound source than microphones with a cardioid pattern (keyword: direct-to-reverberant sound ratio, or 'D/R-ratio') (Fig. 3.1).

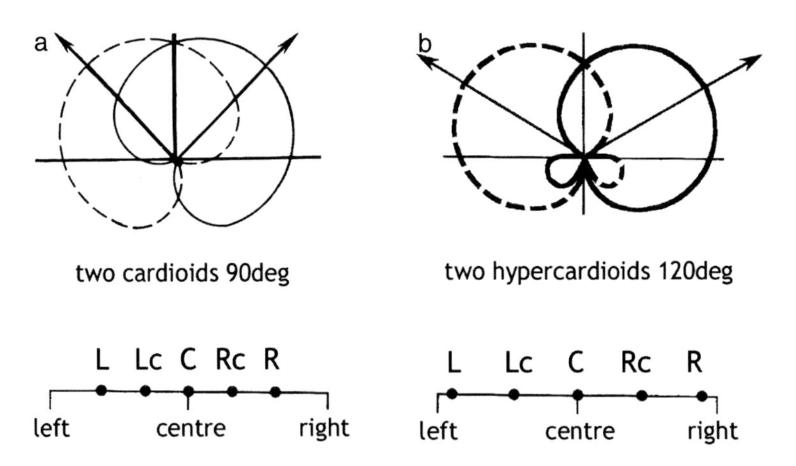

Fig. 3.1 XY-pair with **a** cardioids and **b** hyper-cardioids and associated Franssen's localization analysis (below)

3.1.2 XY with Figure-of-Eights (Blumlein-Pair)

The arrangement with two figure-of-eight capsules at an included angle of 90° to each other goes back to Alan Dower Blumlein's invention in the 1930s (Blumlein 1931). In his patent, he describes two possible configurations of crossed 8's:

(a) with an orientation of −45°/+45° relative to the sound source, the signals thus obtained are routed directly to the L and R channel of a stereo loudspeaker system, and
(b) arranged as an MS system, with one 8 pointing toward the center of the sound-source, while the second one is pointing side-wise at a relative angle of 90°. Signals suitable for replay over a 2-channel stereo system can be obtained via MS-decoding.

In this context, it is worth to mention a system-inherent property of version a) which may—occasionally—cause problems: The microphone which is pointed left also picks up sound from the right rear quadrant, which will then be played back (with inverted polarity!) from the left loudspeaker. This may be problematic, depending on the room acoustics of the recording venue. In order to minimize potential problems, the sound-engineer may decide to use a sound-absorptive panel behind the microphones in order to largely eliminate sound-pickup from the 'out-of-phase' lobe of the figure-of-eight capsule. Strong side-wall reflections in a room for example may sometimes turn out problematic in that respect (Fig. 3.2).

Especially the arrangement under (a) is being described as 'very naturally sounding' by many listeners (see Streicher and Dooley 1985).

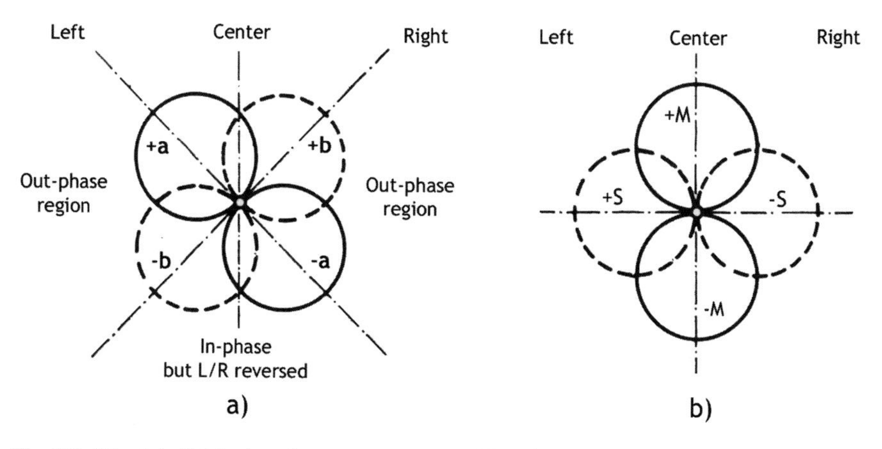

Fig. 3.2 Blumlein-Pair' microphone arrangement: **a** *XY*-orientation of the 8's ($X = -45°/Y = +45°$); **b** MS-orientation of the 8's ($S = -90°/M = 0°$) (after Rumsey 2001)

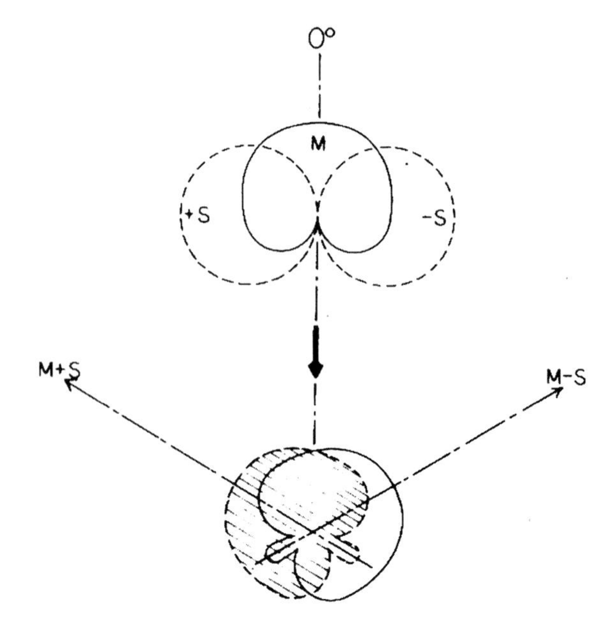

Fig. 3.3 Equivalence of MS and XY arrangements (from Streicher and Dooley 1985)

3.1.3 MS Stereo

The MS arrangement consists of a side-pointing figure-of-eight capsule, combined with a forward-facing microphone, which can have a polar pattern ranging from omni to hyper-cardioid or even figure-of-eight (Blumlein-Pair). The resulting recording angle depends on the volume balance between the M (Mid-) and S (Side-) signal, which is applied during the decoding process. Due to this reason, the MS techniques are appreciated by many sound-engineers, as it leaves options open for post-production, as long as the M and S signals are being kept separate.

For applications in music recording, the S-signal usually contains a lot of diffuse sound, as the 8 points side-wise, while the M-signal contains mainly direct sound. This is also beneficial for mono-compatibility, as for the mono signal mainly the direct signal takes effect (Fig. 3.3).

3.1.4 Coincident Omni-directional Microphones?

This technique—setting up two omnis in XY-mode—was developed by Ron Streicher for use as a soloist pickup (see Streicher and Dooley 1985) and also (Swedien 1997). At first, this arrangement might appear to be senseless, as omni-capsules are equally receptive from all sides for all frequencies. Or should we rather say: the *ideal* omni-capsule … as real-world omni-directional microphones usually exhibit a certain directionality with increasing frequency, due to law-of-physics (which has to do with phase attenuation according to the membrane diameter and the angle of incidence of the sound that arrives at the membrane).

This effect—usually undesired—on the other hand enables us to achieve a very subtle 'stereo'-effect (hint: ASW—apparent source width) for the sound source. The signals of these omni microphones, panned hard L and R, are highly correlated, which has the advantage that a soloist is being captured in stereo, but movements of his/her will not result in disturbing 'image shift'-type of localization distortions. Also, 'the use of pressure capsules eliminates the proximity effect and breath blasting problems associated with gradient microphones' (from Streicher and Dooley 1985).

Coincident Technique with 3 Capsules

3.1.5 BPT Microphone ('Blumlein-Pfanzagl-Triple,' BPT-3.0)

As the name already indicates, the BPT makes use of a third figure-of-eight capsule, which is added to the well-known Blumlein-Pair of two figure-of-eight capsules, which have an included angle of 90° and are oriented at ±45° toward the sound source. The new, third capsule is pointed directly toward the middle of the sound source at 0° (see Fig. 3.4 for schematic orientation of the capsules).

In order to shield off unwanted signals from the rear, it is preferable to place an acoustic barrier (e.g. in the form of a sound-absorptive acoustic panel) behind the BPT microphone. This helps to achieve a better direct/diffuse sound ratio and enables the sound-engineer to move the system further away from the sound-source, thereby usually achieving also a better balance between instruments (or instrumental groups) in case of larger ensembles.

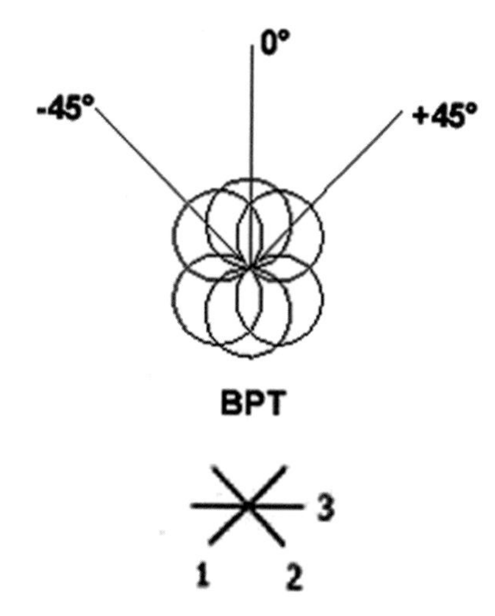

Fig. 3.4 Blumlein-Pfanzagl-Triple schematic (BPT-3.0 arrangement; above: capsules' polar diagrams; below: membrane orientation [plan view])

Why add a third capsule at all, as the Blumlein-Pair already captures 360°? As mentioned in Sect. 3.1.2 the Blumlein-Pair arrangement is appreciated by many as 'very naturally sounding,' which is attributed also to the fact that the two channels are largely de-correlated also at low frequencies, which is vital for good spatial impression. This is a unique feature which is almost exclusively reserved to crossed figure-of-eights (when it comes to 'one-point' or closely spaced microphone techniques), as all other microphone patterns (maybe with hyper-cardioids as an exception) do not provide similar signal-separation for low frequencies. When mentioning the use of the recommended absorptive panel above, some readers may have thought: why not simply use two cardioids crossed at 90°, instead of two fig-8s and the panel behind, as a figure-of-eight (at least in a dual-membrane design) is essentially composed of the (electrical) addition of a forward- and a backward-facing cardioid signal, with a phase reversal of the rear cardioid signal. Shouldn't *one* forward-pointing cardioid do the same job in a much more efficient way? In theory: yes; in practice: no.

Why is that so? Because it is not the same thing, physically. The author has conducted informal listening tests using the identical BPT microphone model of the same manufacturer in order to compare the above-described situation (Rem: the experiment has been conducted twice with microphones by two different manufacturers of high quality). The main difference in signal could be noticed with the correlation at low frequencies: in case of the crossed 8's the sound was 'open' and very natural, in case of the cardioids (which probably started to behave more like omnis with lowering frequency) it started to sound rather 'boxy' and almost slightly 'boomy' due to high signal correlation at low frequencies. In the opinion of the author, in terms of 'naturalness' of sound, figure-of-eights are second best to omnis; all other pressure-gradient transducers—independent of manufacturer or quality—have no chance to even get near due to the almost unavoidable off-axis coloration that seems inherent to their design.

The same thing applies for the sonic difference between MS stereo and the corresponding XY-arrangement of crossed cardioids: even though electrically (and mathematically) one can calculate which ratio of mid- to side-signal should be equivalent to a certain angular orientation of two crossed cardioids, even under the most ideal conditions (same make and model of microphones, same position in the sound field, etc.) the result will not be 100% identical. Why? Because it is not the same thing, physically. To start with, there is a certain degree of 'lack-of-precision' when decoding the signals of two capsules of the MS system, as their membranes are not in the same physical spot, where they should—ideally—be. In addition, in an MS-pair the M-capsule is fully facing the sound source, while the S-capsule points side-wise, while in an XY-pair both capsules are (at least partly) facing the sound-source to the same extent, just 'looking' at the sound source from opposite sides. How sound impinges on the membranes (fully frontal, or at an angle of 45°)—especially at high frequencies—makes a relevant difference: As we know, 'pure' pressure transducers (with one membrane, encapsulated on one side) are not able to maintain their uniform pattern for the whole frequency range. At higher frequencies, they usually exhibit an increase in directionality, as high-frequency sound—impinging with an angle of incidence on the diaphragm—will suffer level attenuation due to partial phase cancelation effects. This simply has to do with the ratio of capsule diameter and the wavelength λ of the sound.

As the wavelength λ is only 17 mm at a frequency of 20 kHz, the diameter of the diaphragm is already in the order of one full cycle of sound, and therefore—for frequencies down to 10 kHz and even below—the angle of incidence of the sound will matter, as the wavelength is in a dimension similar to the diameter of the membrane (Fig. 3.5).

However, it can be noted that such a 'narrowing in' of the omni-capsule's polar pattern with rising frequency occurs mainly with so-called true (or pure) pressure transducers. In contrast to this, an omni-directional microphone, which uses a dual-membrane design like, e.g., the Neumann KM88, is more uniform and does not change its polar pattern as drastically with rising frequency. Nevertheless, a certain frequency dependence of the polar patterns seems inevitable also with the best (dual-membrane) condenser microphones on the market (Fig. 3.6).

The usual tendency in terms of increase in directionality with rising frequency applies—of course—not only to omnis, but essentially to all capsules with directional characteristics.

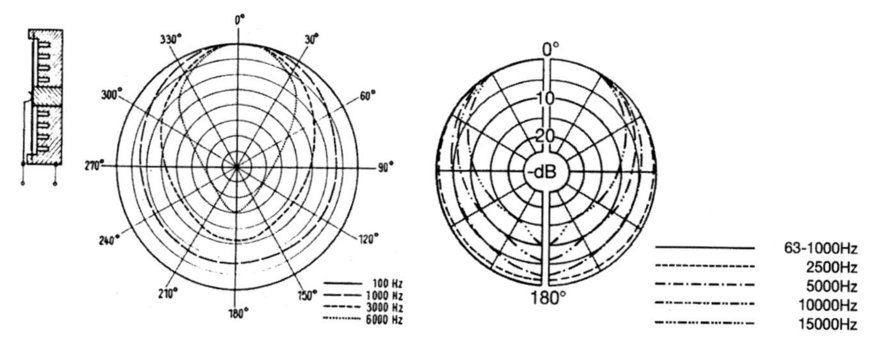

Fig. 3.5 'Pure' pressure transducer (omni): capsule section (far left) and Neumann and Schoeps pure pressure transducer polar patterns (right side)

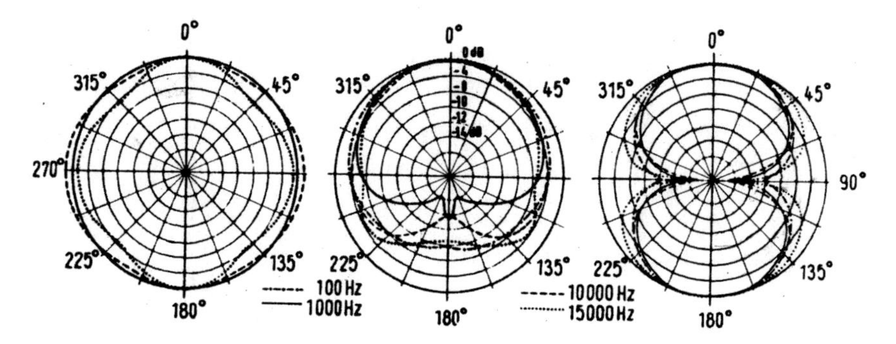

Fig. 3.6 Double membrane transducer Neumann KM88: var. Polar patterns: omni—cardioid—figure-of-8

Tomlinson Holman has described this property, which is also found with figure-of-eight microphones, in (2000, p. 95): '… The system [i.e. Blumlein-Pair] aims the microphones to the left and right of center; for practical microphones, the frequency response at 45° off the axis may not be as flat as on axis, so centered sound may not be as well recorded as sound on the axis of each of the microphones ….' This is of course one of the main reasons why it makes very much (sonic) sense to add a third figure-of-eight microphone or capsule to the 'Blumlein-Pair' arrangement and arrive at the BPT microphone configuration. The addition of the third capsule gives the sound-engineer a high degree of freedom, of how broad/'stereophonic' or narrow/'monophonic' he would like the sound image to be, by simply varying the level-ratio between the L/R capsule and center capsule signals.

Rising the level of C kind of 'focusses' or 'zooms in' to the center of the sound source, while—as a side effect, desirable or non-desirable—increasing the 'monophonic' part of the overall signal. Theoretical calculations have shown that if C is adjusted to a relative level of −10 dB in relation to the L and R capsule signal, almost no emphasis or gain is added to the sounds coming from the center of the sound-source (see Fig. 3.7).

Adjusting the level of the C capsule to a relative level of 0 dB will lead to a gain of +3 dB for the center of the sound-source (Fig. 3.8).

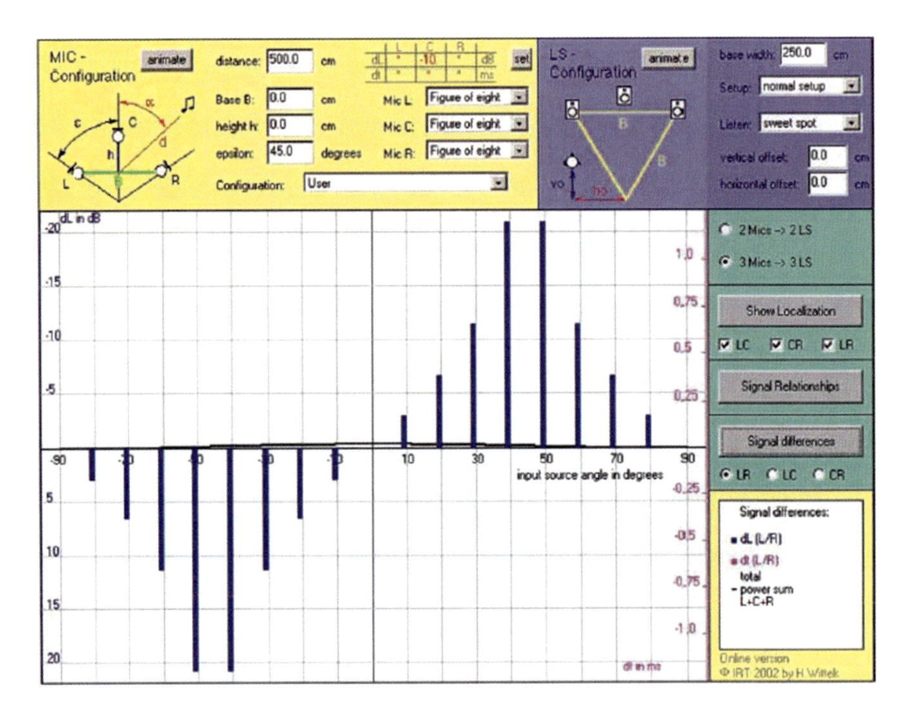

Fig. 3.7 'Image Assistant' graphic (BPT center capsule at −10 dB); (graphic generated from Wittek 2002)

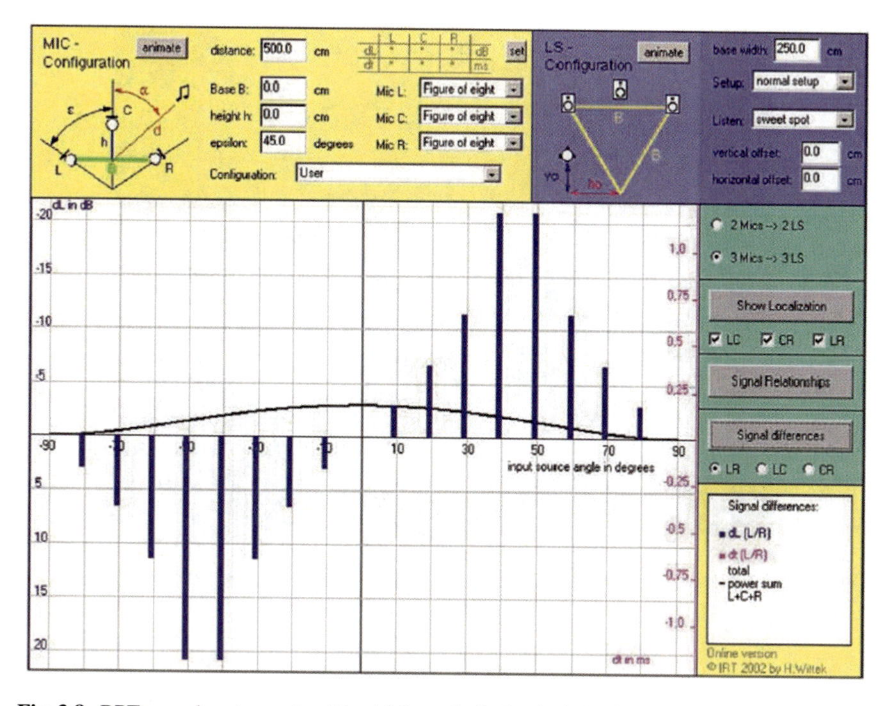

Fig. 3.8 BPT-capsules at same level (multichannel-playback via L, C, R speaker system); (graphic generated from Wittek 2002)

However, in practical applications values between −6 and 0 dB (sometimes even more than 0 dB) have been found to tie in well with the L and R capsule signals, always depending on what kind of sonic image the sound-engineer or Tonmeister wishes to achieve.

Engineers of course have to be even more careful what concerns the room acoustics, as the rear lobes of the figure-of-eight microphones are pointing backwards and are therefore picking up all the diffuse sound or reverb which has been reflected off the side and rear walls.

As Tomlinson Holman describes, with respect to a Blumlein-Pair (see 2000, p. 95): '… The system makes no distinction between front and back of the microphone set and thus may have to be placed *closer* than other coincident types ….' In order to help this problem in terms of direct /diffuse sound ratio an acoustically absorptive panel can be placed behind the BPT microphone so less diffuse sound or room reverb reaches the capsules from behind (Fig. 3.9).

However, in halls where the amount of diffuse sound is not a problem, seasoned sound-engineers like Tony Faulkner have found that: '… the old rule of thumb of coming back half the stage width for the placement of a coincident pair of microphones can often be extended to the whole stage width with the Blumlein-Pair …' (see Faulkner 1981) (Fig. 3.10).

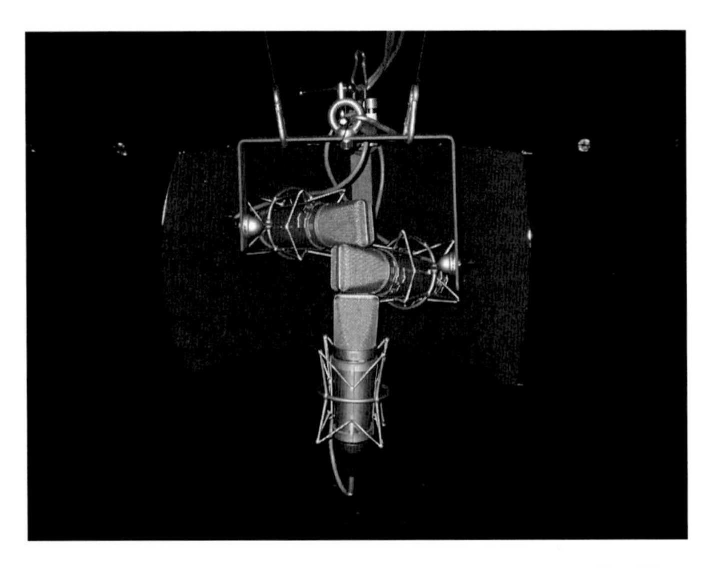

Fig. 3.9 Photograph of suspended BPT-system composed of 3 × Neumann U87 with acoustic panel

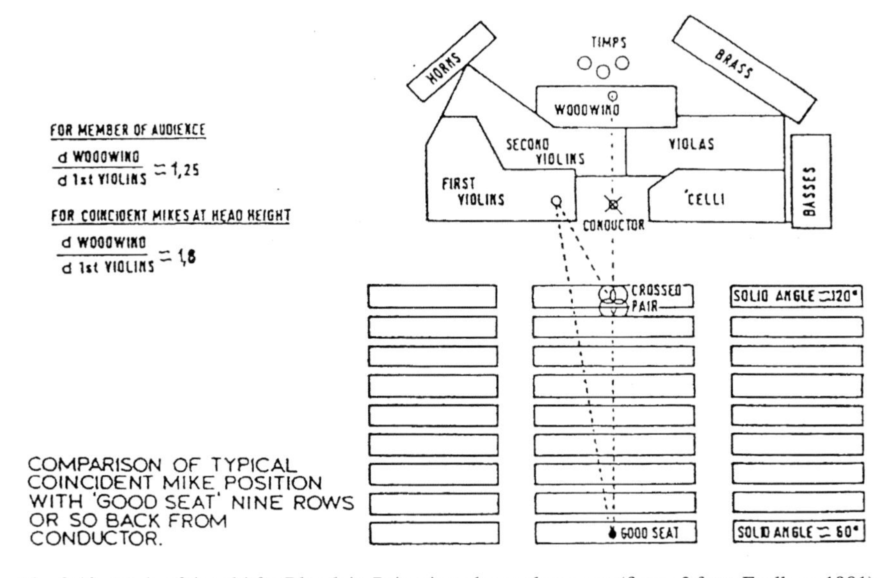

Fig. 3.10 'Rule of thumb' for Blumlein-Pair microphone placement (figure 2 from Faulkner 1981)

This certainly also has to do with the fact that the figure-of-eight capsule's directional characteristics is very 'far reaching'; the 'directivity index' (and hence 'distance factor') is the same for cardioids as well as figure-of-eight microphones (see Figs. 3.11 and 3.12). Only super- and hyper-cardioids exhibit even higher direc-

SUMMARY OF FIRST-ORDER CARDIOID MICROPHONES

CHARACTERISTIC	PRESSURE COMPONENT	GRADIENT COMPONENT	SUBCARDIOID	CARDIOID	SUPERCARDIOID	HYPERCARDIOID
POLAR RESPONSE PATTERN	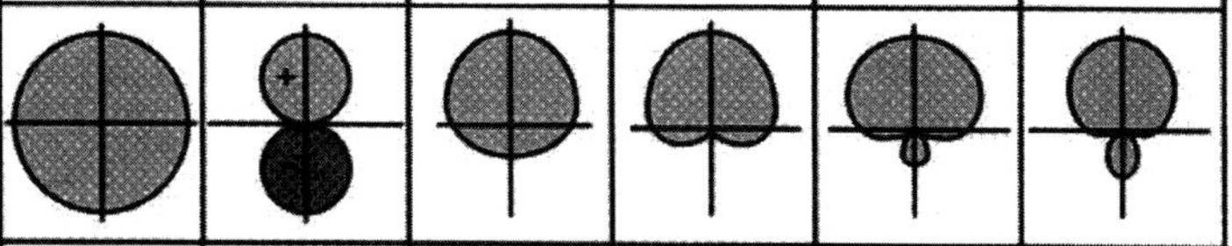					
POLAR EQUATION	1	Cos θ	.7 + .3Cos θ	.5 + .5Cos θ	.37 + .63Cos θ	.25 + .75Cos θ
PICKUP ARC 3 dB DOWN	360°	90°	180°	131°	115°	105°
PICKUP ARC 6 dB DOWN	360°	120°	264°	180°	156°	141°
RELATIVE OUTPUT AT 90° (dB)	0	−∞	−3	−6	−8.6	−12
RELATIVE OUTPUT AT 180° (dB)	0	0	−8	−∞	−11.7	−6
ANGLE AT WHICH OUTPUT = ZERO	–	90°	–	180°	126°	110°
RANDOM EFFICIENCY (RE)	1	.333	.55	.333	.268 (1)	.25 (2)
DIRECTIVITY INDEX (DI)	0 dB	4.8 dB	2.5 dB	4.8 dB	5.7 dB	6 dB
DISTANCE FACTOR (DSF)	1	1.7	1.3	1.7	1.9	2

(1) MAXIMUM FRONT TO TOTAL RANDOM EFFICIENCY FOR A FIRST-ORDER CARDIOID.
(2) MINIMUM RANDOM EFFICIENCY FOR A FIRST-ORDER CARDIOID.
(Data presentation after Shure Inc.)

Fig. 3.11 Characteristics of the family of first-order cardioid microphones (courtesy of Shure Inc.)

tionality, but at the same time their sound is usually considerably inferior with respect to naturalness, due to off-axis sound coloration.

3.2 AB Pair Systems (Δt-Stereophony)

It will most likely be a surprise for many readers that Stereophony is already around 140 years old: On the occasion of the 'International Exhibition on Electricity' in Paris in the year 1881, visitors had the occasion to experience 'spaciousness'—due to 'two-channel sound reproduction'—from a wired sound-transmission of a performance at the Paris Opera house. The arrangement at the opera house consisted of several microphones, positioned along the rim of the stage, which were each connected to a corresponding telephone inside a pavilion on the exhibition site (see Engineering 1981 and more details in Chap. 11). By listening to two adjacent telephones, visitors

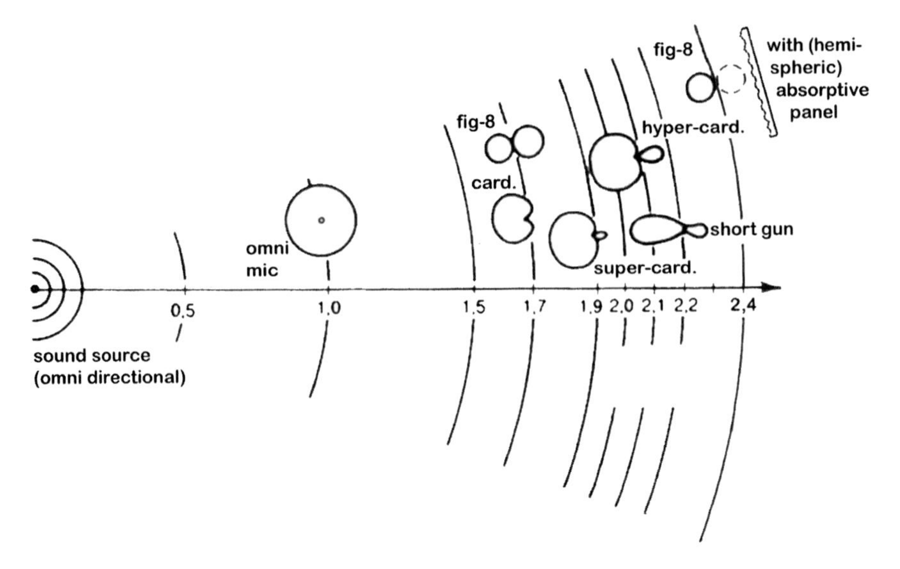

Fig. 3.12 Illustration of distance factor (DF) for the first-order cardioid family (modified from Fig. 1.8.D from Dickreiter 2011, p. 32)

experienced the abovementioned effect of 'spaciousness,' which was caused by the differences in time-of-arrival of the sound impinging on the microphones on stage.

3.2.1 *Small AB—Only a 'Psychoacoustic Trick'?*

In sound-engineering, people in general distinguish between 'small AB' and 'large AB' microphone systems. The first one usually uses capsule spacings in-between approx. 50 and 100 cm (1.5–3 ft.), which aims at creating time-of-arrival differences in the order of 1.5 ms (for soundwaves coming from the side) for human listeners.

A few years ago a study by Wittek and Theile (2002) pointed out that there are huge discrepancies between results derived from listening tests concerning localization. As can be seen in Table 3.1, these differences are—at times—larger than 100%

Table 3.1 Recording angles as found in literature Hugonnet and Walder (1998), Sengpiel (1992), Williams (1987), Wittek and Theile (2000), Wittek (2002); after Wittek and Theile (2002)

Setup	Hugonnet	Sengpiel	Williams	Wittek
AB omnis 50 cm	130°	180°	100°	74°
AB omnis 100 cm	–	62°	–	36°
ORTF cardioids 110°, 17 cm	90°	96°	100°	102°

(compare the values found by Wittek and by Sengpiel for small AB with 50 cm capsule spacing).

In Knothe and Plenge (1978), various studies are listed which indicate that the frequency dependence of sound-source localization between loudspeakers has been known at least since 1934. A more recent study on this topic can be found—among many others—in Griesinger (2002).

Whether the ambiguity of the results found in the study (Wittek and Theile 2002) are caused by differences in the nature of the test signals or varying physical characteristics of the loudspeakers involved (see Gernemann and Rösner 1998), or if it may be related to strong inter-individual differences along with some anomalous localization effects which occur in the mid-frequency range (see Benjamin 2006 and Benjamin and Brown 2007 for details), or in a combination of these factors, may be an interesting subject for further research in the future.

Looking at the differences in the results of study Wittek and Theile (2002) of more than 100%, it seems clear that 'small AB'-based recording techniques apparently cannot provide localization which is 'solid' enough so that inter-individual differences (including loudspeaker quality differences, etc.) can be neglected.

It is commonly known that for sound, a time-of-arrival difference between the ears of a human in the range of 1.1–1.5 ms will result in localizing the sound event entirely in one of the two loudspeakers of a standard 2-channel stereo setup (Wittek and Theile 2002). Therefore, the majority of researchers consider the 'small AB'-based microphone configuration of two omni-directional transducers as the system of choice, as it is able to provide interaural time differences in the requested range.

On the other hand, 'large AB'-based microphone techniques with capsule spacings of several meters (as are commonly used with orchestra recordings) are often all too easily discarded by pointing out their inherent 'hole-in-the-middle' effect, while in practice this acoustic 'hole' can very easily be filled with the help of an appropriate 'centerfill' single microphone or more elaborate microphone 'fill'-system designed for that purpose.

It is important to understand that the presumed 'localization accuracy' of the small AB arrangement [which is not that accurate at all, as put in evidence by the study of Wittek and Theile (2002)] comes at the relevant expense of reduced 'spaciousness' contained in the recording, as low-frequency information below 500 Hz is highly correlated and therefore—essentially—almost 'monophonic' (compare Fig. 3.13).

On the other hand, it is criticized by some engineers (see, e.g., Wuttke 1998) that the large AB configuration sounds better only due to a kind of 'artificial effect,' namely the repeated alternation of in-phase and out-phase signals in neighboring frequency bands, which—supposedly—should lead to an exaggerated effect of apparent source width (ASW).

As can be seen by comparing the measurements presented in Figs. 3.13 and 3.14, this alternation of slight positive and negative correlation of the L and R channel signals is the case for both small AB and large AB.

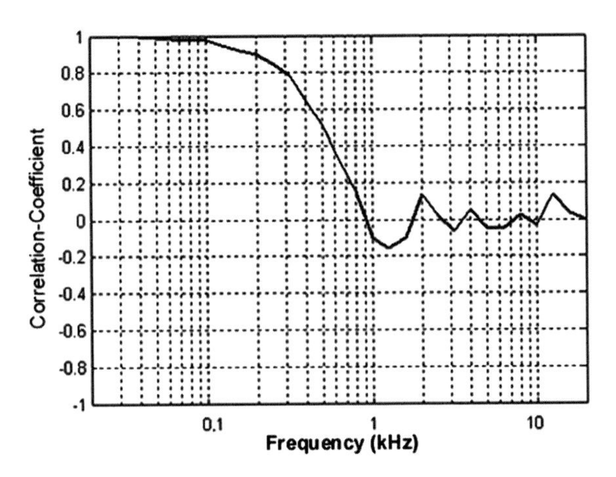

Fig. 3.13 Measured FCC ('Frequency-dependent Correlation Coefficient') for a 'small AB' pair of omni-directional microphones with 20 cm capsule spacing

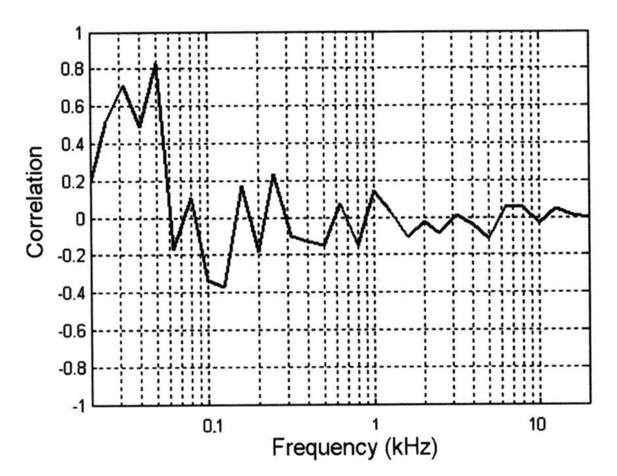

Fig. 3.14 Measured FCC ('Frequency-dependent Correlation Coefficient') for a 'large AB' pair of omni-directional microphones with 1200 cm capsule spacing

Therefore, the argument of 'exaggerated width' would apply also to small AB recordings, with the only difference that the concerned frequency range of alternating in-phase and out-phase information is being shifted to a higher frequency band due to the smaller capsule spacing.

3.2.2 *Faulkner 'Phased Array'*

Developed by British sound-engineer Tony Faulkner (see Faulkner 1981), it is composed of two front-facing parallel oriented figure-of-eight microphones with a capsule spacing of 20 cm. This technique combines a high degree of signal coherence—similar in construction to the Blumlein-Pair—with a certain 'openness' in

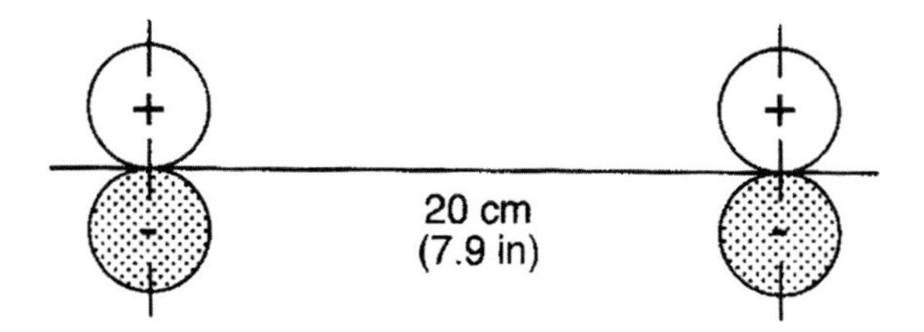

Fig. 3.15 Faulkner,phased array' (from Streicher and Dooley 1985)

terms of sound (Streicher and Dooley 1985, p. 552), which results from the time-of-arrival differences in AB microphone setups. In addition, Faulkner suggests setting up the microphone arrangement further away than usual from the sound-source, in order to achieve a better volume balance between the front- and rear-elements of a large sound source (e.g. orchestra) (Fig. 3.15).

3.2.3 *Faulkner '4-Way Phased Array'*

In recent years, Tony Faulkner has specified and used an 'enhanced' version of his 'Phased Array' technique, which employs two sets of small AB microphones in the following manner (as described by Faulkner himself): '... Phased arrays are very "interesting creatures." The mathematics are complicated in comparison with those for spatial-coincident Blumlein style arrays, but in operation they have one enormous advantage over spatial-coincident arrays. You can place them quite a lot further away from performers and acquire a very helpful sense of presence despite the distance, rather than being overwhelmed by a fog of excess ambience. I work quite a lot recording live concerts for video productions, and video producers are always very happy to see one or two microphone vertical stands rather than thirty or forty boom-stands throughout an orchestra. Quite often for these video jobs in typical modern concert halls I use a 4-way phased array with 67 cm spaced omnis (Schoeps CCM2H) and 41 cm subcardioids (Schoeps CCM21) in between the omnis. The fig8 phased array system is of greatest use recording in massive churches and cathedrals, but some of the commercial models sound a bit colored.' (cited from a personal email-exchange with Mr. Faulkner in 2013)

3.2.4 *Jecklin Disk—or Blumlein Disk? (... and Schneider Disk)*

Described by its inventor, Jürg Jecklin, Tonmeister at the Swiss National Broadcasting Agency as OSS ('Optimal Stereo Signal')-Technique (1981), it is known to most engineers under the name of 'Jecklin disk' (or 'Jecklin-Scheibe' in German): two diffuse-field compensated omni-microphones, with a capsule spacing of 18 cm,

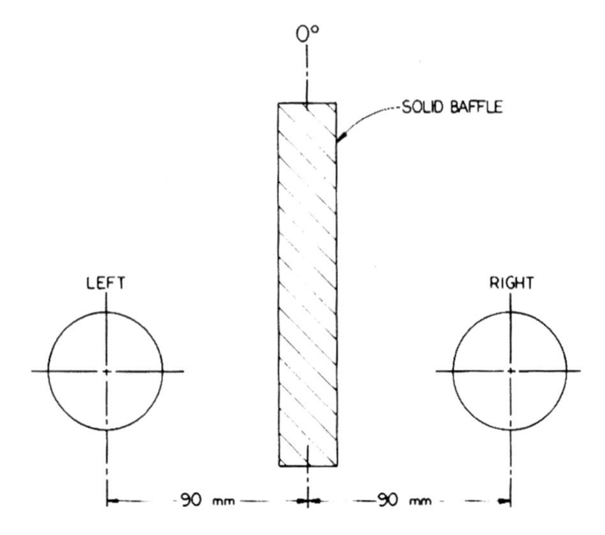

Fig. 3.16 Jecklin disk schematic (from Jecklin 1981)

separated by a disk with a sound-absorptive surface of 28 cm diameter. The microphones should be oriented in a way '... that they slightly "overshoot" the sound-source' (from a personal conversation with Mr. Jecklin in 1998) (Fig. 3.16).

Jecklin describes the sonic characteristics of his system as follows:

> '... 1) In the frequency range below 200 Hz the disk has no influence. Both microphones receive the same signal.
>
> 2) With increasing frequency, diffraction occurs at the edge of the disk with an increasing effect of separation of the two microphones.
>
> 3) The polar response of the OSS microphone is omni-directional, which is satisfactory for rooms with either normal or too short a reverberation time.
>
> To work in rooms with a long reverberation time it is possible to change the omni-directional polar response to a cardioid-like characteristic by using a sound-absorbing screen behind the microphone assembly. This solution does not change the transducer characteristic of the pressure microphones, and the advantages are conserved.' (from Jecklin 1981)

From Jecklin's statement no. (1) it can be understood that for frequencies below 200 Hz the L and R signals are very highly correlated, or—in other words—mono. From the research by Griesinger and others, we know that this is to the disadvantage of perceived good spatial impression, which is instead achieved by high de-correlation at low frequencies (Fig. 3.17).

The Schneider Disk

'...The Schneider disk may be used in OSS technique for stereo recordings of orchestral or chamber music, as well as for stereo miking of solo instruments. In order to properly implement OSS, the microphones should be identical, omni-directional and should be set for equal output to avoid unbalancing of the stereo image (this can be checked using headphones). For the best balance between direct and room sound, the Schneider disk should be located at the distance from the source where the direct

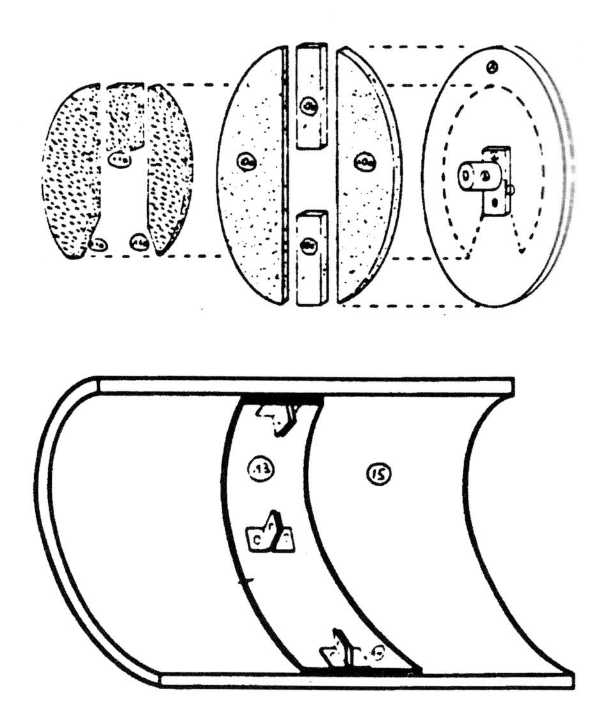

Fig. 3.17 Baffle (above) and absorptive rear panel (below) for 'Jecklin disk'/OSS Technique) (from Jecklin 1981)

sound and the diffuse sound from the room are at equal strength. This distance is known as the "diffuse-field" distance and can be located by careful listening.

The difference between the 'Jecklin' and the 'Schneider' types of disks is the foam sphere at the center of the Schneider disk, which reduces the amount of high-frequency energy reflected from the disk, and results in an increase of stereo separation.

Below approximately 200 Hz, the Schneider disk has little (if any) effect on the stereo signal because the audio wavelength is large enough to bend around the disk, equally reaching both microphones.' [cited from the data-sheet of the 'Schneider-Disk' (MBHO GmbH 2016)] (Fig. 3.18).

Even though the above-described arrangement is known to most people under the name of 'Jecklin disk', strictly speaking, this arrangement has already been invented by Alan Dower Blumlein, presumably in the late 1920s or around 1930, as such an arrangement, using two narrowly spaced 'small AB' omni-directional capsules, separated by an acoustic panel in order to achieve better channel separation, is described in one of his patents (see Blumlein 1931) (Fig. 3.19).

Fig. 3.18 Scheider disk: a variation of the Jecklin disk with a foam sphere added to the center (from MBHO GmbH 2016)

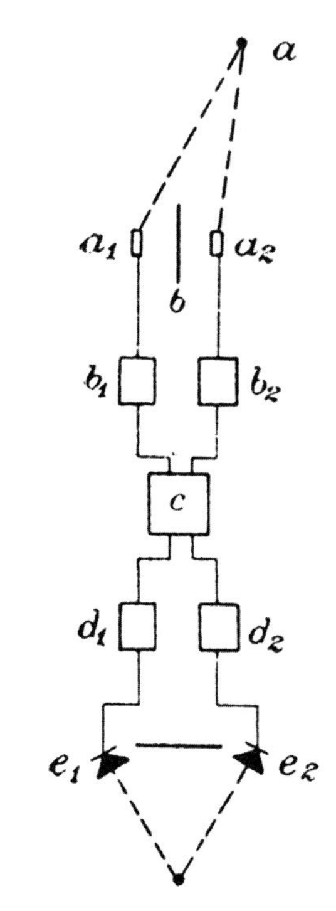

Fig. 3.19 'Blumlein-Disk'—schematic (*a*: sound-source; *b*: baffle; $a_{1,2}$: microphones; $b_{1,2}$: microphone signal amplifiers; *c*: transmission or storage channel; $d_{1,2}$: reproduction amplifiers; $e_{1,2}$: loudspeakers) (from British patent no. 394325; Dec. 14, 1931)

3.3 Equivalence Stereophony

For this group of microphone techniques, the stereo-effect is based on level- as well as time-of-arrival differences of the signals that enter the microphone capsules:

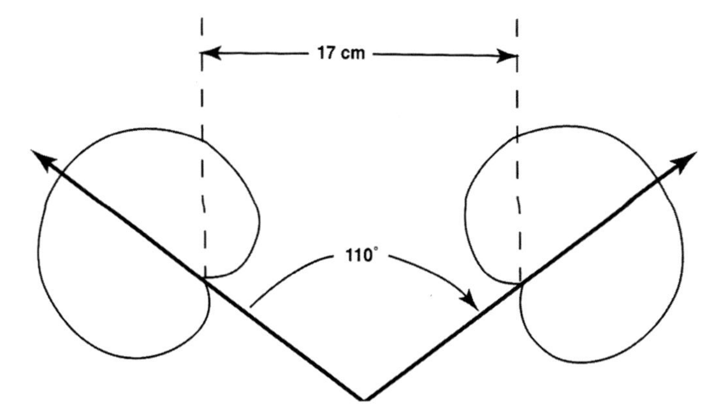

Fig. 3.20 ORTF-technique (from Streicher and Everest 2006)

3.3.1 *ORTF*

This microphone technique has been developed by and is named after the French National Broadcasting Agency (Office de Radiodiffusion-Télévision Française) and consists of an arrangement of two cardioid microphones with a capsule spacing of 17 cm (similar to the average spacing of a pair of human ears) and a physical opening angle of 110°. According to the Williams-Curves, this arrangement is characterized by a resulting recording angle of 96° (Fig. 3.20).

3.3.2 *NOS*

Developed by the Dutch National Broadcasting Agency (Nederlandse Omroep Stichting), this technique uses two microphones with cardioid characteristics, a capsule spacing of 30 cm (approx. 1 foot) and a physical opening angle of 90°, which—according to Williams—results in an effective recording angle of 80° (Fig. 3.21).

3.3.3 *Large AB*

Many readers may think that it is a layout error by the editor or publisher that the 'Large AB' microphone method is showing up in the section of 'equivalence stereophony,' as traditionally one would expect to find this technique in the section of ITD-related recording techniques. I will explain why I think it is more proper to file 'Large AB' among the 'equivalence' techniques, at least as long we are talking about systems with a capsule spacing of several meters.

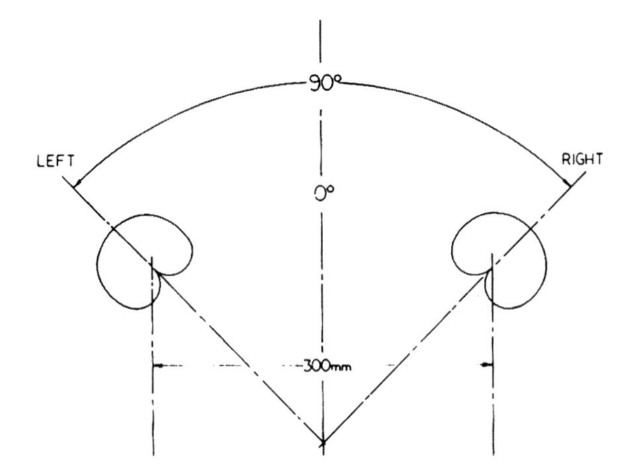

Fig. 3.21 NOS-technique developed by Dutch national radio (from Streicher and Dooley 1985)

It is certainly true for small AB systems that the signals on the two omni-capsules are very similar in terms of amplitude and frequency, but the main difference lies in the time-of-arrival of sound impinging on the membranes. This is not the case for 'Large AB' microphone setups.

As can be seen in Fig. 3.22, the usual capsule spacing in a large AB setup will be in the range of several meters, if it is meant for a large sound-source, for example, an orchestra. Typically the capsule spacing will be one third up to half of the total width of the sound source, which can be up to 10 m (or even more) in case of a large symphony orchestra, which may have a total width of 20 m. However, it needs

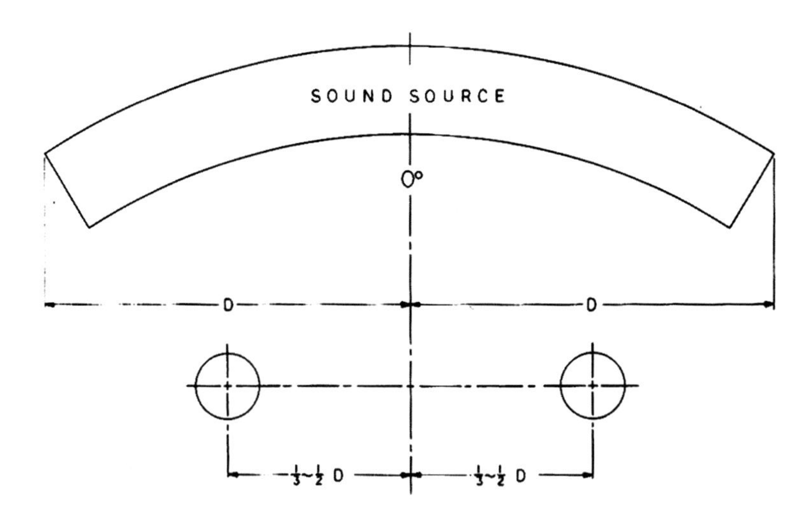

Fig. 3.22 A typical 'Large AB' microphone setup as used for orchestral (or large sound source) recording (from Streicher and Dooley 1985)

to be said that particular care has to be taken with spacings larger than 10 m, as—due to the speed of sound (i.e. 340 m/s at 20 °C)—lateral sound (arriving from + or −90°) would have a time-of-arrival difference of more than 30 ms between the microphones, which might result in echo effects for a sensitive listener.

Due to the large capsule spacing, the resulting ITD cues for the listeners upon playback are of course by far larger than the usual 1.5 ms, which already cause localization of a sound source completely at the side of one speaker (i.e. total L or total R). This is one of the main reasons why some academics simply see recording with a 'large AB' pair as a wrong approach (see Sengpiel 1992; Wuttke 1998), claiming also that large AB was introducing 'phase anomalies' (while small AB wasn't; see Sect. 4.1).

However, it can also be said that even 'exaggerated' ITD cues are apparently valuable indicators for sound source direction as the use of mechanical devices, which deliver such cues on purpose, has a long history (see Figs. 3.23 and 3.24).

The roots of 'acoustic location' extend back to the late nineteenth century, even before the invention of aircraft. The earliest rendition of such a device appeared in the pages of Scientific American magazine from 1880. 'Professor Mayer's Topophone'

Fig. 3.23 An example of the usefulness of exaggerated ITD's is Professor Mayer's Topophone (graphic from 'Scientific American' Magazine, July 3rd, 1880); see also Mayer (1880)

Fig. 3.24 'Ear Radar': two horn 'sound locator' system, as used by the US-Armed Forces in Bollingfield, Washington DC (color painting after a black and white photography from 1921)

invented and patented by Mayer (1880), was intended to assist the wearer in pinpointing the location of sound-sources.

In order to conclude this section on the perception related properties of large AB microphone systems, I would like to justify why they should be considered to be part of the 'equivalence stereophony' techniques. The explanation is that they are not only characterized by the large ITDs which they provide, but—due to the large distance between the capsules in the order of several meters—they will also level-wise pick up quite different sections of the orchestra. In essence: What a pair of cardioid mics in an XY or ORTF arrangement does due to its polar pattern, is achieved by the large AB system in a similar way, due to the very different physical position of the A and B microphones. This is why it is more correct to consider large AB to be an 'equivalence' microphone technique and not to belong to the Δt-stereophony category.

3.3.4 Large AB—and the 'Hole-in-the-Middle' Effect

In contrast to small AB, large AB uses capsule spacings of up to several meters (depending on the size of the sound source) and therefore does not take into account time-of-arrival differences (ITD's—interaural time differences) in an order which would be psychoacoustically 'correct' for human listeners. In this context, it seems important to mention that the stereophonic width (or apparent source width—ASW), as well as the localization of a sound-source depends not only on the Δt-values involved, but—with growing capsule spacing—also on the increasing level difference between the two channels. The result is 'localization zones', which get formed in the vicinity of the microphones. With increasing capsule spacing we also have a growing tendency to perceive the so-called hole-in-the-middle effect, which means that the sound-sources are primarily perceived close to or 'around' the loudspeakers, while there is a 'dead zone' between them.

As already pointed out above, the 'hole-in-the-middle' effect can be avoided by introducing a suitably selected 'Centerfill System'; in the most simple form, this can be achieved by adding a third (omni-directional) microphone (C) in the middle between the A and B microphones (Fig. 3.25).

Such a technique, using three non-directional microphones has first been applied at the Bell Laboratories in the late 1930s (Keller 1981). A similar system has been used later with the legendary 'Living Presence' series of recordings of the 'Mercury' label (see Valin 1994) and more information in Chap. 11, Sect. 11.4). First fundamental research concerning localization distortion in 2- and 3-channel stereo systems (taking into account only level differences between the microphone signals) date back to the 1930s (Steinberg and Snow 1934). In a later publication, one of the authors (Snow 1953) took into consideration also the psychoacoustic effects of time-of-arrival differences in the stereo signal, which led to a correction of the results of the first publication. From today's point of knowledge, it is interesting to notice that

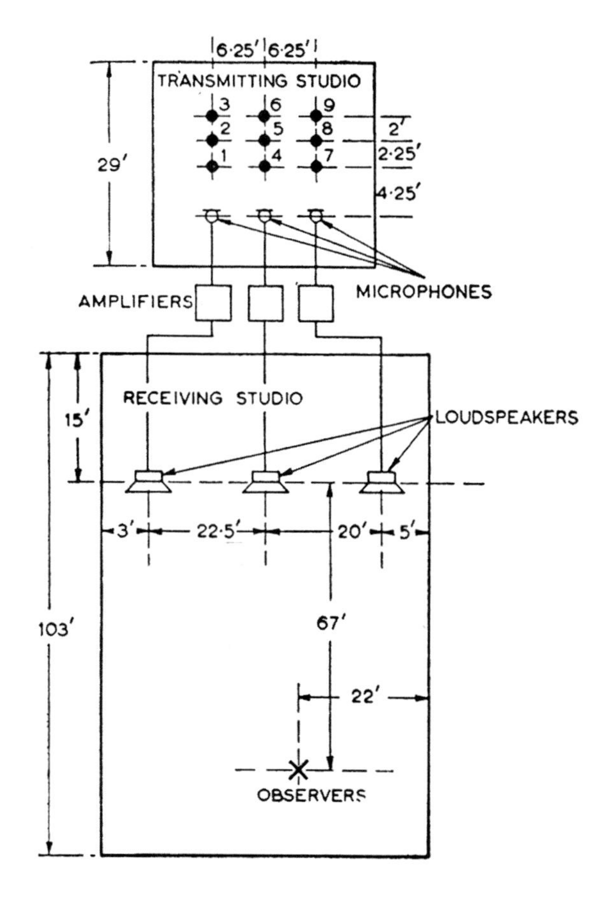

Fig. 3.25 Early 3-channel 'Stereo system' employing 3 microphones and 3 loudspeakers; experimental setup at Bell laboratories in 1934 (from Keller 1981)

already the above-cited research from the 1930s had made it clear that an additional center channel (in the case of a 3-channel system) provides more stable sound-source reproduction.

The rule of thumb with large AB-systems is to have a capsule spacing between one-third and half the width of the sound-source, which—in the case of symphony orchestras—can lead to distances of 10 m (30 ft.) and more, with time-of-arrival differences of up to or even more than 30 ms for sound coming from the side, which is non-negligible due to potential unwanted echo effects. With 3-channel microphone systems, there is the possibility of comb-filtering effects, when summing the signals on a 2-channel stereo bus, which can lead to unwanted sound coloration for the listeners.

In principle, AB arrangements with large, as well as small capsule spacings can be implemented also with directional microphones, as we have seen on the occasion of Faulkner's 'Phased Array' for example, or they can even be combined with other techniques, as is common practice. If we are limiting ourselves to a traditional two-capsule small AB mic-system, the resulting recording angle (based on capsule spacing and included angle in-between) can be calculated on basis of the so-called Williams-Curves (Williams 1987).

As Williams writes: '… The critical point is still −3 dB and we can see in Figs. 3.26, 3.27, and 3.28, the resulting shaded areas that limit our choice of distance and angle between microphones.

Conclusion: In choosing a combination of distance and angle for a given recording angle, we must in general observe two conditions:

(1) choose a combination of distance and angle with a reasonable minimum angular distortion.

(2) avoid the shaded areas as here reverberation "creeps" into the recording angle.

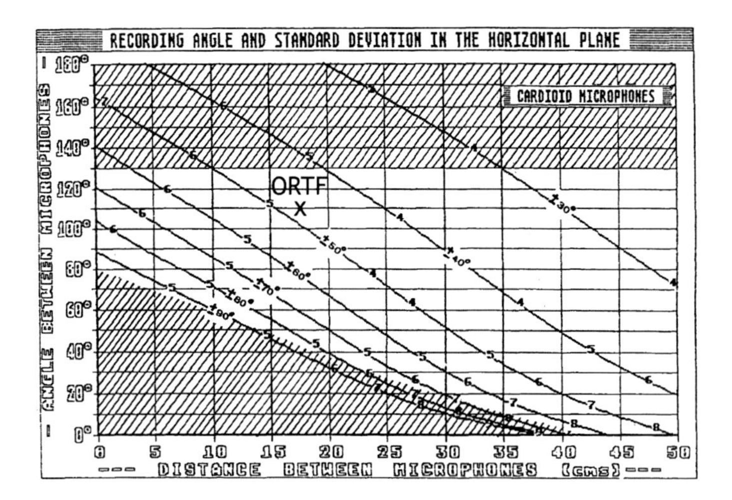

Fig. 3.26 Williams-Curves for spaced and angled Cardioid Microphones (from Williams 1987, ORTF position added by author)

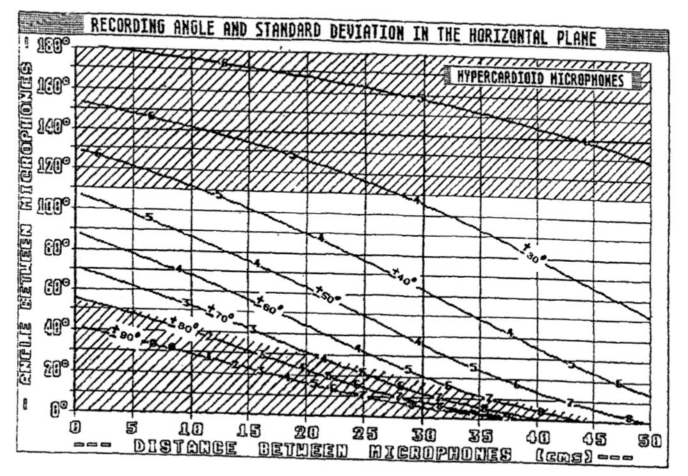

Fig. 3.27 Williams-Curves for spaced and angled Hypercardioid Microphones (from Williams 1987)

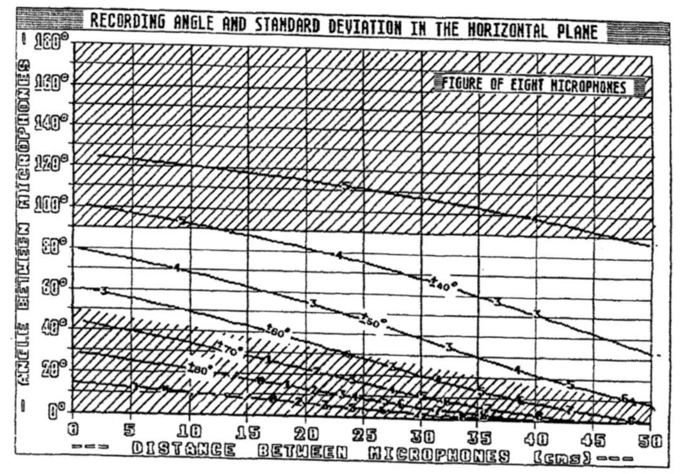

Fig. 3.28 Williams-Curves for spaced and angled figure-of-eight microphones (from Williams 1987)

However, angular distortion can have some useful applications. It is also possible to use the "reverberation effect" in special circumstances (increase in reverberation giving an impression of the source receding).' (from Williams 1987)

3.3.5 *Small AB Versus Large AB: Introducing 'medium AB' and the 'AB-Centerfill' Technique*

As has been reported above, the 'classical' small AB pair which is strictly based on psychoacoustic principles has a capsule spacing in the range of 40–60 cm. Some engineers would probably still consider a spacing up to 1 m (approx. 3 ft.) to be 'small AB,' but above 2–3 m many would refer to it as being 'large AB'.

So—where do we draw the line between 'small AB' and 'large AB'? Should we consider a capsule spacing up to 100 cm to be 'small,' and from 101 cm, it is suddenly to be considered 'large'? This does not seem to make much sense. Everyday work practice shows that many sound-engineers use capsule spacings which are in the range of 1–2 m, which are certainly neither 'small AB' spacings—adhering to the psychoacoustic principles of human hearing, as explained above—nor clear examples of 'large AB.'

My proposal would be to officially introduce the term 'medium AB', with the following definition separating it from both 'small,' and 'large AB' techniques:

'**Small AB**': capsule spacings up to 100 cm,
'**Medium AB**': capsule spacings between 100 cm and extending up to (but not exceeding) the 'critical distance' (measured/calculated for the 1 kHz frequency band) of the recording venue,
'**Large AB**': capsule spacings which exceed the 'critical distance' (or 'reverberation radius') of the 1 kHz frequency band, due to which the signals become more and more de-correlated (over the whole frequency range) with increasing distance.

In this way, the signals provided by AB-systems falling in one of the three categories defined above can easily be distinguished from the others, due to their different acoustic and psychoacoustic properties (rem.: of course—for rather small rooms the 'critical distance' may be way below 100 cm, and therefore, the 'medium AB' distance criterion will be obsolete and 'large AB', by definition, will start right above 1 m. One may also argue that—under such circumstances—the term 'large AB' may already be applied once the 'critical distance' is exceeded between the microphone capsules, which may even be the case below a 100 cm spacing …)

The preference of many academics for the small AB (and also various coincident XY) microphone systems may also have to do with the fact that these are relatively easy to grasp mathematically while already a combination of just two (main-) microphone systems (e.g. large AB with a Blumlein-Pair of crossed figure-of-eights) may not be as easy to calculate in all necessary (psychoacoustic) accuracy.

But what is the everyday reality out there for typical sound-engineers?

Do they restrict themselves to just using two capsules for their main microphone systems?

The answer is no: If we take a look at the microphone setups that are used by practicing sound-engineers, it can be noticed that there is a large number who use not just one main microphone technique, but rather a combination of two (or sometimes even

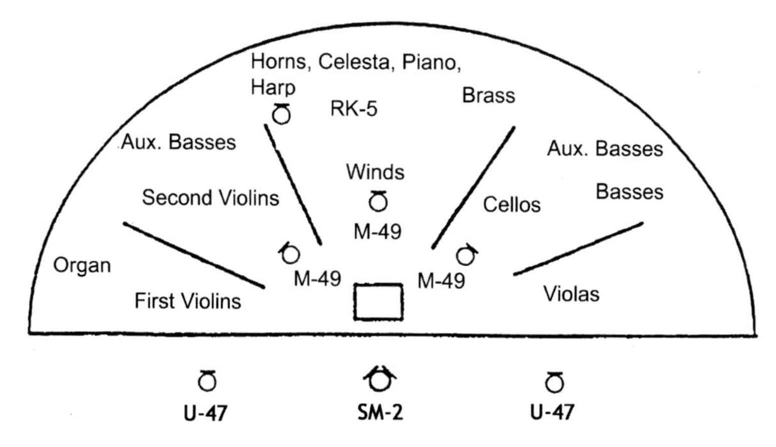

Fig. 3.29 RCA's 'A/B—Centerfill' microphone scheme (1959) (after Valin 1994)

more) appropriate ones. Many times this includes a large AB pair of spaced capsules (see the famous 'DECCA-tree' from the 1950s with omni-directional 'outriggers,' as well as the 'Fukada tree' for surround recording (Sawaguchi and Fukada 1999).

Other famous examples can be found in the highly acclaimed 'Living Stereo' recordings made by the RCA in the 1950s and 60s, of which the original 3-channel recordings have been reissued in a (front signals only) multichannel SA-CD format during recent years. Figure 3.29 shows an example of RCA's 'A/B—Centerfill' microphone scheme, which has been applied on the occasion of the 1959 recording of the 'Pines of Rome,' using a large AB pair of Neumann U-47 microphones in conjunction with a Neumann SM-2 (early stereo microphone) as centerfill system.

Simply put, in order to achieve de-correlation of a 2-channel stereophonic sound signal we need to have either sufficient spacing between two omni-directional transducers or employ coincident or near coincident techniques with highly directional microphones.

Apparently there is a disagreement with respect to the amount of capsule spacing needed for 'correct spatial reproduction' of a sound event: On the one hand side, authorities of the audio community (Rumsey and Segar 2001; Sengpiel 1992; Wuttke 1998) seem convinced that only small microphone spacings ('small AB'), based on psychoacoustic principles, are able to provide correct localization. On the other hand, a large percentage of practicing sound-engineers favor largely spaced AB techniques ('large AB') (see La Grou 1994; Betz 2000), or at least use supplementary 'outriggers' (largely spaced omni-directional microphones in front of the orchestra), due to the more 'open sound' they provide, while they manage to compensate the 'hole-in-the-middle effect' with the use of appropriate centerfill systems or by introducing spot microphones.

Therefore, in practice the better spatial reproduction capability of large AB is preferred by many sound-engineers over the (theoretically) higher localization accuracy of small AB. As we will see later-on in the chapter dealing with listener-comparison of surround recording techniques, there is also a valid reason for this: In various studies, it has been shown that the feeling of 'spaciousness' is highly correlated to listener-preference.

Decca-Tree Style Techniques

The triangular configuration, known to us today as the 'DECCA-tree' has been invented in 1953 by DECCA engineer Roy Wallace, who also had to develop a proper mixing console, capable of handling these signals, as the underlying approach is fundamentally 3-channel.

The usual spacing (or base width) between the A and B (i.e. L and R) microphone lies in the order of about 1–2 m, with the 'Center' C microphone being usually advanced toward the sound source by roughly half the base width. Usually, the DECCA triangle will be set up right above the conductor, but of course it is also possible to position the Decca triangle microphones clearly *in front* of an orchestra—for example—which means they will be positioned 1 or 2 m *behind* the conductor, in order to give a clearer perspective and localization due to larger time-of-arrival differences for sound at the 3 microphones.

The first recording with a DECCA-tree took place in spring 1953 at the 'Victoria Saal' in Geneva with the 'Orchestre de la Suisse Romande.' Three Neumann M-49 (switched to cardioid) were used, placed at a height of about 3.30 m (10 ft.) right above the conductor. The L and R microphone had an included angle of 60° in relation to the Center mic, pointing diagonally across the orchestra, vertically angled by 30°.

From this, it can be seen that the original 'DECCA-tree' (or 'DECCA Triangle,' as it is also called) did not strictly consist of omni-directional microphones, with the intention to mainly make use of the time-of-arrival differences of sound in order to achieve localization and spatial impression in a recording. Instead the main objective was a satisfactory channel separation of signals and a fundamental 3-channel recording approach, which made use of all acoustic means available, such as directional microphones and acoustic absorptive panels to minimize 'crosstalk' between the L, C and R channel signals.

The idea of optimizing channel separation in the pickup of a three capsule microphone systems has been worked on later-on in the OCT (Optimal Cardioid Triangle) system by Theile (Theile 2000), the OCT-Version 2 (see Wittek and Theile 2004 and Sect. 4.2.6), the BPT (Blumlein-Pfanzagl-Triple) System (see Pfanzagl-Cardone and Höldrich 2008 and Sect. 4.1.4), as well as the DHAB, SHAB and CHAB 'hemicardioid AB' systems by Olabe (see Olabe 2014 and sections below).

More details on the historic development of the DECCA triangle technique are available in Chap. 11 (Title: 'A short history of microphone techniques and a few case studies').

3.3.6 OCT—Optimal Cardioid Triangle

An effort in trying to optimize the channel separation (or signal de-correlation) between the three microphones has been undertaken by Theile (2000) with the invention of his OCT (Optimal Cardioid Triangle), for which—initially—he has only defined the directional patterns and spacings of three front capsules, mainly intended for the recording of (small) musical ensembles. He proposes the left and right front microphones as supercardioids, while the center microphone should be a normal cardioid. Later on Theile added two rear-facing cardioids for use in a surround microphone recording setup (for more details, also on capsule spacing, please see the chapter on 'Surround Microphone Techniques', Sect. 4.2.5).

3.3.7 DHAB—Delayed Hemicardioid AB

In Olabe (2014), the author describes an altered version to the classic DECCA-tree arrangement: '… The purpose of DHAB is maintaining DECCA-tree´s psychoacoustic and FDSC (i.e. *Frequency-Dependent Signal Correlation*) properties and improving its main weak point which is found in localization attributes. The left and right omni-capsules are switched to hemicardioid in order to improve localization.

The distances between the capsules of the L-C-R triangle are redefined with respect to the classical DECCA-tree in order to find a positive compromise between localization and all other attributes: Time-of-arrival is a very important topic in order to improve the definition of a sound source. The DHAB system opts for time aligning of the C capsule by applying a delay of 3.5 ms. In this way, the L, R and C capsule are "aligned" within the same timeline but the C (omni) capsule is—of course—physically still in front of the L and R capsule to provide enhanced definition in relation to the L and R hemicardioid capsules' (Figs. 3.30 and 3.31).

Olabe reports on his practical experience with the DHAB System, as follows: (Olabe 2014) '… From a user point of view, the DHAB is a very "forgiving" main system and performs in concert halls, opera houses and auditoriums with different reverb densities and times giving successful results. One of the privileges this system has to offer is the capability of being positioned slightly further away than a spaced omni-system to the limit of the reverb radius of the venue, still maintaining detail, stereo image and tonal quality of the instruments. This distance will vary from venue to venue but you can always focus in with the 2 hemi cardioid L and R capsules if you went further away than desired and had no room for corrections, while still maintaining very consistent results in your main system.

The DHAB is used very frequently by 'El Sistema' orchestras conducted by Gustavo Dudamel touring the world and in the need of a quick setup with high-quality results. Sound-engineer Danilo Alvarez (El Sistema) joined Iker Olabe (U.P. Comillas CESAG and Salzburg Summer Festival) to continue research on the DHAB and two versions for specific applications have been born from the DHAB original

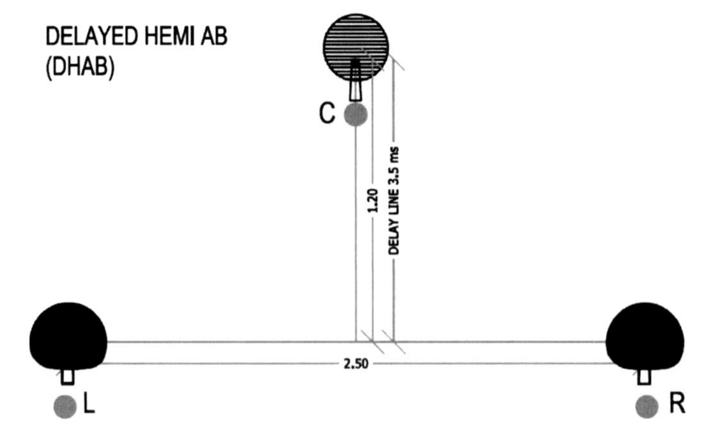

Fig. 3.30 DHAB stereo system microphone layout (from Olabe 2014)

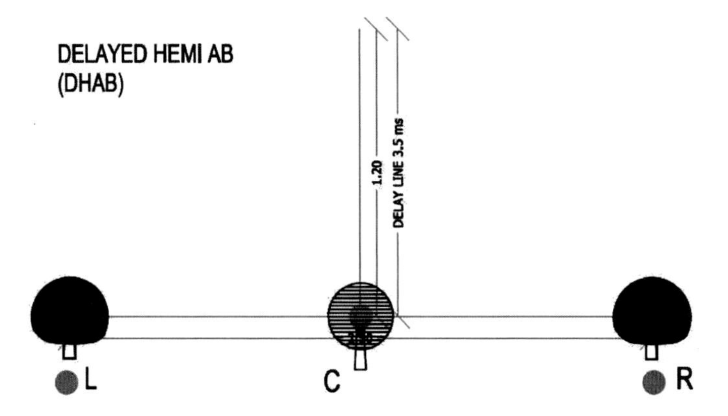

Fig. 3.31 DHAB stereo system with time-aligned C microphone, showing its 'virtual' position, once the time delay is applied (from Olabe 2014)

model. Both versions seem to have a faster rigging process [than the original DECCA-triangle; rem. by the author] and different concepts, which can be applied for specific situations.

3.3.8 SHAB—Side Hemicardioid AB

The SHAB is a DHAB system without the center capsule rigged in the advanced position. Instead it opts for a 'three in a line concept' ideal for broadcast and location recording where positioning the third mic in front could be slightly complex in certain locations. This system is also very useful in opera productions where the main system

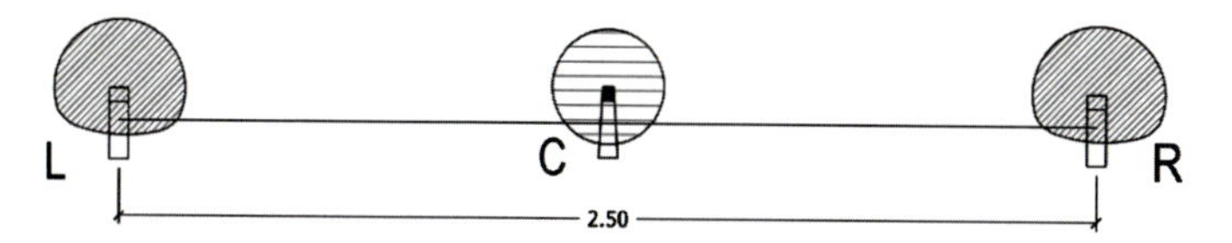

Fig. 3.32 SHAB stereo system 'three in a line concept' (from Olabe 2014)

Fig. 3.33 An SHAB system rigged ready for live broadcast from Palau de la Música de Barcelona, with the Simon Bolivar Orchestra and Conductor Gustavo Dudamel (from Olabe 2014)

will have to be out of the way of lighting and other scenery elements and a balance between intelligibility from the singers and definition from the orchestra will be expected even at a distance. L and R capsules are hemicardioids, spaced 250 cm apart, with an omni center capsule in the middle (Figs. 3.32 and 3.33).

3.3.9 *CHAB—Center Hemicardioid AB*

The CHAB evolved as a different sonic option from the DHAB and SHAB. In large orchestral recordings—if we can control the distance and height at which we rig our system and have an idea of the acoustics of the venue—then we can flip the used polar patterns in SHAB, by using omni-capsules for the L and R mic and a hemicardioid for the C capsule. In this way, a more enveloping and spacious sound can be achieved and more detail can be added by the use of the C microphone.'

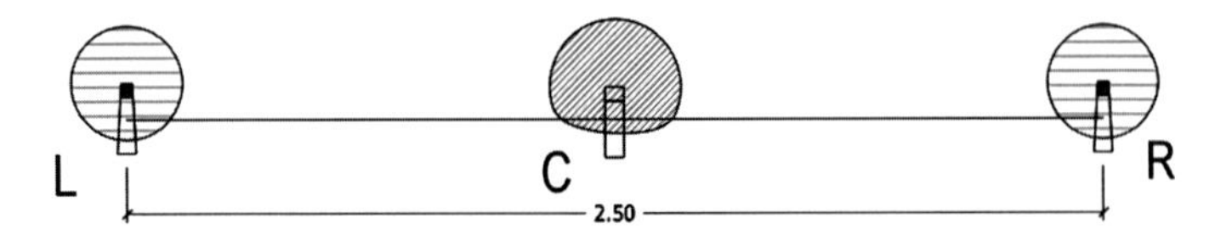

Fig. 3.34 CHAB—Center Hemicardioid AB (from Olabe 2014)

[slightly altered from Olabe (2014)]. Olabe also proposes a 5.0 surround-version of the CHAB system, which uses hemicardioids for the rear channels (for details, see Chap. 4 on 'Surround Microphone Techniques', Sect. 4.2.13) (Fig.3.34).

3.4 Combined Stereo Microphone Techniques

In their everlasting search for 'better sound,' many sound-engineers and Tonmeisters are very happy to try out combinations of various main microphone techniques. Especially combining a coincident or near-coincident stereo main pair with large AB 'outriggers' of omni-characteristics is very commonly found; most likely due to the 'air' which the largely spaced omnis are able to add to the overall sonic picture (see La Grou 1994). This effect is mainly a consequence of the very high degree of de-correlation, which is usually found with large-AB arrangements, for which the capsule spacing is at least 4 m (13 ft.) and more.

3.4.1 The 'AB-Polycardioid Centerfill' (AB-PC)

Two typical arrangements of AB-PC are a combination of Large AB with a DECCA-triangle style arrangement of three cardioids, or Large AB with an ORTF-Triple (ORTF-T) as Centerfill (Pfanzagl-Cardone 2002). In order to avoid redundancy, more details (including its application for surround recording) can be found in the next chapter. The AB-PC system has its roots in the AB-Centerfill Systems which were used by the RCA engineers for recordings of the 'Living Stereo' series, as well as basic DECCA-technique (Decca-triangle + outriggers) (Figs. 3.35 and 3.36).

However, one particular strength of the Polycardioid Centerfill technique should be pointed out: its capability in capturing the sound of solo instruments in a rather 'separated' fashion by making use of the directional characteristics of cardioid microphones.

While the AB-PC technique in general aims at capturing also large sound sources—like an orchestra for example—'as a whole' by using the Large AB spaced microphone pair in combination with an appropriate Centerfill system, the case-specific placement of the Centerfill system allows for better separation of individual or groups of instruments, if necessary.

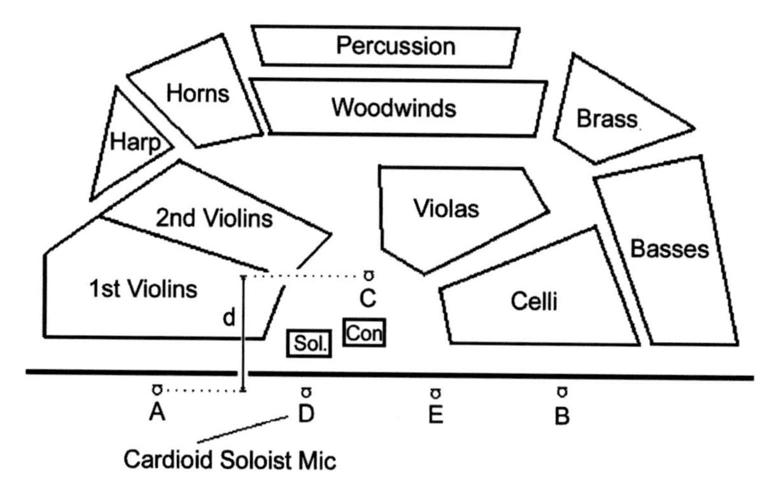

Fig. 3.35 AB-PC using a DECCA-style cardioid triangle (A, B: omni; C–E:cardioids); distance d should be compensated for by use of digital time delay

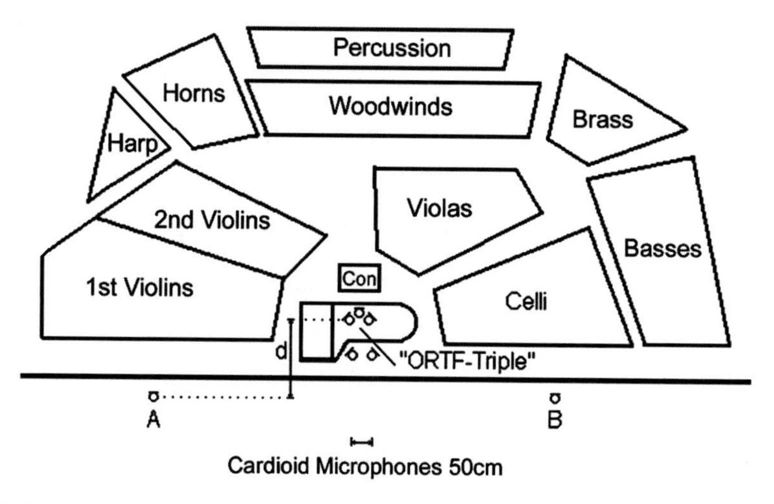

Fig. 3.36 AB-PC using an ORTF-Triple (ORTF-T) as centerfill system; distance d should be compensated for by use of digital time delay

Similar arrangements, using a combination of large AB with a normal ORTF as centerfill have been proposed in Eargle (2004), as can be seen below (Fig. 3.37).

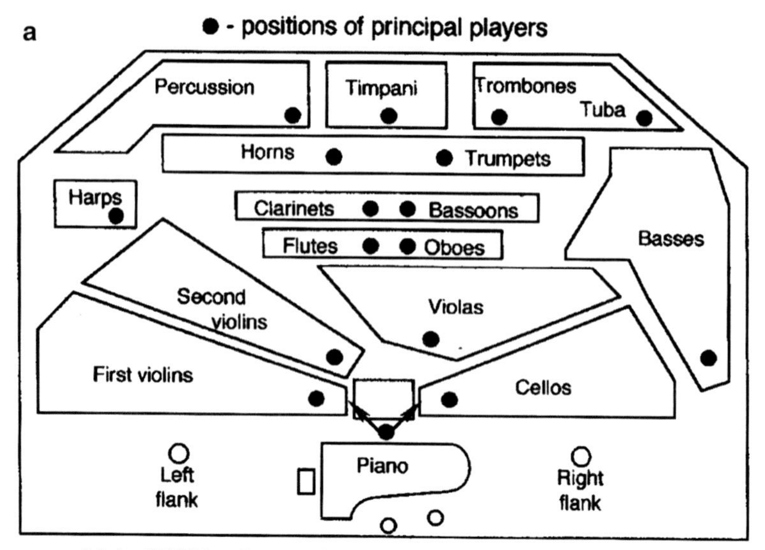

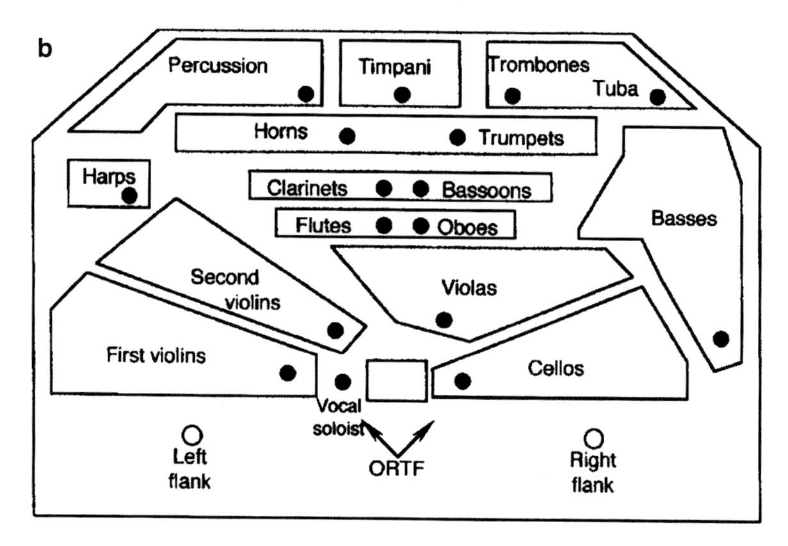

Fig. 3.37 **a** AB + ORTF for orchestra + piano recording; **b** AB + ORTF for orchestra + vocal soloist recording (Fig. 13.15 from Eargle 2004)

3.4.2 'AB-Blumlein Centerfill' and Other Techniques

In a similar way, the combination of a Blumlein style pair of crossed figure-of-eight microphones with large-AB omnis has its sonic merits: while the Blumlein-Pair is renowned for its accuracy in terms of localization, the large-AB pair excels in terms of spaciousness, so—mixed at the right levels, the combination of these two techniques manages to bring together 'best of both worlds.' A suggestion to use this technique can be found on the audiophile 'Chesky' sampler CD (Chesky 1994).

The author has used this technique on many occasions, ranging from small instrumental or vocal ensembles to symphony orchestra with great success. In practical terms, the signals of both microphone systems will usually each be panned hard L and R in order to keep signal correlation to a minimum. As has been pointed out, the signal de-correlation in the frequency band below approx. 500 Hz is of primary importance for achieving good 'spatial impression.' De-correlation in this range will be high for large-AB pairs, but usually a bit less for crossed fig-of-8 microphones (even though—in theory—ideal fig-of-8 microphones are completely de-correlated down to 0 Hz, but in practice deviations from an 'accurate' fig-8 polar pattern and the fact that sound waves easily bend around objects at low frequencies may lead to a slightly less decorrelated stereo signal with lowering frequency). Therefore, the gentle 'roll-off' in terms of amplitude, which is commonly found with fig-8 microphones, is almost welcome when mixing such signals with those of the large-AB pair since the 'narrowing' of the stereo image at low frequencies is overruled by the high degree of de-correlation which the large-AB provides in this frequency range. On the other hand, should a fig-8 microphone be linear in terms of level down to the deepest frequencies, the author would like to suggest to attenuate the signals of such a pair with a second- or even fourth-order high-pass filter (HPF) below approx. 100 Hz (±20 Hz) to maintain a high level of de-correlation for the sake of a convincing spatial impression in the recording (Fig. 3.38).

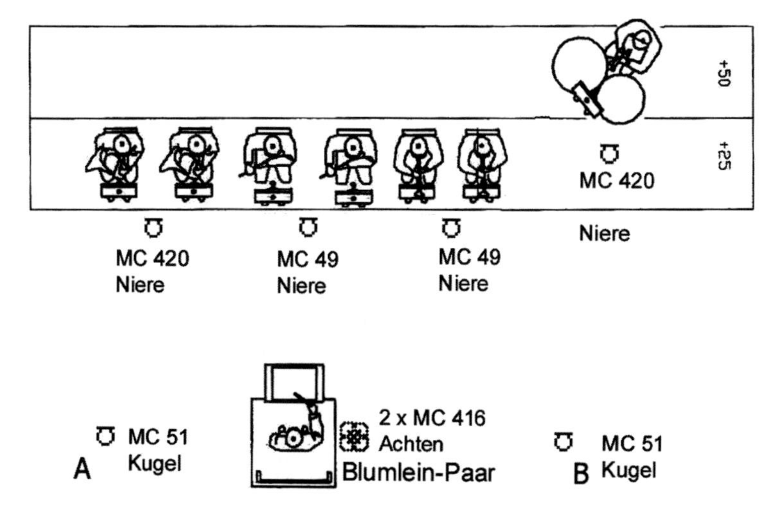

Fig. 3.38 Combination of Blumlein-Pair and large AB for (small) ensemble recording (with small and large membrane condenser microphones with various patterns by manufacturer Nevaton) [rem: 'Achten': fig-8s, 'Kugel': omni, 'Niere': cardioid]

There are similar proposals to combine large-AB with a coincident microphone technique; the following also takes into account the frequency-dependent stereo imaging (i.e. correlation) characteristics of these very different techniques:

In his paper, Preston (1998) proposes a 'frequency-dependent hybrid microphone technique', which combines the signals of a coincident microphone pair with that of a spaced microphone pair. As described in the abstract: 'Frequency-dependent hybrid stereo microphone technique is a method of separately recording intensity and timing information using two stereo microphone arrays and then electrically combining the signals together to create a stereo signal for reproduction over stereo loudspeakers.

The component pairs of a hybrid array (Fig. 3.39) were tested for localization accuracy to verify whether a frequency-dependent hybrid technique resulted in a more stable localization of sound images than conventional coincident and spaced techniques. The results show a useful agreement between the spatial transfer functions of a coincident array (>700 Hz) and a spaced array (<700 Hz).' Preston continues to explain in his paper that '…this combination of intensity and timing information copies, in part, the mechanism of binaural hearing according to the duplex theory of localization (Rayleigh 1907). A frequency-dependent hybrid stereo microphone array can potentially achieve an improvement in stereophonic imaging by combining the strengths of each technique by selectively summing the outputs of a coincident and spaced array according to frequency.'

Preston got inspired to do his test recording upon a suggestion found in Williams (1992):

> '… From the tables of frequency-dependent hybrid arrays listed in this paper, a cardioid microphone array with a stereophonic recording angle of 140° was selected. The array consisted of two coincident cardioid microphones at an included angle of 121°, and a further two cardioid microphones positioned 27 cm apart subtending an angle of 25°. …'

For the test recording carried out with the 'hybrid microphone array' white noise was used as stimulus, the perceived localization angle of which was evaluated in an anechoic chamber by 19 test listeners.

The positive result of the experiment '… indicates that for this particular array there is merit to Williams' theory of stereophonic frequency-dependent hybrid microphone arrays, and to the method of calculating the spacing and angle of the arrays.' In his paper, Preston points out that Williams also suggested a similar idea, using omni-directional microphones for the spaced microphone pair.

Additional Note

I would also like to point the interested reader to a visual documentation series of measurements which are related to frequency-dependent cross-correlation in stereo microphone systems or stereo recordings, which can be found on my Youtube channel 'futuresonic100,' when looking for clips with the term 'mic tech analysis' in the title. These video clips have been produced utilizing MAAT's '2BC multiCORR' VST plugin, which displays the cross-correlation of two audio channels *over frequency* in real time. Several audio samples, based on recordings with well-documented microphone techniques, are analyzed with respect to their signal correlation, which unveils

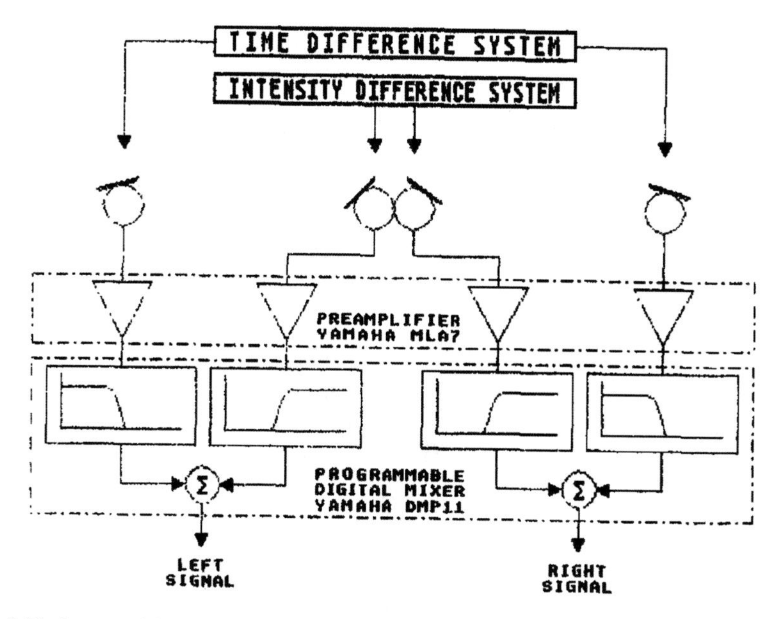

Fig. 3.39 Layout of frequency-dependent hybrid microphone system (after Williams 1992; from Preston 1998)

interesting details concerning their overall sonic character (Rem.: more information on the '2BC multiCORR' plugin can be found in Appendix C).

References

Benjamin E (2006) An experimental verification of localization in two-channel stereo. Paper 6968 presented at the 121st Audio Eng Soc Convention

Benjamin E, Brown R (2007) The effect of head diffraction on stereo localisation in the mid-frequency range. Paper 7018 presented at the 122nd Audio Eng Soc Convention, Vienna

Betz G (2000) Surroundaufnahmen – Praktische Erfahrungen. In: Proceedings to the 21. Tonmeistertagung des VDT, Hannover, pp 485–494

Blumlein AD (1931) Improvements in and relating to sound-transmission, sound-recording and sound-reproducing systems. British Patent 394,325, 14 Dec 1931 (reprinted in: Anthology of stereophonic techniques. Audio Eng Soc, 1986, pp 32–40)

Chesky (1994) Best of Chesky Classics & Jazz and audiophile test disc, vol 3. Tracks 15, 18. Chesky Records, JD 111

Dickreiter M (2011) Mikrofonaufnahme, 4th edn. S. Hirzel Verlag, Stuttgart

Eargle J (2004) The microphone book, 2nd edn. Focal Press

Engineering (1981) 100 years with stereo: the beginning. J Audio Eng Soc 29(5):368–372

Faulkner T (1981) A phased array. Hi Fi News Record Rev, July 1981

Gernemann A, Rösner T (1998) Die Abhängigkeit der stereophonen Lokalisation von der Qualität der Wiedergabelautsprecher. Proceedings to the 20. Tonmeistertagung des VDT, Karlsruhe, p 828

Griesinger D (2002) Stereo and surround panning in practice. Paper 5564 presented at the 112th Audio Eng Soc Convention, Munich, May 2002

Holman T (2000) 5.1 Surround sound – up and running. Focal Press (Elsevier)

Hugonnet C, Walder P (1998) Stereophonic sound recording. Wiley, New York

Jecklin J (1981) A different way to record classical music. J Audio Eng Soc 29(5):329–332
Keller AC (1981) Early Hi-Fi and stereo recording at bell laboratories (1931–1932). J Audio Eng Soc 29:274–280
Knothe J, Plenge G (1978) Panoramaregler mit Berücksichtigung der frequenzabhängigen Pegeldifferenzbewertung durch das Gehör. In: Proceedings to the 11. Tonmeistertagung des VDT, Berlin
La Grou J (1994) Orchestral Recording. Mix Feb 1994, p 32
Mayer AM (1880) Topophone. US Patent 224,199, 3 Feb 1880
MBHO GmbH (2016) Tech-data sheet of Schneider-disc by company MBHO http://www.mbho.de. Accessed 20 Nov 2016
Olabe I (2014) Técnicas de grabación de música clásica. Evolución histórica y propuesta de nuevo modelo de grabación. Dissertation, Universitat de les Illes Balears, http://hdl.handle.net/10803/362938. Accessed 12 Oct 2015
Pfanzagl-Cardone E (2002) In the light of 5.1 surround: why AB-PC is superior for symphony-orchestra recording. Paper 5565 presented at the 112th Audio Eng Soc Convention, Munich
Pfanzagl-Cardone E, Höldrich R (2008) Frequency-dependent signal-correlation in surround- and stereo-microphone systems and the Blumlein-Pfanzagl-Triple (BPT). Paper 7476 presented at the 124th Audio Eng Soc Convention, Amsterdam
Preston C (1998) An analysis of frequency dependent hybrid microphone arrays for stereophonic sound recording. Paper 4793 presented at the 105th Audio Eng Soc Convention, San Francisco
Rayleigh (1907) On our perception of sound direction. Philosophical Magazine, vol 13
Rumsey F (2001) Spatial audio. Focal Press (Elsevier)
Rumsey F, Segar P (2001) Optimisation and subjective assessment of surround sound microphone arrays. Paper 5368 presented at the 110th Audio Eng Soc Convention, Amsterdam, May 2001
Sawaguchi M, Fukada A (1999) Multichannel sound mixing practice for broadcasting. Paper presented at IBC Conference
Sengpiel E (1992) Grundlagen der Hauptmikrophon-Aufnahmetechnik – Skripten zur Vorlesung (Musikübertragung). Hochschule der Künste, Berlin. http://www.sengpielaudio.de. Accessed 5 May 2004
Snow WB (1953) Basic principles of stereophonic sound. SMPTE (Soc Motion Picture and Television Eng) 61:567–589
Steinberg JC, Snow WB (1934) Auditory perspective – physical factors. Electr Eng 53(1):12–15
Streicher R, Dooley W (1985) Basic microphone perspectives – a review. J Audio Eng Soc 33(7/8)
Streicher R, Everest A (2006) The new stereo soundbook, 3rd edn. Audio Eng Associates
Swedien B (1997) Recording with Bruce Swedien. VHS-Videotape, Palefish Enterprises (Inc/AcuNet Corporation), White Salmon, WA
Theile G (2000) Mikrofon- und Mischungskonzepte für 5.1 Mehrkanal-Musikaufnahmen. In: Proceedings to the 21. Tonmeistertagung des VDT, Hannover, pp 348
Valin J (1994) The RCA Bible – a compendium of opinion on RCA living stereo records, 2nd edn. The Music Lovers Press, Cincinatti, Ohio
Williams M (1987) Unified theory of microphone systems for stereophonic sound recording. Paper 2466 presented at the 82nd Audio Eng Soc Convention
Williams M (1992) Frequency dependent hybrid microphone arrays for stereophonic sound recording. Paper 3252 presented at the 92nd Audio Eng Soc Convention
Wittek H (2002) Image Assistant V2.0. http://www.hauptmikrofon.de. Accessed 24 June 2008
Wittek H, Theile G (2000) Investigations into directional imaging using L-C-R stereo microphones. In: Proceedings to the 21. Tonmeistertagung des VDT, pp 432–454

Wittek H, Theile G (2002) The recording angle – based on localisation curves. Paper 5568 presented at the 112th Audio Eng Soc Convention, Munich

Wittek H, Theile G (2004) OCT V2-Info http://www.hauptmikrofon.de. Accessed 24 June 2008

Wuttke J (1998) Das Mikrofon zwischen Physik und Emotion. In: Proceedings to the 20. Tonmeistertagung des VDT, Karlsruhe, p 460

Chapter 4
Surround Microphone Techniques

Abstract In this chapter, over 30 surround microphone techniques are explained in detail and partly analyzed. Starting out with coincident techniques (XY) with various patterns, via MS techniques to special systems such as Ambisonics (SoundField Microphone) and Bauer's 'Phasor Array.' We move on to spaced arrays for two-dimensional sound reproduction, starting from circular arrays, via OCT and ORTF Surround, to the IRT Cross, Hamaski Square and 'Microphone Curtain' designs. A large number of 'Tree'-based techniques is examined, which all have their roots in the famous 'DECCA-tree' arrangement: AB-PC Surround, CHAB 5.0, Fukada tree, Polyhymnia Pentagon, Streicher's 'Surround Sound DECCA-tree' and XY tri, to mention just a few. To finish up, there is a section on 'Baffled and 3D Techniques,' starting out with the KFM360 sphere microphone, various baffled surround-mic systems, the Holophone microphones, DPA 'D:mension 5100,' Sony sphere-arrangement, the 'Eigenmike®,' Jecklin's OSIS-System, the Pan-Ambiophonic 2D/3D system and the BACCH(TM) 3D Sound system. A short look is given to 'Immersive Audio' systems like 'Auro 3D,' 'Dolby Atmos' as well as related microphone techniques.

Keywords Microphone technique · Surround · Coincident · Spaced arrays · DECCA-tree · 3D audio

4.1 Coincident and 'Quasi-coincident' Microphone Arrays

4.1.1 'Soundfield-' or 'Soundpoint-' Microphone?

The combination of 4 subcardioid capsules, arranged in form of a tetrahedron—which the SoundField microphone is based on—was invented by Michael Gerzon (see Gerzon 1973). The idea behind this is to have directional capsules in closest proximity to each other; ideally, they would all be at the same identical point in the soundfield, which is—of course—not possible physically. The tetrahedron represents the closest approximation to this. The orientation of the capsules is as follows:

(Front): Left-up, Right-down, (Rear): Right-up, Left-down

E. Pfanzagl-Cardone, *The Art and Science of Surround and Stereo Recording*,
https://doi.org/10.1007/978-3-7091-4891-4_4

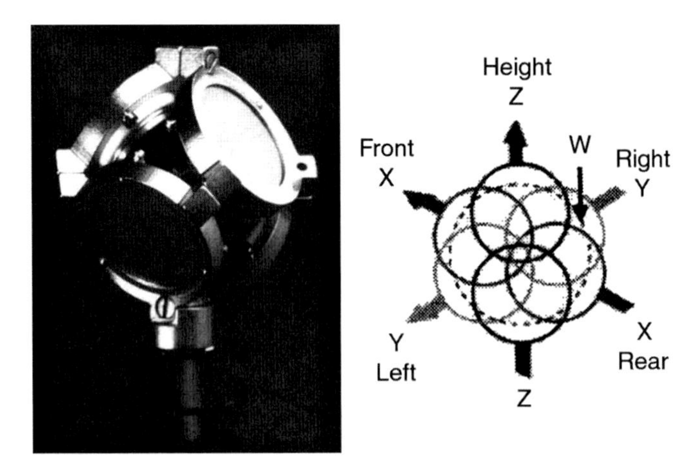

Fig. 4.1 SoundField microphone capsule arrangement and derived directional patterns (courtesy of SoundField)

By means of matrixing, these 4 so-called A-format signals can be transformed into their 'B-format' components, which correspond to the signals of 3 orthogonal fig-8 microphones, as well as a pressure transducer (omni):

W = pressure component = (Lu + Rd + Ru + Ld)
X = front/rear velocity component = (Lu + Rd) − (Ru + Ld)
Y = left/right velocity component (Lu + Ld) − (Rd + Ru)
Z = up/down velocity component = (Lu + Ru) − (Rd + Ld) (Fig. 4.1).

From these 4 B-format signals, first-order directional patterns in any desired direction can be derived with the appropriate processing applied.

For practical applications, SoundField's MK-V processor (or the corresponding software emulation) can be put to use, with the help of which the A-format signals can be transformed into B-format and sound adjustments can be made, such as virtual rotation of the SoundField-mic in all three dimensions (X, Y, Z), and change of capsule directivity.

A very similar capsule arrangement, which seems to have been inspired by the 'SoundField' microphone, but is made up of discrete single microphones has been proposed in (Sokol 2003), and also the 'Ambeo' microphone by Sennheiser, as well as the ZOOM H3-VR, employ the same tetrahedral arrangement of four capsules, which can be found in the original SoundField microphone (Fig. 4.2).

The SoundField-microphone is of interest mainly also due to its 'future-proof' recording format, as—with appropriate algorithms—multichannel reproduction can be realized for any number of desired replay speakers as well as speaker layouts.

Due to the underlying 'Ambisonics' theory and the groundbreaking work by mathematician Michael Gerzon, this system seems very attractive to the more numerically and signal processing minded audio-engineers. Of course, the idea to capture the characteristics of a three-dimensional sound-field with an idealized one-point technique

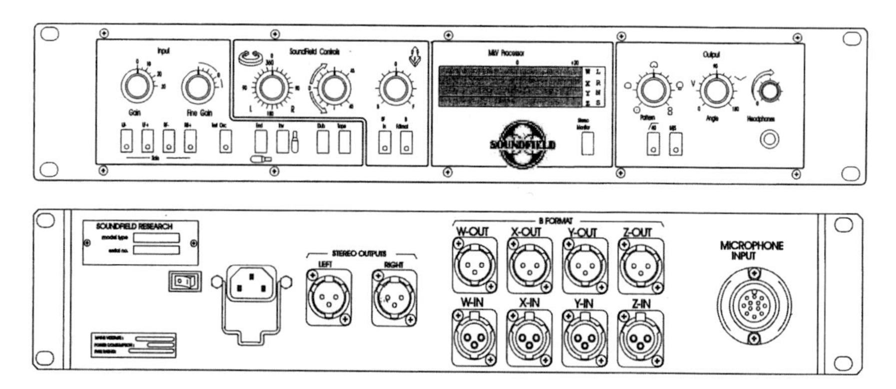

Fig. 4.2 SoundField MK-V processor (courtesy of SoundField)

looks very appealing. One of the drawbacks of this coincident system is that—by principle—it does not capture any information in terms of L/R time-of-arrival differences for the listener and relies solely on representing the surrounding sound sources by the level differences which the capsules (and applied processing) are able to ‘catch.’ While localization—resulting mainly from mid- and high-frequency signal components—is fairly accurate, spatial impression is rather low (manly due to high signal correlation at low frequencies) which is really a shame, as this should be one of the strong points in any surround microphone system.

This is most likely also the reason why the SoundField/Ambisonic-mic has scored low in comparison with the competitor systems in these two comparative listening tests: Camerer et al. (2001), Hildebrandt and Braun (2000) (Fig. 4.3).

The above assumption is backed also by the following analysis, which can be found in Nicol (2018):

‘…The tetrahedral microphone arrangement of the “SoundField” Ambisonic microphone allows the capturing of a full 3D sound field. However, since it is composed only of four capsules, its spatial resolution is low, which means that the discrimination between the sound components is not accurate…

The number of microphones of the spherical array imposes the maximal order that can be extracted. One example is the “Eigenmike®,” which—if composed of 32 capsules—allows HOA (Higher-Order Ambisonics) encoding up to the fourth order. For sound reproduction, a second step of appropriate decoding is needed to correctly map the spatial information contained in the HOA components to the loudspeaker array, in order to compute the loudspeaker input signals.

In its original definition, Ambisonics is based on the spherical harmonics expansion limited to zeroth and first-order components. HOA generalizes this concept by including components of order m greater than 1, as shown in Bramford (1995) and Daniel (2001). …The HOA components convey spatial information as a function of the azimuth and elevation angles (see Fig. 4.4).

Each order m is composed of $(2m + 1)$ components with various directivities. It should be noted that some components are characterized by a null response in

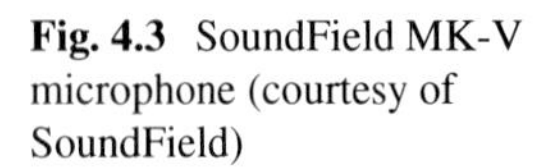

Fig. 4.3 SoundField MK-V microphone (courtesy of SoundField)

the horizontal plane. The consequence is that they do not contribute to any spatial horizontal information. By contrast, the directivity of the remaining components is symmetrical to the horizontal plane. These latter components are referred to as the "2D Ambisonics components," in the sense that, if the sound field reproduction is restricted to the horizontal plane (i.e. the loudspeaker setup is limited to the horizontal plane), only these components must be considered. On the contrary, if a full 3D reproduction is expected, all the components are used and the reproduction setup requires both horizontal and elevated loudspeakers to render spatial height information.

The component of zeroth order corresponds to the spatial equivalent of the DC component and is characterized by no spatial variation. In other words, the zeroth-order component, W, is the monophonic recording of a sound field by a pressure

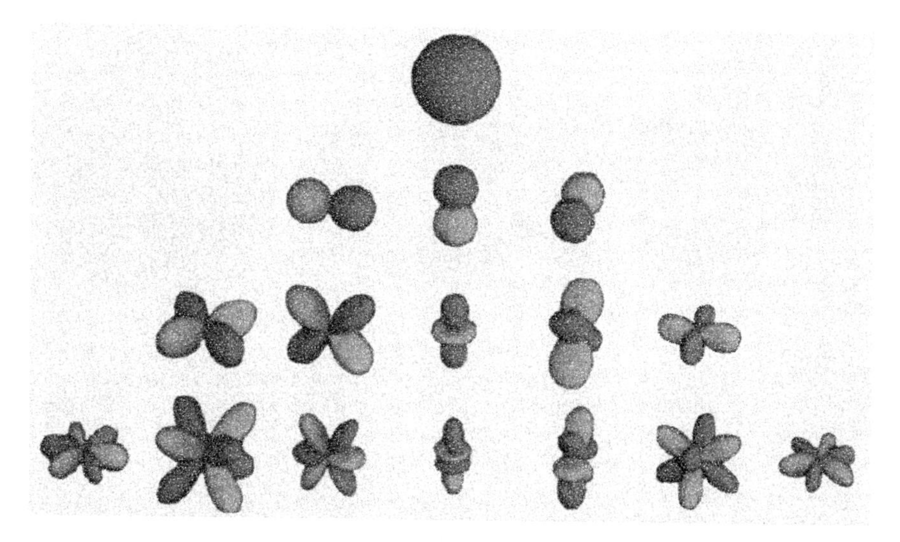

Fig. 4.4 Visualization of spherical harmonics to order three

microphone. The three first-order components are characterized by a figure-of-eight variation (i.e. cosine or sine function). As the order increases, the spatial variation is faster and faster as a function of the angle, as illustrated in Fig. 4.4. A first benefit of including components of higher order is therefore to enhance spatial accuracy and the spatial definition (resolution) of the sound field representation. This is due to the increase of the high-frequency cutoff of the associated spatial spectrum. The resulting effect on the reproduced sound field is complex: Both the size of the listening area and the bandwidth of the "time spectrum" are affected. Indeed, first-order Ambisonics reproduction is penalized by a phenomenon of "sweet spot": The sound field is correctly reproduced only at the close vicinity of the center of the loudspeaker setup. In addition, for a given reproduction area, low frequencies, which are linked to large wavelengths and therefore to slow spatial variations, are better reconstructed than high frequencies. Adding Ambisonics components of order higher than $M = 1$ increases both the size of the listening area and the high-frequency cutoff of the time spectrum. Thus, small movements of the listener are then allowed. In Fig. 4.5, the sound field reproduced by Ambisonics systems of various orders is illustrated in the case of a plane wave. It is observed that a low-frequency plane wave ($f = 250$ Hz) is well-reconstructed over a wide area by a fourth-order system. If the frequency increases up to 1 kHz, the area of accurate reproduction shrinks considerably. An upgrade to a nineteenth-order system is needed to achieve a listening area the size of which is equivalent to that obtained by the first-order system at $f = 250$ Hz. Thus, if the sound field reproduced is observed over a fixed area, the high-frequency cutoff decreases as a function of the maximal Ambisonics order M. In the same way, if the sound field is observed at a fixed frequency the size of accurate reproduction decreases as a function of the maximal Ambisonics order M.

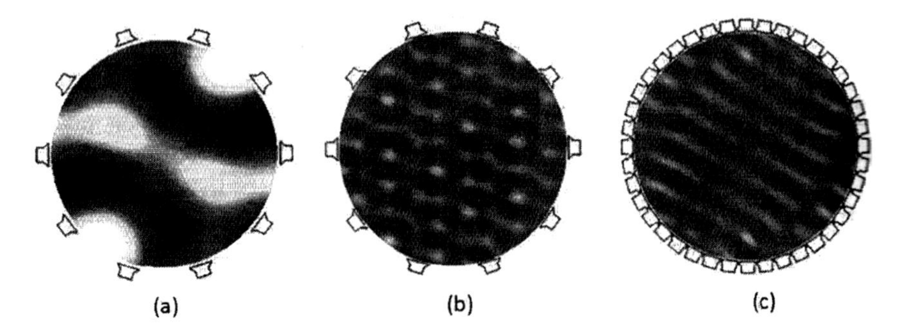

Fig. 4.5 Benefits of adding Higher-Order Ambisonics components: **a** the 250 Hz plane wave is well reconstructed over a fourth-order system; **b** increasing the frequency to 1 kHz yields a poorer reconstruction using the same fourth-order system; **c** to reconstruct the 1 kHz plane wave with the same accuracy as in (**a**), the system must be increased to a 19th order (from Nicol 2018)

In Ward and Abhayapala (2001), a rule of thumb was proposed to estimate the reproduction order as a function of the wave number k and the radius r of the reproduction sphere, to achieve a maximum threshold of the truncation error equal to 4%. The order M is obtained as:

$$M = k\,r \tag{4.1}$$

rounded up to the nearest integer. For instance, if we consider a radius of the reproduction sphere equal to 8.5 cm, which is close to the average radius of a human head, first-order Ambisonics achieves valid reconstruction of the soundfield (i.e. truncation error lower than 4%) only up to 637 Hz.

To increase the frequency cutoff up to 16 kHz for the same area, it is needed to include HOA components up to the $M = 25$th order.' (from Nicol 2018).

Of course one-point systems have their merits for certain applications, and for measurement purposes, the underlying Ambisonics-technique of the SoundField-mic is probably one of the most accurate ways to capture a soundfield in a three-dimensional way. The reproduction with an appropriate multichannel loudspeaker system will most likely be able to 'reconstruct' an almost identical signal in the sweet spot of said arrangement, but this does not mean it is also the best (and most convincing) microphone system to capture signals for replay to humans (or any creatures with two ears, whose detection of sound-source position, envelopment, spatial impression, etc., relies heavily also on L/R time differences at both ears and not only on level differences that are measurable at both ears.).

As a closing note, the author would like to make the following proposal: as the first-order Ambisonics 'SoundField' microphone—in reality—captures sound-information only at *one point* in a three-dimensional soundfield (and it does a good job as such …), it may seem more appropriate to actually refer to it as 'Soundpoint' microphone. If we want to put it in a more 'literary' way: yes, the (renamed) 'Soundpoint' microphone tries to portrait a soundfield, but does so from a 'one point' view of perspective…

The second-order Ambisonics 'Octomic' by Core Sound, as well as the third-order Ambisonics 'ZM-1' microphone by company Zylia and the Higher-Order Ambisonics 'Eigenmike®' by company mh acoustics LLC are presented later in this chapter in Sect. 4.3.7.

A very interesting and up-to-date comparison between several first-order Ambisonics, as well as Higher-Order Ambisonics commercially available microphones including the Sennheiser 'Ambeo,' Core Sound 'TetraMic'; SoundField 'MK-V,' MH Acoustics 'Eigenmike®' and Zoom 'H2n' has been carried out by Enda Bates et al., about which they report in Bates et al. (2016, 2017).

See also Sect. 4.3.7 for details on the Higher-Order Ambisonics Eigenmike®.

4.1.2 DMS (Double-MS Technique)

The initial proposal for a Double-MS arrangement consisting of two cardioids in a back-to-back arrangement, sharing a side-oriented fig-8 microphone came from Curt Wittig and Neil Muncy during the mid-1990s (see Mitchell 1999). Back then is was not yet a coincident arrangement, but the two MS-pairs were split into a 'front-' and a 'rear'-pair, separated by the critical distance (or reverberation radius). For the sake of completeness, it should be mentioned that already more than a decade earlier (Hiroyuki et al. 1980) had been granted a patent for a Double-MS back-to-back arrangement by the Japanese Patent Office.

During the last years, practical Double-MS configurations have been offered by several microphone manufacturers, among them the Schoeps Double-MS (2010), Sennheiser MKH800 Twin (2010), as well as the Sanken WMS-5 (2010).

The Double-MS (surround) system—of course—is characterized by the same flexibility in terms of adjusting the recording angle during the post-production process, as is the case with the normal stereo MS technique. Unfortunately, the outcome of a listening test, comparing a total of 8 surround microphone systems used for orchestra recording, provided the worst grading for the Double-MS System (see Chap. 9, Figs. 9.47–9.50) (Fig. 4.6).

4.1.3 Double-MS + Artificial Head

Michael Bishop, Grammy winning producer and senior sound-engineer for the US-American record label TELARC, uses a surround microphone technique which combines the signals of a Neumann KU100 artificial human head with a Double-MS arrangement which has been turned by 90°: the MS-pairs are pointing side-wise, with a distance of approximately 2 m (6 ft) to each other. They are positioned about 3–8

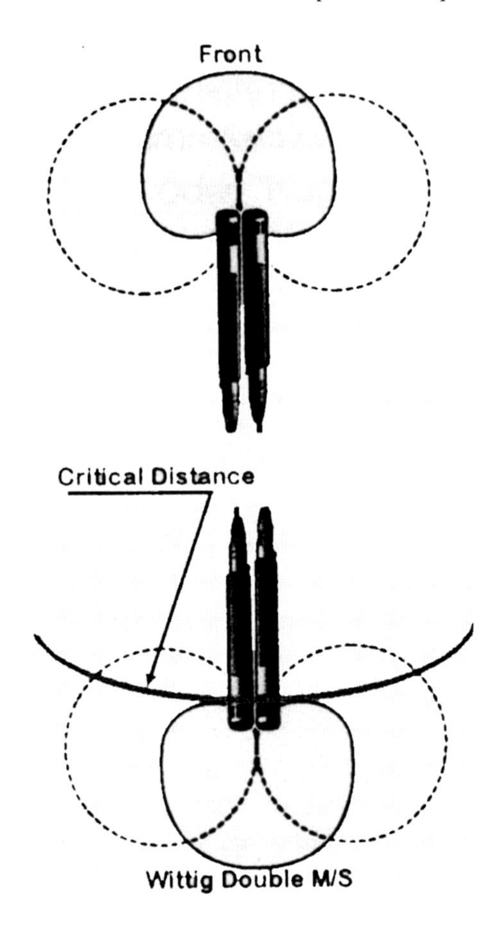

Fig. 4.6 Double-MS arrangement (after Curt Wittig and Neil Muncy; from Mitchell 1999)

ft behind the dummy head, which is facing the sound-source. According to Michael Bishop, the right balance of the MS-signals (in relation to the dummy head signals) needs to be adjusted 'by ear' in order to optimize localization accuracy as well as spatial impression. To achieve this, it may also be necessary to 'pan' the signals of the outfacing MS-cardioid (i.e. M-signal) microphones between the front and rear loudspeakers (L/LS and R/RS, respectively) (see Mitchell 1999) (Fig. 4.7)

4.1.4 BPT (Blumlein-Pfanzagl-Triple)

The BPT microphone (Pfanzagl-Cardone 2005) originally has been developed for the recording of small sound-sources (solo instruments or chamber music); it consists of three capsules with fig-8 characteristics, arranged in close proximity along a vertical axis. The normal orientation of these capsules is $-45°$, $0°$, $+45°$ relative to the sound source, which means that a third, forward-facing fig-8 capsule has been added to the

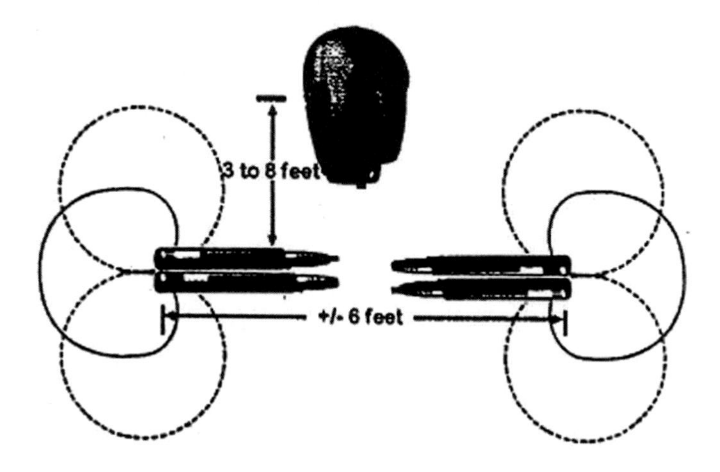

Fig. 4.7 Double-MS with artificial human head (after Michael Bishop; from Mitchell 1999)

well-known 'Blumlein-Pair' arrangement. In 'BPT 3.0'-mode the signals of the three capsules are routed directly to the channels L, C, R of a regular 5.1 system (Fig. 4.8).

In the case of larger sound-sources, like a symphony orchestra, the BPT system can be 'enhanced' by adding a 'large AB' pair of (usually omni) 'outriggers,' thus becoming something like a 'centerfill' system for the 'large AB,' which would otherwise suffer from the 'hole-in-the-middle' effect (Fig. 4.9).

One sonic drawback with figure-8 microphone capsules is the inherent level roll-off for bass frequencies below approximately 200 Hz, due to their pressure-gradient design. In order to achieve a more linear frequency response, this needs to be compensated by making use of an appropriately set EQ or filter. Usually, an LF shelving-type filter with a corner-frequency set between 120 and 180 Hz, using a boost of 2–6 dB should be sufficient (depending on the amplitude-frequency response of the individual microphone used) (Fig. 4.10).

In order to keep the overall resulting signal of the L, C, R capsules as de-correlated as possible for the low-frequency band (approx. 0–200 Hz), it is advisable to apply this filtering only to the signals of the L and R capsule and sometimes even attenuate the corresponding LF-part of the center capsule (e.g. by means of a HPF or attenuating shelving filter).

If it is desired to derive a full 5.1 surround signal from the BPT system, the capsule-anglings need to be rotated by 45° in the horizontal plane. One fig-8 capsule will now be pointing side-wise, the other one front-back, while the third one (on top) needs to be switched to omni characteristics. By using all three signals and applying MS-Decoding, it is possible to generate appropriate signals for all 5 surround, as well as an LFE-channel (see Fig. 4.11).

In order to perform the surround MS-decoding for the three signals of the BPT-microphone arrangement, appropriate software-plugins can be used or hardware MS-matrixes. Also, it is possible to simply use the routing and phase-reversal functions

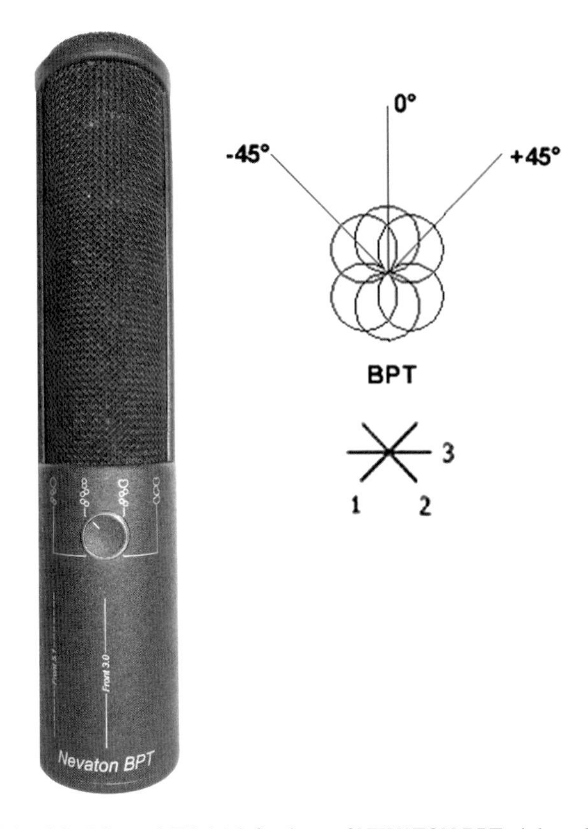

Fig. 4.8 BPT (Blumlein-Pfanzagl-Triple) left: photo of NEVATON BPT; right: schematic

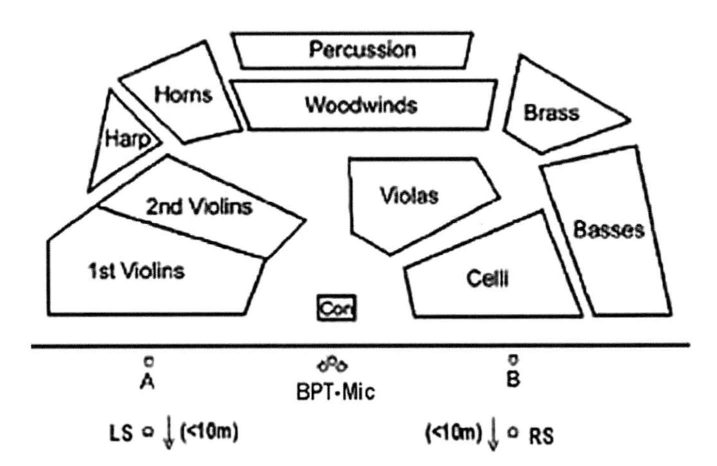

Fig. 4.9 AB-BPT system as main microphone for orchestra recording (rem.: the three capsules of the BPT-mic are pictured as 'non-coincident' only for reasons of clearer visual representation/separation)

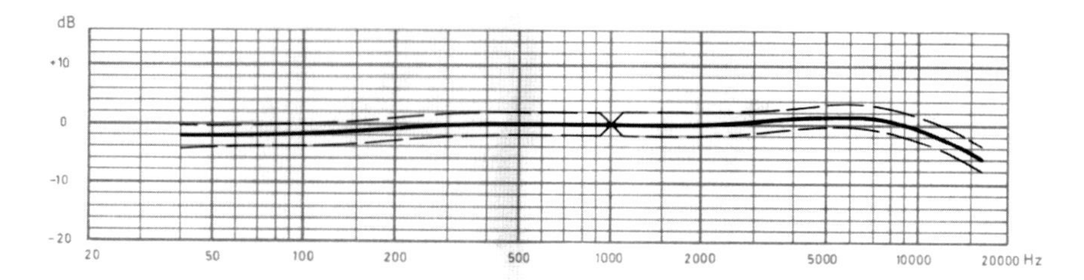

Fig. 4.10 Amplitude-frequency response of a typical (high-quality) figure-8 condenser microphone (Neumann U87) (courtesy of Neumann, Germany)

on a (for ease-of-use preferably 8-bus) mixing desk and perform the necessary signal processing there.

An example of how this can be achieved is shown in Fig. 4.12 (Note: The routing scheme in Fig. 4.12 also makes use of the PAN function on a mixing desk, which is not really needed if the desk allows 'direct routing' to single busses. In that case, all input faders can be set to 0 dB and the PAN-control at center. The −3 dB level attenuation on the fader needs to be applied if the PAN-control needs to be employed in connection with L/R-routing on a stereo-bus, as a signal which is panned in this way profits from a 3 dB gain with respect to signals, which are routed to both bus-channels with the PAN-control set to center (due to the 'panning law' which is usually employed in mixing desks).

Creating an LFE-signal: The signal-routing layout shown in Fig. 4.12 does not provide an LFE-signal; the easiest way to generate such a signal would be to bring up the W (omni-capsule) signal on a second, free input channel, then low-pass filter it with a cutoff frequency set somewhere between 80–120 Hz, route it to bus#4 and balance it in level-wise, according to taste.

BPT Surround Recording—Version 2: 'Back-to-back' Mode

There is a second possibility to achieve a 5.1 surround recording with the BPT microphone, which offers better channel separation (especially between front and rear signals) by use of 2 BPT microphones and—at least one—acoustic baffle or absorptive panel in-between (see Fig. 4.13). In this case, for both—front and rear hemisphere—a BPT-mic in 3.0 mode is employed, each of which directly delivers three signals, e.g. for L, C, R front channels or LS, CS (center surround) and RS rear channels. This version has the advantage that no signal processing (MS-decoding) is needed.

The distance d between the two BPT microphones can be varied to control the amount of spatial impression contained in the recording: If *d* is in the range of 50–100 cm, spatial impression will be less, if *d* is in the order of several meters, the front and rear signals will usually be more de-correlated and spatial impression will be higher. However, a distance of 10 m should preferably not be exceeded between the front and rear BPT in order to keep time-of-arrival differences for sound below 30 ms, as above this limit echo effects might occur, depending on the room acoustics and

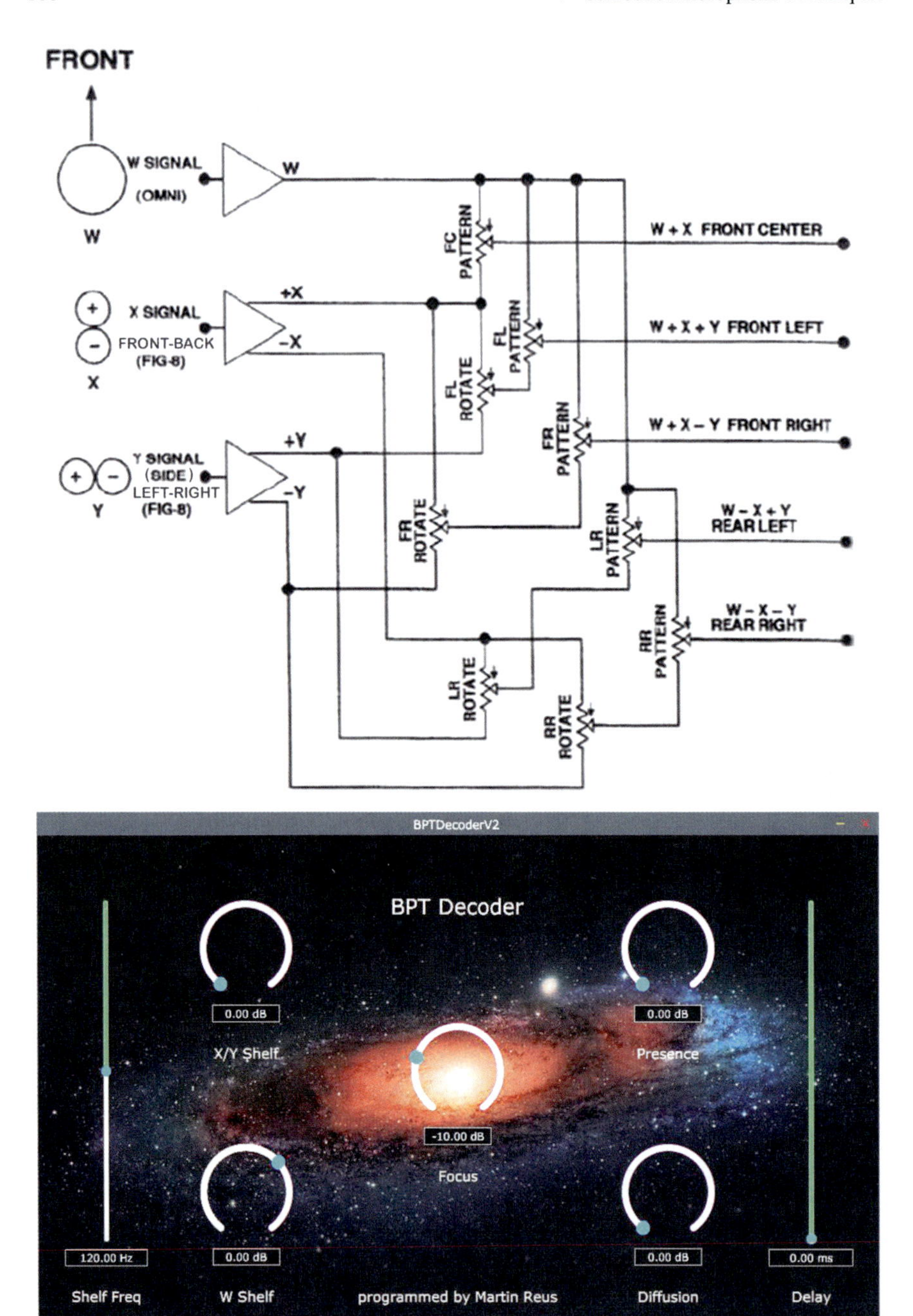

Fig. 4.11 Top: Decoding for 5.0 surround of the BPT-microphone signals (adapted from Josephson Engineering 'Series 7 User Manual'); Bottom: an early version of the surround signal decoder VST-plugin for the NEVATON BPT microphone (© 2020 by Martin Reus)

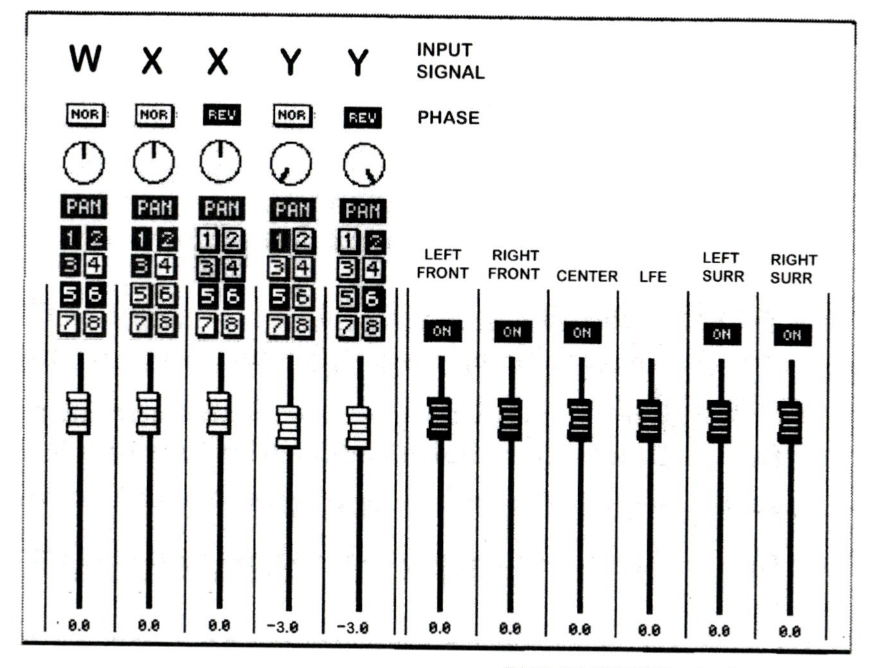

Fig. 4.12 BPT-microphone signal MS-decoding on an 8-bus mixing desk (*Rem.* for this figure, graphical elements from the user manual of the Yamaha DM2000 digital mixing desk have been used by the author)

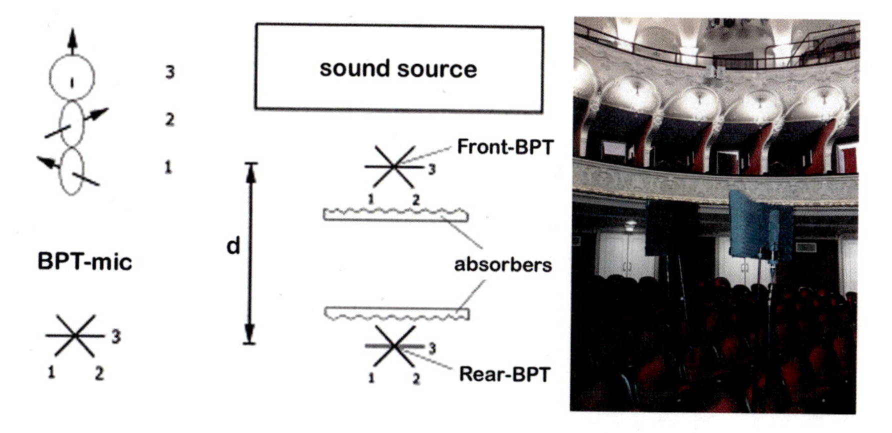

Fig. 4.13 Left: Two BPT microphones with absorptive panels, arranged in 'back-to-back' mode for 6-channel surround recording; Right: BPT-microphones with absorptive panels set up in a theater for a surround recording session

placement of the BPT—microphones (and associated absorptive panels) in relation to the sound source.

Theoretical evaluations on the acoustic effectiveness of the combination of three vertically aligned capsules can already be found in McKinnie and Rumsey (1997). These authors, however, did neither examine the combination of three fig-8 capsules, nor look into the possibility of deriving 5.1 surround signals my means of MS-decoding.

Even ahead of McKinnie and Rumsey, in Cohen and Eargle (1995), a coincident capsule arrangement for deriving three front-channel signals is described. The main focus of that research was the achievement of second-order cardioid patterns, which—due to various technical and qualitative reasons—did not seem feasible at the moment.

The software tool 'Image Assistant V2.0' (Wittek 2002) indicates that there is a 3 dB level-boost for sound-sources toward the middle-axis of the BPT-arrangement, if all three capsule signals (fig-8 mode) are used with the same relative level of 0 dB (see Fig. 4.14). According to the localization simulation of the same tool (in a different mode, not shown), this results in a recording angle of 164°. A reduction of the center capsule level by 10 dB results in an almost even amplitude distribution along the L-R stereo basis and a recording angle of 110°.

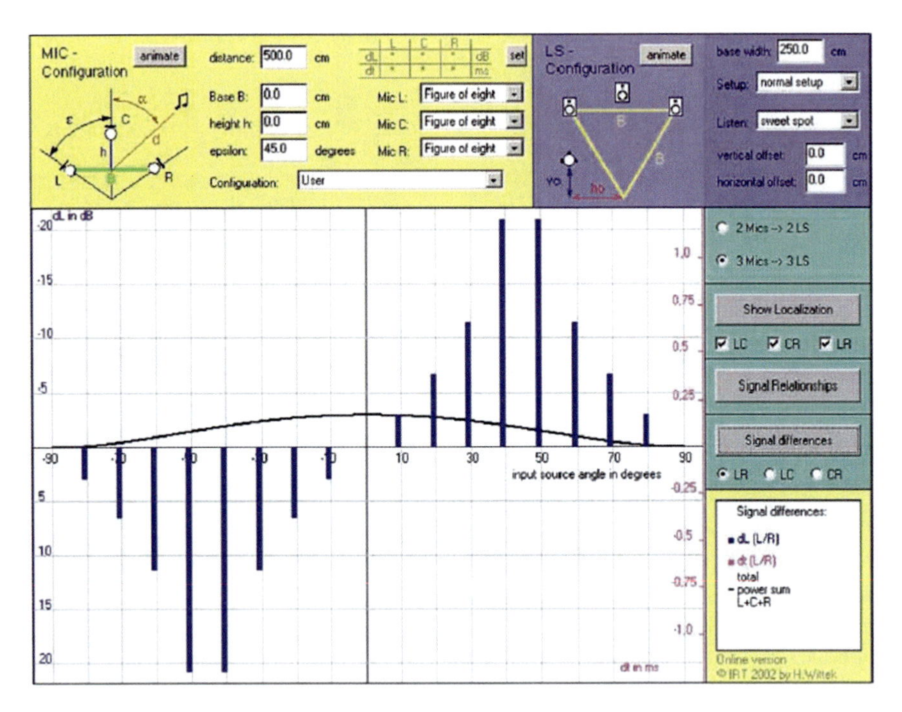

Fig. 4.14 Level distribution on a 3-channel loudspeaker replay system for BPT-capsule signals set to the same level (graphic generated using Wittek 2002)

The accuracy of the calculated effective recording angle has found approval by the author on the basis of informal listening tests. In practical work, the 'correct' (or acoustically useful) level of the center capsule signal usually depends on the distance between BPT and sound source: While for close distances a relative level of −10 dB (in relation to the L and R capsule signal) often turned out to be the 'correct' one, for large distances the level may need to be raised to −6 or −3 dB. On the occasion of a soloist, performing with an orchestra, needing some 'emphasis' volume-wise, the level of the center capsule may even need to be raised to 0 dB or more. Hence, the sound-engineer or Tonmeister has a flexible tool at hand, which can help to re-adjust the sonic picture (and level-balance), as needed. Of course raising the level of the center capsule also makes the overall resulting signal more 'monophonic' (i.e. L- and R-ear signal more highly correlated) for the listener, which is true for playback on a 2-channel stereo, as well as multichannel-surround system.

In this context, it seems interesting to note that Geoff Martin proposes a relative level difference of 9 dB for the center capsule for the rather similar microphone arrangement which he describes in Martin (2005) (see Sect. 4.1.5). Martin also proposes a different tilt for the center capsule in order to minimize overall signal correlation for the front system.

The BPT-microphone system, as used for stereo and surround recording has been described in the following (utility- or innovation-) patents:

Australia: Innovation Patent No. AU2005100255 '3.0 Microphone for Surround-Recording'; Austria: Gebrauchsmuster Nr. 9616 'Hauptmikrofon für Surround-Aufnahmen'; Germany: Gebrauchsmuster Nr. 20 2005 017 198.9 'Hauptmikrofon für Surround-Aufnahmen'; Japan: Utility Model Registration No 3128257 'Microphone for Surround-Recording'; USA: Utility Model Application No. 11279154 'Microphone for Surround-Recording'.

4.1.5 Martin Triple

The Martin Triple has first been presented at the 118th AES convention in Barcelona (Martin 2005) and consists of an arrangement of 3 vertically aligned capsules with fig-8 characteristics. In contrast to the BPT-microphone system (Pfanzagl-Cardone 2005), the center capsule is tilted downwards (at least by 45°), as Martin wants to avoid too high a correlation between the L (or R) and center capsule signals. An alternative proposal of his consists in a vertical displacement of the center capsule by at least 6 cm relative to the L and R capsule, again in order to minimize correlation between the front capsules. For pickup of rear signals, he proposes a Blumlein-Pair of two crossed fig-8 capsules with an included opening angle of 45° to 90°. In order to minimize correlation between the front and rear channels, he proposes to tilt the rear microphones upwards (in a manner, which 'overshoots' the sound-source). Also, he specifies a maximum distance of 70 cm (roughly 2 ft) between the front and rear system, which is—presumably—also not to the advantage of the spatial impression, which this surround microphone system should be able to provide, as we already

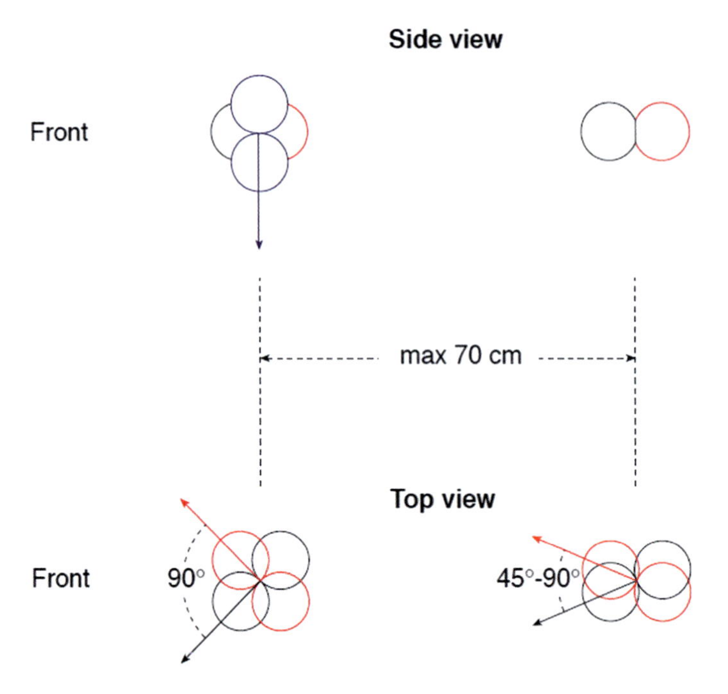

Fig. 4.15 Schematic view of configurations for version 1 of the 'Martin Triple' technique. The L, R, LS and RS microphones are forward-facing. The microphone for the center-channel information is coincident with the front array and is pointing toward the floor (from Martin 2005)

know from research from other sources that spatial impression relies heavily on the de-correlation (especially of the low-frequency components) of all signals involved (Figs. 4.15, 4.16, 4.17 and 4.18).

4.1.6 XY Cardioid Arrangement (Back-to-Back)

A double XY-arrangement with cardioids and physical opening angles between 90° and 120° has already been described in Yamamoto (1973). During the last few years, a Japanese manufacturer has taken up the idea again—with a single arrangement though—and has integrated two cardioid capsules into a card-based audio recorder (Zoom H2) (Figs. 4.19 and 4.20).

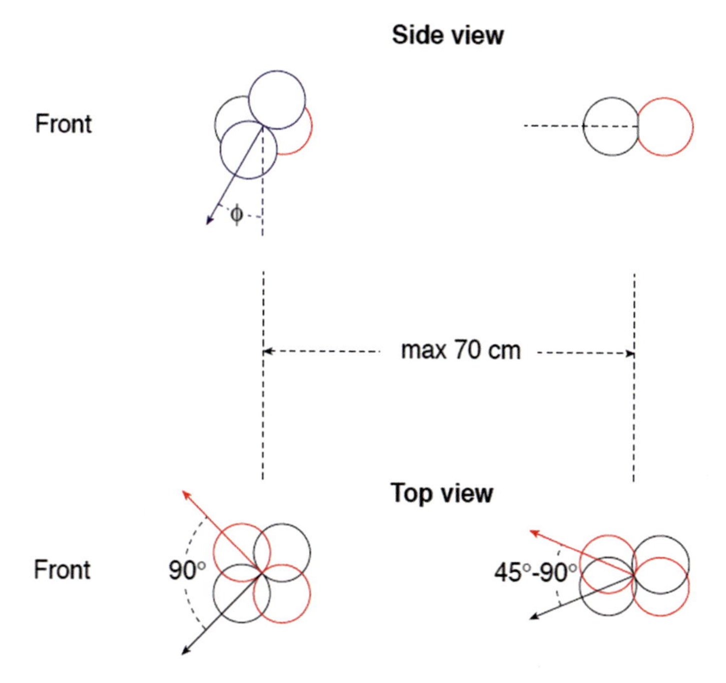

Fig. 4.16 Schematic view of configuration for variation 2 of the technique. This can be used to increase the level of frontal direct sound sources in the center microphone (from Martin 2005)

4.1.7 MILAB 360 (360° Surround Microphone)

The surround microphone SRND 360 of company MILAB contains 3 capsules with cardioid characteristics at included angles of 120° and uses a special matrix to generate signals of virtually 'inserted' cardioid capsules, with the aid of which a variety of surround signals can be generated, among them 6 (virtual) cardioid microphones at included angles of 60°. According to the manufacturer, these output signals can be used in a wide variety of stereo, multichannel and surround formats (Fig. 4.21).

4.1.8 Phasor Arrays and the 'Ghent' Microphone

In 1961, Benjamin Bauer has published a paper on the topic of 'Phasor Analysis of Stereophonic Phenomena' (Bauer 1961). The essence of his findings is that: 'An improved understanding of some stereophonic phenomena may be obtained by the use of acoustical pressure phasors to portray sound pressure at the ears of the observer. With the help of phasors, it is possible to expand and modify certain

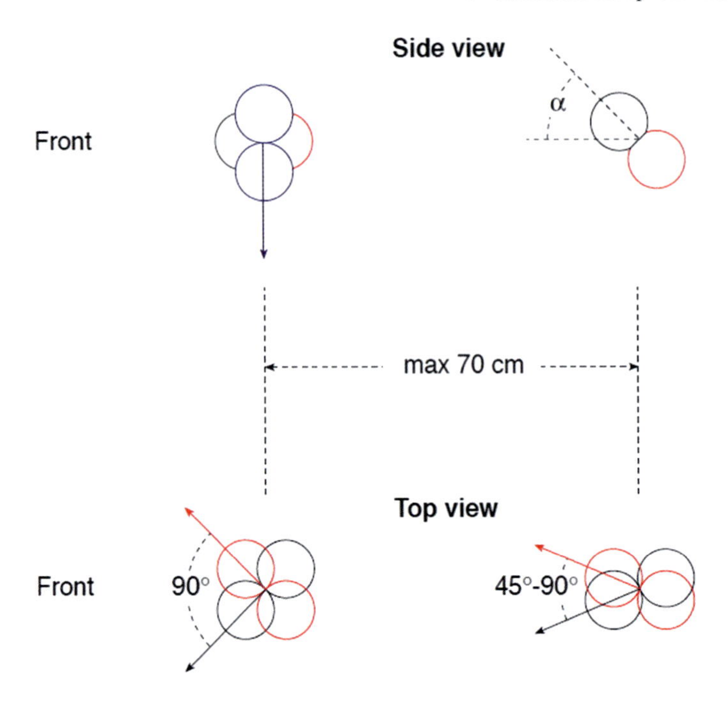

Fig. 4.17 Schematic view for version 3 of the 'Martin Triple' technique. This can be used to decrease the level of frontal sources in the rear microphone array if a problem occurs with front images 'pulling' into the surrounds (from Martin 2005)

conclusions of previous observers and to validate some previously unpublished observations; stereophonic "law of sines" is derived. The existence and location of "out-of-bounds" stereo-phonic image are analyzed and verified. The "allowed maximum out-of-phase ratio" is derived, together with the observation that this maximum is exceeded by certain microphone arrays. The motion and elevation of the center image in stereophonic reproduction are observed and explained.' (from Bauer 1961).

After further research, Bauer came up with the proposal of the 'Ghent' Microphone system, that—in a practically realized version—is made up of 4 cardioid capsules and an elaborated matrixing system to create 4 (virtual) microphone patterns, which capture an angle of 360° and have a pattern similar to hypercardioids. This quadraphonic system is characterized by an all-around inter-channel separation of 9.5 dB, a front-to-back separation of 13 dB and a side-to-side separation of 18.5 dB. More details can be found in Bauer et al. (1978) and Bauer (1979); a practical implementation based on a Neumann QM-69 quadrophonic microphone and a specially constructed microphone encoder can be seen in Fig. 4.22.

While the above-described practical implementation results in 4 coincident hypercardioids, angled at 90° to each other, by use of different coefficients in the signal processing matrix different polar patterns and angling of the virtual microphones can be achieved.

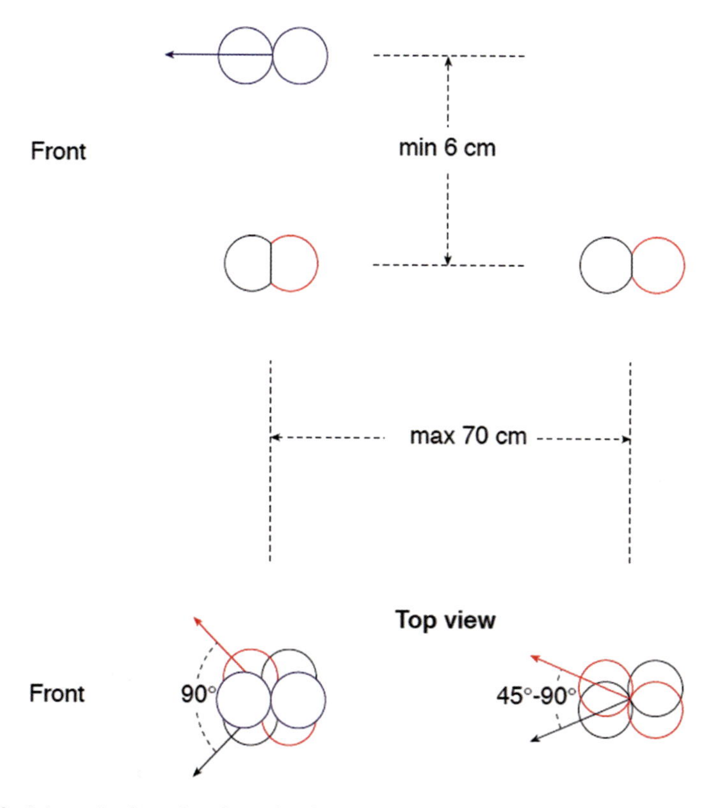

Fig. 4.18 Schematic view of configuration for version 4 of the technique. Not drawn to scale (from Martin 2005)

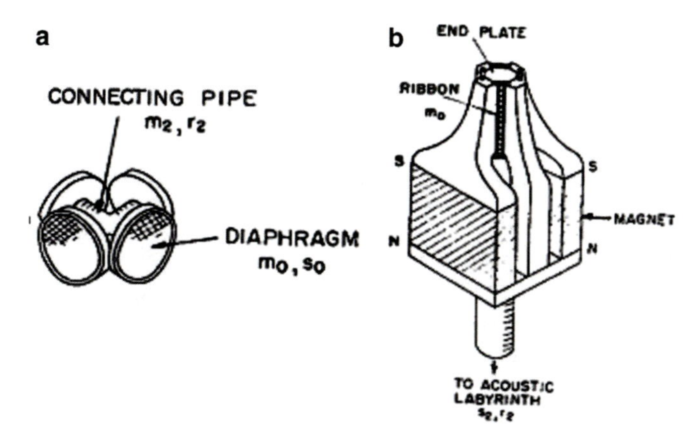

Fig. 4.19 'One point' quad-microphone with four cardioid capsules (**A** = condenser type, **B** = ribbon-type) (from Yamamoto 1973 courtesy of AES)

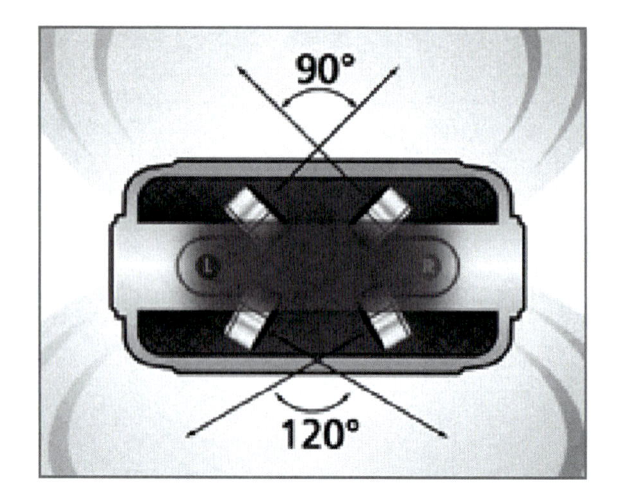

Fig. 4.20 Practical implementation of a 'one point' quad-microphone with four cardioid capsules in the form of the ZOOM 'H2' SD-card recorder with built-in microphones (courtesy of ZOOM Corp., Japan)

Fig. 4.21 MILAB SRND 360° coincident surround microphone (courtesy of MILAB)

In his patent from 1981, Bauer proposes an arrangement which makes use of two pressure-gradient microphones oriented a 0° and 90° relative to the sound-source, as well as a pressure microphone and a considerable amount of signal processing in order to create output signals for quadrophonic sound (see Bauer 1981).

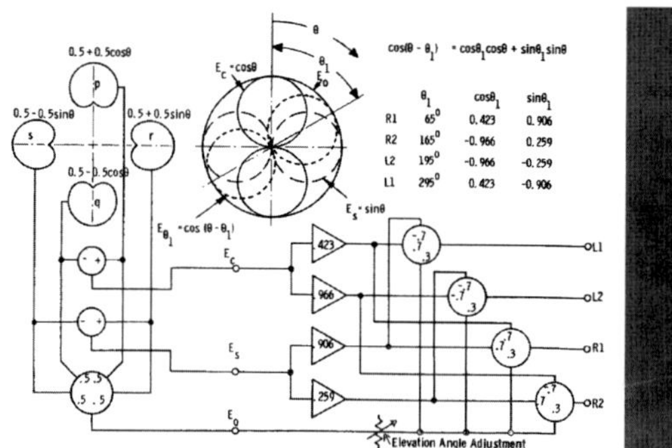

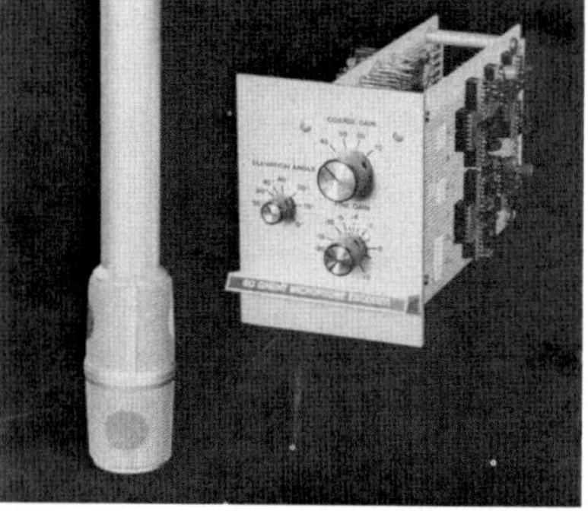

Fig. 4.22 Left: Schematic of Ghent microphone; Right: the Neumann QM-69 microphone and the special microphone encoder used in the Ghent system (partial reproduction of Figs. 7 and 8 from Bauer 1979)

4.2 Spaced Arrays (Two-Dimensional Arrays)

4.2.1 Circular Microphone Array

Mark Poletti has proposed a circular microphone array according to Fig. 4.23 which also employs a digital signal processor, frequency compensation filters and a sum and difference network. The digital signal processor calculates the Fourier transform of sampled output signals from the transducers to produce a plurality of sound wave components specifying the sound field. The frequency compensation network equalizes each component using Bessel functions to flatten the apparent response of the

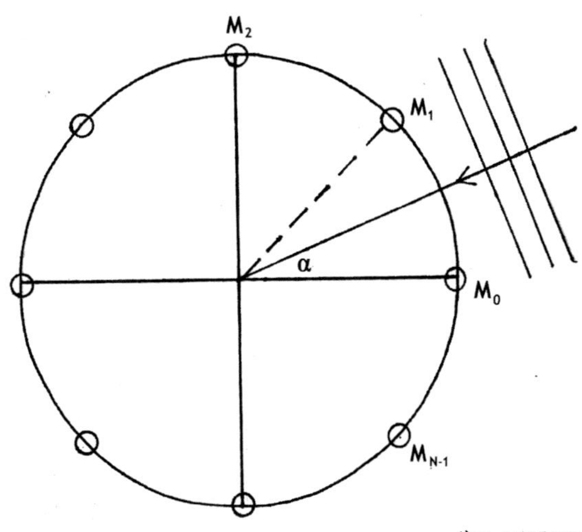

Fig. 4.23 Circular microphone array (after Poletti 2000)

array, and the sum and difference network then combines the equalized components to provide a plurality of audio signals which represent the sound field (see Poletti 2000, as well as international patent WO 01/58209 A1).

4.2.2 *Trinnov-Array*

In Laborie et al. (2004), the authors are describing a compact two-dimensional array, which occupies a space of only about 20 by 20 cm (roughly 0.6 by 0.6 ft) and consists of 8 microphones based on the principle of 'High-Resolution Spherical Harmonics.' The authors claim that a system solution, using directivity of fifth order, should be sufficient to achieve appropriate directional patterns and a surround signal channel separation that conforms to the ITU-R BS.775 recommendation.

According to their research, a fifth-order system is sufficiently flexible to enable calculation of correct signals for the irregular (i.e. asymmetrical—with respect to front/back speaker layout) replay arrangement of a typical 5.1 surround system, using appropriate panning-laws (Fig. 4.24).

Figure 4.25 shows the—acoustically very relevant—differences between the simulated and the actually resulting, measured directional patterns: for low frequencies, fifth order directionality cannot be achieved and at high frequencies 'secondary lobes' start showing up, which certainly cause unwanted coloration of sound (see the 'ripples' at the 1 kHz and 'thorns' at the 5 kHz measurements).

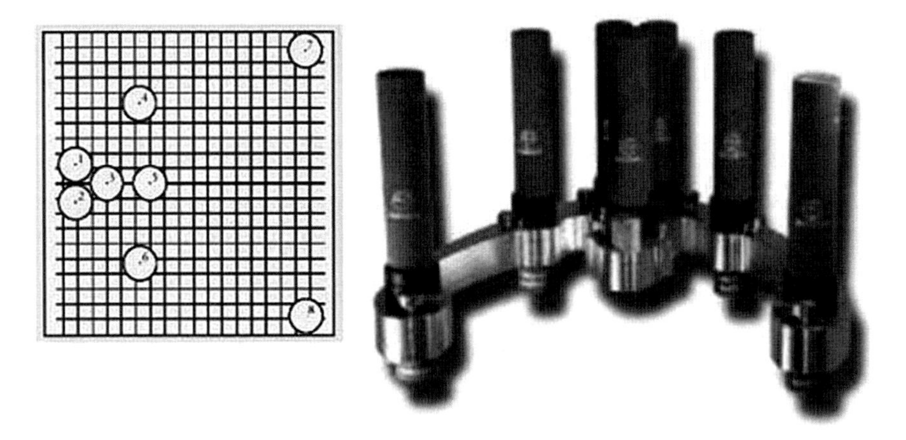

Fig. 4.24 TRINNOV surround microphone arrangement (Left: schematic layout; Right: physical implementation using Schoeps microphones) (from Laborie et al. 2004)

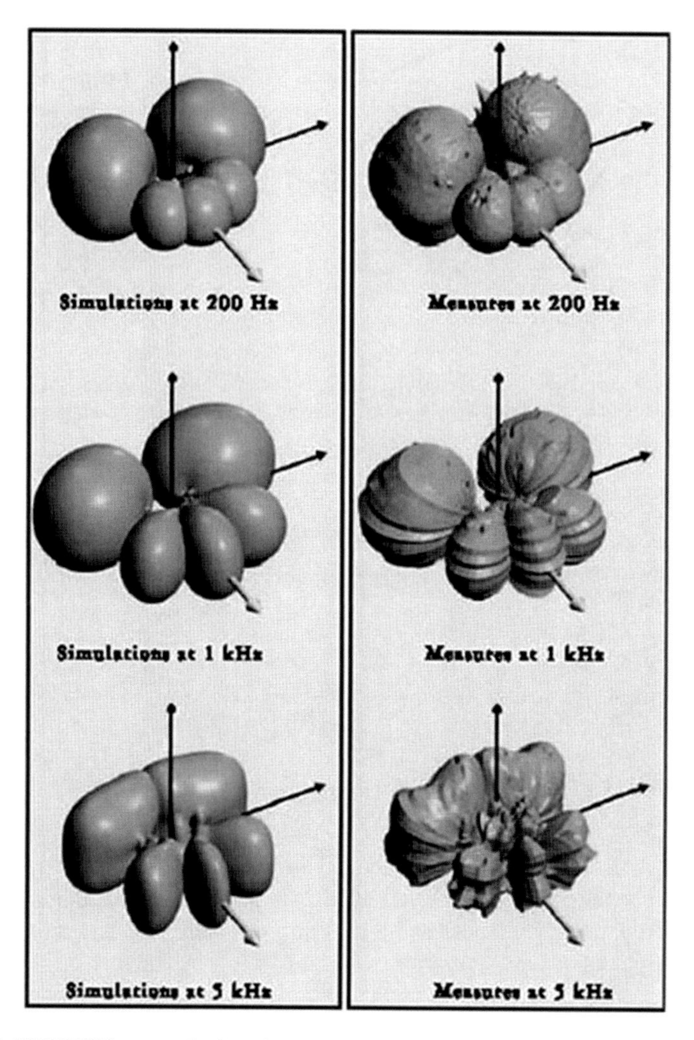

Fig. 4.25 TRINNOV surround microphone array—simulated and measured directional patterns dependent on frequency (from Laborie et al. 2004)

4.2.3 ABC 5 (Omni Array, Decca-Triangle)

The ABC 5 arrangement, which has been proposed in Hermann et al. (1998) also for use as part of a surround-setup has obvious roots in the well-known DECCA technique, first developed and employed by the DECCA sound-engineers during the 1950s. For surround purposes, the ABC 5 needs the addition of (at least) two microphones for the rear channels (not pictured). The localization properties of this arrangement are mainly based on interaural time differences (ITDs), due to time-of-arrival differences of the sound at the omni microphones (Fig. 4.26; Table 4.1).

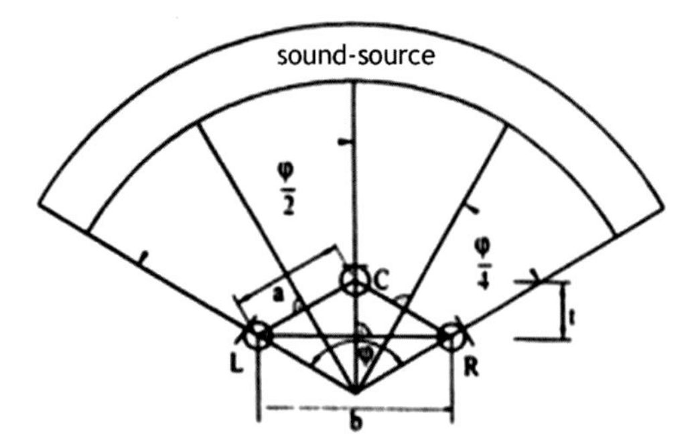

Fig. 4.26 ABC (or DECCA) microphone array; schematic arrangement of the front-channel capsules; calculation of capsule spacing depending on desired recording angle (after Hermann et al. 1998)

Table 4.1 ABC—calculation of effective recording angles based on ITDs (after Hermann et al. 1998)

Recording angle φ (in degree)	Capsule spacing *a* (in cm)	Capsule spacing *b* (in cm)
100	87.5	158.5
120	74	128
140	64.5	105.5
160	57.5	88

4.2.4 INA 5 (Ideal Cardioid Arrangement)

This arrangement has also been described in Hermann et al. (1998) and can be regarded to be the analogy of the ABC/DECCA technique in the domain of equivalence microphone techniques, as the resulting localization is based on both ITDs and ILDs (Interaural Level Differences) (Figs. 4.27, 4.28; Table 4.2).

Fig. 4.27 INA 5 schematic arrangement of the front channels (from Hermann et al. 1998)

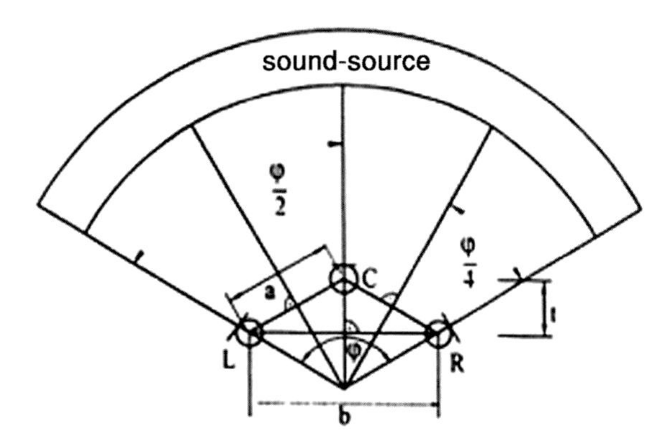

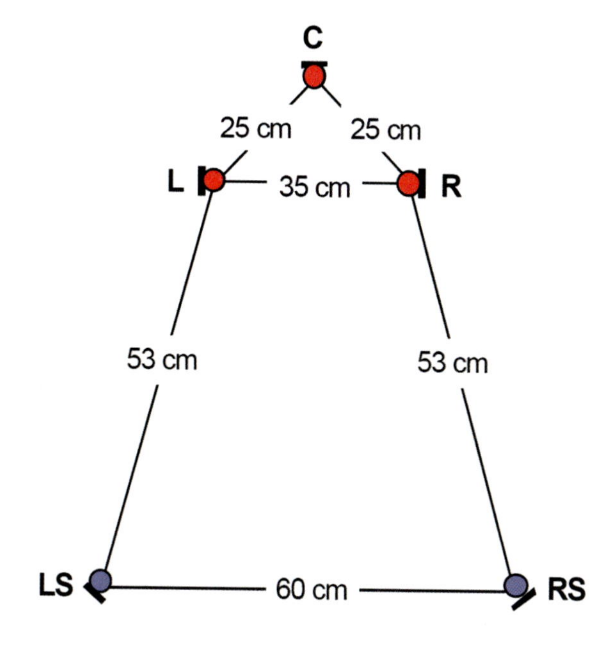

Fig. 4.28 INA 5—schematic layout of capsules for an effective recording angle of 180° (from Theile 2000)

Table 4.2 INA 5—calculation of effective recording angles based on the Williams-curves (after Hermann et al. 1998)

Recording angle φ (in degree)	Capsule spacing *a* (in cm)	Capsule spacing *b* (in cm)	Triangle height *t* (in cm)
100	69	126	29
120	53	92	27
140	41	68	24
160	32	49	21

4.2.5 OCT Surround (Optimal Cardioid Triangle—Surround)

This arrangement has been proposed by Dr. Günther Theile of IRT Munich (Institut für Rundfunktechnik) who, based on his calculations, has concluded that channel separation is not large enough, if only microphones with pure cardioid characteristics (see INA5) are used. This is why he devised the 'Optimal Cardioid Triangle,' for which initially he only defined the directional patterns and spacings of the three front capsules, mainly intended for the recording of (small) musical ensembles (see Theile 2000): left and right front microphone as supercardioids, center microphone as cardioid. Later on Theile added two rear-facing cardioids for use in a surround microphone recording setup. It needs to be added that—due to the relatively small spacing between the two rear capsules, as well as their distance to the front capsules (see Pfanzagl-Cardone and Höldrich 2008)—signal correlation within the OCT system

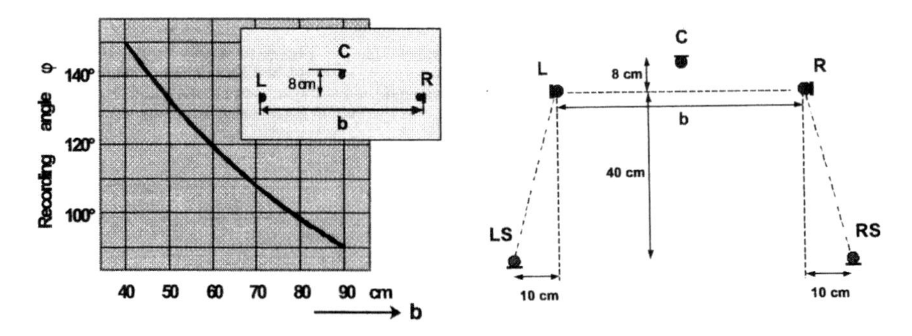

Fig. 4.29 OCT surround: Left—relation between recording angle and capsule spacing; Right—schematic capsule layout (from Theile 2000, 2001)

is unfavorably high for low frequencies (below 200 Hz), which degrades spatial impression for the listener (in this context see also the emails of Cornelis van der Gragt and Jan Korte in the appendix) (Fig. 4.29).

In order to compensate for the lack of bass, which is commonly found with supercardioid microphones, Theile suggests the addition of two pure pressure transducers (omnis), low-pass filtered at 100 Hz, which are to be positioned in close vicinity to the supercardioids (not pictured). By proposing this, he certainly manages to compensate or linearize the frequency response at very low frequencies, but at the same time completely neglects the fact that signal correlation will be very high for the signals of these two omnis, which contributes in a negative way to the impression of spaciousness.

4.2.6 OCT V.2

Despite the very self-confident naming of the original OCT—'Optimal Cardioid Technique,' apparently some ameliorations were still possible, as a few years later Theile proposed a modification (see Wittek and Theile 2004).

As a main change, the cardioid center capsule C has been advanced toward the sound source by 32 cm, which results in a time-of-arrival difference between C and the L or R microphone of 1 ms for frontal sound (Fig. 4.30).

In their paper, the authors describe the '…expected properties of the OCT2:

- Optimized downmix properties in comparison with the OCT1 with respect to the sound color and the spatial impression (due to the reduced correlation of the reverb tail between L/C and C/R).
- The precise and stable directional image of the OCT1 remains unchanged.
- The imaging characteristics with respect to the display of depth and distance remain unchanged.

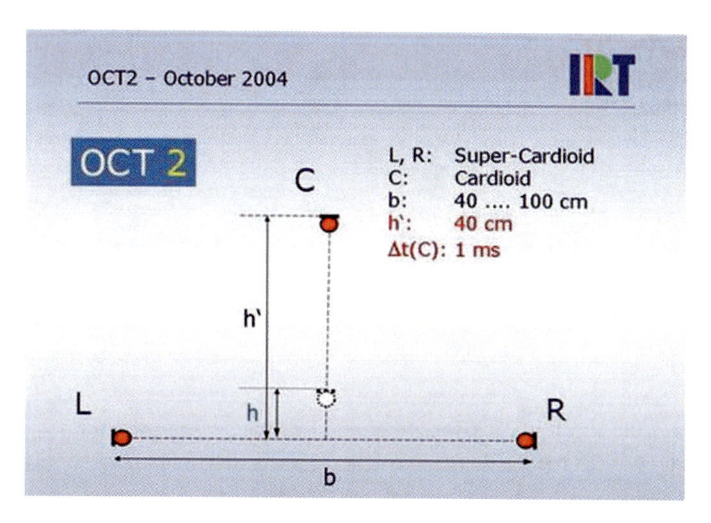

Fig. 4.30 'OCT version 2' (from Wittek and Theile 2004)

- The imaging characteristics with respect to the display of reverb and spaciousness are improved, and the sonic image is now a bit more like the well-known DECCA-tree.' (from Wittek and Theile 2004).

4.2.7 ORTF Surround

The ORTF technique has experienced some development since its inception in the 1970s at the French Broadcasting agency ('Office de Radiodiffusion Télévision Française').

Ceoen (1972) has been able to demonstrate the sonically preferable properties of the ORTF technique in comparison with a number other common stereophonic microphone techniques. For the sake of obtaining a third (center-) channel signal, as required for the three front channels of a 5.1 surround recording, Pfanzagl-Cardone has proposed the addition of a third cardioid capsule to the regular ORTF configuration (see Pfanzagl-Cardone 2002), and later on has been able to demonstrate its beneficial sonic properties, when combined with a large-AB pair of outriggers, through acoustic measurements (FCC—Frequency-dependent cross-correlation Coefficient measurement), as well as subjective listening tests (see Pfanzagl-Cardone and Höldrich 2008).

Probably as early as 2008 Wittek has proposed a 'surround' version of the ORTF microphone technique (see Wittek 2008).

In Wittek and Theile (2017), the system is described, as follows: '...One optimal solution for ambient recordings in multichannel stereophony is the "ORTF-surround" system, in which four supercardioids are arranged in a rectangle with 10 × 20 cm side lengths. Here the distances between microphones help with de-correlation and

Fig. 4.31 ORTF surround microphone array by company Schoeps (from Wittek and Theile 2017)

thereby lend the sonic impression its spatial openness. The microphone signals are routed discretely to the L, R, LS and RS channels. The signal separation in terms of level is approx. 10 dB; thus, the sonic image during playback is stable even in off-axis listening positions.' (Fig. 4.31).

4.2.8 Klepko-Array

The Klepko-Technique seems to be inspired by the ORTF technique as the capsule spacing between the outer supercardioids and the center cardioid microphone is 17.5 cm each (approx. 0.6 ft). With physical opening angles of 30° each, there is a resulting overall recording angle of 150°, according to Klepko. For the capturing of rear signals, a dummy head is employed, positioned 124 cm (roughly 4 ft) behind the front microphones (see Klepko 1997) (Fig. 4.32).

4.2.9 Atmo-Cross (IRT Cross)

The so-called 'Atmo-Cross' consists of an arrangement of 4 cardioid microphones with a capsule spacing of 21–25 cm (approx. 0.7 ft), which has been chosen based on the 'Williams-Curves' (see Williams 1987). According to the inventor, (see Theile 2000) the Atmo-Cross is very well-suited for the recording of atmospheres, sound-effects, noises, etc., as well as for the capturing of diffuse sound in combination with a front microphone system in the context of music recording in surround. In the latter case, the signals of the front-facing microphones of the cross are routed to the front replay channels in order to ensure good 'envelopment' for the listener (Fig. 4.33).

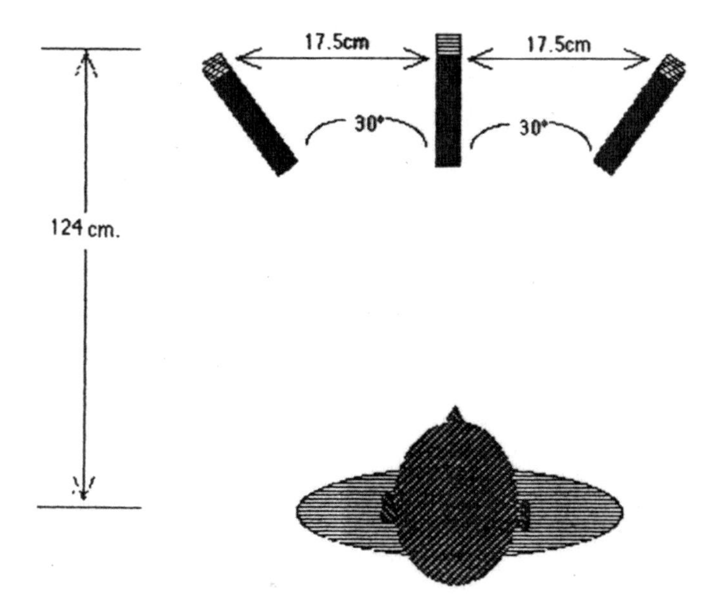

Fig. 4.32 Klepko-array (after Klepko 1997 from Mitchell 1999)

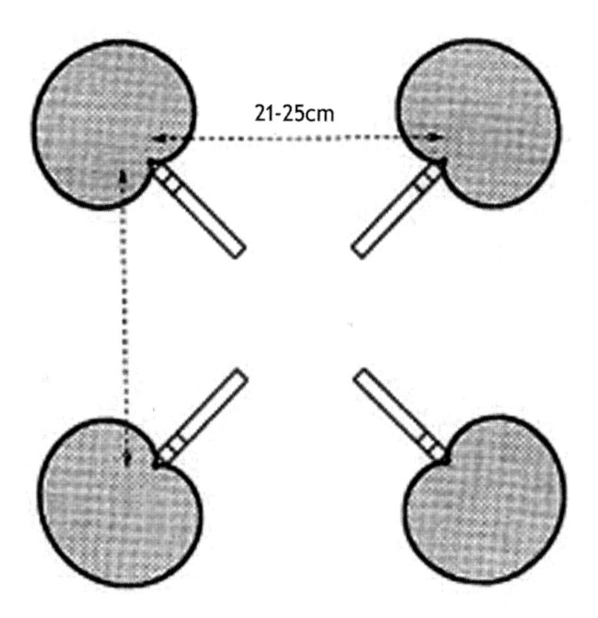

Fig. 4.33 Atmo- or IRT Cross, consisting of 4 cardioids (21–25 cm spacing, 90°)

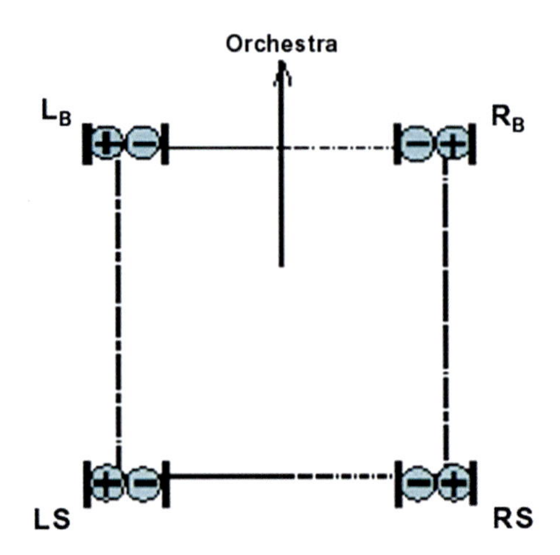

Fig. 4.34 Hamasaki Square (from Theile 2001)

4.2.10 Hamasaki Square

The Hamasaki Square is a rectangular arrangement of 4 side-facing microphones in fig-8 mode, which has also been conceived mainly for the capturing of atmospheres or diffuse sound. Similar to the IRT cross, the signal information of the front microphones can either be routed directly to the front channels, or positioned freely (panned) on the side speaker-bases L-LS and R-RS, respectively. The rear microphones are routed to the channels LS, RS. The fig-8 microphones are angled to the sides in order to minimize pickup of sound from the front. According to the inventor, the array should be placed '…at a very high position in the concert hall where the sound is very diffuse and energy from direct sound is diminished' (Hamasaki et al. 2001) (Fig. 4.34).

While initially the proposed capsule spacing was 1 m (3 ft), later on Hamasaki has revised this to be 2–3 m, for the sake of better de-correlation of the signals, which is preferable for a better spatial impression. Also, it is suggested—as a possible alternative—to replace the rear fig-8 microphones by rear-facing cardioids (see Hamasaki and Hiyama 2003).

4.2.11 Multiple AB ('Microphone Curtain')

Usually, five microphones are arranged in a line along the stage front. This can be considered to be a large L-C-R setup with two microphones added in-between as phantom sound sources for enhanced directional imaging (at times mixed in at −3 dB). The direct-/diffuse sound ratio as well as the balance of the individual instrumental groups within the orchestra can be controlled to a certain extent by microphone positioning. If it is desired to reduce the indirect (diffuse) energy in the

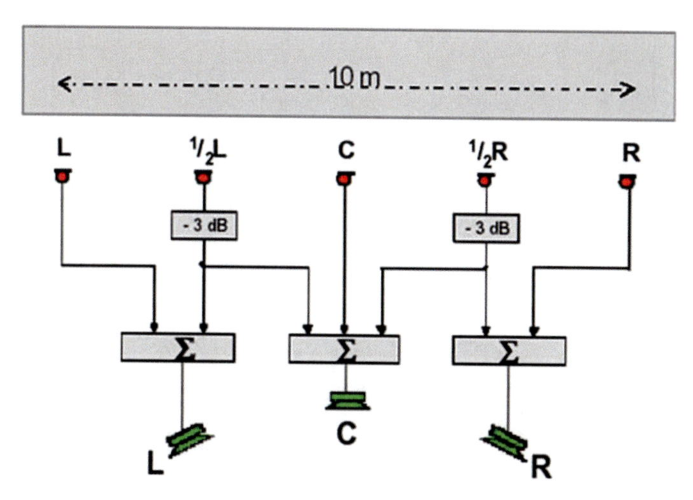

Fig. 4.35 Microphone curtain—schematic view (from Theile 2001)

front channels, cardioid microphones instead of omnis can be used (see Theile 2001) (Fig. 4.35).

The arrangement of 5 omni-microphones along a line is very well-suited to capture front signal information and quite popular among sound-engineers. Among them is also Eberhard Sengpiel, who has been working for TELDEC many years.

Also in the domain of opera-recording employing a 'microphone curtain' can be useful, accompanied by spot-microphones in the orchestra pit and cardioid microphones along the rim of the stage for singer pickup. As can be seen in Fig. 4.36, cardioid microphones along the sidewalls can be added to pickup signals for replay via the rear channels (Fig. 4.37).

The spacing between the fig-8 microphones in the Hamasaki Square was determined based both on calculations of the cross-correlation between the channels at different frequencies and on subjective results. Figure 4.38 shows the cross-correlation at four different frequencies between a pair of omni-microphones in the diffuse sound field; on this basis Hamasaki claims that if the microphone spacing is 2 m, the signals are de-correlated above 100 Hz.

Tree-Arrangements

In principle, all of the following tree-arrangements have their roots in the legacy DECCA-tree or DECCA-triangle technique, which is often accompanied with two additional microphones in a large-AB setup (also called 'outriggers'), in order to capture largely de-correlated signal components, which are vital for a convincing spatial impression. While for the outriggers, omni-microphones have been used almost exclusively, the inner microphones, used for the triangle, have seen quite some experimentation with respect to directional pattern during the 1950s, when the DECCA-tree got introduced first. For some time, during those early years acoustic

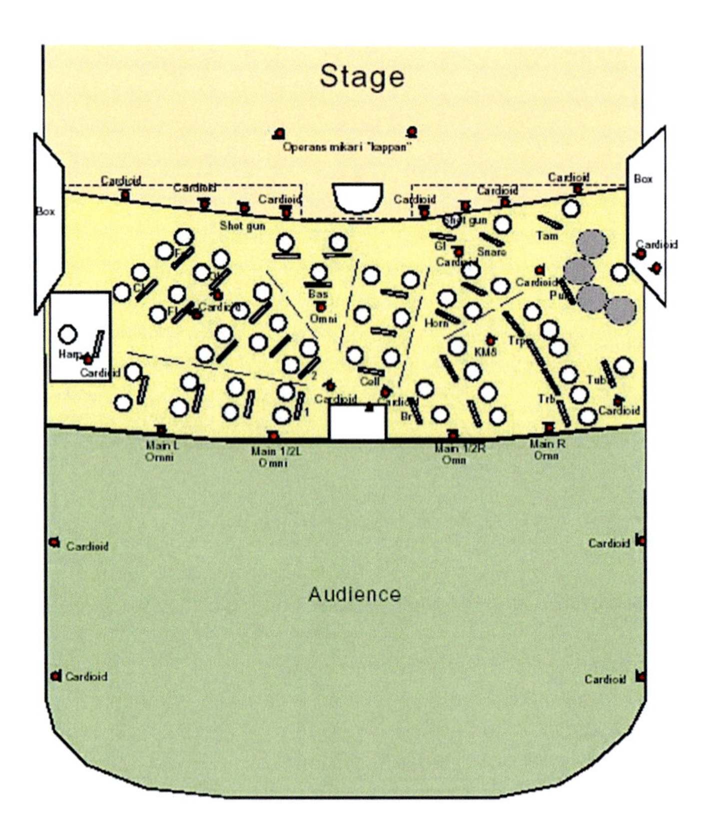

Fig. 4.36 Microphone setup for opera recording (microphone curtain, spot microphones, stage microphones) (from Evers 2003)

baffles were inserted between the triangle microphones to increase channel separation (see Valin 1994). The basic configuration details of the 'classic' DECCA triangle arrangement have already been described in the chapter on stereo microphone techniques. For more detailed information on the historic development of the DECCA-tree please see Chap. 11: 'A short history of microphone techniques and a few case studies.'

In the literature, quite a number of tree techniques can be found, which differ mainly in terms of directional characteristics, as well as capsule spacing. In some cases, these configurations seem to have more to do with the will for experimentation from the side of sound-engineers or Tonmeisters, than with well-founded scientific psychoacoustic considerations, which is—however—a fair approach (Remark: The naming of some of the following microphone techniques has not been chosen by their inventors or the Tonmeisters who have employed them, but are proposed by the author of this book. The tree techniques below are presented in alphabetical order.).

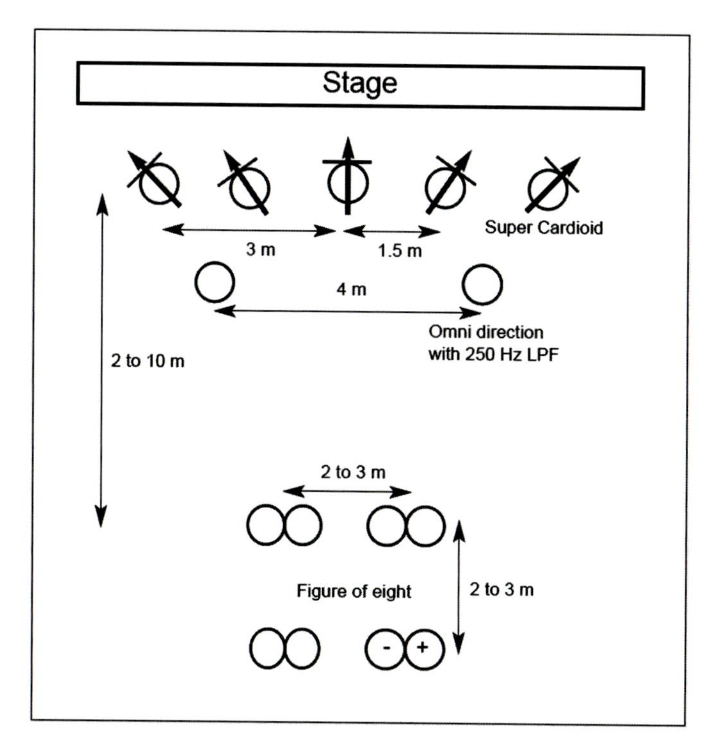

Fig. 4.37 Proposed microphone array for orchestra recording in a concert hall: microphone curtain made up of supercardioids, in combination with large AB omnis for enhanced LF-pickup and Hamasaki Square (from Audio Engineering Staff writer 2004 after Hamasaki and Hiyama 2003)

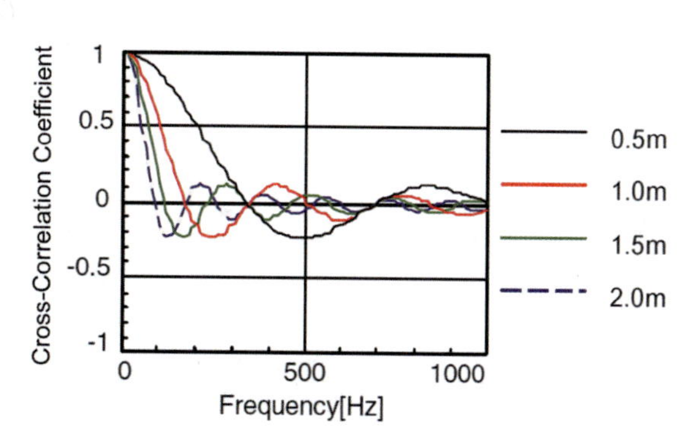

Fig. 4.38 Cross-correlation coefficient over frequency for two omnidirectional microphones in a diffuse sound field (from Hamasaki and Hiyama 2003)

4.2.12 AB-PC (AB-Polycardioid Centerfill)

The AB-PC technique consists of a large-AB arrangement of diffuse-field compensated omnis with a centerfill system in the form of the so-called ORTF-Triple (see Fig. 4.39 right-hand side, which shows a Schoeps CMC6 ORTF microphone with an

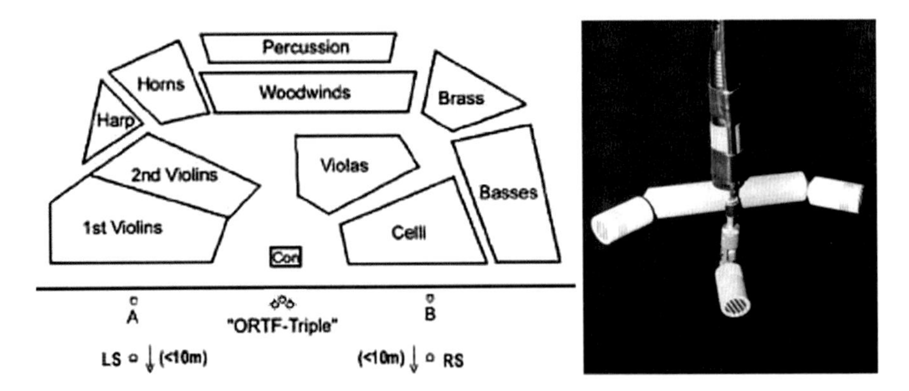

Fig. 4.39 AB-polycardioid centerfill technique (Left) with ORTF-Triple (Right) (from Pfanzagl-Cardone and Höldrich 2008)

added MK4 cardioid capsule for center signal pickup). The L and R signals of the ORTF main microphone are panned on the stereo (or front-surround) bus at 10:30 and 13:30 and in this way are basically used as a (relatively far positioned) accent microphone for the first desks of the string sections left and right of the conductor.

In order to keep signal correlation down for bass frequencies, the LF-part of the ORTF-signal should be attenuated by the use of a high-pass filter set to roughly 150 Hz for all three cardioid microphones of the ORTF-Triple (abbreviated: ORTF-T).

For rear signal pickup of the channels LS, RS, rear-facing cardioids are employed, which can be positioned as far away as 10 m (30 ft) from the line of the front microphones. With a larger spacing, the delay between front and rear microphone system would be more than 30 ms, which could result in unwanted echo effects or a psychoacoustic perception of separate 'front' and 'rear' hemisphere, depending on the acoustics of the venue and the nature of the sound signal (see, e.g., Pfanzagl-Cardone 2002).

Similar main microphone arrangement proposals can be found in Holman (2001) and Eargle (2004), but without the 'enhancement' of the ORTF main microphone by means of adding a cardioid center microphone to become an ORTF-Triple.

As an alternative to the ORTF-T, the BPT microphone system (see detailed information in the appropriate section in Chap. 3) can be employed as centerfill microphone system. When combining the BPT 3.0 and large-AB technique, we get a 'best of both worlds' situation: While the large-AB is able to deliver strong spatial impression due to a high degree of de-correlation also at low frequencies, the coincident BPT-technique is characterized by high localization accuracy (see Muraoka et al. 2007, p. 2).

In Williams (1992), it is proposed to combine the signals of two microphone pairs in the following manner: Frequencies above 700 Hz are picked up by a coincident microphone array, while frequencies below are captured with an AB pair of omnidirectional microphones. In another research paper, based on calculations, Preston

(1998) arrives at the conclusion that the localization properties of this combined microphone system are indeed superior to each single technique.

The abovementioned combination of large-AB with an ORTF-Triple or BP-Triple ('Blumlein-Pfanzagl-Triple') takes a similar approach, even though there is no splitting into a HF- and LF-band (based around the frequency at which the human head becomes effective as a baffle for higher frequencies), as well as no selective summing of frequency components. For the sake of completeness and correctness, it should be mentioned that in Williams (1992) 'extreme cases' (like large-AB and the 'Stereosonic System,' i.e. Blumlein-Pair) are not covered, as Williams adheres to a rather strict approach—based on psychoacoustic principles—for his calculations.

4.2.13 CHAB 5.0—Center Hemicardioid AB for Surround

Rooted in the proposals for 3-capsule stereo recordings DHAB, SHAB and CHAB (see chapter on stereo microphone techniques) in Olabe (2014), the author also describes a 5.0 surround version, which uses hemicardioids for the rear channels.

As with the pure stereo versions, the capsule spacing between the L and R omni microphones remains at 250 cm, with a hemicardioid capsule at the midpoint in between. The backward-facing rear microphones are recessed by 62.5 cm with a slight toe-out by 8°, presumably for enhanced channel separation or better coverage of sound sources at the back (Fig. 4.40).

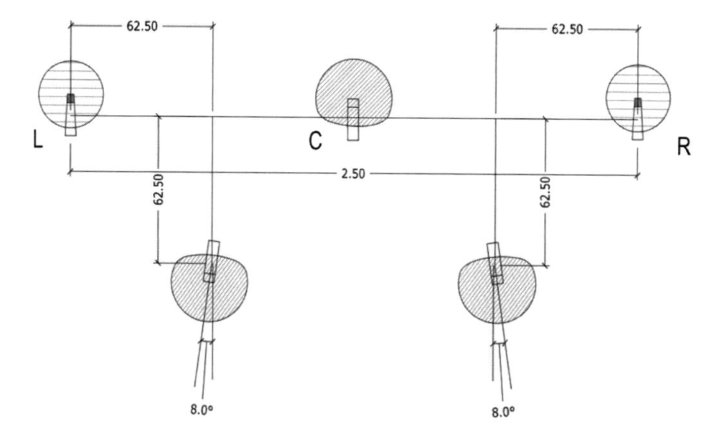

Fig. 4.40 CHAB 5.0—center hemicardioid AB 5.0 surround microphone system (from Olabe 2014)

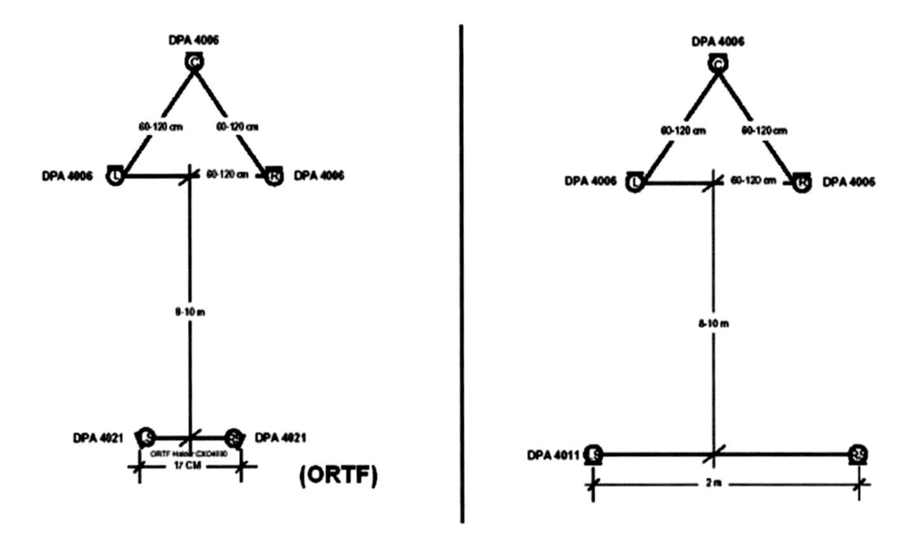

Fig. 4.41 Two versions of tree-arrangement (after Christensen 2003)

4.2.14 *Christensen Tree*

Front: Three omni microphones arranged in an even triangle, with capsule spacings of 60–120 cm (2–4 ft).
Rear: ORTF technique, or large-AB with parallel cardioids and a capsule spacing of 2 m (approx. 7 ft).
Distance front-to-rear system: 8–10 m (27–33 ft); (after Christensen 2003) (Fig. 4.41).

4.2.15 *Corey and Martin Tree*

Front: 3 semi-cardioids with a L-R base width of 120–160 cm (4–5.3 ft), while the center capsule is advanced toward the sound source by 8–15 cm (0.25–0.5 ft).
Rear: two cardioids in parallel, facing to the rear or upwards, with a spacing of (only!) 40 cm (1.3 ft); distance between front and rear system: 60–90 cm (2–3 ft) (see Corey and Martin 2003).

4.2.16 *Fukada Tree*

The Fukada tree has first been proposed in 1999 (see Sawaguchi and Fukada 1999) and has obvious roots in the DECCA-tree technique. It uses cardioid patterns for all microphones, except the outriggers, which are omnis (Fig. 4.42).

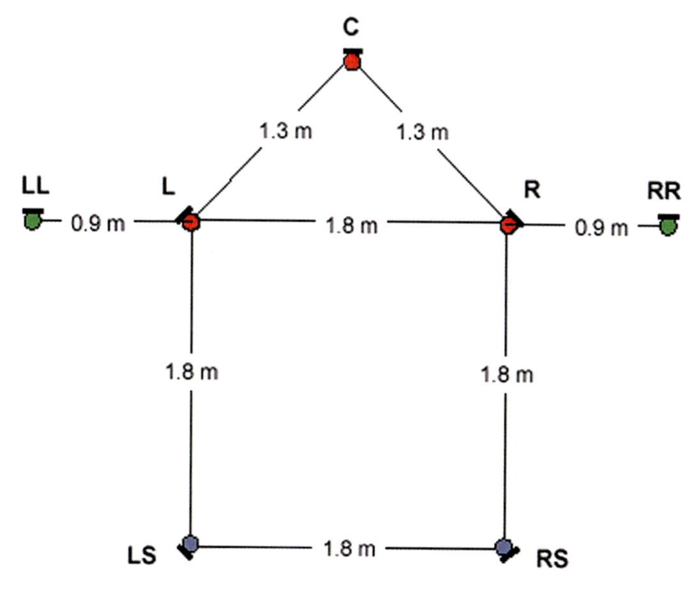

Fig. 4.42 Fukada tree (from Theile 2001)

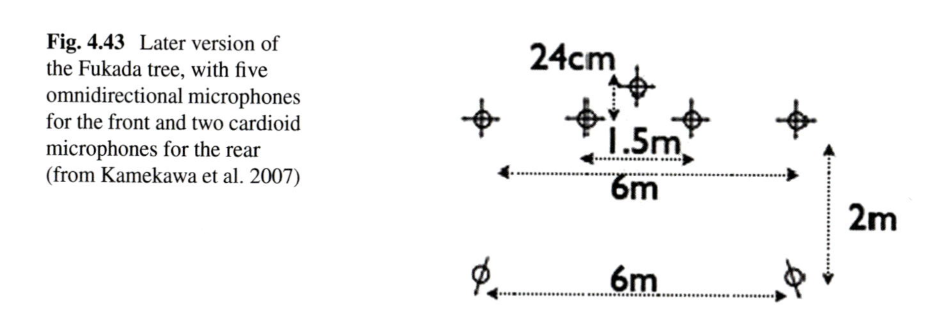

Fig. 4.43 Later version of the Fukada tree, with five omnidirectional microphones for the front and two cardioid microphones for the rear (from Kamekawa et al. 2007)

Similar to the Hamasaki Square, with later versions of the Fukada tree the capsule spacings have been enlarged, most likely with the same background of intended higher signal de-correlation in order to enhance spatial impression (see Fig. 4.43).

4.2.17 King-Tree

R. King (see King 2003) of Sony-Studios has come up with the proposal to use two diffuse-field compensated omnis as rear mics, as well as flying the center microphone lower by 20 cm (0.7 ft) in order to benefit from the 'precedence' effect of the wavefront, which arrives at the center microphone first. All microphones in the King-Tree have omni characteristics, with the two rear mics also being diffuse-field compensated (i.e. HF-boost) (Fig. 4.44).

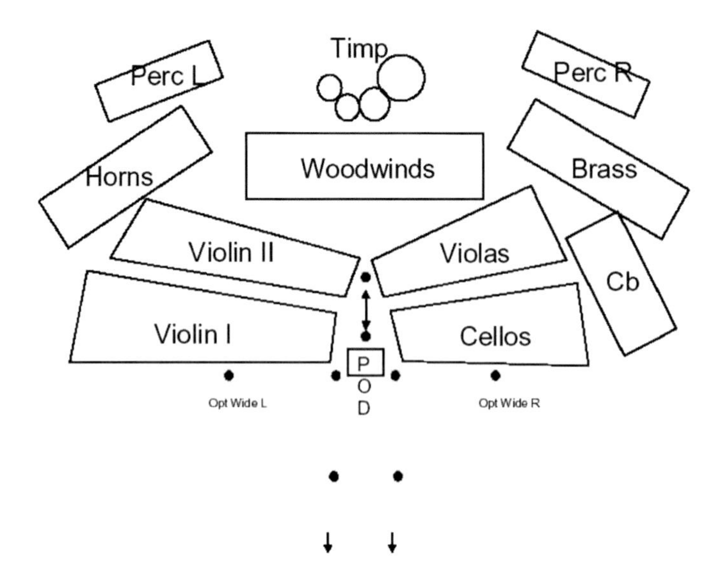

Fig. 4.44 King-tree (from King 2003)

4.2.18 Polyhymnia Pentagon

'Polyhymnia Pentagon,' Philips Classics (see Peters et al. 2007)

Front: three omni-microphones arranged in a triangle, L-R base width approx. 2.5 m (8 ft), center mic advanced by about 60 cm (2 ft).
Rear: two omni-microphones, large-AB with a capsule spacing of about 2.5 m (8 ft).
Distance front-rear: about 3.5 m (12 ft) (Fig. 4.45).

4.2.19 Ron Streicher's Surround Sound Decca-Tree

In Streicher and Everest (2006), Ron Streicher has proposed a significantly altered version of a DECCA tree, in which the Center microphone has been replaced by a SoundField microphone. According to Streicher and Everest (2006) '…on this way, several possible stereo and surround recording options become simultaneously available (Fig. 4.46).

First, of course, is the SoundField microphone itself, which is able to provide a variety of stereo and surround formats. When combined with either or both of the spaced pairs of microphones, additional variations become possible in stereo. Finally, the widely-spaced outer pair of rearward-facing microphones can serve as a surround pickup by themselves, as well as contribute to the stereo or surround perspectives created by the other microphones on the array. Although specific microphones are

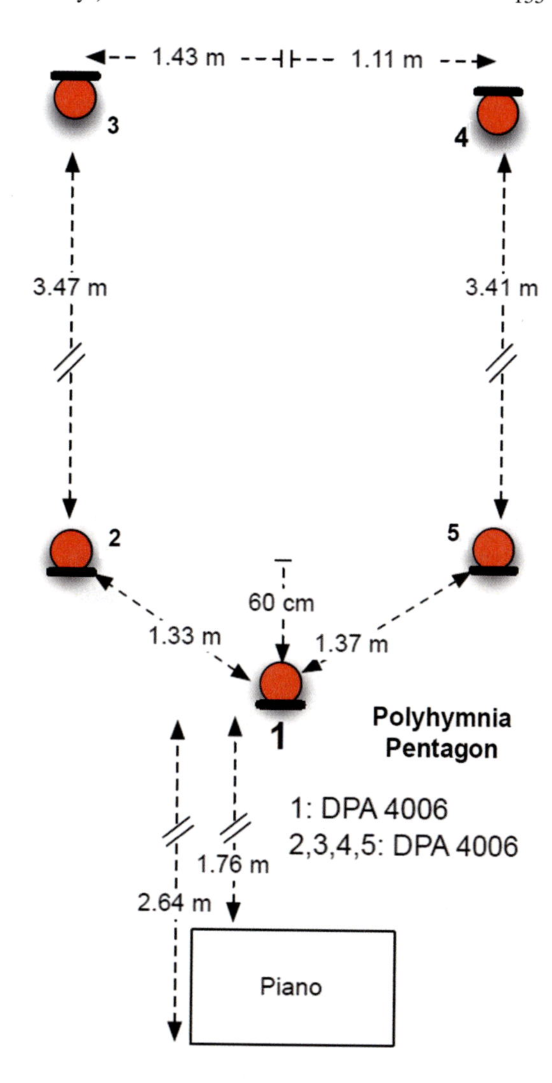

Fig. 4.45 Polyhymnia pentagon tree-arrangement (from Peters et al. 2007)

shown, the choice of microphones is always the creative discretion of the recording engineer…'

According to Streicher, his 'Surround Sound DECCA-tree' consists of—and provides the—following signals:

1. SoundField MK-V Microphone (various XY options),
2. Near-coincident frontal stereo pair: Mics 2, 3,
3. Near-coincident surround stereo pair: Mics 4, 5,
4. Traditional Decca Tree: Mics 1, 2, 3,
5. 'Ambient'Decca Tree: Mics 1, 4, 5.

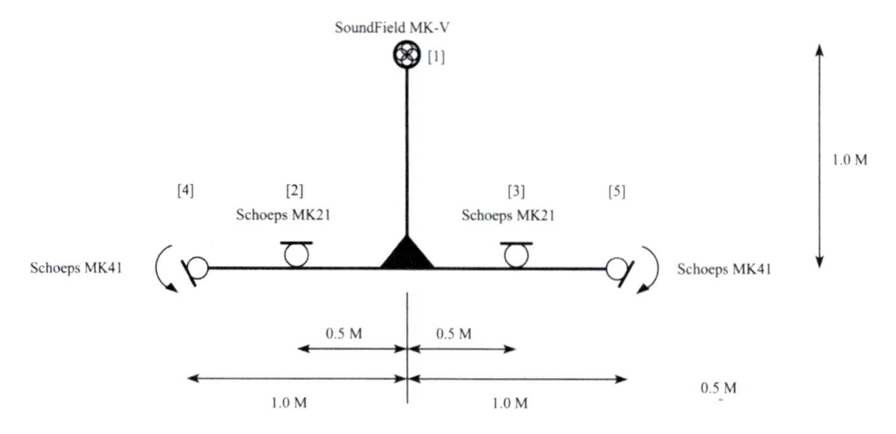

Fig. 4.46 Ron Streicher's 'surround sound DECCA-tree' (from Streicher and Everest 2006)

Streicher refers to his invention as the 'Surround Sound DECCA-tree,' but this may get confused with an ABC 5 arrangement (see the appropriate section above) made up of omni microphones, which would also have its roots in the Decca-Triangle configuration. Therefore, the author of this book would like to propose to use the name 'Streicher-Tree' instead.

4.2.20 XY Tri (by Andrew Levine)

According to Andrew Levine, the 'XYtri is an easy to use, integrated solution born out of the need for accurately verifiable, uncompromised traditional stereophonic recordings when tracking small- to large-scale projects also to be captured in surround.' (see Levine 2008) (Fig. 4.47).

'...The XYtri(angle)-setup consists of three XY-type configurations (with microphones of cardioid characteristic)—subsequently named L-XY, C-XY and R-XY—which result in three coincident and therefore quite mono-compatible stereophonic ranges (facing L, C and R), as well as two connecting runtime-based stereophonic ranges (L-XY_R to C-XY_L and C-XY_R to R-XY_L).

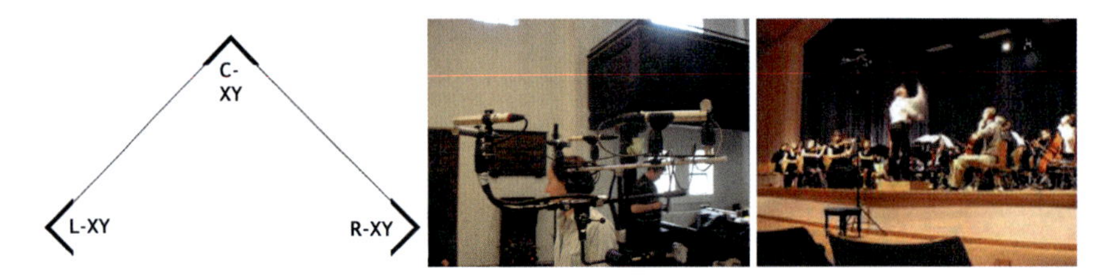

Fig. 4.47 XYtri by Andrew Levine (from Levine 2008)

The original design (XYtri#1) uses three XY-pairs at ±45° arranged in the shape of an isosceles triangle—each covering a range of 196°—, and a distance of 51.5 cm between L-XY/C-XY and C-XY/R-XY, resulting in two AB-ranges of 180° based on a runtime difference of about 1.5 ms [Calculations according to data by Eberhard Sengpiel].

There have been two subsequent modifications:

- 'XYtri#2angle' of L-XY and R-XY has been extended to ±67.5°. While keeping L-XY_R and C-XY_L parallel, there is now a rear-facing AB-configuration between the now also parallel L-XY_L and R-XY_R. The angle of ±67.5° results in a coverage of 143°. The direction of the outward-facing XY-pairs is no longer parallel to the base-axis but facing down by 21.5°.
- 'XYtri#3angle' of all three XY-pairs, placed within an equilateral triangle, has been set to ±60°, resulting in a uniformly coincident coverage of 158° while creating three uniform runtime-based angles. The direction of the outward-facing XY-pairs is 20° down from base-axis.

A modified configuration of XYtri#1 tilts all three XY-pairs down by 45° while moving the direction of both side-facing pairs back by ±30°, forming a tetrahedron. This setup is in many ways equivalent to XYtri#3 and can be applied at a raised elevation, which is of particular use in live recordings of concerts as this set does not obstruct the view as much as the original version. The sonic quality has not yet been tested (Fig. 4.48).

For the horizontally level sets, the advantage of flying them above and slightly behind the conductor's head is that sound approaching from close below is attenuated due to the cardioid characteristic of the capsules. Sound from instrumentalists further

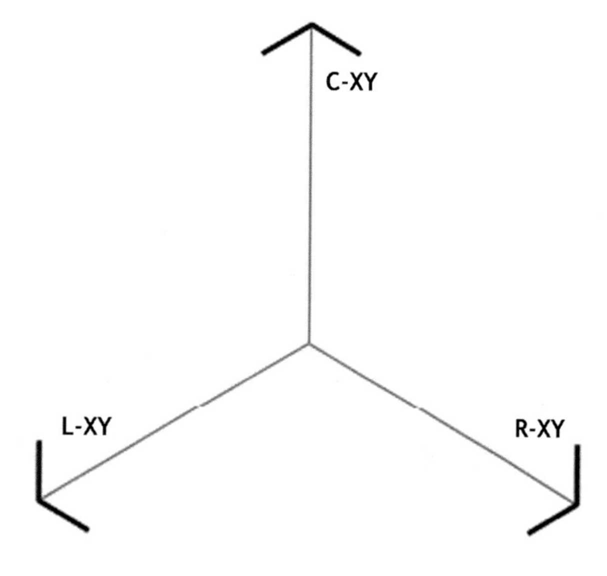

Fig. 4.48 XYtri V1 configuration tilted down 45° (from Levine 2008)

off—and usually raised—, e.g. the woodwinds or the brass, approaches at an angle closer to the 0°-axis.

The skill to assess a stereophonic range with headphones can be acquired easily. We are all more or less used to interpreting binaural cues, and although the impression of the soundstage when using headphone-based monitoring does differ from that observed using loudspeakers it can be intuitively translated and further honed through practice.

From Stereo to Surround

Recordings tracked with only one coincident setup allow for the precise localization of sound sources through differences in sound pressure level between the left and right channel, whereas recordings tracked with only one runtime-based setup contain cues based on time-of-arrival differences between the left and right channel. In both instances, one of the greatest challenges for the listener, when compared to a loudspeaker-based evaluation, is the ability to accurately judge the width of the sound stage.'

Levine reports that '…In the course of many live recordings, I have employed a variety of configurations, from simple to complex, most of which I had to set up, evaluate and monitor using headphones. In most situations I use visual tools to complement the auditory impression by displaying levels, frequency curve and the stereophonic power balance. Still I have found many complex setups, e.g. ones based on the DECCA-tree, or sessions intended for surround, distinctly difficult to judge on location. XYtri gives me confidence in this respect.'

Downmixing the XYtri

The configuration maps (nearly) discretely to a 7.0 monitoring environment can be folded down to 5.0, LCR (analogue to OCT-2), combined or discrete directional stereo as well as directional mono.

The following two screenshots give an overview of the primitives used to process the six inputs to create a 5.1/2.0 downmix (Fig. 4.49).

For folddown to 5.1, the following processing is applied:

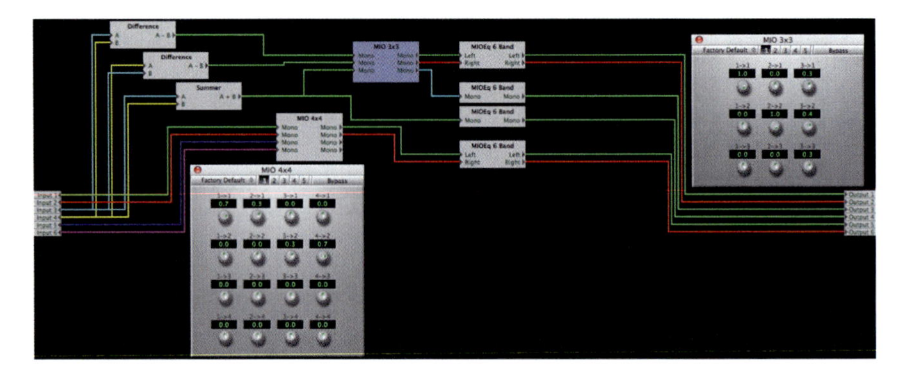

Fig. 4.49 Xytri—folddown to 5.1 (from Levine 2008)

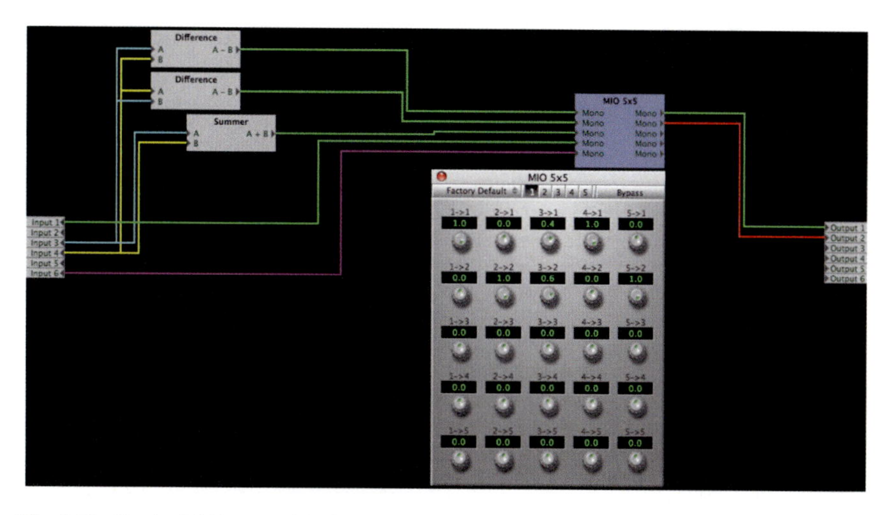

Fig. 4.50 Xytri—folddown to 2.0 (from Levine 2008)

- the difference of C-XY_L minus C-XY_R is low-cut and fed to the front L speaker.
- the difference of C-XY_R minus C-XY_L is low-cut and fed to the front R speaker.
- a small part (0.3) of the sum of C-XY_L plus C-XY_R is low-cut and fed to the C speaker.
- a small part of the low-cut-filtered sum is added to the front L (0.3) and R (0.4) speakers.
- the same sum is low-passed and fed to the LFE.
- 2/3 of L-XY_L and 1/3 of L-XY_R are low-cut and fed to the rear L speaker.
- 2/3 of R-XY_R and 1/3 of R-XY_L are low-cut and fed to the rear R speaker.

For folddown to 2.0, the following processing is applied (Fig. 4.50):

- the difference of C-XY_L minus C-XY_R is low-cut and fed to the front L speaker.
- the difference of C-XY_R minus C-XY_L is low-cut and fed to the front R speaker.
- a small part of the sum of C-XY_L plus C-XY_R is low-cut and added to the front L (0.4) and R (0.6) speakers.

To increase the sense of ambience in the stereophonic representation:

- L-XY_L is fed to the front L speaker.
- R-XY_R is fed to the front L speaker.

Matrix-Processing for 7.x

Matrixing can be employed to increase channel separation in 7.0 and 7.1. The concept currently being explored uses differences in SPL between coincident pairs and compensates for runtime differences when multiplexing signals from spatially distinct inputs.

$$L = \left((\text{C-XY\_L} - \text{C-XY\_R})^{d1} + \left(\text{C-XY\_L} - \text{L-XY\_L}^{d1}\right)\right)^{d3}$$

$$R = \left((\text{C-XY\_R} - \text{C-XY\_L})^{d1} + \left(\text{C-XY\_R} - \text{R-XY\_R}^{d1}\right)\right)^{d3}$$

$$C = (\text{C-XY\_L} + \text{C-XY\_R})^{d2}$$

$$Lm = \left((\text{L-XY\_R} - \text{L-XY\_L})^{d1} + \left(\text{L-XY\_R} - \text{C-XY\_R}^{d1}\right)\right)^{d3}$$

$$Rm = \left((\text{R-XY\_L} - \text{R-XY\_R})^{d1} + \left(\text{R-XY\_L} - \text{C-XY\_L}^{d1}\right)\right)^{d3}$$

$$Ls = (\text{L-XY\_L} - \text{L-XY\_R})^{d2} + \left(\text{L-XY\_L} - \text{R-XY\_R}^{d2}\right)$$

$$Rs = (\text{R-XY\_R} - \text{R-XY\_L})^{d2} + \left(\text{R-XY\_R} - \text{L-XY\_L}^{d2}\right)$$

^{d1} Delay between C-XY_L and L-XY_L/C-XY_R and R-XY_R
^{d2} Delay between L-XY and R-XY; this constitutes the total delay of all processed signals
^{d3} Delay d2—Delay d1 (Fig. 4.51)

Usability of the XYtri
With large ensembles, there is always the question of how to capture the width of the sound stage convincingly in a stereophonic representation—perhaps also one that can potentially be extended. Another concern is that of discovering new ways to maximize the enfolding nature of a surround monitoring setup—without compromising the possibility to fold down to an 'optimal' stereophonic representation. XYtri, a composite of traditional stereophonic techniques, is useful in these kinds of recording situations, preserving the option to 'expand' the image by allowing the mixer to extend the 'edges' to beyond the L and R speakers, enfolding the listener more so than can usually be achieved even in a live performance.

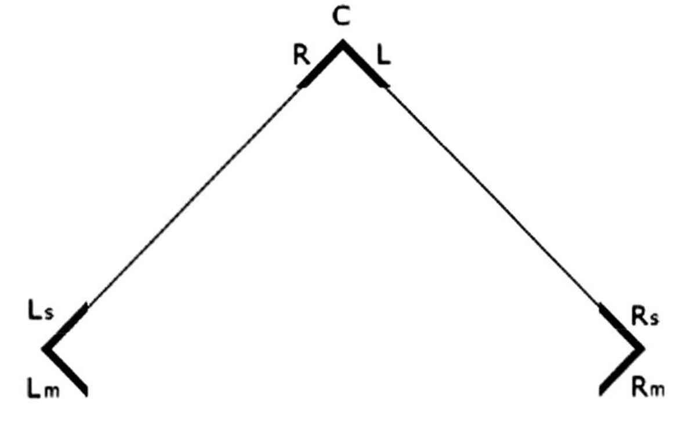

Fig. 4.51 Xytri—7.0 matrixing (from Levine 2008)

As a live location recordist I have certain requirements:

- Setup time is precious and I require immediate feedback, especially regarding my decisions concerning microphone placement.
- Currently, most of my clients ask for a stereophonic recording only. Still I often attempt to capture a wider than usual image in case I am subsequently asked to do a surround remix.
- If I can set up in an adjoining but separate space I usually employ a stereophonic near-field-monitoring solution only, then again I like to be present in the performance space.
- I often monitor using only headphones—using software Spectafoo for visual confirmation of critical variables.
- Evaluating a stereophonic panorama (AB, XY, Blumlein, ORTF) with headphones is relatively easy, whereas the image from complex configurations in my experience has to be assembled in a controlled environment and can not be reliably judged 'on the fly.'

The screenshot (see Fig. 4.52) displays the setup for a live recording of a harp recital. I am easily able to monitor each of the five stereophonic ranges while tracking and for subsequent 'peace of mind' confirmation.

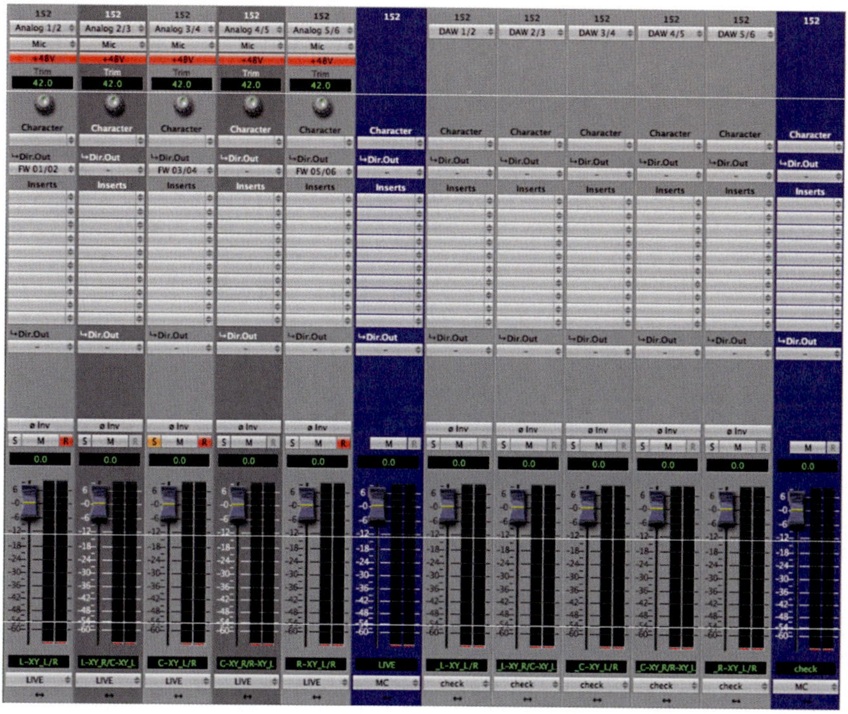

Fig. 4.52 Matric Halo 2D-mixer for monitoring (from Levine 2008)

There is an overlap of monitored channels. Inputs 1 + 2, e.g., produce the leftward-facing XY-image, inputs 2 + 3 the forward-and-left-facing AB-image.

There are many spaces that bring with them a very special ambience. In my work as a live location recordist, I have tracked uncorrelated ambience many times by 'sampling' with microphones of omni-directional characteristic at two or more points. This gives me a return on the reverb that I can use to sweeten the more direct sound of the recording or rebalance it in the mix.

I have consistently found that these 'sampling points' primarily have to be spatially distinct from the main setup while avoiding extraneous noise, e.g. from an audience. They can be behind the musicians, on the second tier of a concert hall, in the far wings of a church, etc. Discovering them is a matter of creativity.

While remixing many basically quadraphonically tracked sessions to surround I have found the (true sound of the) ambience feeds to again be very helpful. I believe that at least part of their applicability to sculpt the rear sound stage is due to the distance between the sampling points and the main setup that introduces specific runtime/phase differences to the mix. This can only be achieved by a complex setup, tailored to the requirements of the location and musical event.

Personally I am 'into runtime differences.' I enjoy the spaciousness of AB and often employ ORTF as the configuration of my main microphone. I use XY primarily as a spot when I either want to directionally cover a wider range than I can with one microphone only, while retaining the option to manipulate the width of the image at a later stage, or when I want to conserve the movement of the musician in space (Fig. 4.53).

My first step toward XYtri came when doing a stereo and surround production with the ensemble 'Wavegarden' featuring mainly crystal bowls and flutes.

I had placed a Blumlein-setup with ribbons in the central position, with two musicians sitting opposite one another on both sides of the main axis. Since the crystal bowls are arranged in the shape of a horseshoe I complemented the main setup with two XY-pairs on both sides of the ring of bowls, facing inward and aligned to be on the diagonal axis 'of the Blumlein-Pair. Seen in retrospect it was kind of like XYtri in reverse…' (from Levine 2008).

4.3 Baffle and 3D Techniques

4.3.1 KFM 360 ('Kugelflächenmikrofon')

The 'side-angled' MS technique of Michael Bishop (covered in a previous section) bears some similarity with the use that Jerry Bruck makes of the Schoeps KFM 'Kugelflächenmikrofon,' to which two forward-facing fig-8 microphones are added at the sides (see Bruck 1996). Also, the resulting two MS-pairs are used to derive front- and rear-channel information, but the M-signals come from the pressure transducers built into the sides of the KFM-sphere, which acts as a baffle (similar to a human

Fig. 4.53 Xytri + large AB with Norddeutscher Kammerchor, 2008 (from Levine 2008)

head) for mid- and high-frequencies, while Michael Bishop's technique makes use of cardioids for the M-signals and the MS-pairs are separated by a larger distance.

Based on the invention of Theile (1986), the first version by the name of KFM 6U had a slightly larger diameter of 20 cm (0.7 ft) and was meant for stereo recording. The next version, KFM 360, had a smaller diameter of 18 cm in order to enlarge the effective recording angle to 120°. Two fig-8 microphones are added left and right of the pressure transducers which are built into the sphere at the sides. The signals of the fig-8 and pressure transducers (omnis) are used for MS-matrixing, and thus at least four surround signals for L, R, LS and RS can be obtained (Figs. 4.54, 4.55).

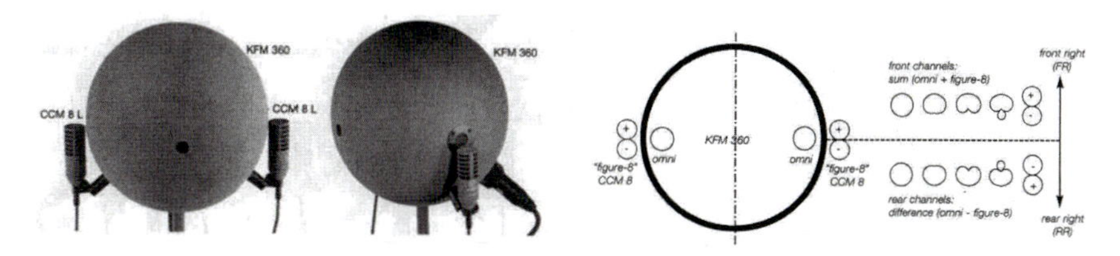

Fig. 4.54 Schoeps KFM360—sphere microphone ('Kugelflächenmikrofon'); generation of signals R and RS by means of MS-matrixing (© Schoeps Mikrofone)

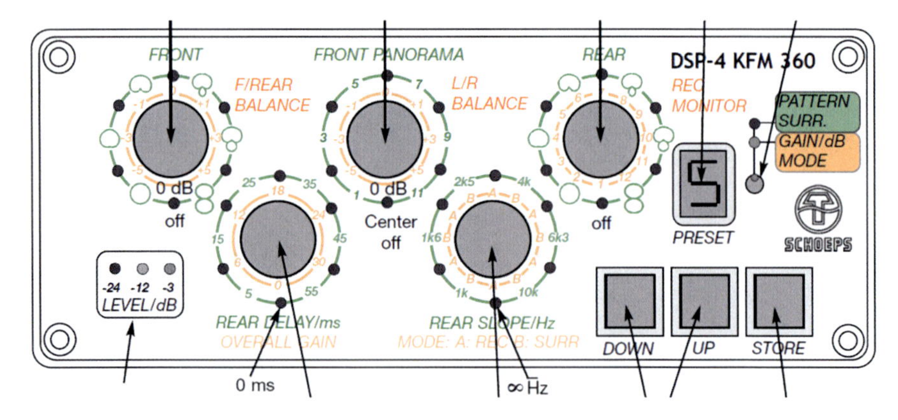

Fig. 4.55 Schoeps DSP-4 signal processor: front panel schematic (© Schoeps Mikrofone)

With the implementation of the so-called Gerzon-Matrix in the Schoeps DSP-4 Processor, it is possible to derive the proper signals for 5.1 surround. In addition to this, the DPS-4 offers further options for signal processing, among them adjustable directional patterns for the dematrixed microphone signals, delay for the rear-channel signals (in order to keep localization to the front-plane, due to the 'first wavefront' principle or 'Haas'-effect), etc.

4.3.2 Various Baffled Surround Microphone Systems

The following techniques are somehow similar to the sphere microphone, but—since they employ several built-in microphones—some of them do not need additional signal processing in the form of matrixing, or similar. Most of the following techniques come from the area of film- and video-recording.

- modular mic array for surround recording (after Fox and McGreor 2002)
- 4-capsule surround-arrangement (after Iredale and Keller 2003)

The arrangement in Fig. 4.56 is based on 4 capsules: 3 are positioned to capture front signals Left, Front (i.e. Center), Right plus one capsule for a (mono-) rear channel. In order to make the recording of signals easier, the down-matrixing of the 4 signals to 2 is proposed, in order to remain compatible with standard decoders (which used to be 'Dolby Surround' in times of analogue tv).

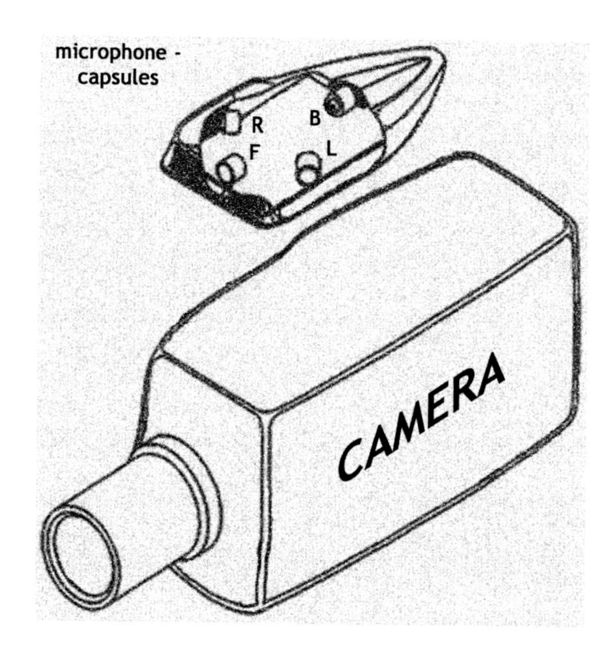

Fig. 4.56 4-capsule baffle surround microphone (after Iredale and Keller 2003)

4.3.3 Holophone 'H2-Pro' and 'PortaMic 5.1'

For these two surround microphone systems, several miniature capsules have been integrated into the surface of a more or less sphere-shaped body. In this manner, no signal processing is necessary and the larger model 'H2-Pro' (approx. 19 by 15 cm diameter) is in the range of the dimension of the human head and can deliver up to 7.1 surround plus 'height information' due to an additional capsule integrated on top. The much smaller 'PortaMic' is capable of delivering 5.1 Surround and can even be mounted on a consumer video camera. Due to its much smaller diameter, it must be assumed that it becomes effective as a baffle only at much higher frequencies, and therefore, channel separation will be rather weak for low frequencies, which is to the disadvantage of good spatial impression (Holophone 2008) (Fig. 4.57).

4.3.4 DPA—'D:Mension 5100' Mobile Surround Microphone

Apart from the surround microphone solutions offered by DPA, which use discreet microphones in an ABC or INA 5 style arrangement mounted on a metal support, the company has come up with a very compact, lightweight, baffled solution, which—in some way—resembles the saddle of a bicycle:

The 'D:mension 5100 Mobile Surround Microphone' uses five miniature pressure transducers, which are pre-polarized condenser elements of 5.4 mm diameter.

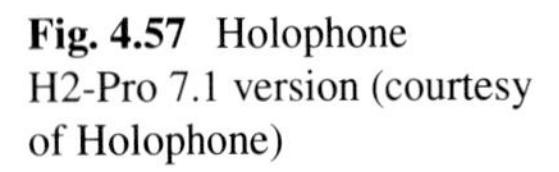

Fig. 4.57 Holophone H2-Pro 7.1 version (courtesy of Holophone)

'...Appropriate channel separation and directionality are achieved by using interference tubes on the L/C/R omni-directional capsules. Acoustic baffles preserve the accuracy of levels between the discrete analog output channels.

The three front microphones in this system are time-aligned to eliminate comb filtering, which—according to the manufacturer—also ensures frequency consistency when downmixing to stereo or mono. In contrast, the rear microphones, with standard omni-directional patterns, are spaced from both each other and the front array to simulate the most natural time arrival differences. The LFE (0.1) channel is comprised of a L/R sum, which is then attenuated 10 dB in comparison with the signal from the main channels, in accordance with the 5.1 format...' (from DPA 2018).

The individual polar patterns of the single miniature microphones show a directional pattern down to the range of about 1–2 kHz, after which the microphone patterns become more omni-directional, as can be expected, according to the dimensions of the baffle in which the microphones are mounted (24 cm × 14 cm × 19 cm) (Fig. 4.58).

Fig. 4.58 DPA 5100 mobile surround microphone mounted on a video-camera (© DPA-Microphones)

4.3.5 Sony Sphere-Arrangement

Somehow similar to Holophone's 'H2-Pro' is an arrangement by Akitaka Ito, which he has patented for Sony (see Ito 2001) and consists of 8 microphones which have been arranged along the surface of a sphere. Six of them are positioned as a horizontal circle, and one microphone is on top of the sphere, while the other one is at the bottom. With the help of an additional signal processor, horizontal as well as vertical panning of a sound source can be achieved (Fig. 4.59).

4.3.6 Multichannel Microphone Array (Johnston-Lam)

Very similar to the Sony/Ito arrangement is a proposal by Johnston and Lam: a 7-channel array with 5 hyper-cardioid microphones in a horizontal circle at included angles of 72°. Together with a 'top' and 'bottom' microphone, all capsules are arranged like on the surface of a virtual sphere with 29 cm diameter (approx. 0.9 ft). For music recording, this 'Johnston-Lam Array' should normally be positioned in a 'beast seat of the house' position at a height of about 3 m (9 ft) (see Eargle 2004, p 261) (Fig. 4.60).

According to the authors, the microphone array layout has been chosen in order to optimize ILD and ITD for listeners in the sweet zone of a regular 5.1 surround loudspeaker arrangement according to ITU-R 775.1, with the option that—ideally—the loudspeakers should also be placed at included angels of 72°.

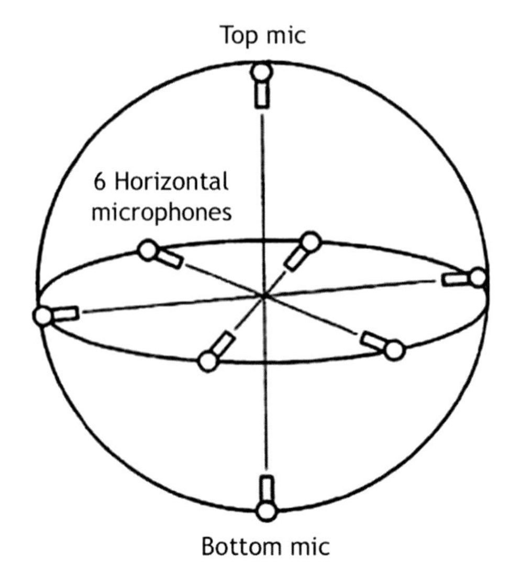

Fig. 4.59 Sphere-arrangement, patent of Sony (after Ito 2001)

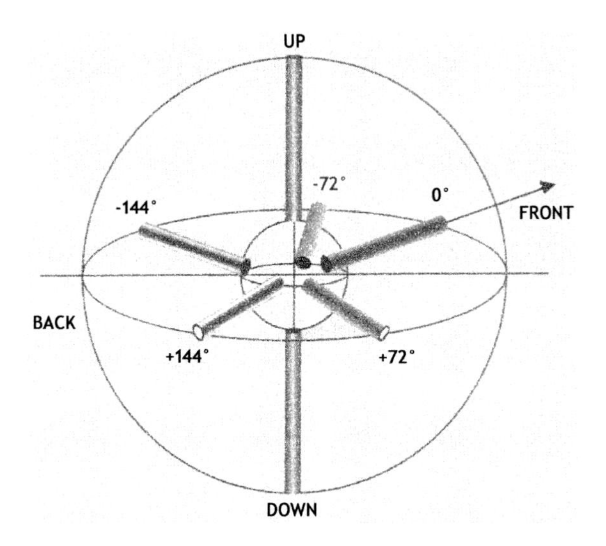

Fig. 4.60 Multichannel microphone array by Johnston and Lam (after Eargle 2004, Figs. 15–19, p. 261)

4.3.7 HOA—Higher-Order Ambisonics: The 'Eigenmike®', Octomic and ZM-1

In (Meyer and Agnello 2003) a sphere microphone with 24 miniature capsules and a diameter of only 7.5 cm (roughly 0.25 ft) is described. With post-processing, a microphone directivity of third order can be achieved (Fig. 4.61).

With the Eigenmike®, the number of microphones (single transducer elements) n defines the order which can be achieved:

Fig. 4.61 Eigenmike® Left: older model with 24 capsules; Right: new version 'Eigenmike® m32' with 32 capsules (courtesy of mh acoustics LLC)

$$n = (M + 1)^2 \tag{4.2}$$

with M being the order (rem.: therefore with the 4 capsules of the SoundField-Microphone only first-order microphone characteristics can be realized).

With the 24 transducers of the Eigenmike®, above 1.5 kHz third-order directional characteristics can be achieved, below this second order and below 700 Hz only first order (see Eargle 2004).

The 'Eigenmike® m32,' with its 32 capsules, is able to achieve fourth-order Ambisonics. An arrangement similar to the Eigenmike® is registered as international patent WO 03/061636 A1 by Elko et al. (2003) 'Audio system based on at least second-order Eigenbeams.'

In 2018 US-company Core Sound, which also produces the 'TetraMic,' has come up with the first commercially available 'Second-Order Ambisonic' (SOA) microphone: the 'OctoMic' is made up of 8 capsules with cardioid characteristics, which apparently allows not only for a broad range of first- and second-order microphone patterns but also delivers more accurate spatial impression and localization due to better signal separation between the capsules. The enhanced microphone pattern flexibility of the OctoMic in terms of post-production is undoubtedly an advantage; however, it still needs to be seen if some of the new second-order microphone patterns (e.g. with three-dimensional '4-leaf flower' characteristics, looking a bit like a crossed figure-of-eight Blumlein-Pair, but as *one unified* microphone pattern; for reference see the two left ones in the middle row of patterns in Fig. 4.4) will find a practical application as part of new microphone technique concepts.

Based on the estimate proposed in Ward and Abhayapala (2001) and taking a look at the 'spatial resolution' or accuracy, which second-order Ambisonics is able to achieve in terms of soundfield reproduction, we can use Eq. 4.1 (with the wave number k substituted as in Formula 2.6) and for a 'reproduction sphere' of 17 cm diameter (similar to the size of a human head) we will arrive at a frequency of approximately only 1300 Hz, up to which the OctoMic should be able to achieve a valid reconstruction of the soundfield (i.e. truncation error lower than 4%).

Around the same time, Polish company Zylia has released their third-order Ambisonics (TOA) microphone 'ZM-1,' based on 19 omni-directional capsules. Accompanying software enables beamforming and the creation of useful microphone patterns for 2D and 3D Audio use, including binaural signal generation.

Repeating the calculation, which has been made above for the OctoMic: based on the estimate proposed in (Ward and Abhayapala 2001) and an assumed 'reproduction sphere' (i.e. the 3D analogy to the 'sweet zone' in 2D-surround sound) of 17 cm diameter, we arrive at a frequency of approximately 1900 Hz, up to which the ZM-1 should achieve a valid reconstruction of the soundfield, according to theory.

Zylia states a recording resolution of 48 kHz, 24 bit which would result in a theoretical dynamic range of 24 * 6 dB = 144 dB (rem.: as each bit is the equivalent of an amplitude resolution/dynamic range of 6 dB). However, most likely due to the quality of the transducers and the electronics involved, in their technical specifications Zylia states a dynamic range of 105 dB and a signal-to-noise ratio of only 69 dB(A), which is similar to what well-aligned analog tape machines (without a noise reduction system) were able to achieve in previous times; all this results in an effective (usable) dynamic range equivalent to approximately 12bit, because the audio signal in the amplitude range of the lower 12bits (of the total of 24bits) will be drowned in noise.

A very interesting and up-to-date comparison between several first-order Ambisonics, as well as Higher-Order Ambisonics commercially available microphones including the Sennheiser 'Ambeo,' Core Sound 'TetraMic'; SoundField 'MK-V,' MH Acoustics 'Eigenmike®' and Zoom 'H2n' has been carried out by Enda Bates et al., about which they report in Bates et al. (2016, 2017).

4.3.8 *OSIS-System (Optimal Sound Image Space—'Jecklin Surround')*

Derived from his 2-channel stereo recording system OSS (Optimal Stereo Signal), Tonmeister Jürgen Jecklin in 2002 proposed the OSIS-system, which makes use of a Jecklin disk also for the rear microphones. Similar to the OSS system, also the OSIS uses a 32 cm (1 foot) diameter dampened disk for the front (i.e. 'Image' disk), which is situated between the two diffuse field-compensated omni-microphones, with a capsule separation of 17 cm and an azimuth set in a way to slightly ‘overshoot’ the sound source. In addition, a directional microphone is integrated into the disk, from which the center signal is derived.

The diameter of the ‘Space’ disk (rear) is defined with 28 cm and—instead of the omni microphones—a pair of rear-facing cardioids is employed. According to Jecklin’s motto ‘No image in space, no space in image’ the rear system is meant to pick up only diffuse sound (‘no image in space’), while the front arrangement should pick up only direct sound (‘no space in image’) (Fig. 4.62).

As Jecklin had already stated for the OSS system Jecklin (1981) low frequencies bend around the disk, with the consequence that frequencies below approx. 200 Hz are captured essentially in mono. The same is true at least for the frontal ‘Image’ arrangement of the OSIS system, which is certainly to the disadvantage of spatial impression for which sufficient de-correlation of frequencies below 500 Hz and even more so of frequencies below 200 Hz is necessary (see Hidaka et al. 1995, as well as Morimoto and Maekawa 1988).

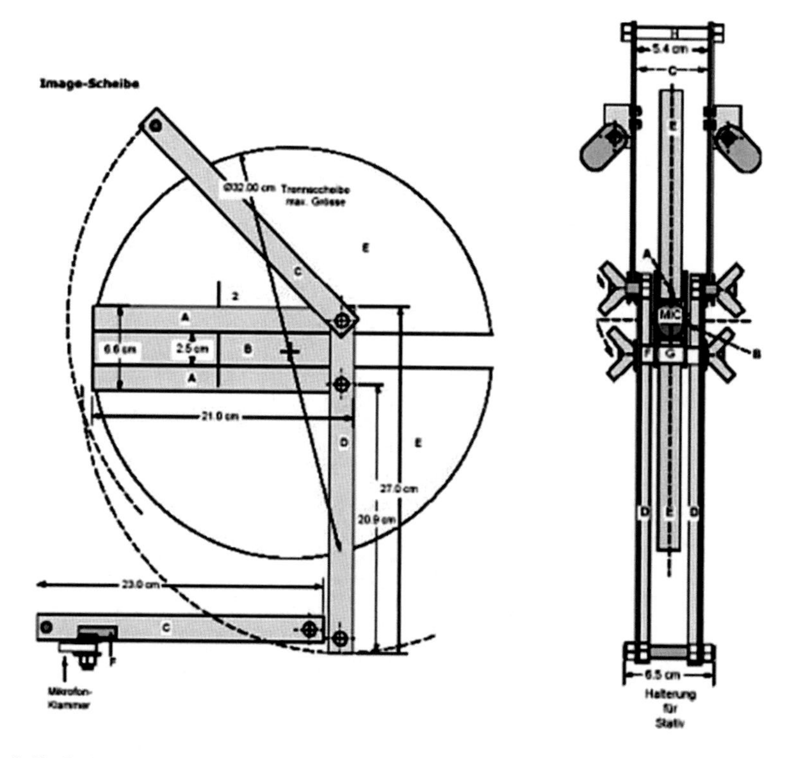

Fig. 4.62 Side (Left) and frontal view (Right) on the ‘image disk’ with in-built center microphone, as used for the OSIS—‘optimal sound image space’ surround microphone (adapted from Jecklin 2002)

4.3.9 Pan-Ambiophonic 2D/3D System

Miller combines the signals of a discreet Ambiophonic arrangement built of single microphones with those of two omni-transducers, which are situated at 'ear positions' of the head-related baffle structure which can be seen in Fig. 4.63.

With 'Ambiophonic 2.0,' the resulting stereo signal is replayed on 2 speakers which have an included angle of only 10°–20° (base width) and are claimed to have a correct reproduction angle of 120°. According to Miller playback via the stereo system can be enhanced by use of additional ambience speakers, which reproduce signals which have been generated by means of impulse-response convolution (see Miller 2004a, b) (Fig. 4.64).

As an alternative, in the PerAmbio 2D/3D System, the abovementioned signals are replayed via a total of 10 loudspeakers according to Fig. 4.65, a setup in which the exact placement of the overhead speakers can be taken into account via appropriate position programming in the decoder.

Miller notes that if the listener moves back from the sweet spot by 26%, this results—relative to the new position—in a loudspeaker layout according to the 5.1 surround standard.

Fig. 4.63 'PanAmbiophone'—a total of 8 recording channels for 'PerAmbio 3D' (from Miller 2004a)

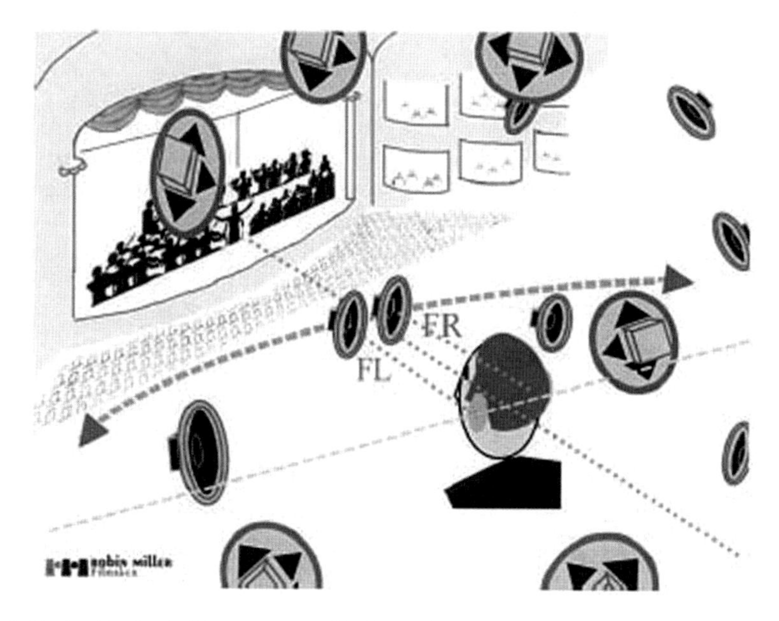

Fig. 4.64 Ambiophonic 2.0 turns stereo 'inside-out,' reproducing a 120° wide front stage, using 10–20° spaced speakers (from Miller 2004a)

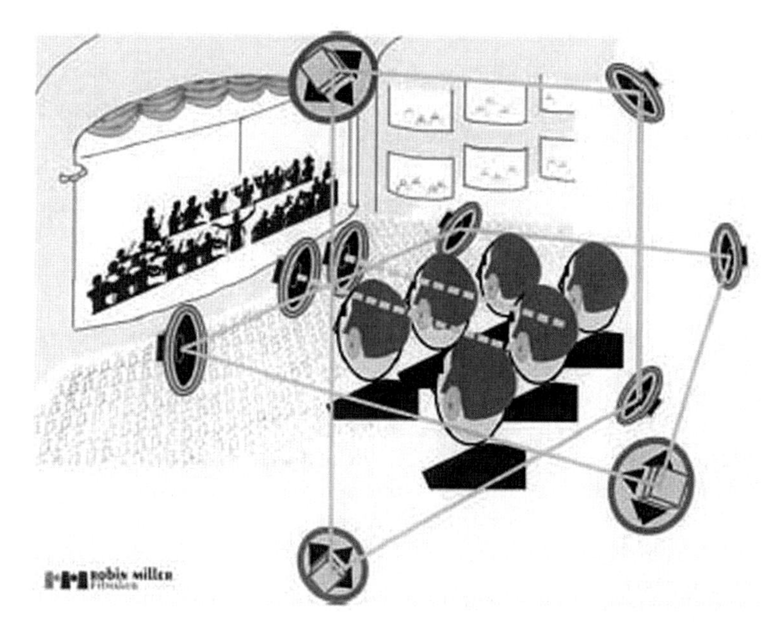

Fig. 4.65 'PerAmbio 3D/2D' system (pat. pend.) plays both ambiophonic and 3D (with height) recordings using 10 speakers (from Miller 2004a)

4.3.10 BACCH$^{(TM)}$ 3D Sound

Back in 1961, Benjamin Bauer has proposed to enhance the quality of music reproduction—listened to via headphones—by electronically adding a 'crosstalk-effect' (sound from the left channel of a stereo recording will arrive both at the left and at the right ear, but with some alterations—level, frequency and time delay-wise—which is characteristic for HRTFs), which happens 'naturally,' when listening to the same signal via loudspeakers (see Bauer 1961).

A somehow reversed technique has been proposed by Edgar Choueiri as the basic principle behind the BACCH$^{(TM)}$ 3D sound system: The replay of a binaurally recorded signal (via an artificial human head or 'dummy head') can be optimized for loudspeaker representation, if a crosstalk cancelation filter is applied which prevents sound from the left speaker to arrive at the right ear, by essentially adding a modified 'inverse' version of the left-channel signal to the right-channel signal (which will effectively 'cancel out' the left-channel signal at the entrance to the right ear canal), and vice versa.

The BACCH$^{(TM)}$ 3D sound system has been proposed by Professor Edgar Choueiri, head of the '3D Audio and Applied Acoustics (3D3A) Lab' of Princeton University.

It is based on the 2-channel Ambiophonic principle, which has been presented in the previous section, with the difference that it is strictly limited to replay via 2 loudspeakers, at least as long as only one listener needs to be reached in the sweet spot.

In analogy to the 3D capabilities of stereoscopic photographs, this approach relies on optimizing the acoustic 3D effect for a sweet spot-listening position by utilizing a crosstalk cancelation filter (XTC) (Fig. 4.66).

3D imaging and 3D audio through two loudspeakers share the same *stereo* principles. In 3D imaging, two lenses of a camera separated by the typical human interocular distance are used to record two separate images intended for the left and right eyes separately. Such a 'stereoscopic' recording contains the visual 3D cues needed to see in 3D.

In 3D audio, the same applies for the interaural distance between the two ears and the crosstalk cancelation filter, which is applied.

A crosstalk cancelation (XTC) filter is a digital filter that allows the signals D_L and D_R (see Fig. 4.67) on the left and right channels of a stereo (or binaural) recording to be manifested as air pressure signals P_L and P_R at the left and right ears, respectively, so that the left ear hears only what is on the left channel and vice versa.

The XTC filter, which has been known for some time, consists of simply inverting the transfer matrix (shown schematically and symbolically for the idealized case of point sources in free space) that describes the wave propagation from two loudspeakers to two ears.

In order to let the system work under optimized conditions, replay has to happen in an anechoic space. The XTC filter can be implemented as a digital filter that is used to process (in real time or offline) digital audio on a DSP chip or computer (Fig. 4.68).

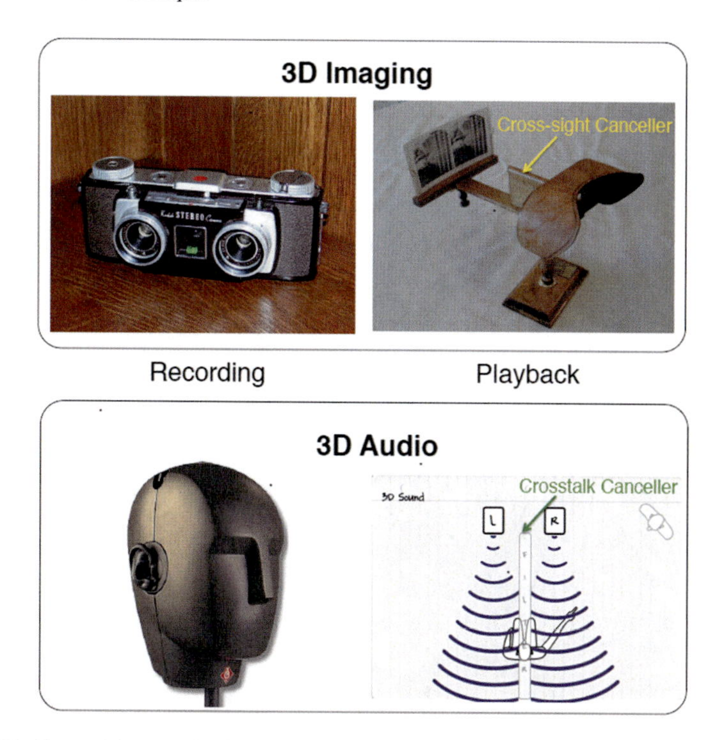

Fig. 4.66 Cross-sight canceler in optical 3D and cross-talk canceler techniques in 3D audio-soundfield (from Choueiri 2010)

According to the inventor, the BACCH(TM) 3D sound can be implemented in a way to function for multiple sweet spots simultaneously: 'DynaSonix' is a combination of Princeton's BACCH(TM) 3D Sound technology with the phased array speaker technology developed by an industry partner, Cambridge Mechatronics Ltd., (Cambridge, UK), which allows up to 6 listeners sitting anywhere in a room to get simultaneously a 3D audio image. The multiple sweet spots are adjusted dynamically by steering the 12 sound beams from the phased loudspeaker array to the ears of the listeners, located with a head tracking camera (Fig. 4.69).

4.4 '3D' or 'Immersive Audio'

Currently, there seem to be mainly three widely accepted systems—or industry-defined 'standards'—for the representation of 3D audio, at least for music and film reproduction:

- Auro 3D
- Dolby Atmos, and

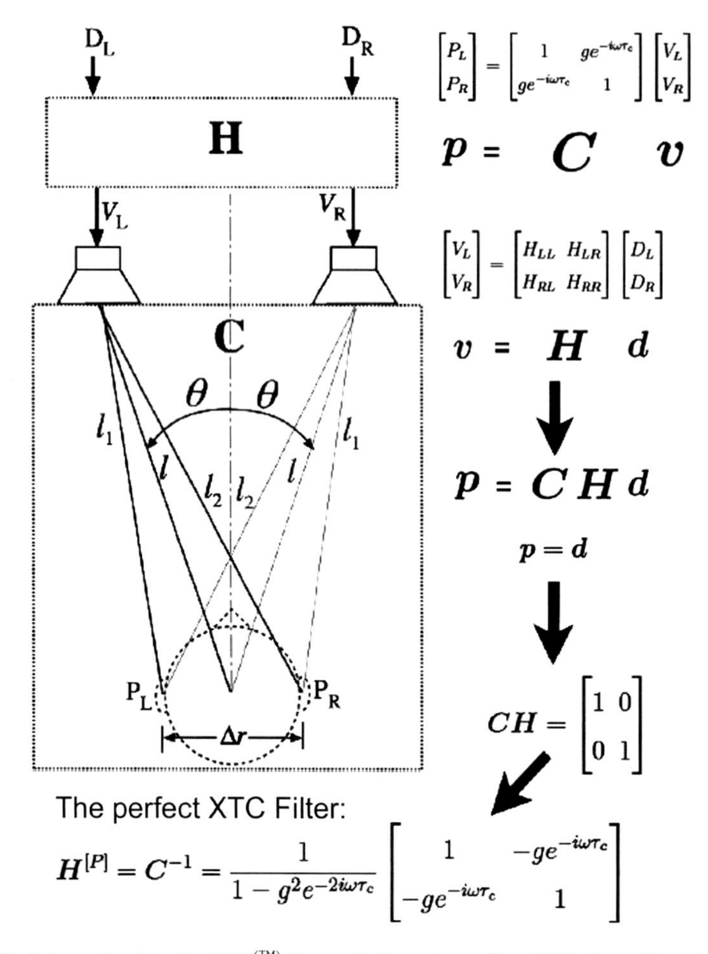

Fig. 4.67 Schematic of the BACCH(TM) Crosstalk Cancelation filter XTC (from Choueiri 2010)

- the Ambisonic Format (First-Order Ambisonics FOA or Higher-Order Ambisonics HOA), as an appropriate means for capturing—or representing—3D audio in the form of a recording (or rendering) from a 'one point' perspective.

In Corteel et al. (2016, p. 2), the following distinction between the three principally differing approaches is made (citings are slightly altered):

'…If we look at the current technical situation, 2D and 3D audio will essentially be implemented using one of the three following principles, in which each of the following 3D audio formats uses multiple sound streams:

– **Channel-based**: Each audio stream is assigned to a fixed loudspeaker. This is the case for the traditional and well-established delivery formats of 2-channel stereo and 5.1 surround. Sound positions of objects are a consequence of their representation through the use of the main microphone system (with the possibility of the

Fig. 4.68 Playback of demonstration of 3D audio teleconferencing using BACCH$^{(TM)}$ 3D sound (from Choueiri 2010)

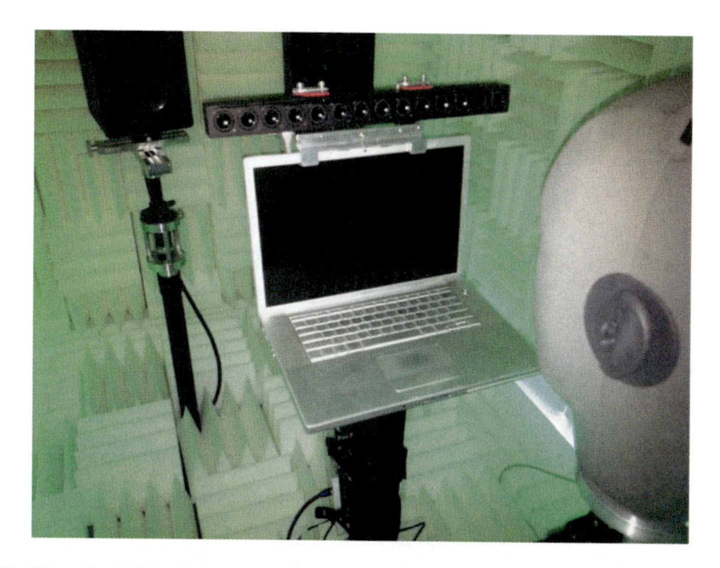

Fig. 4.69 'DynaSonix' implemented on a phased loudspeaker array mounted on top of a laptop computer tested in the anechoic chamber of the 3D3A Lab (from Choueiri 2010)

additional use of spot-microphones, included during the mixdown process), or a kind of pre-rendering (through bus allocation and mixing) of the sound-sources during the mixing stage as well as a well-defined loudspeaker setup (number and position of speakers) for replay.

- **Scene-based**: directional encoding of a sound scene, based on a set of eigen-functions of radiation/directionality (spherical harmonics in the case of Higher-Order Ambisonics or "HOA").
  This is a scalable format which offers increasing spatial accuracy with the order (number of HOA components/streams) it uses. It requires some matrixing on the rendering side since each channel reproduction is neither a source nor attached to a direction where loudspeakers could be located.
- **Object-based**: descriptions in which each audio stream is described as a sound object with associated metadata (position, spatial extent, etc.) that may evolve over time. Apart from its complete independence from the reproduction system, shared with the scene-based approach, the advantage of the object-based principle is the possibility to manipulate each object independently from the others. Theoretically, the same content can be used on different devices and different platforms…'

The next-generation audio formats—according to the aim of the EDISON 3D audio project (see Corteel et al. 2016) will be supporting either all or a subset of the three principles explained above.

The film and music industry has recently introduced various 3D audio systems to the movie theater (and home theater) market, with proprietary approaches, which we will be only shortly outlined in the following sections.

4.4.1 Auro 3D

In some way, German Tonmeister Werner Dabringhaus has anticipated part of the basic principle of Auro-3D's height loudspeakers by proposing his '2 +2+2' system of surround sound playback in 2000 (see Dabringhaus 2000). Instead of adhering to the regular 5.1 surround layout he remained with only 4 speakers for the horizontal layer (by leaving out the center speaker) and added two height speakers—above the front speakers L, R—lifted up half the distance of the front loudspeaker's base width.

As stated in the 'AURO-3D Home Theater Setup Installation Guidelines' (Auro Technologies NV 2015):

'…The Auro-3D System was introduced by the inventor Wilfried Van Baelen (CEO Galaxy Studios and Auro Technologies) with demonstrations of his Auro 9.1 and Auro 10.1 speaker layouts at the AES convention in Paris and San Francisco during the "Surround Sound with Height" workshops in 2006. A fundamental element of the concept behind the format is compatibility. All Auro-3D listening formats are mutually compatible, but they are also completely accordant with any existing workflows, delivery formats and available bandwidth, as well as with the existing standards for stereo and 5.1 Surround. Using the three orthogonal axes (x, y, z), the speaker setups reproduce true three-dimensional sound as a hemisphere all-around the audience. Auro-3D literally introduces the third dimension "Height" to movies and music, which enhances and augments their emotional impact.

Instead of using the term "Surround Sound with Height," Van Baelen introduced the term 'Immersive Sound' as a more all-round description. Therefore, it is not dependent on the technology being either channel- or object-based. The use of an object-based technology provides no guarantee to great immersive sound; it all depends on the way it is has been implemented by creatives. The Auro-3D format is a hybrid format, which uses both channel- and object-based technology (the latter called AuroMax), depending on the final goal.'

4.4.2 Dolby Atmos

The 'DOLBY ATMOS' technique also makes use of audio 'beds' and 'objects.' Beds are effectively channel-based submixes or stems, which can be retained as separate bed stems through the mixing process; they are combined into a single bed as part of the print-master process. These beds can be created in different channel-based configurations, such as 5.1, 7.1 or 9.1 (including arrays of overhead loudspeakers).

As Dolby Atmos allows up to 128 tracks to be packaged,—in addition to the 9.1 bed—you can have 118 simultaneous mono objects, each of which can be placed individually within a virtual 3D space as automation metadata. This 3D model is then rendered down to produce a mix specifically for the individual monitoring environment, based on the number and position of speakers in your room and the metadata of the objects.' (from DOLBY 2014a, b).

The possibilities of Higher-Order Ambisonics (HOA)—see the 'scene-based' approach above—have already been outlined in the above Sects. 4.3.7. During the last years, the Ambisonics/HOA format is embraced by the VR community, not least due to the support it gets by video platforms and social media like Youtube and Facebook.

4.4.3 3D Audio: Psychoacoustic Considerations and Comparative Tests

In Gribben and Lee (2014), a short summary on the current state of 3D Audio and related psychoacoustic knowledge can be found:

'...The recently proposed multichannel audio formats such as 22.2 (ITU-R BS.2159-4 2012) and Auro-3D (Van Daele and Van Baelen 2012) employ height channels to provide the auditory sensation of a "three-dimensional (3D)" space. For cinema sound or pop music production, the height channels could be used for creative panning of source image in the vertical domain as well as for providing extra ambience. On the other hand, for acoustic recordings made in a concert hall, the use of height channels is likely to be focused on extra ambience since source images would not need to be elevated in most cases (an exception of which could be choir singers on high stands).

In recent years, a few main microphone techniques employing height channels have been introduced (see Theile and Wittek 2011, Williams 2013, Geluso 2012). For example, (Theile and Wittek 2011) proposed a technique called "OCT-9" that employs four upward-facing cardioid microphones that are placed above the front left, front right, rear left and rear right microphones of the main microphone array "OCT-5." The recommended spacing between the main and height microphones for this technique is 1 m or wider (Fig. 4.70).

Williams (2013) also designed a 3D microphone array with four height microphones that are vertically spaced from the main microphones. The proposed spacing between the lower and upper layers is 1 m, and the polar pattern of the height microphones is figure-of-eight. On the other hand, Geluso (2012) proposed using a "coincident" microphone technique as a method to capture height information; a vertically oriented figure-of-eight "side" microphone is configured with a front-facing "mid" microphone without any spacing between the two.

To date, however, no formal experimental data have been provided on the effect of spacing between main and height channel microphones on perceived spatial impression. In the context of horizontal stereophony, it is widely known that a more spaced microphone pair would produce a greater spatial impression in reproduction (see Lipshitz 1986, Hamasaki 2003, Rumsey and Lewis 2002). This is due to the fact that a larger spacing between the microphones would lead to a lower degree of inter-channel correlation between the signals (Hamasaki 2003). However, research suggests that the principles of horizontal stereo might not be directly applicable to vertical stereo. In terms of localization, it is well known that vertical localization relies on spectral cues rather than interaural cues (see Roffler and Buttler, 1968, Blauert 1997). The amplitude panning of phantom image in vertical stereophonic reproduction has been reported to be unstable (Barbour 2003). It has also been found

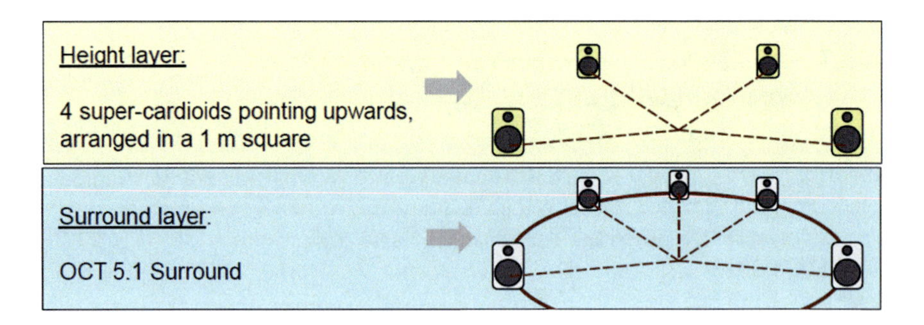

Fig. 4.70 'OCT-9' proposal by Theile and Wittek (graphic from Theile and Wittek 2012)

that the precedence effect does not fully operate between vertically arranged loudspeakers regardless of the time difference applied to them (see Lee 2011, Wallis and Lee 2014), and that time panning in the vertical plane is ineffective (Wendt et al. 2014). With respect to spatial impression, Lee and Cribben investigated the effectiveness of inter-channel de-correlation for controlling the perceived image spread of band-passed pink noise, using two loudspeakers arranged vertically in the median plane as well as those horizontally arranged (Gribben and Lee 2014). It was found that the effectiveness of vertical de-correlation was not as strong as that of horizontal de-correlation, depending on frequency. However, the perceptual mechanism of vertical spatial impression has not been fully explored yet and therefore needs further investigation…'

In connection to the abovementioned 'standards' of AURO 3D and DOLBY ATMOS, which define mainly the mixing and reproduction side of 3D audio recordings, there are several proposals concerning microphone techniques for capturing audio in a suitable format, among them:

- 'Bowles-Array with Height Layer' (see Bowles 2015)
- Ellis-Geiger Triangle (see Ellis-Geiger 2016),
- Paul Geluso's 'MZ-Microphone,' 'Blumlein + Z' and 'Spaced 5.0Array + Z' techniques (Geluso 2012),
- The 'Howie-Tree' (see Howie and King 2015, Howie et al. 2016, 2017)
- Lee's 'ESMA near-coincident quad-array' (after Williams), (see Lee 2016)
- the 'Magic Array' and Isoceles Triangle Structure, as proposed by Michael Williams (see Williams 2012a, b, 2013, 2014),
- the 'OCT-9' (or OCT 3D) technique by Theile and Wittek (see Theile and Wittek 2011),
- the 'ORTF-3D' Technique by Wittek and Theile (see Wittek and Theile 2017) (Fig. 4.71).

An interesting comparison of two 3D audio microphone techniques ('Twins-Square' vs. 'Double-MSZ') used for the recording of a pop-style music ensemble can be found in Ryaboy (2015).

A more extensive comparative 3D audio microphone array test has taken place at Abbey Road Studios in 2017, led by Hashim Riaz and Mirek Stiles (see Riaz et al. 2017 and Stiles 2018), which has involved a total of 11 different microphone systems:

in Position A:

- mh acoustics EM32 'Eigenmike®'(HOA)
- SoundField ST450 MKII
- ESMA (Equal Segment Microphone Array)
- ORTF-3D Surround
- and Neumann KU100 dummy head (as reference)

in Position B:

- Sennheiser AMBEO (FOA)

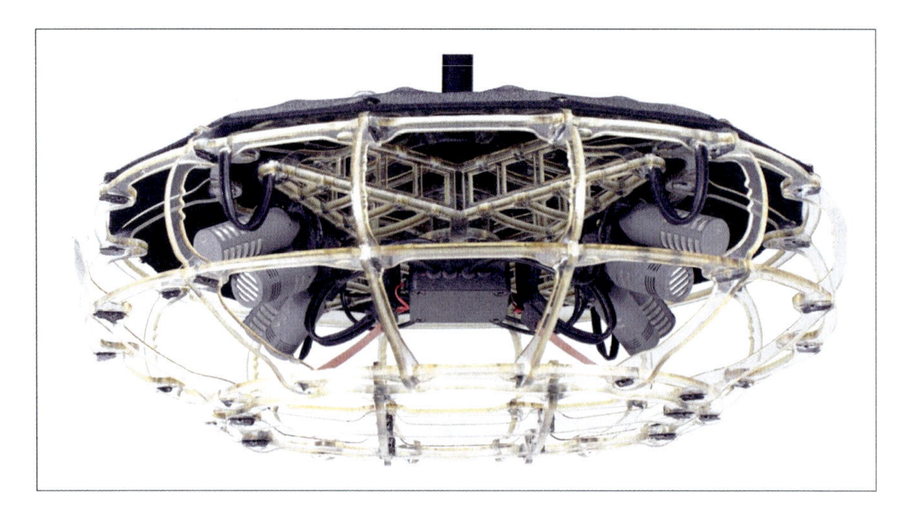

Fig. 4.71 'ORTF-3D' proposal by Wittek and Theile (from Wittek and Theile 2017)

- OCT-9 Surround
- PCMA 'Perspective Control Microphone Array'
- Stereo XY-pair

in Position C:

- IRT Cross
- Hamasaki Cube
- SoundField ST450 MKII

The objective was to undertake a live popular music recording while using dedicated VR (virtual reality) multichannel microphone arrays (Fig. 4.72), as the performance of a band was also captured by a Samsung Gear 360° camera. During post-production, binaural processing was applied to the audio material in Reaper (audio recording software), using additional third-party plugins.

'The 360 video was previewed using Kolor's GoPro VR Player, SpookSyncVR Software was employed to synchronize the video to the audio in Reaper. … Using SpookSyncVR, it is possible to gather the X (yaw) and Y (pitch) positional data from a headset such as an Oculus Rift and transfer the information to Reaper so that the soundfield can be rotated accordingly to create an enhanced listening experience.' (from Riaz et al. 2017).

More detailed information on these—and other 3D audio techniques, not mentioned above—shall be found in an upcoming publication of the author of this book, which will focus only on techniques for 3D audio capturing and reproduction.

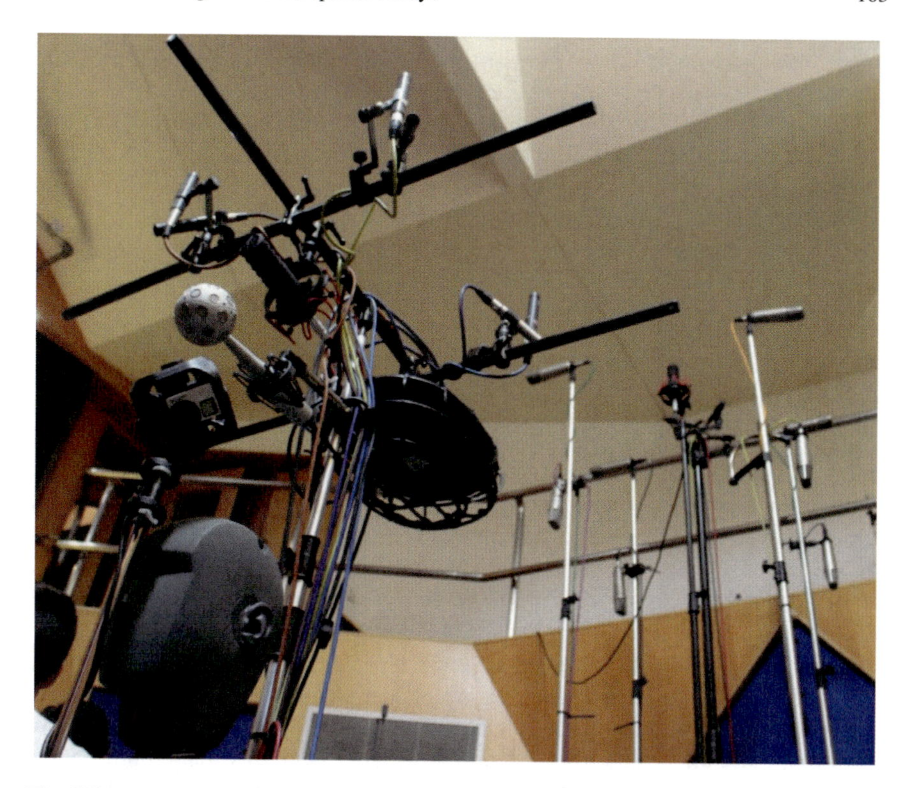

Fig. 4.72 Arrangement of several 3D audio microphone systems (and a Neumann KU100 artificial human head) as used in the comparative recording at Abbey Road Studios in London (from Riaz et al. 2017): front (bottom to top): Neumann KU100, mh acoustics M32 'Eigenmike'®, SoundField ST450, ORTF-3D (in basket), ESMA; rear (bottom to top): Hamasaki-Cube, SoundField ST450

4.5 General Thoughts on Microphone Arrays

In Woszczyk (1992) thoughts on microphone arrays, optimized for music theater use can be found: coincident, as well as microphone techniques with sufficiently large capsule spacings are suited to avoid unwanted sound coloration (comb-filtering effects) due to phase cancellation. Signals that are picked up by a 'spaced' microphone array can be considered statistically 'independent' above the frequency for which the capsule spacing equals half the wavelength. In this context, the terms 'correlation length' and 'coherence interval' are used.

Stereo microphone arrays can be seen as mono-arrays, the microphones of which are panned hard left and right. Such an array can be enlarged by the addition of further mono-arrays, which are panned between the L and R bus.

With a combination of coincident and spaced arrays, whose properties are optimized in certain frequency ranges, positive results with respect to desired directivity or evenly distributed pickup sensitivity can be achieved, which would otherwise not be possible with a conventional single microphone technique. Woszczyk further notes that microphone arrays are particularly well adapted for the pickup of large sound sources—like orchestras—for example.

But—of course—also the technical properties of the replay side have a strong influence on how the signal information, captured by a microphone array, 'translates' to the listener:

Among others, in Burkhard et al. (1991) it is analyzed how the sound quality and response characteristic of the replay loudspeakers influence spatial impression, localization, envelopment and clarity for the listener.

4.6 Conclusion

As has already been pointed out, with stereo microphone techniques, coincident ('one-point') techniques are often characterized by very good localization properties. However, their signals—obviously—do not carry any time-of-arrival information. The above described 'SoundField Microphone' manages to capture the sound signal at one point in a room in a three-dimensional manner and—via matrixing and dematrixing—further down the signal chain the information can be played out to the desired loudspeaker arrangement, which can be comprised of any number of speakers, according to need. However, even under ideal conditions, this will lead to the perfect reconstruction of the original sound-field only at one point in the playback room and the results from listening tests (see Chap. 9) strongly indicate that there are other surround microphone systems or techniques, which do provide better sonic results outside of that 'sweet spot' or 'reproduction sphere' (in the case of 3D audio), since they may be better suited for the capturing and reproduction of sound for a selected surround sound format (in our case: 5.1-Surround).

While the SoundField microphone may really be ideal for monaural listening, the normal listener will usually make use of both ears, and it is therefore to be excluded that the one-point Ambisonic principle (FOA or HOA)—in the absence of any time-of-arrival differences included in the signal—can provide realistic impression to both ears in that respect. This may very well explain the discrepancy between the theoretically 'perfect' signal reproduction, which the Ambisonic principle is able to provide, and the clearly inferior judgement it has received in listening comparisons (see Camerer et al. 2001, Hildebrandt and Braun 2000).

However, it also needs to be said that coincident microphone techniques are usually of advantage when it comes to capturing sound-sources with small diameters (e.g. solo instruments) or sound-sources with complex (i.e. direction- and frequency-dependent) radiation patterns. Under such circumstances microphone techniques which use several dispersed microphones run the risk to cause a 'wandering' or 'jumping' effect of the sound-source (i.e. localization distortion) upon replay.

The majority of the microphone systems described in Sect. 4.2 'Spaced Arrays' is based on the psychoacoustic principle of time-of-arrival differences according to the ear-spacing of human heads (see the 'ABC 5' system, for example), in many cases supported by level differences due to the use of directional microphones (see 'INA

5,' 'OCT,' 'Fukada tree' as well as other 'Tree-Systems'). Some of these are more meant for the pickup of diffuse sound ('IRT-/Atmo-Cross' and 'Hamasaki Square'), others are instead using large capsules spacings, aimed at capturing the sound-source in its full width (e.g. 'Microphone Curtain' or 'Multiple AB' technique, as well as AB-PC—'AB-Polycardioid Centerfil' technique) and are therefore also adequate for a use as surround microphones for front-channel pickup.

In contrast to 'spaced arrays' (with small capsule spacings, i.e. small AB), the signals of the two latter systems are characterized by much larger 'time-of-arrival' differences, as well as relevant level differences, which—in case of orchestra recordings, for example—are caused by the relative proximity to the various instrumental groups of each of the microphones involved. This is very much in favor of the distinctness and precision of signal localization, as ITD and ILD will 'pull' the psychoacoustic impression in the same direction. In fact, techniques with such large capsule spacings are more appropriately situated in the group of 'Equivalence Microphone Techniques,' in which both physical parameters (time component and level component) contribute to localization and spatial impression for the listener.

In case of 'large AB' arrangements, a 'Centerfill' system has to be used in order to avoid the 'hole-in-the-middle' effect. This is the case already with a the simple extension of large AB to large ABC by adding a C (Center) capsule in the middle, or by adding more elaborate centerfill systems like a DECCA-triangle (Δt technique), ORTF-Triple (equivalence technique) or BPT 'Blumlein-Pfanzagl-Triple' (coincident technique) and thus arriving at the sonic advantages of so-called combined main microphone systems (see Chap. 3, Sect. 3.4). What must be taken into consideration is the fact that 'large-AB' (and 'Microphone Curtain') techniques aim at 'capturing the sound-field along a line' and replaying it (usually in a 'downscaled' form) through the loudspeakers in the (usually much smaller) playback room, while preserving important sonic aspects of the underlying 'wavefront' principle. (For more details on the wavefront principle see Chap. 11.)

The 'Circular Microphone Array' instead takes a completely different approach: With a relatively large number of single microphones—arrayed in circular fashion at an appropriate place in the recording room—the signal of interest is captured in order to be processed and optimized before replay on a multichannel loudspeaker system. This microphone technique seems a bit unpractical for use in music recording, due to the involved mechanical setup of a large number of microphones.

Similarly, the rather complex 'Trinnov-Array' needs signal processing before playback in 5.1 surround format.

In Sect. 4.3, various baffle and 3D techniques have been described, of which most are designed to record the sound-field in a manner which allows for the discrete capturing of sound from 'above' and 'below.' Some of these techniques use sound-transducers integrated into a surface (usually similar to a human head in terms of diameter), in order to make use of sound-shadowing effects which are effective at higher frequencies (see 'Eigenmike®' and 'Holophone H2-Pro'), others are simply arranged along a (virtual) sphere (see 'Sony-Sphere Microphone' and 'Multichannel Microphone Array'). With these techniques for localization and spatialisation, time-of-arrival differences, as well as level differences due to the use of directional microphones (as in the case of the 'Multichannel Microphone Array') are used, sometimes accompanied by signal processing (see 'Sony-Sphere Microphone' and 'Eigenmike®').

A reduction of the three-dimensional 'head-related baffle technique' to a quasi-two-dimensional disk can be found in Jecklin's OSIS System (Optimal Sound Image-Space), which uses two different systems for frontal and rear pickup of sound signals. Another rather individualistic recording approach has been implemented in the 'Pan-Ambiophonic 2D/3D System,' which combines the signals of a discreetly assembled Ambisonics-array with that of two pressure transducers at ear positions of a human-head-like baffle.

During the last years the 'Auro 3D,' 'Dolby Atmos' and 360° VR (mainly related to the Ambisonics HOA format), industry standards have managed to establish themselves as formats for multichannel 3D music and film-sound related reproduction formats with height information, which has opened doors for great creativity among sound-engineers in coming up with new, suitable microphone techniques (see the OCT-9, ORTF-3D and ESMA proposals, among many others in Sect. 4.4), as well as challenging scientists in researching this new 'psychoacoustic (reproduction) reality.'

References

Audio Eng Society-Staff writer (2004) Novel surround sound microphone and panning techniques. J Audio Eng Soc 52(1/2):74–80

AURO-Technologies NV (2015) AURO-3D home theater setup—installation guidelines. Rev 6. http://www.auro-3D.com. Accessed 28 Oct 2015

Barbour JL (2003) Elevation perception: phantom images in the vertical hemisphere. In: Proceedings to the 24th international conference: multichannel audio, the new reality. Audio Engineering Society

Bates E, Gorzel M, Ferguson L, O'Dwyer H, Boland FM (2016) Comparing ambisonics microphones: part 1. Paper presented at the conference on sound field control, Audio Engineering Society, Gilford, 18–20 July 2016

Bates E, Doonery S, Gorzel M, O'Dwyer H, Ferguson L, Boland FM (2017) Comparing ambisonics microphones—part 2. Paper presented at the 142nd audio engineering society convention, Berlin, 20–23 May 2017

Bauer B (1961) Phasor analysis of some stereophonic phenomena. J Acoust Soc Am 33:1536–1539

Bauer B (1979) A unified 4-4-4-, 4-3-4, 4-2-4 SQ(TM)-compatible system of recording and FM bradcasting (USQ). J Audio Eng Soc 27(11):866–880

Bauer B (1981) Microphone system for producing signals for surround-sound transmission and reproduction. US Patent 4,262,170

Bauer B, Abbagnardo L, Gravereaux D, Marshall T (1978) The ghent microphone system for SQ(TM) quadraphonic recording and broadcasting. J Audio Eng Soc 26(1/2):2–11

Blauert J (1997) Spatial hearing. The MIT Press

Bowles D (2015) A microphone array for recording music in surround-sound with height channels. Paper 9430 presented at the 139th audio engineering society convention, New York, Oct 2015

Bramford JS (1995) An analysis of ambisonics systems of first and second order. Dissertation, University of Waterloo, Ontario, Canada

Bruck J (1996) Solving the surround dilemma. In: Proceedings to the 19 Tonmeistertagung des VDT, pp 117–124

Burkhard M, Bray W, Genuit K, Gierlich HW (1991) Binaural sound for television. Proceedings to the audio engineering society, 9th international conference on television sound today and tomorrow, Detroit, Feb 1991

Camerer F, Sodl C, Wittek H (2001) Results from the Vienna listening test. http://www.hauptmikrofon.de/ORF/ORF_und_FHD.htm. Accessed 1 Dec 2016

Ceoen C (1972) Comparative stereophonic listening tests. J Audio Eng Soc 20(1)

Choueiri E (2010) Optimal crosstalk cancellation for binaural audio with two loudspeakers. http://www.princeton.edu/3D3A/Publications/BACCHPosterV4.pdf. Accessed: 27 Feb 2018

Christensen L (2003) A surround microphone setup for classical music. DPA microphones workshop on Mic techniques for multichannel audio. Paper presented at audio engineering society 24th international conference in Banff, Canada, 2003

Cohen E, Eargle J (1995) Audio in a 5.1 channel environment. Paper 4071 presented at the 99th audio engineering society convention, New York

Corey J, Martin G (2003) Description of a five-channel microphone technique. Paper presented at the DPA microphones workshop on Mic techniques for multichannel audio of the audio engineering society 24th international conference in Banff, Canada, 2003

Corteel E, Pesce D, Foulon R, Pallone G, Changenet F, Dejardin H (2016) An open 3D audio production chain proposed by the Edison 3D project. Paper 9589 presented at audio engineering society 140th international convention in Paris, France, 2016

Dabringhaus W (2000) 2 + 2+2—kompatible Nutzung des 5.1 Übertragungsweges für ein System dreidimensionaler Klangwiedergabe klassischer Musik mit drei stereophonen Kanälen. In: Proceedings to the 21. Tonmeistertagung des VDT

Daniel J (2001) Représentation de champs acoustiques, application á la transmission et á la reproduction de scènes sonores complexes dans un contexte multimédia. Dissertation, University of Paris VI, France

DOLBY (2014a) DOLBY ATMOS—next-generation audio for cinemas. White paper by Dolby Laboratories, Inc. https://www.dolby.com/us/en/technologies/cinema/dolby-atmos.html. Accessed 27 Feb 2018

DOLBY (2014b) Authoring for Dolby atmos—cinema sound manual (issue 3 for software 1.4). White paper by Dolby Laboratories Inc., https://www.dolby.com/us/en/technologies/cinema/dolby-atmos.html. Accessed 27 Feb 2018

DPA (2018) D:mension 5100—mobile surround microphone. http://www.dpamicrophones.com/microphones/dmension/5100-mobile-surround-microphone. Accessed 27 Feb 2018

Eargle J (2004) The microphone book, 2nd edn. Focal Press

Elko G, Kubli R, Meyer J (2003) Audio system based on at least second-order Eigenbeams. Int Patent WO 03(061636):A1

Ellis-Geiger J (2016) Music production for Dolby atmos and auro 3D. Paper 9675 presented at the 14st audio engineering society convention, Los Angeles

Evers H (2003) Swan lake in 5.1. Presented at the DPA microphones workshop on Mic techniques for multichannel audio, audio engineering society 24th international conference in Banff, Canada

Fox C, McGregor W (2002) A modular microphone array for surround sound recording. Paper 5566 presented at the 112th audio engineering society convention, Munich, May 2002

Geluso P (2012). Capturing height: The addition of Z microphones to stereo and surround microphone arrays. Paper 8595 presented at the 132nd audio engineering society convention

Gerzon M (1973) Periphony: with-height sound reproduction. J Audio Eng Soc 21(1)

Gribben C, Lee H (2014) The perceptual effects of horizontal and vertical interchannel decorrelation using the Lauridsen decorrelator. Paper 9027 presented at the 136th audio engineering society convention

Hamasaki K (2003) Multichannel recording techniques for reproducing adequate spatial impression. In: Proceedings to the audio engineering society 24th international conference on multichannel audio—the new reality, Banff, Canada, 2003

Hamasaki K, Hiyama K (2003) Reproducing spatial impression with multichannel audio. Paper presented at the audio engineering society 24th international conference on multichannel audio—the new reality, Banff, Canada, 2003
Hamasaki K, Shinmura T, Hiyama K (2001) Approach and mixing technique for natural sound recording of multichannel audio. Paper 1878 presented at the audio engineering society 19th international conference, June 2001
Hermann U, Henkels V, Braun D (1998) Comparison of 5 surround microphone methods. In: Proceedings to the 20. Tonmeistertagung des VDT, Karlsruhe, p 508
Hidaka T, Beranek L, Okano T (1995) Interaural cross-correlation, lateral fraction, and low- and high-frequency sound levels as measures of acoustical quality in concert halls. J Acoust Soc Am 98 (2)
Hildebrandt A, Braun D (2000) Untersuchungen zum Centerkanal im 3/2 Stereo-Format. In: Proceedings to the 21. Tonmeistertagung des VDT, p 455
Hiroyuki N, Michio M, Katsunori F, Isanaga Y (1980) Microphone. JP Patent 55–117538:26
Holman T (2001) Mixing the sound (part 2): perspective—where do the sounds go? Surround Prof 35
Holophone (2008) Available via www.holophone.com. Accessed 10 June 2008
Howie W, King R (2015) Exploratory microphone techniques for three-dimensional classical music recording. Convention E-brief presented at the 138th audio engineering society convention, Warsaw, Poland, May 2015
Howie W, King R, Martin D (2016) A three-dimensional orchestral music recording technique, optimized for 22.2 multichannel sound. Paper 9612 presented at the 141st audio engineering society convention, Los Angeles
Howie W, King R, Martin D, Grond F (2017) Subjective evaluation of orchestral music recording techniques for three-dimensional audio. Paper 9797 presented at the 142nd audio engineering society convention, Berlin, 20–23 May 2017
Iredale JJ, Keller R (2003) Microphone apparatus for producing signals for surround reproduction. US Patent 6,507,659, Jan 2003
Ito A (2001) Surround sound field reproduction system and surround sound field reproduction method. EU Patent EP 1,259,097 A2
ITU-R Recommendations BS.2159-4 (2012) Multichannel sound technology in home and broadcasting applications. International Telecommunications Union
Jecklin J (1981) A different way to record classical music. J Audio Eng Soc 29(5):329–332
Jecklin J (2002) Surround-Aufnahmetechnik OSIS 321. In: Proceedings to the 21. Tonmeistertagung des VDT, Hannover, Nov 2002
Kamekawa T, Marui A, Irimajiri H (2007) Correspondence relationship between physical factors and psychological impressions of microphone arrays for orchestra recording. Paper 7233 presented at the 123rd audio engineering society Convention, New York, Oct 2007
King R (2003) A five-microphone technique for music recording in a large venue. DPA microphones workshop on Mic techniques for multichannel audio, audio engineering society 24th international conference on multichannel audio—the new reality, Banff, Canada, 2003
Klepko J (1997) 5-channel microphone array with binaural head for multichannel production. J Audio Eng Soc 45:127
Laborie A, Montoya S, Remy B (2004) High spatial resolution multichannel recording. Paper 6116 presented at the 116th audio engineering society convention, May 2004
Lee H (2011) The relationship between interchannel time and level differences in vertical sound localization and masking. Paper 8556 presented at the 131st audio engineering society convention, Oct 2011
Lee H (2016) Capturing and rendering 360° VR audio using cardioid microphones. Paper presented at the conference on audio for virtual and augmented reality, Los Angeles, 30 Sept–1 Oct 2016
Levine A (2008) XYtri—from stereo to surround, and back! In: Proceedings to the 25. Tonmeistertagung des VDT, Leipzig, Nov 2008

Lipshitz SP (1986) Stereo microphone techniques: are the purists wrong? J Audio Eng Soc 34:717–743

Martin G (2005) A new microphone technique for five-channel recording. Paper 6427 presented at the 118th audio engineering society convention, Barcelona 2005

McKinnie D, Rumsey F (1997) Coincident microphone techniques for three channel stereophonic reproduction. Paper 4429 to the 102nd audio engineering society convention, Munich, March 1997

Meyer J, Agnello T (2003) Sperical microphone array for spatial sound recording. Paper 5975 to the 115th audio engineering society convention, New York

Miller R (2004a) Spatial definition and the panambiophone microphone array for 2D surround and 3D fully periphonic recording. Paper presented at the 117th audio engineering society convention, San Francisco

Miller RE (2004b) System and method for compatible 2D/3D (full sphere with height) surround sound reproduction. US Patent 2004/0247134

Mitchell D (1999) Tracking for 5.1—surround-recording techniques. Audio Media

Morimoto M, Maekawa Z (1988) Effects of low frequency components on auditory spaciousness. Acustica 66

Muraoka T, Miura T, Ifukuba T (2007) Ambience sound recording utilizing dual MS (mid-side) microphone systems based upon frequency dependent spatial cross correlation (FSCC). Paper 6997 to the 122nd audio engineering society convention, Vienna

Nicol R (2018): Sound field. In: Geluso P, Roginska A (eds) Immersive sound. Focal Press, Routledge

Olabe I (2014) Técnicas de grabación de música clásica. Evolución histórica y propuesta de nuevo modelo de grabación. Dissertation, Universitat de les Illes Balears. http://hdl.handle.net/10803/362938. Accessed 12 Oct 2015

Peters N, McAdams S, Braasch J (2007) Evaluating off-center sound degradation in surround loudspeaker setups for various multichannel microphone techniques. Paper 7197 presented at the 123rd audio engineering society convention, New York

Pfanzagl-Cardone E (2002) In the light of 5.1 surround: why AB-PC is superior for symphony-orchestra recording. Paper 5565 presented at the 112th audio engineering society convention, Munich

Pfanzagl-Cardone E (2005) 3.0 microphone for surround-recording. AU patent AU2005100255

Pfanzagl-Cardone E, Höldrich R (2008) Frequency-dependent signal-correlation in surround- and stereo-microphone systems and the Blumlein-Pfanzagl-Triple (BPT). Paper 7476 presented at the 124th audio engineering society convention, Amsterdam

Poletti MA (2000) A unified theory of horizontal holographic sound systems. J Audio Eng Soc 48(12):1155–1182

Preston C (1998) An analysis of frequency dependent hybrid microphone arrays for stereophonic sound recording. Paper 4793 presented at the 105th audio engineering society convention, San Francisco

Riaz H, Stiles M, Armstrong C, Chadwick A, Lee H, Kearney G (2017) Multichannel microphone array recording for popular music production in virtual reality. E-brief presented at the 134rd audio engineering society convention, New York, Oct 2017

Roffler SK, Buttler RA (1968) Factors that influence the localization of sound in the vertical plane. J Acoust Soc Am 43(6):1255–1259

Rumsey F, Lewis W (2002) Effect of rear microphone spacing on spatial impression for omnidirectional surround sound microphone arrays. Paper 5563 presented at the 112th audio engineering society convention, April 2002

Ryaboy A (2015): Exploring 3D: a subjective evaluation of surround microphone arrays catered for Auro-3D reproduction system. Paper 9431 presented at the 139th convention of the audio engineering society, New York, Oct 2015

Sawaguchi M, Fukada A (1999) Multichannel sound mixing practice for broadcasting. Paper presented at IBC conference

Sokol M (2003) FLRB-array. Paper presented at the 'DPA microphones workshop on Mic techniques for multichannel audio' audio engineering society 24th international conference, Banff, Canada
Stiles M (2018) Recording spatial audio. Resolution 17(2):49–51
Streicher R, Everest A (2006) The new stereo soundbook, 3rd edn. Audio Eng Associates
Theile G (1986) Das Kugelflächenmikrofon. In: Proceedings to the 14. Tonmeistertagung des VDT, Munich, p 277
Theile G (2000) Mikrofon- und Mischungskonzepte für 5.1 Mehrkanal-Musikaufnahmen. In: Proceedings to the 21. Tonmeistertagung des VDT, Hannover, p 348
Theile G (2001) Multichannel natural music recording based on psychoacoustic principles. In: Proceedings to the 19th international conference of the audio engineering society, pp 201–229
Theile G, Wittek H (2011) Principles in surround recordings with height. Paper 8403 presented at the 130th audio engineering society convention, May 2011
Theile G, Wittek H (2012) 3D Audio natural recording. In: Proceedings to the 27. Tonmeistertagung des VDT, Cologne, Nov 2012
Valin J (1994) The RCA Bible—a compendium of opinion on RCA living stereo records, 2nd edn. The Music Lovers Press, Cincinatti, Ohio
Van Daele B, Van Baelen W (2012) Productions in auro-3D: professional workflow and costs. White paper by Auro-Technologies, Feb 2012
Wallis R, Lee H (2014) Investigation into vertical stereophonic localization in the presence of interchannel crosstalk. Paper 9026 presented at the 136th audio engineering society convention, April 2014
Ward DB, Abhayapala TD (2001) Reproduction of a plane wave sound field using an array of loudspeakers. IEEE Trans Speech Audio Process 9(6):697–707
Wendt F, Florian M, Zotter F (2014) Amplitude panning with height on 2, 3, and 4 loudspeakers. Proceedings to the 2nd international conference on spatial audio
Williams M (1987) Unified theory of microphone systems for stereophonic sound recording. Paper 2466 presented at the 82nd audio engineering society convention
Williams M (1992) Frequency dependent hybrid microphone arrays for stereophonic sound recording. Paper 3252 presented at the 92nd audio engineering society convention
Williams M (2012a) Microphone array design for localization with elevation cues. Paper 8601 presented at the 132nd audio engineering society convention, Budapest, April 2012
Williams M (2012b) 3D and multiformat microphone array design for the GOArt project. In: Proceedings to the 27. Tonmeistertagung des VDT, Cologne, Nov 2012, p 739
Williams M (2013) The psychoacoustic testing of the 3D multiformat microphone array design, and the basic isosceles triangle structure of the array and the loudspeaker reproduction configuration. Paper 8839 presented at the 134th audio engineering society convention, May 2013
Williams M (2014) Downward compatibility configurations when using a univalent 12 channel 3D microphone array design as a master recording array. Paper 9186 presented at the 137th audio engineering society convention, Los Angeles, Oct 2014
Wittek H (2002) Image Assistant V2.0. http://www.hauptmikrofon.de. Accessed 24 June 2008
Wittek H (2008) Stereophonie in Theorie und Praxis. https://www.hdm-stuttgart.de/~curdt/Wittek_HdM_Stereo_12_2008.pdf. Accessed 15 May 200
Wittek H, Theile G (2004) OCT V2-Info http://www.hauptmikrofon.de. Accessed 24 June 2008
Wittek H, Theile G (2017) Development and application of a stereophonic multichannel recording technique for 3D Audio and VR. Paper presented at the 143rd audio engineering society convention, New York, Oct 2017
Woszczyk WR (1992) Microphone arrays optimized for music recording. J Audio Eng Soc 40(11)
Yamamoto T (1973) Quadraphonic one-point pickup microphone. J Audio Eng Soc 21(4)

Chapter 5
Artificial Head Recordings

Abstract In this chapter, the importance of the use of an artificial human head (dummy head) as a 'human reference' for acoustic measurements in connection also with 5.1 surround recordings is pointed out. Also, the results of two studies on the quality of artificial human heads, carried out at Aalborg University, Denmark and IRT (Institut für Rundfunktechnik), Germany are analyzed. Aspects taken into account for these studies were mainly related to localization accuracy (out-of-cone error, within-cone error, median-plane error, distance error): while the Aalborg University study was only dealing with localization perception, the IRT study was also evaluating sound coloration. Dummy heads used for these two studies were by Bruel & Kjaer, Head Acoustics (HMS I, II and III), Techn. Univ. of Aachen HUGO, Knowles Electronics Manikin for Acoustic Research KEMAR, Neutrik-Cortex Instruments 'MANIKIN MK1,' Neumann KU80, KU81, KU100 and University of Toronto Dummy Head. As a result, researchers of the Aalborg University study were unsatisfied with the overall localization accuracy of the dummy heads tested, since there was none that would stand out quality wise, which was also attributed to the dummy heads—in general—maybe not being similar enough to human heads in all relevant aspects. Part of the conclusion was that 'both the results from the listening experiments and the analysis of the transfer functions indicate that the difference between artificial heads and humans are larger than differences between humans.'

Keywords Artificial head · Dummy head · Localization accuracy · HRTF · Binaural · Human reference

5.1 Use of an Artificial Head as a 'Human Reference'

For the surround test recordings, which have been carried out by the author (described in Pfanzagl-Cardone and Höldrich 2008) and are analyzed in detail in this book, an artificial human head (Neumann KU81) has been used as a reference. The binaural signals recorded with this dummy head were used both for an acoustic signal analysis (FIACC—Frequency-dependent InterAural Cross-Correlation), as well as a reference signal in listening tests.

E. Pfanzagl-Cardone, *The Art and Science of Surround and Stereo Recording*,
https://doi.org/10.1007/978-3-7091-4891-4_5

At least at the time of conducting the experiments in 2003 (and maybe up until today), there was an open question in the academic word concerning the optimum degree of signal cross-correlation between the channels of 5.1 surround microphone signals. While Günther Theile pleads for generally high correlation of at least 0.35 (see Chap. 2), David Griesinger is an advocate for de-correlation of signals in the front and rear channels, as well as between the rear channels, in order to achieve a 'natural' reproduction of a sound source. Due to this uncertainty, it seems to make very much sense to introduce an artificial human head in the evaluation chain as a reference element.

This move is certainly justified by the fact that the ears (and the perceptual part of the brain related to hearing) of the human listener are the undisputed 'final goal' at the receiving end. (Rem.: also the 'correct' spacing between omni-directional microphone capsules in case of 'small AB' technique (or even the 17.5 cm spacing with the ORTF microphone technique) is based on the average spacing of the human ears; for birds or elephants a very different spacing would need to be employed to achieve proper Inter-Aural Time Differences (ITDs) at the ears of these animals…).

Also, the introduction of an artificial head has the advantage of reducing the amount of measurements needed in the context of frequency-dependent cross-correlation (FCC) measurement for 5.1 surround microphone techniques: instead of measuring the 5 full-range channels pair-wise (which means a total of 10 measurements, excluding the LFE channel signal) it is sufficient to do only *one* FCC measurement for the binaural dummy head signals.

Examples of the use of a dummy head as reference can be found in Toole (1991), Olive et al. (1994), Braun et al. (1996), Pfanzagl-Cardone (2002) and Peters et al. (2007).

The 'human head as reference'-principle could also have been employed for the listening tests with 5.1 surround signals, but this would have forced the test listeners to switch between listening to at least one or two 5.1 surround microphone techniques being replayed via loudspeakers and the artificial head reference recording from the concert hall, listened to via headphones.

A constant change between headphone- and loudspeaker-listening is a distraction which is surely not desirable for a listening comparison situation, which certainly also would have had its effects on the accuracy of the results. As an alternative, one could have considered to do an EQ-compensation for the binaural dummy head signal and replay it over the 5.1 surround loudspeaker systems. There are hardware units (see e.g. the Lexicon 'Logic 7' processor) as well as software algorithms (see Griesinger 1988) available, but it can also be assumed that this kind of processing introduces changes in the character of the binaural replay-material, which would lead to a relevant change in the perception of the signal.

Due to this consideration, the author has decided to let the test listeners perform a headphones-only-based listening comparison for part of the test and refrain from having them to switch between headphone and loudspeaker reproduction.

5.2 Studies Concerning the Quality of Artificial Heads

In a study carried out at the Acoustics Laboratory of the Danish Aalborg University (Moller et al. 1999), eight artificial heads have been rated in respect to their localization attributes by 20 test listeners.

Test A ('Real-Life') had been carried out as an 'identification experiment' in which listeners had to assign their impression of sound-source localization to one of 19 loudspeakers (14 of which were positioned on the surface of a virtual sphere with a radius of 1 m; out of these, 7 speakers were arranged along the median plane and the remaining 5 (not on the sphere) were arranged at distances of up to 5 m from the listener position). The test signal was female speech (Fig. 5.1).

In tests B and C, the listeners were presented a binaural recording, made with a dummy head in the sweet spot of the abovedescribed arrangement played back via Beyerdynamic DT990 (or STAX, for the recordings with the HMS artificial head). For part B of the test, the built-in microphones of the dummy heads were used, while in test part C miniature microphones were placed right at the entrance of the ear canals.

In order to keep the signal chain as neutral as possible for the test listeners, the combined transfer function of the signal path from headphone to dummy head microphone was measured beforehand (from the headphone input to the microphone output), and an amplitude correction (inverse EQ) was then applied to the test signals.

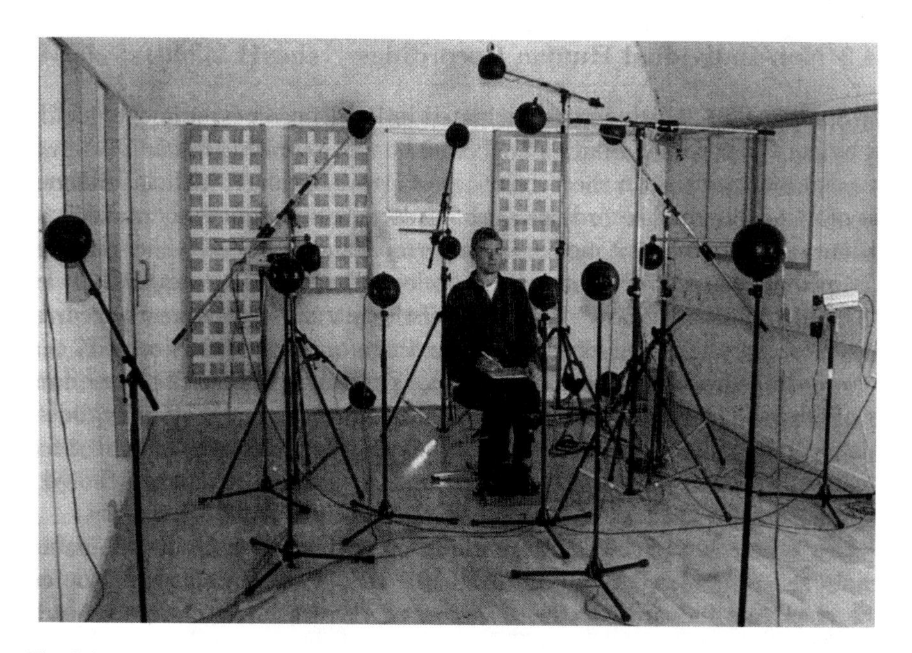

Fig. 5.1 Photograph of setup in standard listening room (originally published in Minnaar et al. 2000, reproduced from Hammershoi and Moller 2005, Fig. 9.8)

For the tests, the following artificial heads or head-torso combinations were used:

- KEMAR (Knowles Electronics Manikin for Acoustic Research) Head-and-Torso Simulator (to which 4 different pinnae of different size can be attached; used in conjunction with Bruel & Kjaer 4134 microphones),
- Neumann KU 80i (dummy head),
- Neumann KU 81i (diffuse-field compensated) (dummy head),
- HMS I (company HEAD Acoustics/University of Technology, Aachen, Germany) (head-and-torso; used in conjunction with B&K 4166),
- HMS II (company HEAD Acoustics/University of Technology, Aachen, Germany) (head-and-torso; used in conjunction with B&K 4165),
- Artificial Head of the University of Toronto (on KEMAR-Torso, used in conjunction with Sennheiser MKE-4-211-2 microphones).

As part of the experiments, the various dummy heads were tested in respect to localization distortion, both in the horizontal, as well as the vertical plane. Test stimuli were assigned to the following main areas (cones of perception) LEFT, FRONT, ABOVE, BEHIND, RIGHT. If a stimulus had been localized in a wrong position within the right cone, this was registered as a 'within-cone error'; if the stimulus had been perceived by the listener in a wrong cone, this was marked as an 'out-of-cone error.' Distance errors in the listener evaluation were another aspect taken into account, as well as 'median-plane errors'—like 'front/back' localization errors—which are quite commonly found with dummy head sound reproduction.

In order to respect the complexity of artificial head quality evaluation, the outcome of this test was not simply listed in a numerical fashion, but great care was taken to give detailed information regarding the various parameters, which had been evaluated (Fig. 5.2).

The following results appear to be of interest:

Even though the HMS I turned out to be the 'winner' with the lowest percentage of 'median-plane errors,' and finds itself also at the second best position regarding distance errors, at the same time it was responsible for the highest percentage of 'out-of-cone errors,' but also the lowest percentage of 'within-cone errors.'

The HMS II was second to last or last in respect to 'median-plane errors' within the different parts of the test, but scored best in terms of distance errors at least in one part of the test (i.e. Section B).

The big differences in the results regarding the two models HMS I and HMS II can be explained with the fact that the HMS I was modeled after a specific human being, while the HMS II is a human head model with a rather simplified head-geometry. The over-proportionally high 'out-of-cone' error rate with HMS I was attributed to the above average ear-to-ear spacing (which had been verified by an inter-aural time delay measurement), which apparently had led to wrong localization for the majority of test listeners (for details see Moller et al. 1999, p. 93).

The 'out-of-cone' and 'within-cone' error rate is more or less in the same order for all dummy heads tested, apart from the Neumann KU80i and KU81i: For those two these rates are clearly above average, but—according to the research—there seems to be a plausible cause: the Neumann dummy heads were the only test objects for which

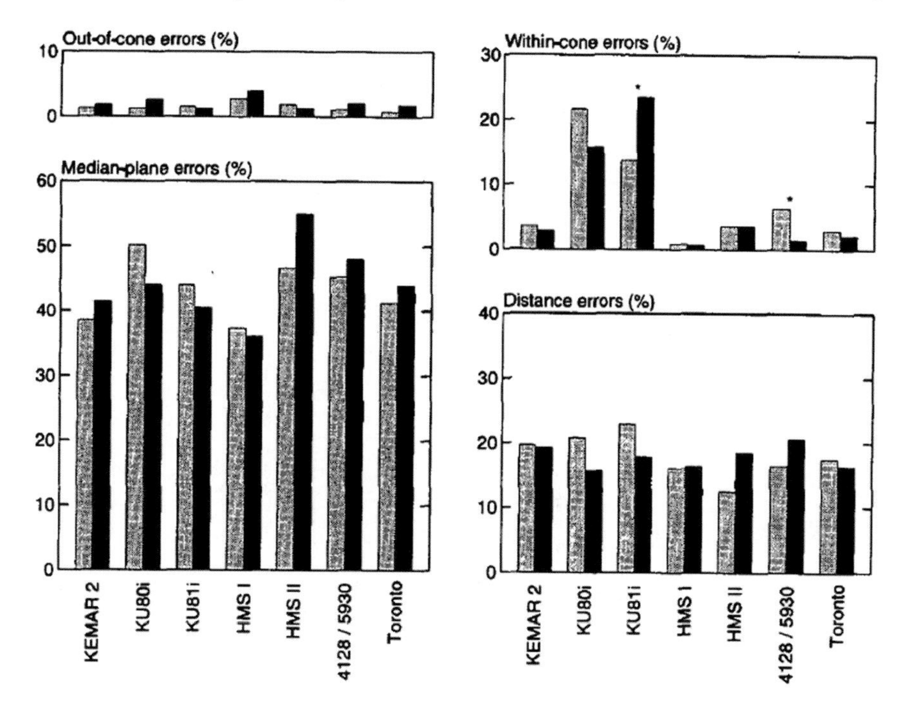

Fig. 5.2 Comparison of recording types: light columns—built-in microphone (experiment B, 912 stimuli for each head); dark columns—blocked ear canal microphones (experiment C, 380 stimuli for each head). Statistical tests compared the two recording types for each of the artificial heads (two-sided Fisher–Irwin tests). *Significance at 5% level (*rem.* the column-chart above is showing only a partial result from the paper of Moller et al. 1999)

no torso was available; therefore the absence of torso-specific sound-deflection and reflection seems to be the reason (rem.: stimuli from the directions DOWN LEFT, as well as DOWN RIGHT had often been identified as UP LEFT and UP RIGHT by the test listeners).

On the other hand, the KU80 was a winner in respect to distance errors (for the second half of test section C), and the KU81 was at the 4th position, which should result in a more convincing reproduction of 'stage depth' for these dummy heads.

In respect to 'median-plane errors,' the KU80 scored last (in test section B), but achieved 4th position for the same aspect in test section C. For comparison: the KU81 achieved position 4 (test section B) and position 2 (test section C) for this parameter.

Now for a comparison of the results of KU80 and KU81 with the other 4 dummy heads: KEMAR 2 ('equipped' with outer-ear version 2), as well as the B&K 4128 and 5930 (rem.: the dummy head is the same one, the difference is only that with the model 4128 the microphones are built into the dummy head, while with the 5930 system, the miniature microphones are positioned right outside of the ear canal) and the 'Toronto'-head.

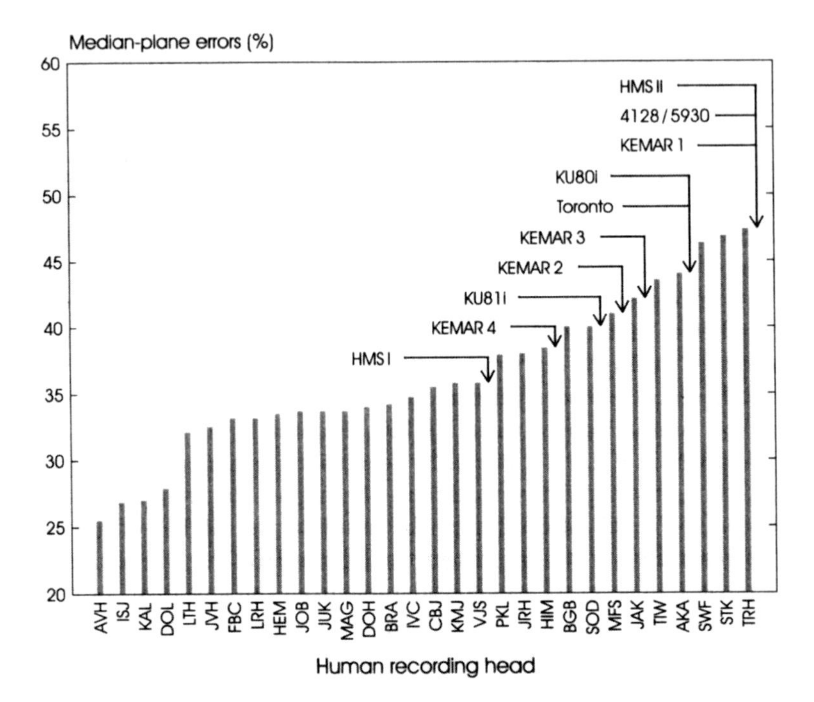

Fig. 5.3 Average median-plane error percentage for a panel of 20 listeners for the recording head indicated on the abscissa. Arrows indicate the performance with artificial head recordings (originally published in Moller et al. 1999, reproduced from Hammershoi and Moller 2005, Fig. 9.10)

In test section B, 'median-plane errors,' the rating is as follows (from better to worse):
HMS I, Kemar 2, Toronto, KU81, B&K, HMS II, KU80

In test section C, 'median-plane errors,' the rating is as follows:
HMS I, KU81, Kemar 2, KU80, Toronto, B&K, HMS II.

In test section B, 'distance error,' the rating is as follows:
HMS II, HMS I, B&K, Toronto, Kemar 2, KU80, KU81.

In test section C, 'distance error' the rating is as follows:
KU80, HMS I, Toronto, KU81, HMS II, Kemar 2, B&K.

Looking at the qualitative criteria 'median-plane error' and 'distance error,' the HMS I seems best, achieving two times position 1, as well as two times position 2. But at the same time, it is the dummy head with the highest 'out-of-cone' error rate, which is almost double as high as with all the other heads.

The HMS II almost always scored very low, except in test section B, distance error.

Also, the B&K never scored higher than position 5, except in test section B, distance error.

The Toronto head received an 'average' quality rating with two times 3rd and two times 4th position. The KU80 is not performing well in respect to 'median-plane errors.' Apart from the lack of a torso, another reason for this may lie in the frequency response of this dummy head, which is characterized by a relevant attenuation in the high-frequency band, according to the study. For this reason, the KU80 was tested only with audio stimuli up to a frequency of 6.5 kHz.

Nevertheless it turned out best in test section C, 'distance error,' which is rather surprising, since the perception of distance of a sound source is usually—at least partly—related to the *absence* of high-frequency information contained in the sound; therefore, a general lack of (or attenuation in) the high-frequency band of a dummy head should rather conceal or at least hinder appropriate distance perception for a listener.

The KU81 achieved two times position 4, once position 2 and once position 7. Of all dummy heads tested, it was the only one with a 'diffuse-field' compensation, which—according to the description of the manufacturer—results in no need for extra signal processing as long as the binaural recordings of the KU81 are listened to with diffuse-field compensated headphones as well.

The HMS I and HMS II are built for free-field conditions and therefore should be listened to with a STAX SR Lambda headphone in conjunction with filter banks delivered by the manufacturer of the dummy heads.

The final outcome of the above study is that none of the dummy heads under test results as a clear winner. The conclusion of the authors is therefore that target-oriented improvements with the various dummy heads could lead to a significant gain in overall quality, as—in comparison—much better test results have been achieved with binaural recordings, which had been made with miniature microphones placed appropriately at the ear canals of selected test persons.

Cited from Hammershoi and Moller (2005): 'The difference between human and artificial heads may be assessed objectively by comparison of their HRTFs. Frontal HRTFs for the artificial heads are given in Fig. 5.4. It can be seen that the human HRTFs are well grouped, though with higher variations the higher the frequency is. It can further be seen that the HRTFs of the artificial heads do generally not well represent the human HRTFs. The artificial head HRTFs deviate in several cases considerable from the 'average' or 'typical' structure, and often do not even fall within the range of humans. This is disappointing, since the general design goal for the artificial heads is to replicate humans.

The deviations in Fig. 5.4 do, however, not directly predict the ranking shown in Fig. 5.3. The Toronto head has a very human-like transfer function, but it has a relatively high percentage of median-plane errors even so. The HMS I, on the contrary, has a fairly non-human-like transfer function, but has a lower percentage of median-plane errors, at least when compared to the other artificial heads. ...Generally, both the results from the listening experiments and the analysis of the transfer functions indicate that the difference between artificial heads and humans are larger than differences between humans.'

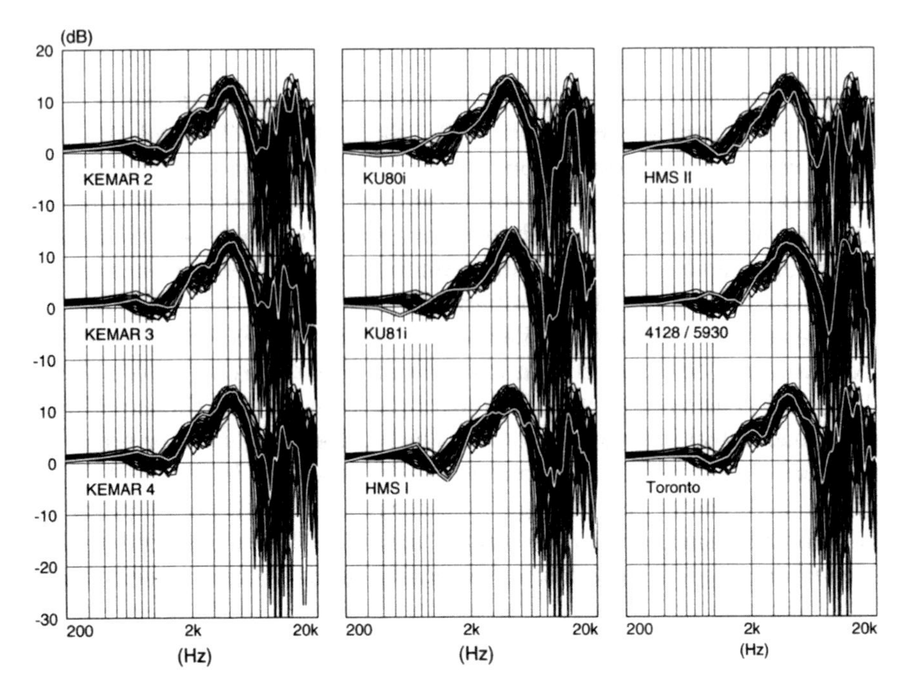

Fig. 5.4 Blocked-entrance HRTFs for frontal sound incidence for 40 humans (thin black lines) and manikins (white lines) (originally published in Hammershoi and Moller 1992; reproduced from Hammershoi and Moller 2005, Fig. 9.11)

As part of a study regarding 'Binaural Room Scanning' (Rathbone et al. 2000) 5 dummy heads were rated at the IRT (Institut für Rundfunktechnik), Munich by a group of 18 test listeners in respect to localization accuracy and sound coloration:

- HMS III (Head Acoustics) head-and-torso,
- HUGO (Institute for Technical Acoustics of the Technical University of Aachen); head-and-torso
- MANIKIN MK1 (Neutrik-Cortex Instruments); head-and-torso,
- KU 81 (Neumann); head only
- KU 100 (Neumann); head only,
- KU 100 (Neumann); head combined with the torso of the MK1

The study arrives at the conclusion that in respect to localization accuracy no statistically significant differences between the dummy heads were found. Surprisingly, basically with all dummy heads, the listening tests revealed a perceived elevation of about 7° (mean value).

In respect to sound coloration, all dummy heads (except for the KU 100) showed satisfactory results with 'tonal' sound examples (i.e. classical music, pop music, speech). When noise-like signals (applause) were used, none of the dummy heads provided satisfactory results (but also with this kind of signal, the KU 100—without torso—got the worst rating).

In this context it is interesting to note that the KU100 has received ratings similar to the other dummy head systems, when being combined with the torso of the MK1, but has received much inferior ratings when used on its own (rem.: in this case the ratings of the KU 100 were significantly worse that the ratings for the KU 81, which was also used without torso).

Even thought the KU 100 was introduced as a successor model to the KU 80 and KU 81, the visible simplification of the face- and head-geometry (see photograph) apparently has relevant negative effects on the perceived 'naturalness' of its performance (Fig. 5.5).

Fig. 5.5 Various artificial head systems as used in the study of Rathbone et al. (2000) and Hammershoi and Moller (1992) (graphic modified from Rathbone et al. 2000)

5.3 Summary and Conclusion Concerning the Studies of Artificial Heads

The research carried out at the IRT investigated less aspects than the study at Aalborg University, as no attempt was made to analyze median-plane errors or reliability in respect to distance perception.

The fact that the IRT study unveiled the presence of (minor) vertical localization errors for basically all dummy heads tested is certainly of interest, but should not present a major obstacle for regular music recordings, for example, as the musicians are usually seated on a relatively plane stage, with occasional risers for woodwind, lower strings, percussion or choir.

It seems much more relevant how well horizontal localization perception works for the listeners, utilizing the various dummy heads. In this respect, the Neumann KU80 and KU81 dummy heads turned out favorites in the Aalborg University test. The HMS I and HMS II artificial heads seem rather unsteady in their overall performance (HMS I highest 'out-of-cone' error rate, HMS II very high 'median-plane error rate' in test sections B and C).

The Toronto head turned out to have slightly above average quality, but is built only on order and is therefore less widely in use.

The Bruel & Kjaer (B&K) artificial head performed below average.

For the acoustic measurements made by the author and presented in this book, a Neumann KU81 has been used. Then main reason for this choice was that the KU81 is characterized by diffuse-field compensation, which makes evaluation of its recordings easy when diffuse-field compensated headphones are used upon playback. Even though the KU81 does not have an accompanying torso on its own, it managed to achieve good results in comparison with the competitor models in the study of Rathbone et al. 2000. For the music recording situation in which it was mainly used, the lack of the torso may also be considered less relevant since the KU81 was positioned at 'normal' height for the recording of orchestral music in a hall, a situation in which a torso would have largely been covered by the seats of the front rows anyway.

References

Braun D, Gutzke K, Wönicker C (1996) Ein objektives Verfahren zur Beurteilung der räumlichen Abbildung von Lautsprechern. In: Proceedings to the 19. Tonmeistertagung des VDT, Karlsruhe, pp 630–641

Griesinger D (1988) Equalization and spatial equalisation of dummy head recordings for loudspeaker reproduction. Paper 2704 presented at the 85th audio engineering society convention, Los Angeles

Hammershoi D, Moller H (1992) Artificial heads for free field recording: how well do they simulate real heads? Paper H6-7 presented at 14th international conference on congress acoustics, ICA, Beijing

Hammershoi D, Moller H (2005) Binaural technique—basic methods for recording, synthesis, and reproduction. In: Blauert J (ed) Communication acoustics. Springer

Minnaar P, Olesen SK, Christensen F, Moller H (2000) Localisation with binaural recordings from artificial and human heads. J Audio Eng Soc 49:323–336

Moller H, Hammershoi D, Jensen CB, Sorensen MF (1999) Evaluation of artificial heads in listening tests. J Audio Eng Soc 47(3):83–100

Olive SE, Schuck PL, Sally SL, Bonneville ME (1994) The effects of loudspeaker placement on listener preference ratings. J Audio Eng Soc 42(9):651–669

Peters N, McAdams S, Braasch J (2007) Evaluating off-center sound degradation in surround loudspeaker setups for various multichannel microphone techniques. Paper 7197 presented at the 123rd audio engineering society convention, New York

Pfanzagl-Cardone E (2002) In the light of 5.1 surround: why AB-PC is superior for symphony-orchestra recording. Paper 5565 presented at the 112th audio engineering society convention, Munich

Pfanzagl-Cardone E, Höldrich R (2008) Frequency-dependent signal-correlation in surround- and stereo-microphone systems and the Blumlein-Pfanzagl-Triple (BPT). Paper 7476 presented at the 124th audio engineering society convention, Amsterdam

Rathbone B, Fruhmann M, Spikofski G, Mackensen P, Theile G (2000) Untersuchungen zur Optimierung des BRS-Verfahrens (binaural rooms scanning). In: Proceedings to the the 21. Tonmeistertagung des VDT, Hannover, Nov 2000, pp 92–106

Toole FE (1991) Binaural record/reproduction systems and their use in psychoacoustic investigations. Paper 3179 presented at the 91st audio engineering society convention

Chapter 6
Some Thoughts on Subjective Listening Tests

Abstract The chapter opens with a research evaluation of commonly used sound attributes in listening tests. A reasoning for the selection is made along with ideas how to 'calibrate' the test listeners. Music recorded in 5.1 surround at the Salzburg Festival Hall, by use of five different microphone techniques DECCA, KFM, OCT, AB-PC and BPT, is the program material of the presented listening tests (seven-step scales, ANOVA and also analyzed in respect to acoustical parameters in Chaps. 7 and 8), the results of which are presented in this chapter. The music material (orchestral, as well as a chamber music duo) has been evaluated through loudspeaker reproduction, as well as in a binaural listening comparison via headphones. For the latter, the music has been re-recorded via a Neumann KU81 dummy head ('human reference') while being played back through a 5.1 loudspeaker setup. A short look at headphones with FEC properties ('freeair-equivalent coupling'), suited for listening tests, is taken. Finally, a correlation analysis is made for the 15 sound attributes which have been used for the listening tests. A high correlation between preference, naturalness, sound color, as well as spaciousness and source width is revealed.

Keywords Human reference · AB-PolycardioidCenterfill · Blumlein-Pfanzagl-Triple · DECCA · Kugelflächen-Mikrofon · Optimal-cardioid-triangle

6.1 Basic Considerations Concerning Listening Tests

An overview concerning the fundamentals of listening test design can be found in Bech (1999) and an extensive analysis in the excellent book "Perceptual Audio Evaluation—Theory, Method and Application" by Bech and Zacharov (2006). In this chapter, the author will try to give only a very short outline of some important aspects, which should be taken into consideration for the realization of a listening comparison.

In an article titled "New Horizons in Listening Test Design" (Staff-Writer 2004), the author gives a timely view on the state and tendencies in the area of listening test design. Independent of whether test listeners are asked to describe their impression or if they are supposed to do a rating according to a scale, in either case it is necessary

E. Pfanzagl-Cardone, *The Art and Science of Surround and Stereo Recording*,
https://doi.org/10.1007/978-3-7091-4891-4_6

to have a common vocabulary. The methodology usually applied with listening tests stems from various areas, such as psychology and quality rating in the food industry, to name just a few.

Applied methods are 'multidimensional scaling' (MDS), 'preference mapping', and—in connection to the development of attributes adequate for the purpose—'descriptive analysis' (DA) [see Bech (1999), 'repertory grid technique' (RGT) Berg and Rumsey (1999) and 'free choice profiling' Bech (1999)].

The method of multidimensional scaling offers the possibility to discover underlying structures in relatively simple data sets. The test listeners have the task to compare paired stimuli by rating their dissimilarity or individual preference for one over the other.

A study concerning the resulting difference depending on whether 'expert' or untrained listeners are used can be found in Olive (2003). It came to the conclusion that, in general, results will coincide independent of the level of training of the listeners, but expert listeners will differentiate more and give lower overall preference ratings.

The following studies are dealing with the quality of spatial sound reproduction: Berg and Rumsey (2006), Choisel and Wickelmaier (2006), Mason and Rumsey (2000) as well as Camerer and Sodl (2001), which we will look at in more detail later on.

6.2 Requirements for the Reproduction System; Consideration of Qualitative Parameters

Acoustic requirements for the listening room as well as qualitative requirements regarding the 5.1 surround loudspeaker system are analyzed in the following documents:

- ITU (1992–1994) Int. Telecommunications Union—Recommendation ITU-R BS.775-1: "Multichannel stereophonic sound system with and without accompanying picture" (Geneva)
- SMPTE (1991) Soc. of Motion Pictures and Television Engineers—Recommended Practice SMPTE RP-173: "Loudspeaker placement for audio monitoring in high definition electronic production" (SMPTE N 15.04/152-300B)
- EBU (2000) Recommendation R22: "Listening conditions for the assessment of sound programme material" (Details see EBU Tech 3276 with Supplement 1)
- ITU (1997) Int Telecommunications Union—Recommendation ITU-R BS.1116-1: "Methods for the subjective assessment for small impairments in audio systems including multichannel sound systems" (Geneva, Supplement 1 to Volume 1997)
- SSF (Surround Sound Forum) Empfehlungen für die Praxis SSF—01 (Recommendations for practical work):
  "Hörbedingungen und Wiedergabeanordnungen für Mehrkanal Stereofonie" (Listening conditions and reproduction setup for multichannel stereophony)

– SSF (Surround Sound Forum) Empfehlungen für die Praxis SSF—02 (Recommendations for practical work):
  "Mehrkanalton-Aufzeichnungen im 3/2 Format—Parameter für Programmaustausch und Archivierung, Einstellung von Wiedergabeanlagen" (Multichannel-recording in 3/2 format: parameters for program exchange, archiving and alignment of reproduction systems)

In connection with sound recording and reproduction, the signal of interest is exposed to several conversions which almost inevitably leads to a distortion of the signal in various respects:

First, the acoustic sound is converted into an analog electrical signal by means of a microphone, then—after analog/digital conversion—the signal is usually stored either in a linear (e.g. PCM 16 or 24 bit integer) or 32-bit floating point, or other high resolution format like DSD or the like (Rem.: for the research of the author a PCM 48 kHz/16bit recording format was used). Usually, for the preparation of listening examples, some kind of signal processing (level correction, audio file editing, EQing) may need to be applied in the analog or digital domain. For replay to the listener, another conversion of the signal from the electrical to the mechanical domain by use of an electroacoustic transducer (i.e. loudspeaker) is needed.

While today it is possible to produce microphones with a very flat frequency response and low total harmonic distortion (THD), most loudspeakers still exhibit THD that is much larger and lies in the range of several percent.

In respect to qualitative parameters that are related to signal distortion especially in regards to spatial impression (and therefore potentially very relevant for the results of listener evaluation), only a few examples of studies shall be mentioned:

– Gernemann A, Rösner T (1998) Die Abhängigkeit der stereophonen Lokalisation von der Qualität der Wiedergabelautsprecher. (The dependence of stereophonic localization on loudspeaker quality), proceedings of the 20th convention of the VDT, Karlsruhe, pp 828–846
– Braun D, Gutzke K, Wönicker C (1996) Ein objektives Verfahren zur Beurteilung der räumlichen Abbildung von Lautsprechern. (An objective procedure for the evaluation of spatial reproduction of loudspeakers), proceedings to the 19th convention of the VDT, Karlsruhe, pp. 630–641
– Olive SE, Schuck PL, Sally SL, Bonneville ME (1994) The Effects of Loudspeaker Placement on Listener Preference Ratings. J. Audio Eng. Soc. 42(9):651–669

In respect to psychoacoustics, there are more factors—apart from the physical position of the replay loudspeakers—which will influence listener perception (mainly in respect to localization for what concerns the papers mentioned below); among them, the spectral distribution of the signal of interest (see, e.g. Chap. 1, Figs. 1.16, 1.17, 1.21 and 1.22).

– Knothe J, Plenge G (1978) Panoramaregler mit Berücksichtigung der frequenzabhängigen Pegeldifferenzbewertung durch das Gehör. (Panorama-control with consideration of the frequency dependent level-perception by the human ear), proceedings to the 11th convention of the VDT, Berlin, pp. 136–143

– Griesinger D (2002) Stereo and Surround Panning in Practice. Paper 5564 presented at the 112th Audio Eng Soc Convention, Munich, May 2002.

6.3 Further Factors Influencing Listener Perception

In general, it can be said that a listening test should be composed in a way in order to ensure that

(a) The reliability and validity of the listener responses can be analyzed.
(b) The succession (sequence) of listening examples (audio files) has to be structured in a way, so that a biasing of the test results due to 'habitude'/conditioning effects (e.g. example B is always followed by example F) will be excluded (e.g. by randomization of the sound examples).
(c) An A/B comparison (without a reference stimulus) should therefore be structured in such a way that the potentially biased rating of example A due to a pairing with example B is evenedout over the complete test set of samples. Therefore, the exact order in which sound examples are presented is of great importance in order to ensure that such a biasing will evenout over the course of each complete set of sound examples.

As an example: for the paired comparative listening tests performed by the author and presented in this book, four surround microphone techniques (DECCA, KFM, OCT and AB-PC) had to be evaluated by the test listeners. In principle, there were three factors which should have an influence on the rating of the sound examples or microphone techniques:

1. Individual listener preference (inter-individual differences in 'taste'), as well as the individual capability of test listeners in judging the difference between (paired) sound examples, as well as the grading of various aspects, which define their overall sonic character,
2. Stimulus B, to which stimulus A should be compared (in the case of 'paired' listening comparisons),
3. The sonic differences, by which the microphone systems under test are characterized (which are the real 'objects of interest'…)

ad (1) It can be assumed that the sound examples will be rated differently by various test listeners. In order to ensure proper conditions to have a chance to obtain statistically significant results from the listening tests, it is necessary that the panel of test listeners is both large enough and made up of listeners, which are adequate for the purpose.

In the context of microphone technique evaluation, it may be of interest to analyze, if part of the listeners may be biased in their taste by individual 'preferences' (like, for example: a more direct sound, which might favor recordings that use a lot of spot microphones instead of a more purist 'one-point' recording technique that might use just one main microphone, positioned further away from the sound source and therefore will have a quite different direct-to-diffuse sound ratio, but may also provide a more pronounced impression of stage depth).

ad (2) On the occasion of A/B comparisons (without a reference stimulus), it can be assumed that the listener evaluation of sample A will be influenced by the sonic properties of sample B. Therefore, it is of utmost importance that each sample is compared (paired) with all of the other samples. Combined with a randomization regarding the order of presentation of the pairs, it should be possible to exclude the danger of biasing the results (e.g. presenting a dry sample with a rather reverberant sample may result in the first sample being rated 'very dry' in comparison with the second and vice versa). If the listening test is structured in such a way, a potential biasing of the evaluation of the each pair should 'even out' over the whole set of test data.

ad (3) as already mentioned above, the real question of interest is the sonic difference, by which microphone systems under test are characterized.

6.4 Selection of Appropriate Sound Attributes

In the context of the research carried out by the author, several research papers have been consulted, which also deal with listening tests:

In Nakayama et al. (1971), the multichannel reproduction of music surround recordings has been evaluated by ten test listeners, using preference and similarity ratings. The rating of the sound examples was done by use of 'multidimensional scaling'. As part of the preference rating, the test listeners had to rate the samples along a seven-step scale ('very good' to 'very bad').

For the similarity rating, each A/B-pair had to be rated along a seven-step scale (from 'just the same' to 'quite different').

As a result of the multidimensional analysis (MDA) of the 'similarity rating', it was possible to show that the multichannel reproduction was characterized mainly by the following three attributes:

- Fullness,
- Clearness and
- Depth of the image sources.

In the context of the study by Nakayama et al. (1971), 'fullness' was characterized by a well-balanced spectral distribution, pleasing stereophonic width, an appropriate direct/diffuse sound ratio; 'clearness' was an equivalent to 'clarity' and characterized by good localization of single sound-sources, as well as a high direct/diffuse sound ratio; 'depth of the image sources' was an equivalent to 'stage depth'.

One result of the study was that 'fullness' turned out to be directly related to listener preference. Also, it was possible to prove that 'clearness' is a function of the measurable parameter of acoustic 'definition' (German: 'Deutlichkeit').

A study concerning the sonic characteristics of stereo main microphone techniques, conducted by Ceoen (1972) with the participation of 64 test listeners (who were visitors to an AES-convention held in Cologne in 1971), used the following attributes for evaluation:

- Liveness (the spatial impression contained in a recording: 'diffuse sound field'),
- Intimacy (impression of 'directness'; acoustical parameter: 'definition'),
- Perspective (stereophonic perspective [stereo reproduction angle, distance to sound-source, etc.])
- Stage continuity (continuity in respect to localization [no stretching or skewing of the sound image in respect to L/R stage width, as well as stage depth; no 'hole in the middle' effect, etc.])
- Extra width (impression of an extra-wide stereo basis),
- Dynamic range
- Warmth (individual impression of 'warmth' in respect to sound color; high amount of low frequencies, lack of high frequencies),
- Brilliance (the equivalent of 'warmth' for the high-frequency band, usually achieved by having enough 'presence' in the sound).

A study which was part of the MEDUSA research project ("Multichannel Enhancement of Domestic User Stereo Applications", EUREKA 1652) funded by the European Union (see Rumsey 2000) dealt with a qualitative ranking of "Virtual Home Theater Algorithms" (see Mason and Rumsey 2000) and used the following attributes:

- source width, depth, envelopment and naturalness.

In addition, individual listener preference was measured.

Further attributes, which are commonly used in studies of the MEDUSA project are 'source focus' (related to localization accuracy) and 'room impression' (spatial impression).

The—by far—biggest differences in terms of listener evaluation occurred with the attribute 'naturalness', which also had the highest correlation to listener preference.

A research on five different surround microphone systems (Hildebrandt and Braun 2000) used A/B comparisons with 59 test listeners, who had to rate along a five-step scale. The sound examples had been taken from orchestral and string quartet recordings, which had to be graded in respect to the following parameters:

- Spatial impression (envelopment),
- Localization and
- Sound color (regarding naturalness, as well as the 'overall sound')

In a listening comparison carried out at the National Austrian Radio Broadcaster ORF ("Österreichischer Rundfunk") in May 2001 regarding 7 different 5.1 surround techniques, 18 test listeners had to rate the following attributes against a five-step scale (see Camerer and Sodl 2001; Rem.: later-on a second panel of 14 test listeners at the Fachhochschule Düsseldorf evaluated the same samples):

- Expansion of the orchestra image: wide/narrow,
- Distance of the orchestra image: close/distant,

- Depth of the orchestra image: deep/flat,
- Stability of the image: stable/instable,
- Precision of the image: precise/blurred
- Sound color: satisfactory/unsatisfactory,
- Room impression: perfect/imperfect,
- Presence of room information: too much/too little,
- Surround signals: identifiable/not identifiable.

The results of these two listening tests (ORF and FH-D) were printed separately on basis of the mean values and 95% confidence interval (see Chap. 9, Sect. 9.2.1 for more details).

For his own research (see Sect. 6.5 and following sections), the author of this book has decided to use the same attributes which had been used for the ORF test, as they seem very well suited for the purpose of 5.1 surround microphone system evaluation. With the last question 'surround signals: identifiable/not identifiable', it seemed a bit unclear whether the possibility of identification of the surround signals by the test listeners was to be seen as positive or negative. In addition, the research of the author was focused on the aspect of 'spatial impression' in surround sound reproduction; therefore, it was decided to split this question in two parts in order to enable a more in-depth analysis of the listener impression regarding the rear-channel signals.

Usually, these signals should simply enhance the 'stereophonic' (or spatial) impression for the listener, and by doing so, adding to the perceived 'naturalness' in the reproduction of the sound event, without drawing too much attention from the listener (which can easily happen due to inappropriate signal content or wrong—usually too high—level of the rear signals).

In addition to the attributes of the ORF-listening test, the author decided that also the overall perceived 'naturalness' of a recording should be of interest, as well as the listener preference in respect to the recordings (representing the various microphone techniques) under test (Rem.: this choice was made, as these two attributes can also be found among the attributes of the MEDUSA project).

In this way, it should also be possible to get an answer to the question of whether the 'most natural' sounding recording is also the recording of 'preference' for most listeners.

This should enable one also to see if there is any biasing among the test listeners in respect to a certain 'recording aesthetic', meaning that some of the listeners could already be very much used to the sound aesthetics of (industry standard) multimicrophone or spot microphone techniques and may have learned to rate them higher than any other technique.

For the double-blind listening test carried out by the author—taking into account various considerations above—the following attributes have been selected for listener evaluation:

1. Difference between the listening examples (i.e. microphone systems)

This is a 'true' AB comparison in the sense that you need to have sample B available for your rating, while for the attributes below it would be possible to rate each one individually on the scale without the need of a B stimulus.

2. Subjective preference (like very much–do not like at all):
   This rating has been put to the beginning of the questionnaire on purpose, as the test listeners were supposed to give a very spontaneous, immediate reply which may not have been achievable, if they had already dealt with the ratings for all the other parameters that follow.
3. Naturalness (very natural–very unnatural):
   Also this rating was put among the first questions in order to have a rather spontaneous answer.
4. Sound color (detail: naturalness):
   First, the naturalness of the 'overall sound color' was asked for.
   (semantic opposites: natural–unnatural)
5. Sound color (detail: balance of highs/lows):
   The listeners were asked to rate whether they felt that the recording was in tendency 'too bright', 'too dull' or 'well balanced' (this was a rating in a mere quantitative—not qualitative—sense, as the brightness [or—on the opposite—warmth] of a recording could qualitatively be rated very different by various test listeners).
6. Localization (very good–not good):
   The question asked to the test listeners was how well they were able to distinguish the positioning of the individual instruments or instrumental groups within the orchestra. To facilitate orientation for the listener, the order of the instruments or groups along the L/R stereo base was listed.
7. Balance (more–less):
   This question referred to volume balance between the instruments or instrumental groups.

According to the main topic of interest of the author's research, several questions and attributes followed, which regard aspects of spaciousness and spatial impression:

8. Spatial impression in respect to the orchestra (close–far):
   'Impression of distance', i.e. how close or far do you perceive the orchestra as a whole to be?
   In the layout of the questionnaire 'close' was put on the left (usually 'positive') side, while 'far' was labeled on the right side. This assignment (and therefore relative association with 'positive' and 'negative') was also chosen because in the field of sound engineering, as a recording in which diffuse sound dominates, would be rated as 'unprofessional' by most.
9. Spatial impression in respect to the orchestra (deep–flat):
   'Stage depth': the equivalent of 'stereophonic base with' (or ASW—"apparent source width") along the x-axis, is 'stage depth' along the y-axis of a three-dimensional sonic image.
10. Spatial impression in respect to the orchestra (stable–instable):
   'Stability' of the spatial impression had to be rated here.

11. Spatial impression in respect to the orchestra (precise–blurred):
    'differentiation' (in the sonic image): in the case of an abundance of diffuse sound being picked up by the microphone systems under test it could be assumed that the recording will be perceived as 'blurred' or 'unclear' by the listener. This should then have a negative impact on the localization rating (or impression of 'located-ness' of the orchestra) given by the listener.
12. Spatial impression in respect to the room (convincing–not convincing):
    The question was, if the microphone technique under test managed to convey a convincing spatial impression of the recording room to the listener.
13. Spatial impression in respect to the room (too much room/too reverberant–too little room/too dry):
    Essentially, the listeners had to rate the direct/diffuse sound ratio. In combination with the previous question, this question should also serve to better evaluate the way and quality with which a surround microphone technique under test was able to capture and reproduce diffuse sound in a room or given recording situation. The rating of this parameter will most certainly be closely related (i.e. highly statistically correlated) to the impression of 'distance' from the sound source (orchestra).
14. Rear-channel signals (integrate seamlessly–stand out):
    'sonic character': this question serves to determine how well the rear-channel signal information integrates into the 'overall sonic picture'.
15. Rear-channel signals (too soft–too loud):
    Together with the previous question, it can be determined from the results of the listener ratings whether it is the rear-channel signal-*content* (i.e. sonic nature or character of the signal) and/or the signal-*level* which might be responsible, in case the listener rates the signal to 'stand out' in an negative way.

In addition to the ratings of the 14 aspects (along a seven-step scale) explained above, the test listeners also had the opportunity to add a short comment or explanation on the lower part of the questionnaire.

For completeness, it should be mentioned that no questions regarding 'envelopment' had been included in the questionnaire, because this aspect was not of main interest for the study.

Levels of the rear-channel signals were either set according to the description or recommendation of the inventors of the respective microphone techniques, or at a 'non-intrusive' level (i.e. they were balanced-in at such a low level that their presence would be noticed only, if these channels were muted). This was done also in order to prevent the perception of 'intrusive rear-channel signals', mainly due to too much level. Intrusive-rear channel signals (i.e. non-proper in terms of character/signal content) should therefore be singled out more easily, as they would probably also become apparent at lower levels.

6.5 A Few Thoughts on 'Calibrating' Test Listeners and ANOVA

Calibration in respect to verbal attributes

In order to ensure a common understanding and agreement on the meaning of the acoustic aspects (attributes) which had to be evaluated by the test listeners, an introductory session was held in which the intended meaning of the vocabulary used in the questionnaire was explained. Nevertheless, the statistical evaluation—which was carried out on the data later-on—seems to indicate that for some (or at least *one* of the attributes) there was no clear understanding from the side of the test listeners (or a complete disagreement in opinion…) which sonic aspect was to be rated in respect to the attribute 'differentiation'.

Calibration in respect to scale range

As it was desirable for the context of the test that all listeners would make use of the full range of a seven-step scale in their ratings, it was necessary to let them have an impression of the gamma of sonic differences between the sound examples beforehand. For this purpose, an introductory 'calibration-audiofile' was replayed for the listeners before they started with the actual listening session. The calibration file contained excerpts of 20 s duration each, of the same passage from the four different microphone techniques, which were presented one after the other. The calibration file was played for each test listener at least two times or more often, if requested.

ANOVA

For statistical evaluation, an analysis of variance (ANOVA) was carried out on the data set obtained through the listening tests. In order to obtain valid results from the ANOVA, it must be ensured that the following four criteria are met:

1. independence of observations (i.e. the ratings obtained from the test listeners must be independent from each other),
2. homogeneity of variance across tested combinations of independent variables,
3. normal distribution of repeated observations and
4. detection of outliers.

A more detailed analysis on these four basic requirements (and on statistic evaluation methods in general) can be found—among others—in Bech and Zacharov (2006).

6.6 Recording of Program Material for 5.1 Surround at the Salzburg Festival Hall

Taking up the proposal from a previous publication of the author (Pfanzagl-Cardone 2002), recordings were undertaken using the following five surround microphone techniques:

- OCT surround (Optimal Cardioid Triangle, as proposed by Günther Theile of IRT, Munich with two parallel spaced cardioids pointing backward for rear signal pickup,
- a DECCA-triangle style arrangement with three omni-directional capsules, spaced according to psychoacoustic principles, utilizing interaural time differences for localization (Hermann et al. 1998); rear: two cardioid microphones at a distance to the front and separated by approx. 10 m (same as with the AB-PC system below),
- The KFM 360 (Kugelflächenmikrofon, i.e. "sphere microphone", invented by Theile (1986) and built by company Schoeps) with two figure-of-eight microphones at the sides to enable de-matrixing of the sound to 5.1 surround with an appropriate processor (DSP-4) (see Langen 2000) (see Figs. 4.54 and 4.55),
- The SOUNDFIELD microphone with the appropriate processor SP451, based on the invention by Gerzon (1973); (see Figs. 4.1–4.3),
- The AB-PC (AB-Polycardioid Centerfill) System as proposed by the author (Pfanzagl-Cardone 2002): with a spacing of the large AB omni-directional capsules of 12 m, and the use of a so-called ORTF-Triple (abbr.: ORTF-T) as centerfill system. For the rear channels, the signals of two cardioid microphones—pointing toward the rear wall of the concert hall—were used at a distance of slightly less than 10 m from the front microphones and a capsule separation larger than the critical distance of the hall (Rem.: with a more than 10 m capsule spacing this is effectively almost double the critical distance or reverberation radius, which is 5.5 m at the recording venue) (mic-placement as in Fig. 4.9, replace BPT with the ORTF-T centerfill microphone).

As the KFM 360 with its 18 cm diameter provides a recording angle of 120° for stereo (based on to the manufacturer's specifications), the recording angles of the other systems were set accordingly, if possible (i.e. via capsule spacing for the OCT and DECCA Systems, and by means of the SP451 processor for the Soundfield microphone) (Fig. 6.1).

All microphones were suspended from the ceiling of the concert hall, between the second and third row of seats at a height of approx. 3.5 m (12 ft) above stage level (except for the SoundField microphone, which was moved approx. 2 m (7 ft) further out in the hall, due to its tendency to suppress diffuse-field components and—therefore—sound much drier) (Fig. 6.2).

In order to enable an objective comparison of the microphone systems under test and not get influenced by manufacturer-specific sonic differences, only microphones of one manufacturer (Schoeps) were used (Rem.: this was possible, as the recordings of the SoundField microphone were excluded from the listening tests, which was due to time constraints, and also because this system had already scored low with previous listening tests (see Hermann et al. 1998; Camerer et al. 2001) this decision was taken).

Along with the surround microphone systems, the signals of a Neumann KU81 artificial head were recorded. This dummy head was placed at an appropriate position (in respect to the direct/diffuse sound ratio) in the hall (sixth row, floor area), a position for which at least the instruments of the first rows of musicians were still inside the

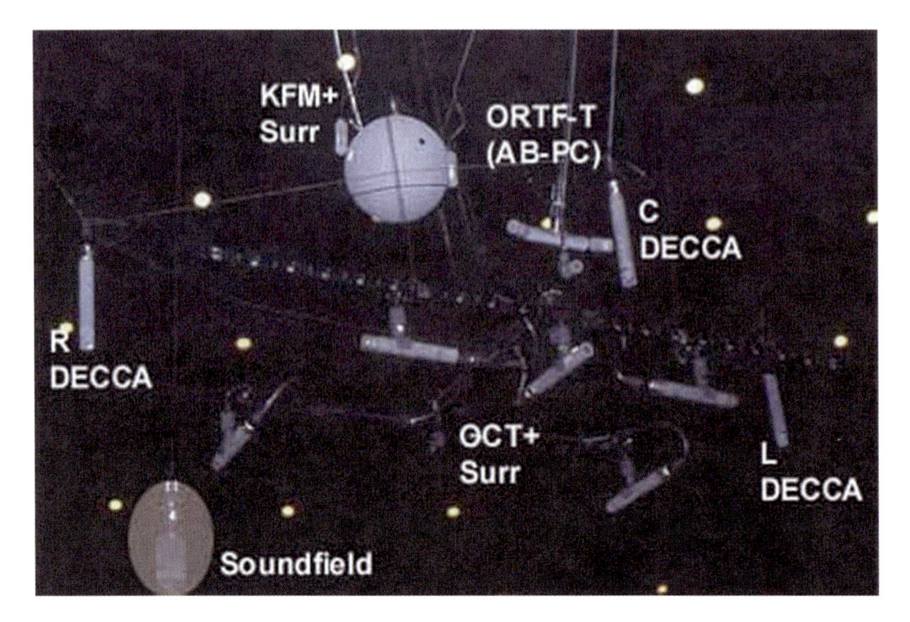

Fig. 6.1 Five surround microphone systems under test at the Salzburg Festival Hall (*photo* Pfanzagl-Cardone)

Fig. 6.2 Suspended surround microphone systems in the "Grand Hall" (for the test: hall 1) of the Salzburg Festival concert and opera-house (to the far right: iron curtain closed)

critical distance (or reverberation radius, which equals approx. 5.5 m or 16.5 ft at mid-frequencies at the Grosse Festspielhaus, Salzburg) (Fig. 6.3 and Table 6.1).

Fig. 6.3 Neumann KU81i artificial human head placed in the sixth row at the Salzburg Festival hall for binaural recording

Table 6.1 Reverb time and corresponding reverberation radius ('critical distance') measured for several octave bands at 'Grosses Festspielhaus Salzburg'

Octave band (Hz)	RT_{60} (s)	Crit. distance (m)
8000	–	–
4000	1.54	5.72
2000	1.73	5.40
1000	1.75	5.36
500	1.85	5.22
250	1.88	5.18
125	1.78	5.32
63	1.53	5.74

Rem. RT_{60} values are originally RT 5–35 ms measurements from the study of Takenaka Research & Development Institute, Japan: 'acoustic measurements of Grosses Festspielhaus, Salzburg (1996)'

6.6.1 *Program Material and Venue Acoustics for the Orchestral Recording Listening Test*

The program material recorded was a symphonic orchestra piece by Lutoslavsky. The 'Grosses Festspielhaus Salzburg' hall holds roughly 2250 seats, a volume of approximately 15,500 m^3 and is characterized by an $RT_{60} = 1.75$ s (at 1 kHz, unoccupied), corresponding to a reverberation radius of 5.4 m (see (Table 6.1)).

The acoustic quality of this concert hall has been classified B+ (i.e. good to excellent) in an international study (see Hidaka et al. 1995).

6.6.2 *Program Material and Venue Acoustics for the Duo Recording Listening Test*

The recording of the duo cello–piano (romantic period) took place in a rehearsal stage with the following dimensions: 20.5 m × 11.35 m × 7.8 m, with an approximate volume of 1815 m^3 (Figs. 6.4 and 6.5; Table 6.2).

6.7 Subjective Listening Test with 5.1 Surround Loudspeaker Reproduction

As already described in more detail in the sections above, the double-blind listening test (see Pfanzagl-Cardone and Höldrich 2008) was set up as a randomized sequence of A/B comparisons of the four surround microphone systems under test. According to the number of microphones systems to be rated and including a few special cases of

Fig. 6.4 Duo recording piano and cello at a rehearsal stage of the Salzburg Festival

A/B comparisons which made it possible to evaluate the reliability of any individual test listener (redundant pair, identical stimulus pair, etc.), this resulted in 10 A/B pairings of surround samples to be judged, with 15 attributes (+one similarity rating).

The group of test listeners consisted mainly of sound engineering students of the IEM (Institute of Electronic Music and Acoustics at the KUG—University of Music and Performing Arts, Graz, Austria), as well as professional sound-engineers and one loudspeaker-designer. The disadvantage with such a non-trained, non-expert panel of test subjects may be that the results are characterized by a higher random error than with expert listeners.

On the other hand, the results obtained with such a group are likely to be closer to the results of average consumers (or 'naive listeners') and is therefore probably more similar or applicable to the group of potential listeners (i.e. the buying public) at large.

In this context, it is also interesting to note that—as already mentioned above—a study concerning the resulting difference depending on whether 'expert' or untrained listeners are used, can be found in Olive (2003). Olive came to the conclusion that in general results will coincide independent of the training of the listeners, but expert listeners will differentiate more and give lower overall preference ratings.

Even though the current test does not stand out in terms of a very high number of test listeners—which would of course be desirable—it never-the-less has a total number of 50 participants, split into three groups for an orchestra, duo recording (both presented in 5.1 surround), as well as a 'binaural' listening group, which evaluated only the orchestra recording.

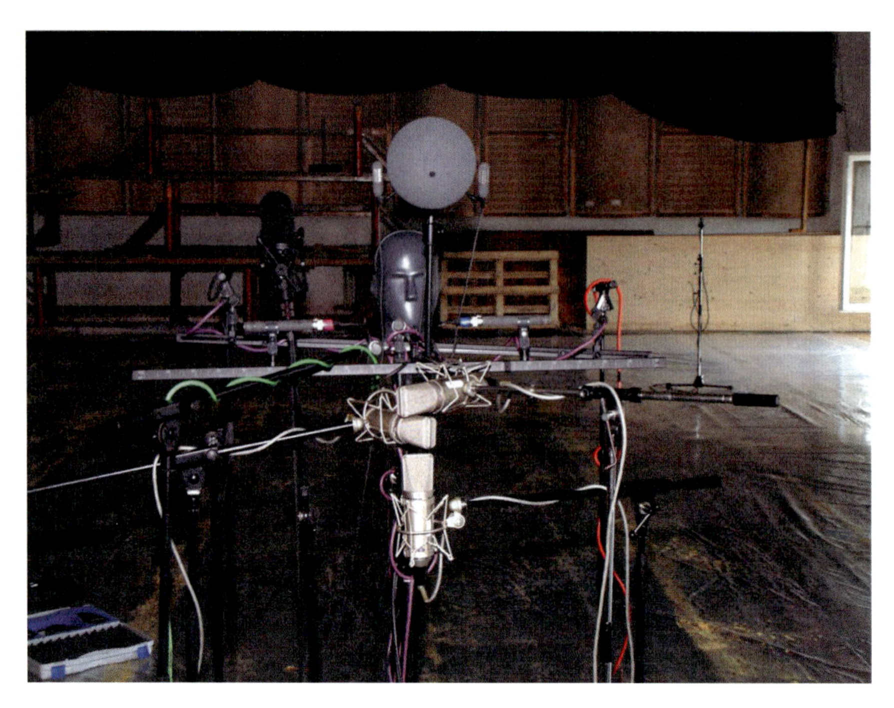

Fig. 6.5 Surround microphone systems (BPT, DECCA, KFM, OCT and soundfield) as well as dummy head (Neumann KU81i) setup at rehearsal stage 'room 447' (Salzburg Festival building) for duo recording of cello and piano (rem.: the LS-rear microphone, as used with the BPT and DECCA techniques, is seen on a microphone stand at the rear of the room)

Table 6.2 Reverb time and corresponding reverberation radius ('critical distance') measured for several octave bands at the rehearsal stage 447 at 'Grosses Festspielhaus Salzburg'

Octave band (Hz)	RT_{60} (s)	Crit. distance (m)
8000	0.58	3.19
4000	1.03	2.39
2000	1.43	2.03
1000	1.54	1.96
500	1.34	2.10
250	1.17	2.25
125	1.22	2.20
63	1.98	1.73

Rem. measurement of RT_{60} values performed with Neutrik 'Acoustilizer' (class 2 instrument, IEC 60651)

What makes this test special is the fact that the same recordings are analyzed through acoustic measurements, *and* in addition these results are compared with the outcome of the subjective listening tests. For this purpose, an artificial human head (Neumann KU81i) has been introduced as a kind of 'human reference'. More details can be found in this and the following chapters.

6.7.1 Listening Test ORCHestral Recording in 5.1 Surround

Twenty-five listeners rated the four surround microphones techniques under test in the form of a double-blind A/B comparison. After an initial 'listener calibration intro,' they were allowed to switch between the A and B sample at any time. The two 80 s surround samples of the respective microphone techniques were looping simultaneously to enable instantaneous switching between A and B without losing the musical context. The employed loudspeakers were two-way systems, which can be found quite frequently in classical music production studios (Genelec 1032A), with a frequency range from approx. 42 Hz–21 kHz, according to the manufacturer's specification.

6.7.2 Listening Test DUO Recording in 5.1 Surround

A group of 15 listeners had to rate the following four surround microphone techniques which were presented to them via surround loudspeaker replay: OCT surround, DECCA + surround, KFM and BPT + surround (see the chapter on surround microphone techniques for a more detailed description of these microphone systems).

During this test—both for the DECCA, as well as the BPT mic system—for the capturing of rear signals two parallel cardioid capsules facing the back wall have been used, at a distance of several meters from the front arrangement and with a capsule separation larger than the reverberation radius.

In the rating of the duo recording, the attribute 'localization' also became a quantitative (not qualitative) judgment, as the listeners had been asked to mark *where* they localized the cello within the L-R panorama.

6.7.3 Duration of the Orchestral and DUO Listening Tests and Listening Room Acoustics

Listening times for the ORCH as well as the duo microphone A/B comparison test were about 60–90 min for the average test listener.

The acoustically treated listening room at the IEM ('Institute of Electronic Music and Acoustics' at the University of Music and Performing Arts in Graz, Austria) normally serves as a control room for surround production. This non-rectangular room has the following properties: Approx. dimensions: 7.5 m (w) × 6.5 m (d) × 3.8 m (h) with an RT_{60} of approx. 0.25 s and conforms to ITU-R BS.116-1 down to 90 Hz.

6.8 Subjective Listening Test (Binaural) with Headphone Reproduction

A different group of ten listeners rated the four microphone systems under test (DECCA surround, KFM surround, ABPC surround and OCT surround), being presented a binaural recording via headphones, which had been obtained in the following way: the surround recordings of the four microphone systems had been replayed via the surround loudspeaker setup at the control room of the IEM and re-recorded with the same Neumann KU81 dummy head which had been used for the original recording at the concert hall (Fig. 6.6).

Fig. 6.6 Re-recording of the surround microphone signals by means of a Neumann KU81i artificial human head in the sweet spot of the control room of the IEM (rear-channel loudspeakers outside of the picture; rem: clothes placed on and around the mixing desk to avoid unwanted reflections)

The listeners were then asked to rate the four surround microphones techniques through a double-blind A/B/REF comparison (REF being the original dummy head recording from the concert hall.). Similar to the surround replay situation, they were able to switch between the three microphone techniques (which were looping in a synchronized fashion) at any time. At the digital mixing desk, with which the switching was performed, the A, B and REF stimuli were clearly labeled, so the listeners knew which acoustic stimulus was the one (i.e. REF) they had to compare the two unknown techniques (A, B) to. Before the actual rating process, the listeners were exposed to the abovementioned (in this case: binaural) calibration intro. Due to the different replay method (headphone instead of loudspeaker reproduction), some of the questions related to acoustical attributes did no longer make sense: The judgments related to the rear channels only (signals non-intrusive/intrusive and signals loud/soft) were therefore excluded. Due to this reason, the number of pairings was also reduced, since two special cases were no longer applicable: The total number was reduced to 8.

One additional question however was included at the beginning of the test: The subjects were asked to make a judgment on how different the two presented microphone techniques A and B were in relation to the REF recording. Also, when giving their ratings for the 13 remaining acoustic attributes (+2 similarity ratings), the listeners were asked to also give a rating for the REF sample (where applicable). This meant that the number of necessary ratings was increased by 50% in comparison to the non-headphone listening tests. Therefore—even though the rating should have been made easier than for the group of listeners with loudspeaker reproduction—due to the presence of the reference stimulus, the rating procedure took much longer, which was probably not only attributable to the higher number of ratings, but also due to the fact that the binaural representation of the recordings via headphones—of course—differs much more from natural hearing than listening to loudspeakers and therefore requires more concentration and causes a greater effort. The average listening time was usually close to 90 min per subject.

6.8.1 Choice of Headphones for the Binaural Listening Test

The Beyerdynamic DT990 dynamic headphones with diffuse-field compensation were chosen, as they are characterized by almost FEC properties (FEC freeair equivalent coupling; see Hammershoi and Moller 2005 for more details). Due to these properties and according to research by Moller et al. (1995) for some high-quality headphones, further compensation (equalization) is not necessary, which also applies to the DT990 (Fig. 6.7).

In order to explain the FEC properties in more detail, we cite the following paragraph from Hammershoi and Moller (2005), p. 232: 'If the pressure division in the headphone situation equals the pressure division in a free-air situation, PDR (pressure-division ratio) reduces to unity. The term 'free-air-equivalent coupling', FEC, was introduced in Moller et al. (1995) to describe headphones for which this is

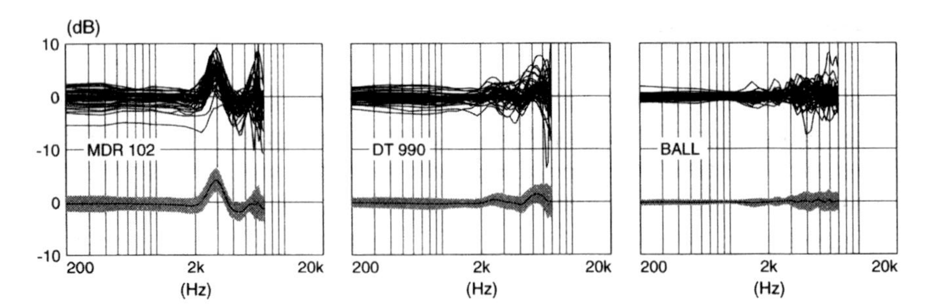

Fig. 6.7 Examples of pressure division ratios (PDRs) for the Sony MDR 102, the Beyerdynamic DT990 Professional headphones and for a freestanding ball loudspeaker. Data from 40 human subjects, individual curves (above), mean value ± one standard deviation (below). Each PDR is computed on basis of four measurements, i.e. open and blocked entrance measured in the free field and open and blocked entrance measured with the headphone as the source. The validity of the results for higher frequencies is doubtful; therefore, no results are shown for frequencies above 7 kHz (Fig. 9.6 from Hammershoi and Moller 2005)

the case. For headphones that have FEC properties, the equalization filter thus only needs to compensate for the headphone transfer function itself, even if recordings are made at the blocked ear canal. For headphones that do not have FEC properties, the equalization filter also needs to compensate for the PDR.'

The reasons for choosing the Neumann KU81i as dummy head for the research of the author have already been outlined in Chap. 5.

6.9 Analysis of the Listening Tests

6.9.1 Statistical Analysis of the Orchestral Recording '5.1 Surround' Listening Test

The data obtained through the listening test were subject to analysis of variance (ANOVA) using MATLAB software. A *p*-value of $p < 0.05$ was chosen; concerning the ratings of the 15 acoustic attributes (or variables) 13 turned out to be statistically significant ($p < 0.05$); (exceptions: 'differentiation' and 'rear channels non-/intrusive').

In Fig. 6.8, the mean values and 95% confidence intervals of the attributes are displayed. For most attributes, the differences between the microphone systems are not statistically significant (see overlapping error bars or confidence intervals). For more than half of the attributes however, at least one mic technique differs significantly from all others. In many cases, it is the KFM microphone, which is singled out: This system seems to be characterized by providing a significantly drier recording than the others, which also resulted in less spaciousness. Its stage depth, width, balance, sound color and naturalness are rated low. Not surprisingly, the mean value

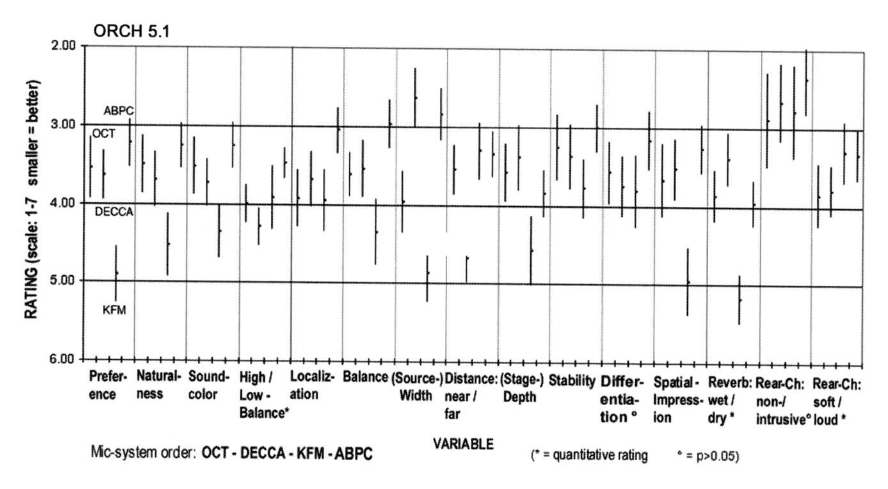

Fig. 6.8 'ORCH 5.1'—listener ratings of four surround microphone techniques (mean value and 95% confidence interval) (from Pfanzagl-Cardone and Höldrich 2008)

of preference for the KFM microphone is therefore considerably lower than for the competitive surround microphones, which coincides well with the findings of other studies (see Camerer et al. 2001; Hermann et al. 1998).

The mean values for the OCT and DECCA system are in close proximity to each other for almost all acoustic attributes (except 'source width' and 'distance near/far'). For both, preference appears to be above average (i.e. smaller than the scale middle value of 4), which coincides well with the findings of other studies (Camerer and Sodl 2001).

The AB-PC system received the best mean value ratings for two-thirds of all attributes. Consequently, it also scored best in terms of listener preference (even though not statistically significant; see overlapping error bars). The sonic picture it provides seems to give very good results in terms of localization, stability and balance. It also scored best concerning spatial impression, sound color and naturalness.

6.9.2 *Statistical Analysis of the Orchestral Recording 'binaural' Listening Test*

The data obtained through the listening test were subject to analysis of variance (ANOVA) using an appropriate function in MATLAB software. A p-value of $p < 0.05$ was chosen, and the results are displayed in the appendix in Table A2. Concerning the ratings of the 14 acoustic attributes (or variables), ten turned out to be statistically significant ($p < 0.05$); (exceptions: 'high/low balance', 'distance near/far', 'stability' and 'differentiation').

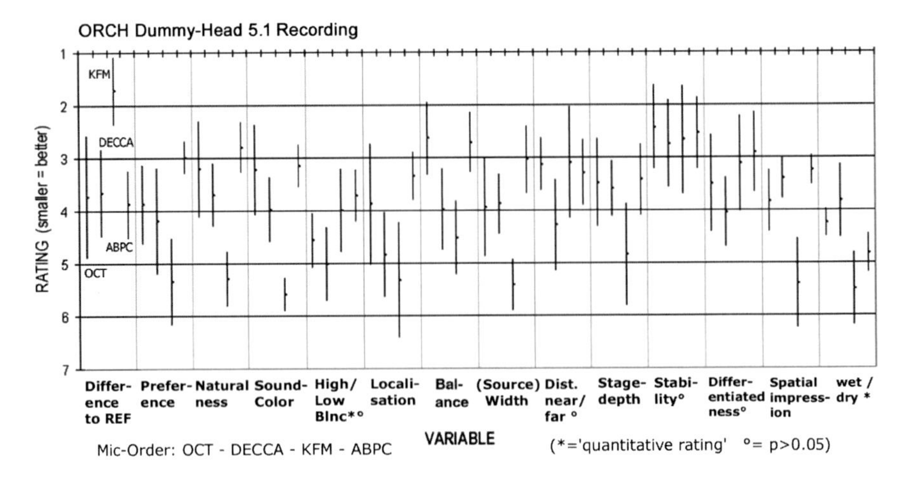

Fig. 6.9 'ORCH binaural'—listener ratings of four surround microphone techniques (mean value and 95% confidence interval) (from Pfanzagl-Cardone and Höldrich 2008)

In Fig. 6.9, the mean values and 95% confidence intervals of the attributes are displayed. For many attributes, the differences between the microphone systems are not statistically significant (overlapping error bars or confidence intervals). For more than one-third of the attributes, however, at least one microphone technique differed significantly from all others. In many cases, it was again the KFM microphone which was singled out: This system seems to be characterized by providing a significantly drier recording than most of the others, which also resulted in poor spatial impression. It sounds considerably flatter (stage depth) and narrower (source width) than its competitors and is rated inferior in terms of localization. In addition, its sound color and naturalness have been rated low.

Again, in line with all the low attribute ratings received, also the mean value of preference of the KFM microphone is, therefore, much lower than for the other surround techniques.

Concerning the "difference to REF" ratings, it should be noted that the rule "smaller = better" (as applicable to all other ratings of this research study) has to be inverted, which is due to a mistake in the questionnaire (Rem.: the adjectives assigned to the min. and max.values of the scale on the form should have been swapped). Therefore, the fact that the KFM has been rated above value 2, in fact means that it is sonically quite different from the reference (original dummy head recording).

The AB-PC technique has received the highest mean values for nine out of 12 qualitative attribute ratings. Not surprisingly, it has again received the highest preference rating.

6.9.3 *Comparison of the Analysis Results 'ORCH 5.1' and 'ORCH Binaural'*

Overall, the results of the two double-blind listening tests coincide very well: One surround microphone technique (KFM) has clearly been singled out due to low ratings. The OCT and DECCA techniques usually have received quite similar ratings (exceptions: balance, localization, distance, source width). Concerning the rating for the attribute 'source width,' it is interesting to note that in respect to the DECCA technique there is a significant difference between the ratings received upon loudspeaker playback and headphone playback. As the DECCA technique is based on the psychoacoustic principle of localization due to time-of-arrival differences (with rather small level differences between the three front-channel signals), it seems that these relatively subtle cues got lost for the listeners of the ORCH binaural test (headphone representation), as they did not have the possibility to move (or turn) their head freely in the sound field, as was the case with the ORCH 5.1 listeners. Therefore, the attributes source width and localization got both downgraded in the binaural test in comparison with the 5.1 test for the DECCA system. It is also interesting to note that under headphone reproduction, the DECCA signal is now rated to sound significantly less far away than with 5.1 loudspeaker reproduction. The ABPC technique—in both listening tests—has received the highest mean values for the majority of the qualitative attribute ratings. It has consistently scored best in terms of preference, naturalness, sound color, localization and spatial impression.

Some readers may criticize that the ORCH 5.1 listening test is not complete in the respect that only an ideal listening position in the sweet spot has been evaluated by the test listeners. Albeit this being true, the author would like to offer the proposal that maybe the ORCH binaural test, which certainly puts the test listeners in a more difficult acoustic situation—as they have to listen to the surround recordings by means of 'someone else's' ears (i.e. those of the artificial head)—and, in addition, through headphones, represents a sonic degradation which may be comparable to the disadvantages of an off-center seat position in a surround loudspeaker setup. As has been pointed out above, the DECCA technique (based on psychoacoustic small time-of-arrival differences, i.e. ITD) seems to be much more vulnerable in respect to non-ideal listening conditions than the OCT and ABPC techniques, which rely on strong level differences (and also much larger time-of-arrival differences in the case of the ABPC) in the signals of the (front-) surround channels.

This conclusion agrees with the results of other studies, which have been undertaken in the field of two-channel stereo (see Gernemann and Rösner 1998).

One could have expected that the re-recording of the surround microphone signals with the dummy head and subsequent replay over headphones to the test listeners may have resulted in a considerable sonic degradation, which could have 'blurred' the sonic impression and therefore also make differences in the ratings smaller. This does not necessarily seem to be the case, maybe also due to the fact that a reference stimulus was presented. Instead it looks as if the 'roughness' introduced due to the transformation process has 'amplified' the perceived differences between the surround techniques under test (similar to a 'grayscale' picture which has been converted into a 'black and white' image.).

The good consistency between the results of the 'ORCH 5.1' and 'ORCH binaural' listening tests supports the correctness of this assumption.

6.9.4 *Statistical Analysis of the DUO Recording '5.1 Surround' Listening Test*

In addition to the ORCHestral recording also a recording of a musical duo (piano and cello) was carried out with 3 of the abovementioned surround microphone systems: OCT, DECCA and KFM.

The AB-PC technique was not used for this recording, as its application makes sense only with large sound-sources. Instead, another surround microphone technique, which is also based on the principle of low signal correlation, but suited for the recording of small sound-sources has been used as fourth technique: the 'Blumlein-Pfanzagl-Triple' (BPT)—see both chapters on stereo, as well as surround microphone techniques for further details.

Concerning the ratings of the 15 listeners regarding the DUO recording, 13 of the 15 attributes provided statistically significant results (except: 'differentiation', 'rear-channels non-/intrusive').

Overall, it can be noted that for most attributes the confidence intervals are much larger than with the 'ORCH 5.1' listening test, independent of microphone technique. The two ratings that stick out the most are the perceived acoustic vicinity of the KFM system (see: 'distance near/far') and the high amount of diffuse sound (see: 'reverb wet/dry') picked up by the DECCA system.

Concerning localization, which has been redefined to give a quantitative (instead of qualitative) rating, confidence intervals are much smaller and the OCT system deviates the most from a 'slightly right of center' position, where the cello was physically situated.

With the DECCA recording, informal listening indicates that it is most likely the high amount of diffuse sound (which results also in a highly enveloping room acoustics) which is of disadvantage for a more precise localization.

As far as the BPT system is concerned, informal listening (as well as the listener ratings; even though not statistically significant) indicates that it provides the highest stage depth, but—as a consequence—this also results in a poorer volume balance, than with the other surround systems.

It is also interesting to note that even though the BPT system has achieved the highest mean value only at one of the (qualitative) attributes (i.e. stage depth), it has nevertheless also received the highest mean value in terms of listener preference (Fig. 6.10).

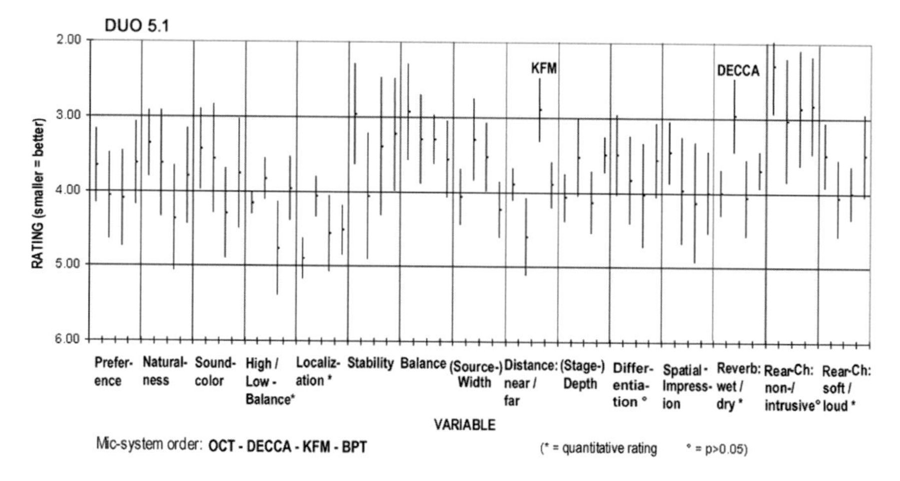

Fig. 6.10 'DUO 5.1'—listener ratings of four surround microphone techniques (mean value and 95% confidence interval) (from Pfanzagl-Cardone and Höldrich 2008)

6.10 Correlation Analysis of the Listening Test Data

The results of the 'ORCH 5.1' listening test were subject to correlation analysis, based on the Pearson product moment correlation coefficient, using MATLAB software. It seems of primary interest to note which attributes have the highest correlation to (listener-) preference: These are naturalness ($r = 0.84$), balance ($r = 0.69$), spatial impression ($r = 0.68$); sound color ($r = 0.67$) and width ($r = 0.63$).

The attributes, which have the highest correlation to naturalness, are sound color (0.75) and spaciousness (0.64). Stability, balance and localization follow with correlation values of around 0.6.

Apart from preference and naturalness, sound color is most strongly correlated with spaciousness (0.57), followed by localization and width, both with values around 0.51.

Spaciousness is (apart from preference and naturalness) strongly correlated with width (0.65). Not surprisingly, it is also quite strongly correlated with reverberation (wet/dry balance) ($r = 0.51$).

Below a listing of the six strongest correlations, as found in the test (Table 6.3).

For completeness, the matrix of all 15 aspects can be found in the appendix (see Table A3).

The author is relieved to notice that 'naturalness' has the highest correlation to preference, as this seems to indicate that the listeners still have a strong sense for (and clear idea about) what 'natural sound' is like. Together with the results of the acoustic measurements, it will become clear that 'listener preference' actually goes toward those surround microphone systems, which are also the best in approximating the binaural signal correlation characteristics (FIACC) of the original sound field in the concert hall (for details see Pfanzagl-Cardone 2012).

Table 6.3 Highest attribute correlations found for the listening test ORCH 5.1

Attribute combination	Corr.-coefficient
Preference—naturalness	0.8383
Naturalness—sound color	0.7539
Preference—balance	0.6879
Preference—spaciousness	0.6829
Preference—sound color	0.6712
Spaciousness—width	0.6522

The outcome of the correlation analysis finds its parallels in other studies (Berg and Rumsey 2001; Gabrielsson and Lindström 1985).

Among others, the correlation found between the attributes 'preference' and 'envelopment' in Berg and Rumsey (2001) coincides well with values for the correlation between 'preference' and 'spaciousness' of the current study.

Also, the high correlation between 'spaciousness' and 'naturalness' that was found matches well with the results of previous research by Toole (1985).

References

Bech S (1999) Methods for subjective evaluation of spatial characteristics of sound. Paper 16-044 presented at the audio engineering society 16th international conference on spatial sound reproduction, pp 487–504

Berg J, Rumsey F (1999) Identification of perceived spatial attributes of recordings by repertory grid technique and other methods. Paper 4929 presented at the 106th audio engineering society convention

Berg J, Rumsey F (2001) Optimization and subjective assessment of surround sound microphone arrays. Paper 5368 presented at the 110th audio engineering society convention, Amsterdam

Berg J, Rumsey F (2006) Identification of quality attributes of spatial audio by repertory grid technique. J Audio Eng Soc 54(5)

Bech S, Zacharov N (2006) Perceptual audio evaluation—theory, method and application. Wiley, Hoboken

Braun D, Gutzke K, Wönicker C (1996) Ein objektives Verfahren zur Beurteilung der räumlichen Abbildung von Lautsprechern. In: Proceedings to the 19. Tonmeistertagung des VDT, Karlsruhe, pp 630–641

Camerer F, Sodl C (2001) Classical music in radio and TV—a multichannel challenge. The IRT/ORF surround listening test. http://www.hauptmikrofon.de/stereo-3d/orf-surround-techniques. Accessed 1 Dec 2016

Camerer F, Sodl C, Wittek H (2001) Results from the Vienna listening test. http://www.hauptmikrofon.de/ORF/ORF_und_FHD.htm. Accessed 1 Dec 2016

Ceoen C (1972) Comparative stereophonic listening tests. J Audio Eng Soc 20(1)

Choisel S, Wickelmaier F (2006) Extraction of auditory features and elicitation of attributes for the assessment of multichannel reproduced sound. J Audio Eng Soc 54(9)

EBU (2000) Recommendation R22: "listening conditions for the assessment of sound programme material" (details see EBU Tech 3276 with supplement 1)

Empfehlungen für die Praxis SSF—01 (Recommendations for practical work) Hörbedingungen und Wiedergabeanordnungen für Mehrkanal Stereofonie (Listening conditions and reproduction setup for multichannel stereophony)
Empfehlungen für die Praxis SSF—02 (Recommendations for practical work) Mehrkanalton-Aufzeichnungen im 3/2 Format—Parameter für Programmaustausch und Archivierung, Einstellung von Wiedergabeanlagen (Multichannel-recording in 3/2 format: parameters for program exchange, archiving and alignment of reproduction systems)
Gabrielsson A, Lindström B (1985) Perceived sound quality of high-fidelity loudspeakers. J Audio Eng Soc 33(1/2):33
Gernemann A, Rösner T (1998) Die Abhängigkeit der stereophonen Lokalisation von der Qualität der Wiedergabelautsprecher. In: Proceedings to the 20. Tonmeistertagung des VDT, Karlsruhe, p 828
Gerzon M (1973) Periphony: with-height sound reproduction. J Audio Eng Soc 21(1)
Griesinger D (2002) Stereo and surround panning in practice. Paper 5564 presented at the 112th audio engineering society convention, Munich, May 2002
Hammershoi D, Moller H (2005) Binaural technique—basic methods for recording, synthesis, and reproduction. In: Blauert J (ed) Communication acoustics. Springer
Hermann U, Henkels V, Braun D (1998) Comparison of 5 surround microphone methods. In: Proceedings to the 20. Tonmeistertagung des VDT, Karlsruhe, p 508
Hidaka T, Beranek L, Okano T (1995) Interaural cross-correlation, lateral fraction, and low- and high-frequency sound levels as measures of acoustical quality in concert halls. J Acoust Soc Am 98(2)
Hildebrandt A, Braun D (2000) Untersuchungen zum Centerkanal im 3/2 Stereo-Format. In: Proceedings to the 21. Tonmeistertagung des VDT, p 455
ITU (1992–1994) Int telecommunications union, recommendation ITU-R BS.775-1: "Multichannel stereophonic sound system with and without accompanying picture"
ITU (1997) Int telecommunications union, recommendation ITU-R BS.1116-1: methods for the subjective assessment for small impairments in audio systems including multichannel sound systems (Geneva, Supplement 1 to Volume 1997)
Langen C (2000) Signal processing for the KFM 360. In: Proceedings to the 21. Tonmeistertagung des VDT, Hannover
Mason R, Rumsey F (2000) An assessment of the spatial performance of virtual home theatre algorithms by subjective and objective methods. Paper 5137 to the 108th audio engineering society convention, Paris, May 2000
Moller H, Hammershoi D, Jensen CB, Sorensen MF (1995) Transfer characteristics of headphones measured on human ears. J Audio Eng Soc 43:203–217
Nakayama T, Miura T, Kosaka O, Okamoto M, Shiga T (1971) Subjective assessment of multichannel reproduction. J Audio Eng Soc 19(9):744–751
Olive S (2003) Differences in performance and preference of trained versus untrained listeners in loudspeaker tests: a case study. J Audio Eng Soc 51:808–825
Olive SE, Schuck PL, Sally SL, Bonneville ME (1994) The effects of loudspeaker placement on listener preference ratings. J Audio Eng Soc 42(9):651–669
Pfanzagl-Cardone E (2002) In the light of 5.1 surround: why AB-PC is superior for symphony-orchestra recording. Paper 5565 presented at the 112th audio engineering society convention, Munich
Pfanzagl-Cardone E (2012) 'Naturalness' and Related Aspects in the Perception of Reproduced Music. In: Proceedings to the 27. Tonmeistertagung des VTD, Köln, Nov 2012
Pfanzagl-Cardone E, Höldrich R (2008) Frequency-dependent signal-correlation in surround- and stereo-microphone systems and the Blumlein-Pfanzagl-Triple (BPT). Paper 7476 presented at the 124th audio engineering society convention, Amsterdam
Rumsey F (2000) Surround sound: placing European research in an international operational context. In: Proceedings to the 21. Tonmeistertagung des VDT, Hannover

SMPTE (1991) Soc of motion picture and television engineers—recommended practice SMPTE RP-173: "loudspeaker placement for audio monitoring in high definition electronic production" (SMPTE N 15.04/152-300B)
Staff-Writer (2004) New horizons in listening test design. J Audio Eng Soc 52(1/1)
Theile G (1986) Das Kugelflächenmikrofon. In: Proceedings to the 14. Tonmeistertagung des VDT, Munich, p 277
Toole FE (1985) Subjective measurements of loudspeaker quality and listener performance. J Audio Eng Soc 33(1/2):2–32

Chapter 7
Analysis of Frequency-Dependent Signal Correlation and Coherence in Surround Microphone Systems

Abstract An enhanced form of audio signal cross-correlation analysis is applied by measuring the frequency-dependent cross-correlation (FCC) as well as the frequency-dependent inter-aural cross-correlation (FIACC) (by use of a Neumann KU81 artificial head, introduced as a 'human reference') for the surround microphone techniques AB-PC (AB-Polycardioid Centerfill), DECCA, KFM (Kugelflächen-Mikrofon), Optimal Cardioid Triangle (OCT), and SoundField (Ambisonic). In addition, also the frequency-dependent signal coherence is estimated. These analyses were performed for paired channel combinations of the surround signals, as well as for the binaural (KU81) signal. It was possible to prove that A) determining the FCCs or FIACC provides a characteristic 'fingerprint' for each surround microphone technique, and B) that the signal correlation provides more sonically relevant information, than estimating the coherence. A comparison of the acoustic measurements with the results from the subjective listening test, concerning the same surround microphone techniques (as presented in Chap. 6) is made. Furthermore the results of frequency-dependent cross-correlation of paired surround channels are presented for the BPT (Blumlein-Pfanzagl-Triple) and SoundField MK-V microphone systems, as well as Hamasaki Square, the IRT-cross, ORTF-surround, the 'Theile-Trapezoid', ESMA (Equal Segment Microphone Array) and Coincident Microphones (Double-MS). Some thoughts on the importance of a main microphone's diffuse-field correlation (DFC) pickup characteristics for the overall 'spatial impression' are shared and conclusions are drawn for related optimum microphone patterns and capsule configurations (i.e. suitable microphone systems).

Keywords Signal correlation · Signal coherence · Frequency-dependent cross-correlation (FCC) · Frequency-dependent inter-aural cross-correlation (FIACC) · Diffuse-field correlation (DFC) · Diffuse-field image (DFI)

The technically most objective way to measure how well a particular microphone technique is able to translate an acoustic event to the listener at home with the least amount of alteration may be the following:

1. make a recording of the sound event in a concert hall with the microphone technique of your choice.

E. Pfanzagl-Cardone, *The Art and Science of Surround and Stereo Recording*,
https://doi.org/10.1007/978-3-7091-4891-4_7

2. at the same time, make a second recording at the 'best seat in the house' position of that concert hall with an artificial human head (dummy head)
3. while reproducing the first recording through a stereo or 5.1 loudspeaker system, make a third recording (using the same dummy head) in the sweet spot of the listening room.
4. measure the correlation (over frequency) between the two dummy head recordings.

Of course one has to be aware that with this method signal distortion (concerning amplitude, frequency and phase) introduced by the replay system and the acoustics of the listening room will be included in the evaluation process and might bias the results.

However, the microphone technique, which produces the highest correlation (regarding the entire frequency range) between the original sound event and the re-recorded reproduction would appear to be the one with the highest fidelity.

7.1 Measurement of the Frequency-Dependent Cross-Correlation (FCC) and Frequency-Dependent Inter-aural Cross-Correlation (FIACC)

In order to be able to measure the frequency-dependent cross-correlation coefficient, it was necessary to program an appropriate tool (function) within MATLAB software. The two channel signals of interest were split into 31 independent frequency bands (with center frequencies according to ISO) which were then subject to correlation analysis.

In addition, the correlation of the low-frequency signal content (below 400 Hz; important in respect to spaciousness) and high-frequency signal content (above 1200 Hz) was measured.

For the measurements, appropriate sections of the audio material recorded for the listening tests were selected:

orchestral recording (hall 1):
sample #1, 7 s (loud),
sample #2, 16 s (soft),
sample # 3, 60 s (soft to loud).
Duo recording (hall 2):
sample #4, 16 s.

In Figs. 7.1, 7.3 and 7.4 the FIACCs of the original dummy head recordings (KU81) from the concert hall (dotted line) are compared with the dummy head re-recording of the selected surround microphone (solid line).

In Fig. 7.1, it is interesting to note that the microphone techniques which employ also omni-directional microphones (i.e. DECCA, ABPC) are both characterized by

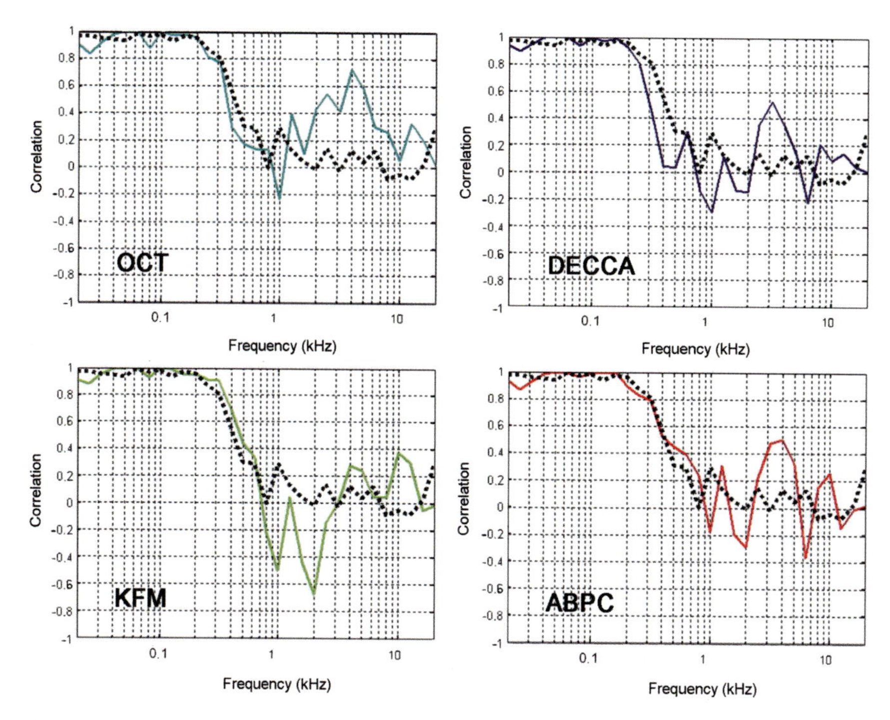

Fig. 7.1 FIACC: T. U. T ('Technique under Test', solid line) and KU81 (dotted line) for orchestral sample #1 (7s, loud, hall 1) (from Pfanzagl-Cardone and Höldrich 2008)

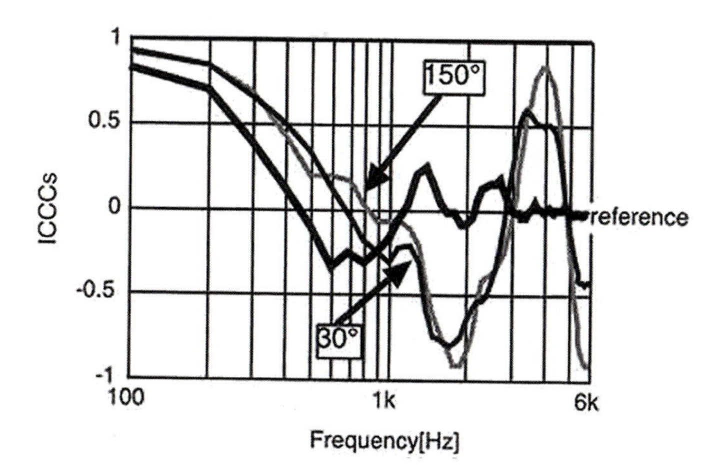

Fig. 7.2 ICCCs—inter-aural cross-correlation coefficient (=FIACCs) for symmetrical arrangement of a pair of loudspeakers at indicated angles (i.e. at ±30° and ±150°) (adapted from Hiyama et al. 2002)

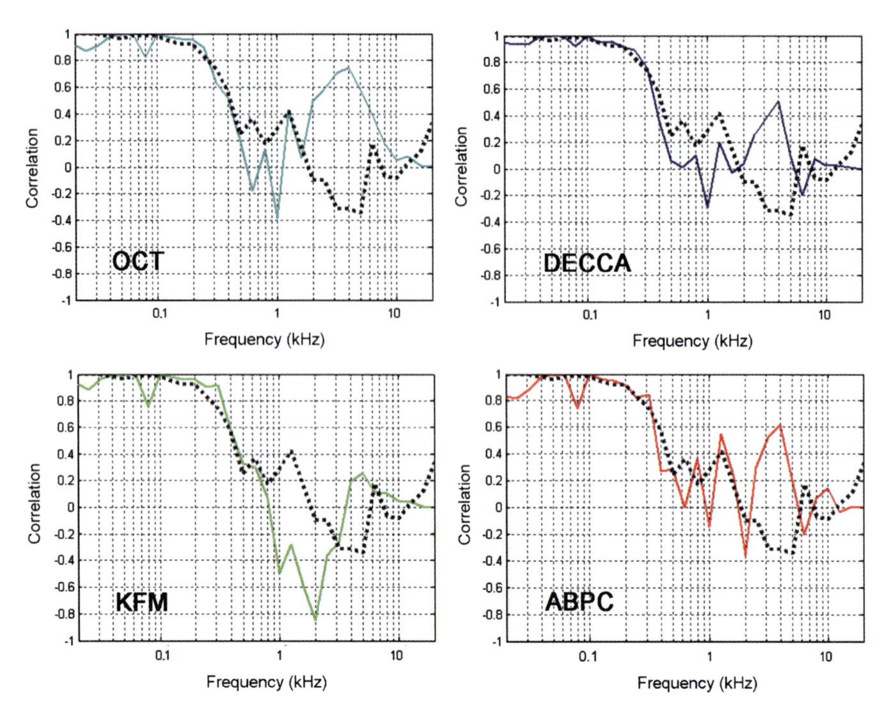

Fig. 7.3 FIACC: T. U. T ('Technique under Test', solid line) and KU81 (dotted line) for orchestra sample #2 (16 s, soft, hall 1) (from Pfanzagl-Cardone and Höldrich 2008)

moderately pronounced correlation (max. +0.5) in the frequency range around 3–4 kHz.

As was already shown in (Hiyama et al 2002), this is a simple consequence of the standard 5.1 loudspeaker layout, in interaction with the HRTFs of a dummy or human head. (see Fig. 7.2; functions for the 30° (R) and 150° (RS) loudspeakers). More evidence on this and a calculation of the resulting filter, based on a spherical head model, can be found in a more recent study (Shirley et al. 2007).

In comparison, the OCT system is characterized by an even higher correlation (max. +0.7) in the same frequency range, while the KFM displays prominent *negative* correlation (max. −0.7) in the frequency range around 2 kHz.

With the 16 s, soft orchestra sample #2, it is interesting to see that the peak of correlation at the OCT system has become more pronounced, and so has the negative correlation of the KFM system (which ranges now from 800 Hz to 3.5 kHz, with a peak of around −0.8).

In Fig. 7.4, a much longer sample of 60 s duration was analyzed leading to quite astonishing results: While correlation has increased over the entire frequency range for the OCT, DECCA and KFM system, the ABPC system is the only one which still 'meanders' around the non-correlation '0'-axis in relative proximity to the reference

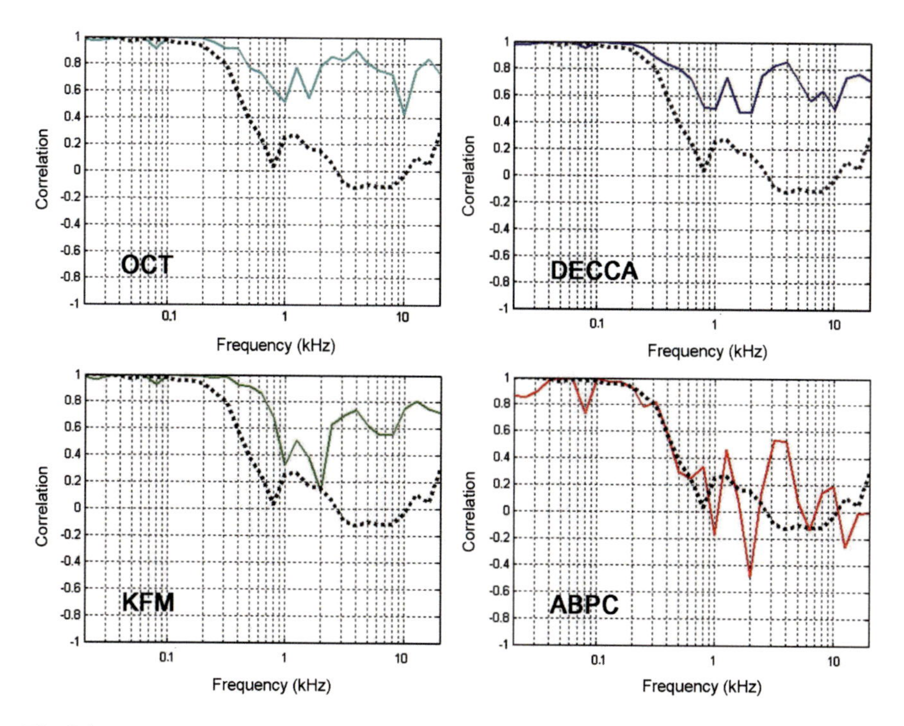

Fig. 7.4 FIACC: T. U. T ('Technique under Test', solid line); KU81 (dotted line) for orchestra sample #3 (60 s, soft to loud, hall 1) (from Pfanzagl-Cardone and Höldrich 2008)

(original KU81 dummy head recording from the concert hall, which now displays a much smoother curve, due to the higher sample length).

Also, it is very interesting to note that the ABPC system is the only one which follows the KU81-reference function almost exactly below approximately 700 Hz (which is the frequency range above which the human or dummy head starts to be acoustically effective as a baffle). As has been pointed out already in previous chapters, faithful reproduction of de-correlated signals in the frequency range below 500 Hz is highly important for spatial impression with human listeners.

According to the FIACC-measurements in Fig. 7.4, the ABPC system is the technique under test (T.U.T.), which approximates the original sound field much better than its competitors.

Table 7.1 shows the numeric values of the correlation coefficients for the four microphone systems under test and the three samples of different duration: Apart from the fact that the ABPC system is usually closest to the KU81 dummy head reference in terms of correlation, it is most interesting to note that it does so—almost always—for the LF-part as well as for the HF- part of its signal (and not only for the overALL correlation). As can be seen from the values of the 'ALL' correlation coefficients, (loud) sample #1 is characterized by much lower overall correlation than (soft) sample #2, which is most likely due to the much higher amount of diffuse

Table 7.1 Correlation coefficients for three music samples and four surround microphone techniques, as well as one dummy head (reference); (from Pfanzagl-Cardone 2011)

Correlation coefficient:	LF(400 Hz)	HF(1200 Hz)	ALL
ORCH-sample 1: (*loud*, 7 s)			
OCT	0.87	0.28	0.41
DECCA	0.85	-0.01	0.23
KFM	0.88	-0.30	0.39
AB-PC	0.89	0.15	0.48
KU81	**0.92**	**0.11**	**0.55**
ORCH-sample 2: (*soft*, 16 s)			
OCT	0.94	0.36	0.85
DECCA	0.96	0.11	0.89
KFM	0.98	-0.46	0.88
AB-PC	0.92	0.37	0.78
KU81	**0.85**	**0.24**	**0.74**
ORCH-sample 3: (soft+ *loud*, 60 s)			
OCT	0.98	0.71	0.88
DECCA	0.98	0.66	0.90
KFM	0.99	0.44	0.87
AB-PC	0.88	0.27	0.54
KU81	**0.87**	**0.20**	**0.57**

sound or reverb (stochastic signal) in the hall, due to the room being excited by the high-level music. With the ‘higher correlation’ (soft) sample #2, the ABPC system ‘follows foot’ and hence also displays higher overall correlation.

Therefore, the ABPC system is not characterized by higher or lower correlation ‘as such,’ it simply seems to manage to capture the original sound field in the hall in a physically more correct way, as can be deducted from the results.

In this context, a study by Kamekawa et al. (2007) should be mentioned, in which the temporal dependence of inter-aural signal cross-correlation has been analyzed for several surround microphone techniques. In Fig. 7.5, the IACC of a total of 8 different surround techniques is displayed for a time-span of 60 s. It can be noted that for the dummy head signals the resulting IACC can have very differing values, depending on the surround microphone technique used (compare e.g. techniques 5CH, DTO, DMS, DTH and 3OI in the time segment from 20 to 30 s).

Looking at the—at times—very high *negative* values of inter-aural signal correlation for some of the techniques (see e.g. 5CH, 3OI, DTH, DTO, DMS with high negative values during the time segment 20–30 s, while Fukada, INA and OM8 display only positive values for the same passage of music. Interestingly these latter three techniques are again the only ones—with the addition of 5CH—to exhibit negative correlation for the time segment 8—12 s, while in this case all other techniques display positive signal correlation), it becomes clear that it is of high importance to use a sample length of a certain minimal duration (e.g. 60 s) for frequency-dependent

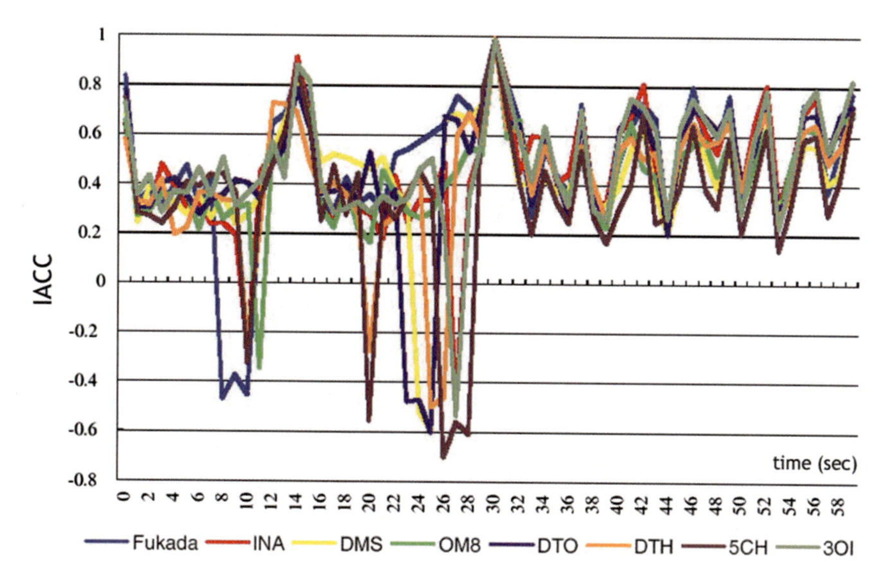

Fig. 7.5 IACC values for a 60 s excerpt of the orchestral work 'Wellington's Victory' (adapted from Kamekawa et al. 2007) (Abbreviations: *DMS* double-MS, *OM8* omni + 8, *DTO* Decca-Tree + Omni-square, *DTH* Decca-Tree + Hamsaki-square, *5CH* 5 cardioids + Hamsaki-square, *3OI* 3 Omnis + IRT-cross)

cross-correlation analysis to arrive at more objective results. In case of music samples it also makes sense to use excerpts which contain both soft (piano) and loud (forte) parts of a performance, as—depending on the volume of the sound-source—this tends to excite the room in different ways and can therefore result in different effective direct-/diffuse sound ratios at the microphones (Fig. 7.6).

Even though the surround recordings of the duo cello-piano have taken place in an acoustically very different room, the FIACC of the KFM in Fig. 7.7 again exhibits pronounced out-of-phase components in the mid- to high-frequency range (up to more than −0.8), as has been the case already with the ORCHestra recordings (compare with FIACC of KFM in Fig. 7.1).

On the other hand, the binaural signals of the other three surround microphone techniques OCT, DECCA, BPT display out-of-phase components only in very small frequency bands and at rather moderate values (with a maximum of −0.4) and in general follow the reference function of the KU81 much better.

7.2 Measurement of the Frequency-Dependent Signal Coherence

Some researchers are convinced that it is preferable to measure signal coherence instead of signal correlation (see e.g. Jacobsen and Roisain 2000 and Martin 2005, among others). The most basic differences between correlation and coherence have been outlined in Chap. 2.

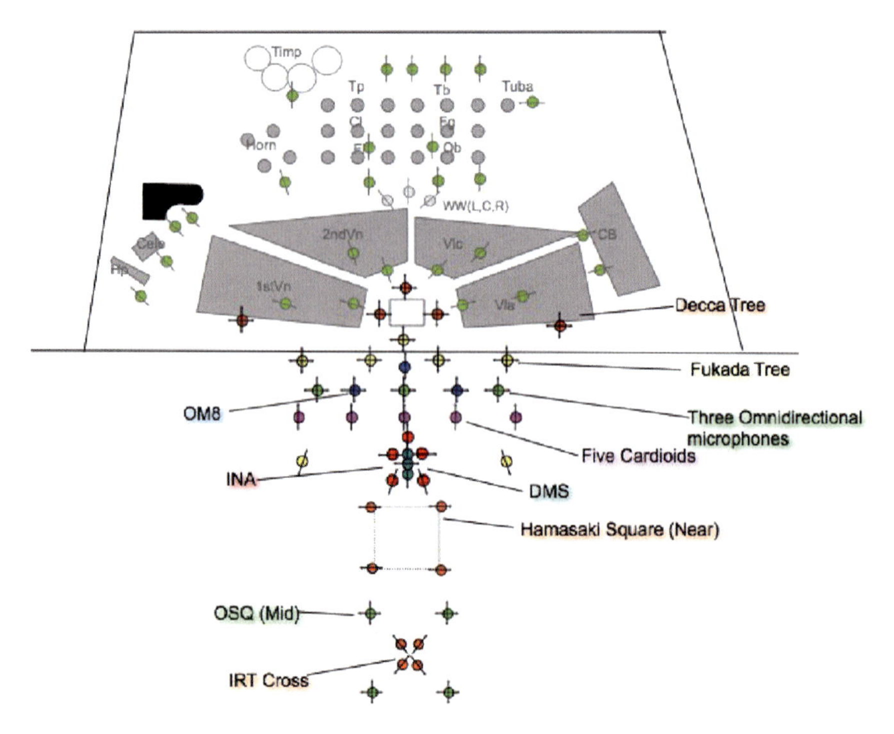

Fig. 7.6 Layout of main and ambient microphones for the experiment of Kamekawa et al. (2007) (reproduced from Kamekawa et al. 2007)

For the analysis in this research, the 'mscohere' function in MATLAB was used, which finds the magnitude squared coherence estimate of the input signals. As this signal processing algorithm is based on the discrete Fourier transform or DFT, an adequate 'computational resolution' has to be chosen in order to achieve sufficient resolution in the frequency domain. With the sample rate at 48 kHz, an appropriate 2048 'DFT-point' resolution was chosen to achieve approximately 20 Hz spacing for the frequencies at which coherence would be calculated, in order to obtain reliable results also for the lowest frequency bands.

Looking at Fig. 7.8, the graph evokes the impression of a much higher accuracy than with the (band-filter bank based) correlation algorithm. This impression is not necessarily correct, as one should bear in mind that:

1. the computation of the signal coherence (however accurate it may be) is only an *estimate*,
2. the DFT-algorithm calculates coherence at many points along the frequency range of interest, but no matter how high the resolution, there will always be way more frequencies for which the DFT-algorithm did not (or better: could not) calculate coherence, than frequencies for which it did…

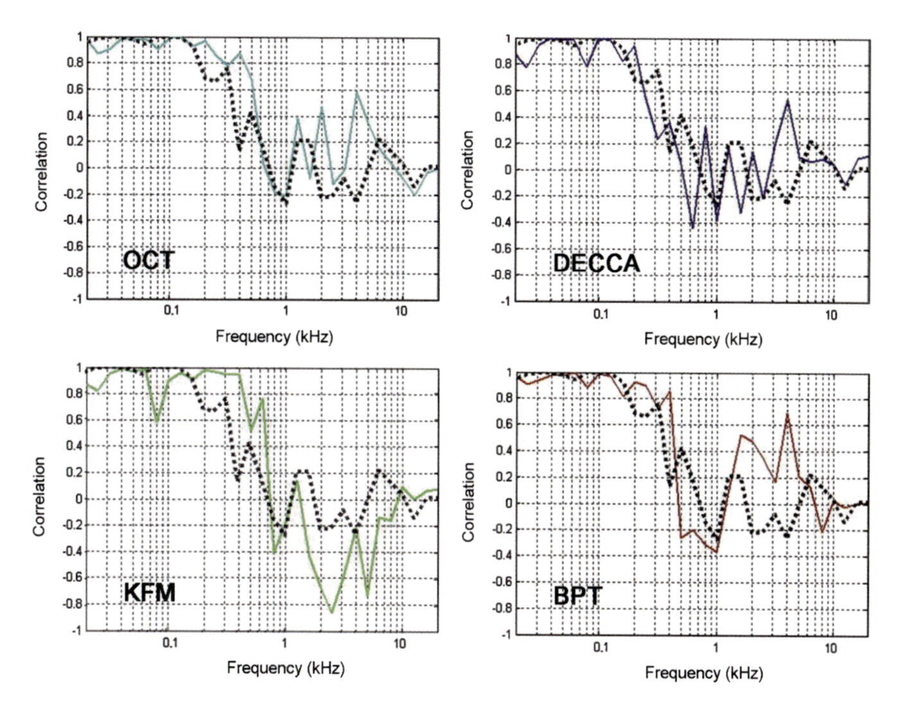

Fig. 7.7 FIACC: T. U. T ('Technique under Test', solid line); KU81 (dotted line) for duo recording; sample #4 (16 s, hall 2) (from Pfanzagl-Cardone 2011)

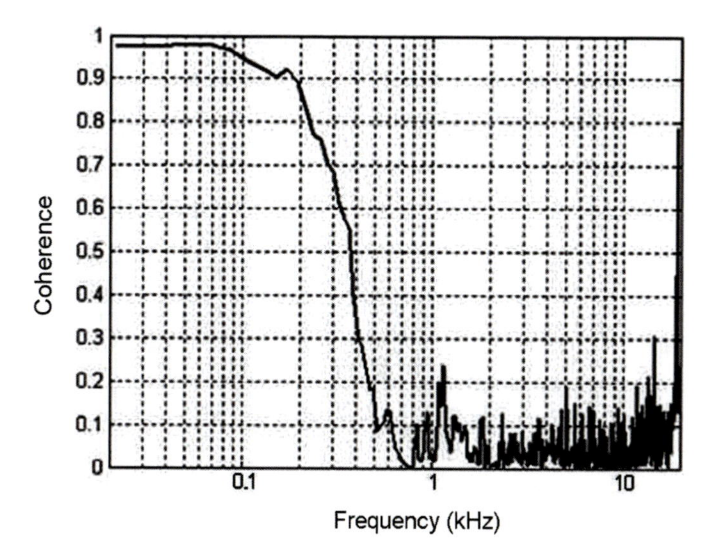

Fig. 7.8 Coherence for the 'ear signals' of the Neumann KU81 dummy head recording from the concert hall (sample 3, hall1) (from Pfanzagl-Cardone and Höldrich 2008)

Therefore, while the correlation coefficient gets *measured*, the coherence gets only *estimated*; and the sometimes quite high spikes along the coherence-'curve' within an otherwise relatively 'calm' part of the curve shows that there can be vast differences for coherence values in a rather small frequency range.

Apart from these 'system-inherent' properties of coherence estimation the main point of criticism becomes evident when taking a closer look at the results, which the coherence estimation provides for the surround microphone systems under test (Fig. 7.9):

Examining the coherence functions of the microphone systems it is evident that they are all characterized by higher coherence in the frequency band around 3–4 kHz (this is very pronounced for the OCT, DECCA and KFM system, much less pronounced for the ABPC). In addition, the KFM exhibits a relevant 'notch' in terms of coherence mainly in the frequency range of 1–2 kHz which must be related to the relevant phase changes in that frequency band which were also evident with the previously performed correlation analysis.

However, this is—in comparison—only a very 'faint echo' of what one was able to see clearly with correlation analysis before…

Also with coherence analysis, the appearance of the ABPC system is quite different from the three competitive systems, and nevertheless, it seems clear to the

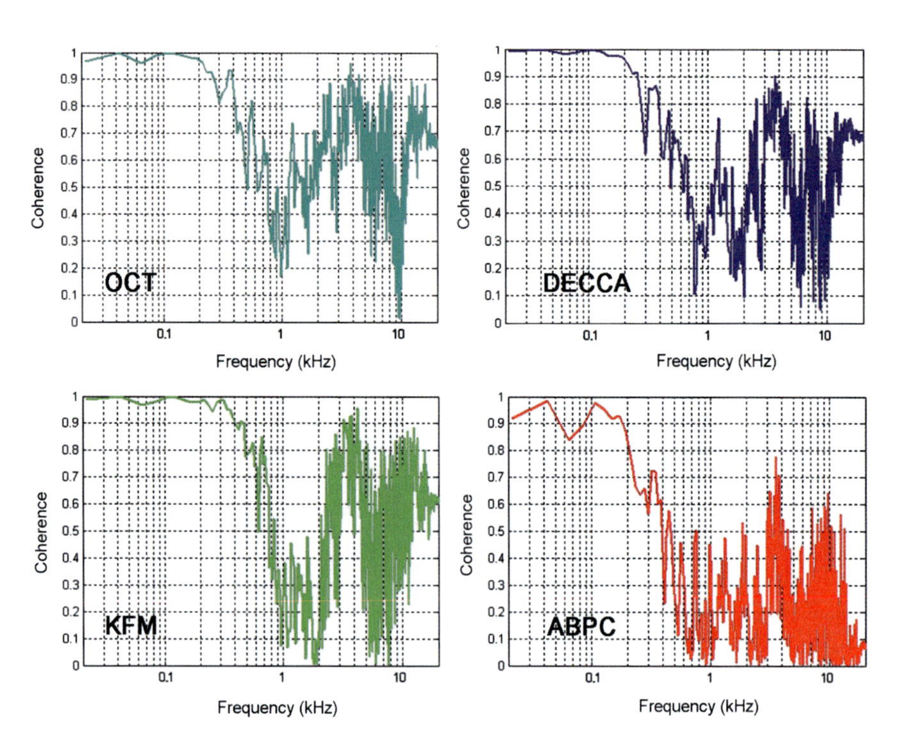

Fig. 7.9 Estimated coherence over frequency for four different surround microphone techniques (2048-point DFT, sample 3, hall 1) (from Pfanzagl-Cardone and Höldrich 2008)

author that measuring signal correlation is superior to measuring signal coherence if a detailed analysis of the T.U.T. is desired:

The strong out-of-phase signal components that were found through FIACC analysis with the KFM system in a frequency range (1–3 kHz) which is of primary importance in relation to human hearing, are most likely responsible for the bad sound color ratings, which this system has received throughout the listening tests.

The fact that these out-of-phase components have been detected by means of the cross-correlation measurement, but would not have been noticed if only the coherence estimate had been performed, is a strong indicator for the superiority of the cross-correlation measurement approach.

An in-depth view on the problem of sound coloration due to comb filtering upon loudspeaker signal summation can be found in a publication dealing with wave field synthesis (Wittek et al. 2007).

For the sake of a clearer visual representation in order to allow that the coherence function can also be compared with the corresponding correlation function, the data obtained by the coherence estimation have been grouped in 1/3rd octave bands (with center frequencies according to ISO), the result of which is displayed in Fig. 7.10. Also in this form of representation, it is evident that the curve of the ABPC technique is closest to the reference (i.e. KU81), especially in the low-frequency region which is of vital importance for good spatial impression.

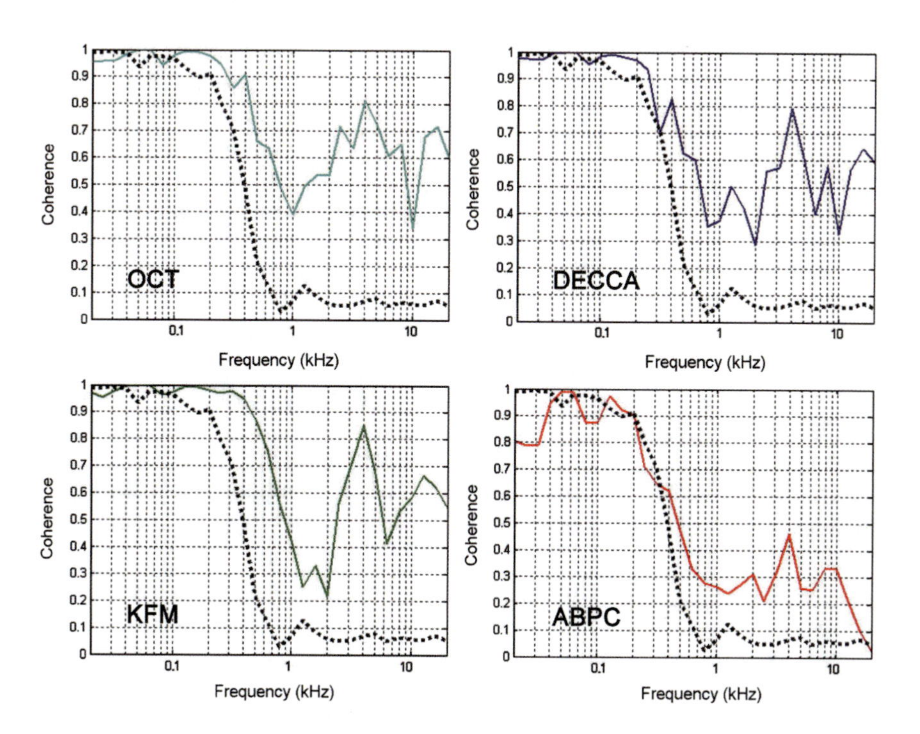

Fig. 7.10 Binaural signal coherence over frequency for four different surround microphone techniques (solid line); KU81 concert hall (dotted line); (ORCHestral recording; sample #3, 60 s, hall 1); 1/3rd-octave resolution, 31-band (from Pfanzagl-Cardone 2011)

7.3 Pair-wise Analysis of Correlation and Coherence with Surround Microphone Signals

Even though correlation and coherence analysis of the binaural dummy head signals has provided quite meaningful results which make it possible to clearly distinguish the reproduction accuracy of the surround microphone systems under test, it seems nevertheless worthwhile to also examine the signal relationships between each of the five surround channel signals in order to gain a better understanding.

Examining the OCT surround system in Figs. 7.11 and 7.12, it becomes clear that coherence for the L/C/R front channels is rather low for the entire frequency range, as can be expected from the two hyper-cardioids that are pointing side-wise. However, coherence is reasonably high between the front channels and the center channel below 500 Hz with peaks reaching values of 0.75. Interestingly, there seem to be some 'spurious' peaks in the frequency region near 10 kHz which—at certain frequencies—reach values of 0.8.

Coherence of the rear-channel pairing LS, RS is very low for most of the frequency range, but below 200 Hz there is a steep rise which totals in high coherence of 0.8.

Also, the coherence between the L (front) and LS (rear) channel is characterized by a relevant rise below 200 Hz, which means that in this frequency range signal coherence is overall significantly high for the OCT system. This high coherence in the low-frequency range is to the disadvantage of good spatial reproduction, as we already know from Morimoto and Meakawa (1988) and Hidaka et al. (1995).

The visual differences in the coherence functions for L,C and C,R as well as L,LS and R,RS—especially in the low-frequency band—could have their roots in an (non-intentional) slight asymmetry in the installed microphone layout or individual deviation of capsule characteristics from the ideal pattern.

We get a more balanced impression (which is also more easily comparable with the results of the frequency-dependent cross-correlation measurement) by 'smoothing' the visual display of the coherence function due to grouping the estimated coherence values in 1/3rd-octave bands (see Fig. 7.12).

Comparing the coherence function for the OCT system in Fig. 7.12 with the correlation function in Fig. 7.13, we arrive at similar conclusions: The L,C signals are de-correlated above 1 kHz, while the same is true for the LS, RS signals already from 300 Hz upwards. Also, neighboring channels front/rear (see e.g. L,LS) are de-correlated from 300 Hz upwards. Between 400 and 500 Hz, there is slight out-phase behavior, which arrives at a maximum negative value of about −0.2 which is in the order of the typical out-phase correlation between two omni microphones in a diffuse-sound field.

In Fig. 7.14 signal correlation over frequency for an OCT system has been calculated, based on a L,R capsules spacing of 80 cm, resulting in an claimed effective recording angle of 106° (see 'Image Assistant' graphic on the left side of Fig. 7.14), as opposed to the effective recording angle of 120° as applied for the OCT system in Figs. 7.12 and 7.13 with a L,R capsule spacing of 60 cm.

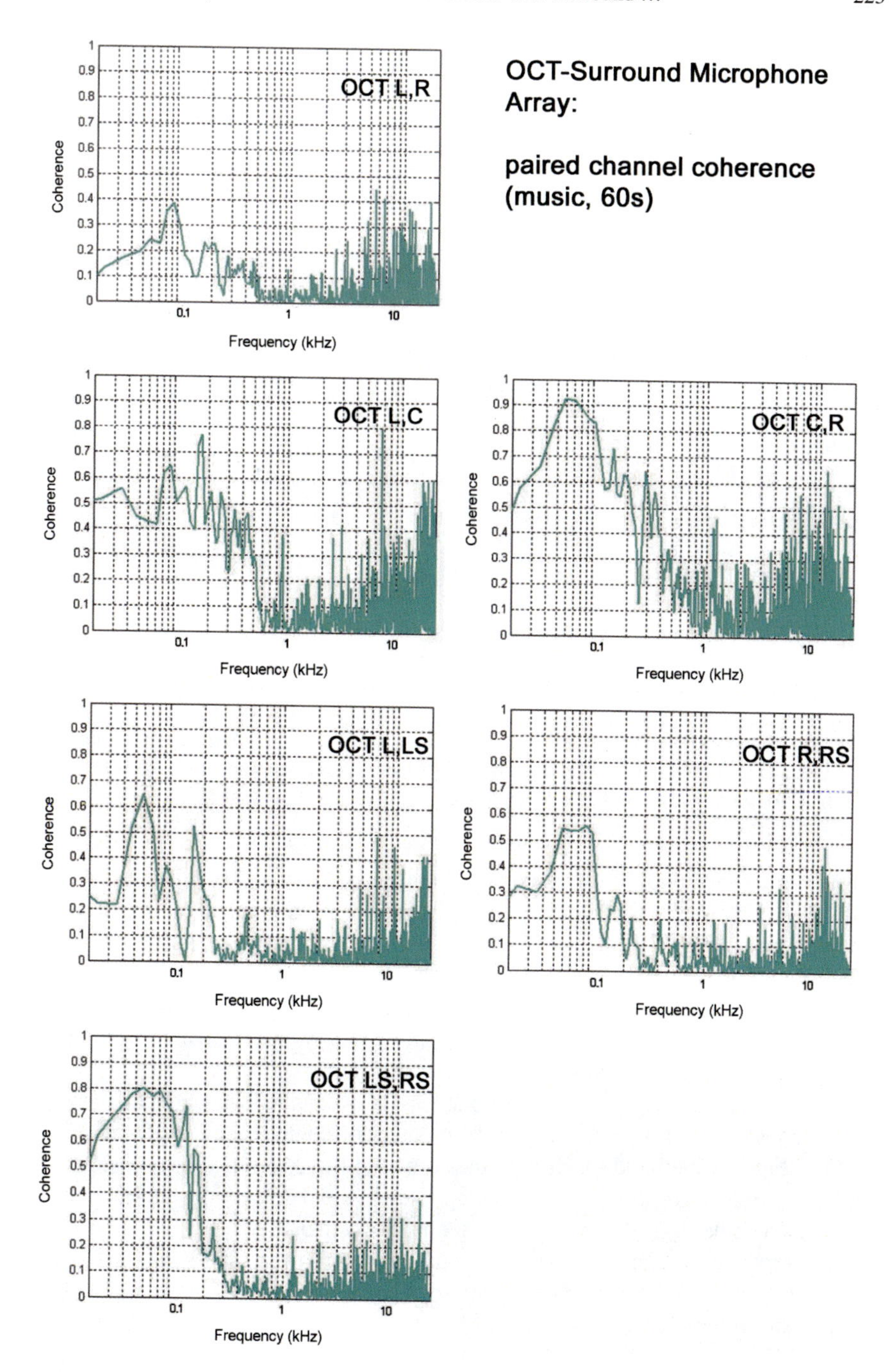

Fig. 7.11 Paired channel *coherence* for OCT surround microphone system signals (2048-point DFT; music, 60 s, hall 1) (from Pfanzagl-Cardone 2011)

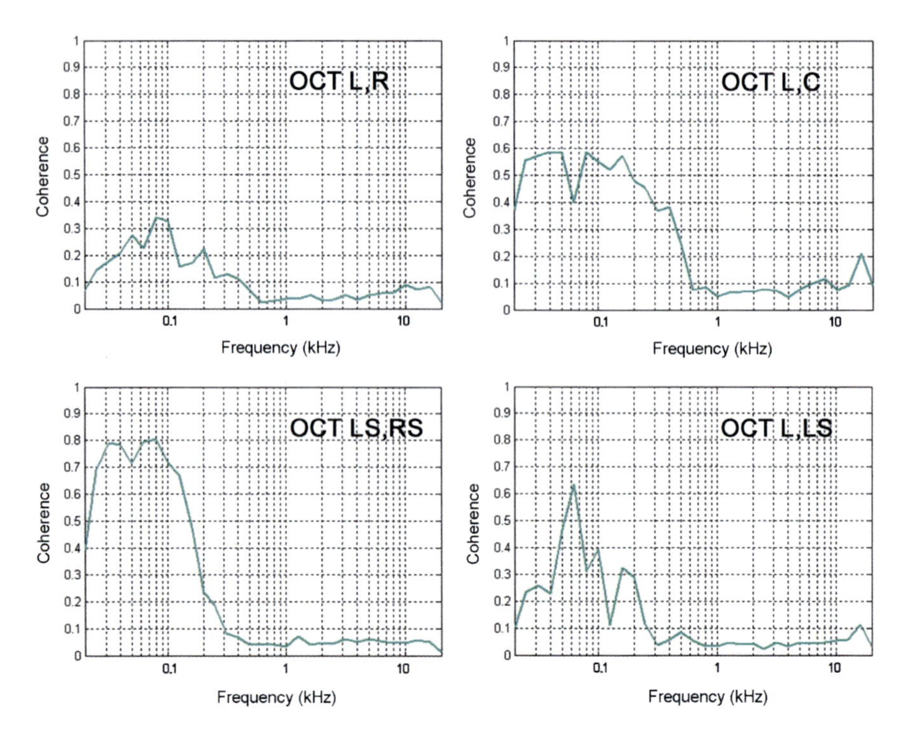

Fig. 7.12 Paired channel *coherence* for OCT surround microphone system signals (2048-point DFT grouped to 31 frequency bands, 1/3rd octave, center frequencies according to ISO; music, 60 s, hall 1) (from Pfanzagl-Cardone 2011)

As can be expected, low-frequency coherence is much higher with the DECCA system, at least between the three front channels, which employ omni-directional microphones. Again a steep rise below 200 Hz (similar capsule spacing as with the L and R capsule of the OCT system), but to even higher values of close to 0.9.

Due to the fact that the backwards-pointing cardioid microphones for the rear channels are positioned quite far away (approximately 10 m), there is almost no coherence between them, as well as to the front channels.

This enables good spatial impression at least between the front and rear channels.

The conclusions drawn from the analysis of the correlation function of the DECCA array in Fig. 7.16 coincide well with the results of the coherence measurements in Fig. 7.15: Below 200 Hz, there is a steep rise toward 1 for all signals of the front system. The other signal combinations are de-correlated.

A comparison can be made with the findings of Wittek (2012) in Fig. 7.17, a calculation which simulates L/C signal correlation for a DECCA triangle with a capsule spacing of 111 cm (roughly 4 ft).

The KFM system in Fig. 7.18, on the other hand, is characterized by much higher coherence between all channels, mainly at low frequencies, but not only: It can be seen that L/R coherence is almost down to 0 around 500 Hz, and then has a steep

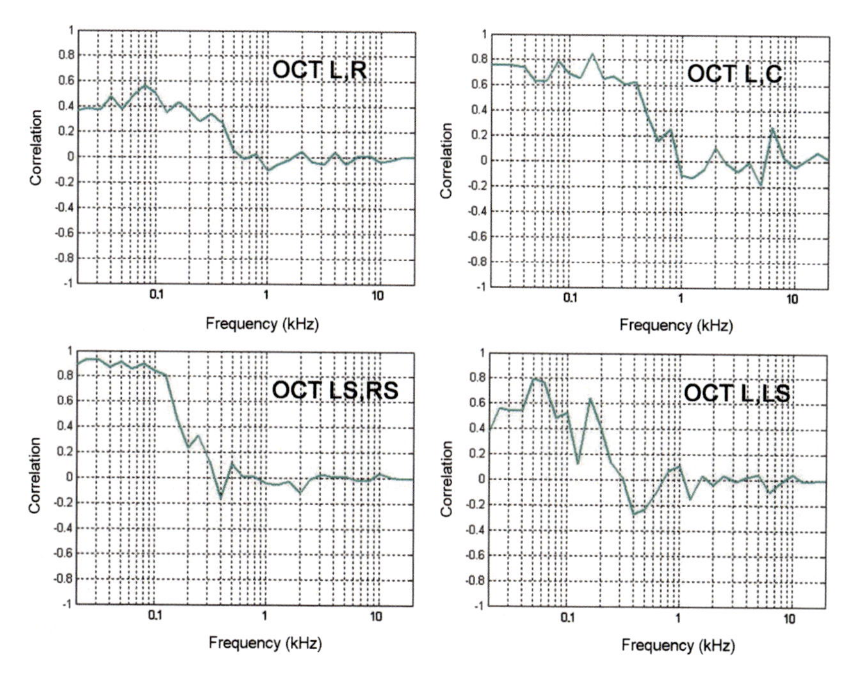

Fig. 7.13 Paired channel *correlation* over frequency for OCT surround microphone system signals (2048-point DFT grouped to 31 frequency bands, 1/3rd octave; center frequencies according to ISO; music, 60 s, hall 1); capsule spacing for L/C and C/R is 31 cm (from Pfanzagl-Cardone 2011)

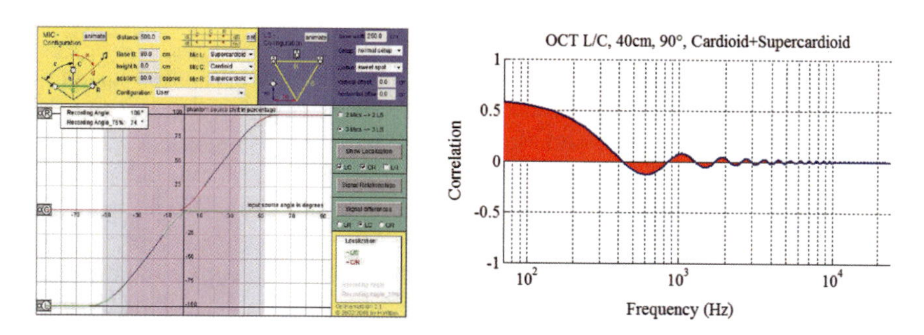

Fig. 7.14 Signal *correlation* over frequency for L/C for OCT surround microphone; (capsule spacing 40 cm, 90°. Cardioid + Supercardioid) (from Wittek 2012); calculated recording angle: 106°. (according to 'Image Assistant' by Wittek 2002)

rise again toward the high frequencies, where max. values of up to 0.65 are reached. Also, coherence between L and C channel is very high for low frequencies up to 500 Hz and also above is higher than with the other three techniques.

For the rear channels, LS/RS high coherence is limited to the low-frequency band up to approximately 350 Hz.

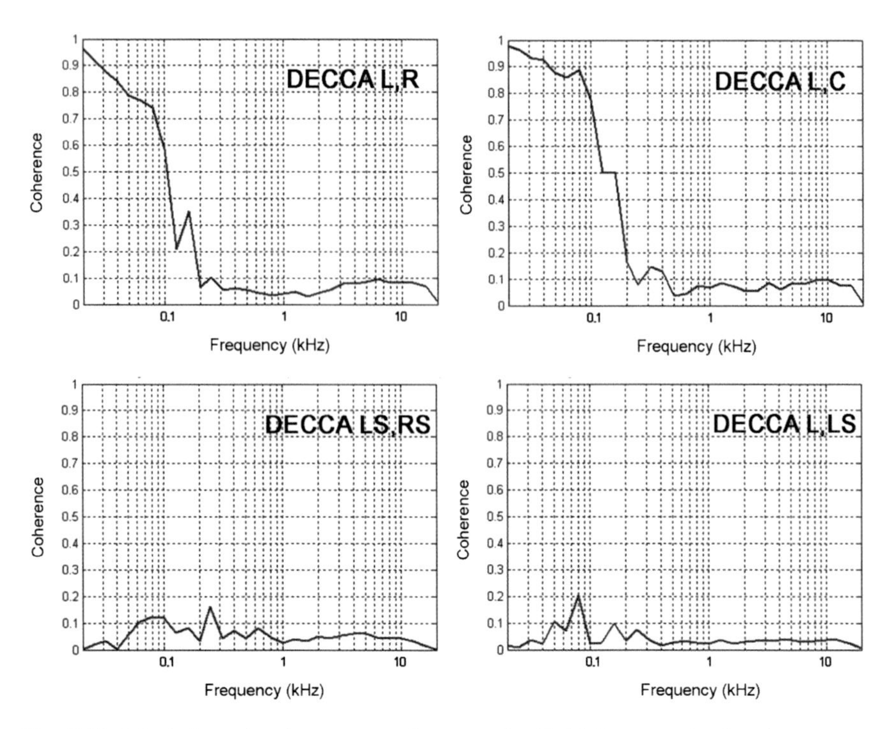

Fig. 7.15 Paired channel *coherence* over frequency for DECCA-surround microphone system signals (2048-point DFT grouped to 31 frequency bands, 1/3rd octave, center frequencies according to ISO; music, 60 s, hall 1) (from Pfanzagl-Cardone 2011)

Coherence between L and LS is very high as well for low frequencies, and almost 0 from approximately 1 to 2 kHz and rises again toward high frequencies.

Taking a look at the correlation functions of the KFM system in Fig. 7.19, it is quite evident that for some of the channel pairings there are strong out-of-phase relationships in some frequency bands, which we have already noticed in the FIACC measurements (compare Figs. 7.1, 7.3 and 7.7). Now with the much more detailed analysis by means of frequency-dependent channel-pair cross-correlation measurements, it becomes clear what happens: For the front-channel combination L,R, there is a broadband out-of-phase relation in the frequency range from 500 Hz to 9 kHz, as well as with the rear channels L,LS, which are basically out-phase for almost the whole frequency range of interest up to 10 kHz, apart from the region around 1 kHz which is close to 0 and therefore largely de-correlated (Rem.: for the de-matrixing of the MS-signals of the KFM system the Schoeps DSP-4P signal processing unit has been used).

The ABPC system in comparisons shows a very different characteristic: As can be seen in Fig. 7.20, coherence is very low in the low-frequency band. For the L,R front channels, there is a slight rise toward high frequencies, but coherence stays usually below 0.35 up until almost 10 kHz. Also, coherence between the L and C

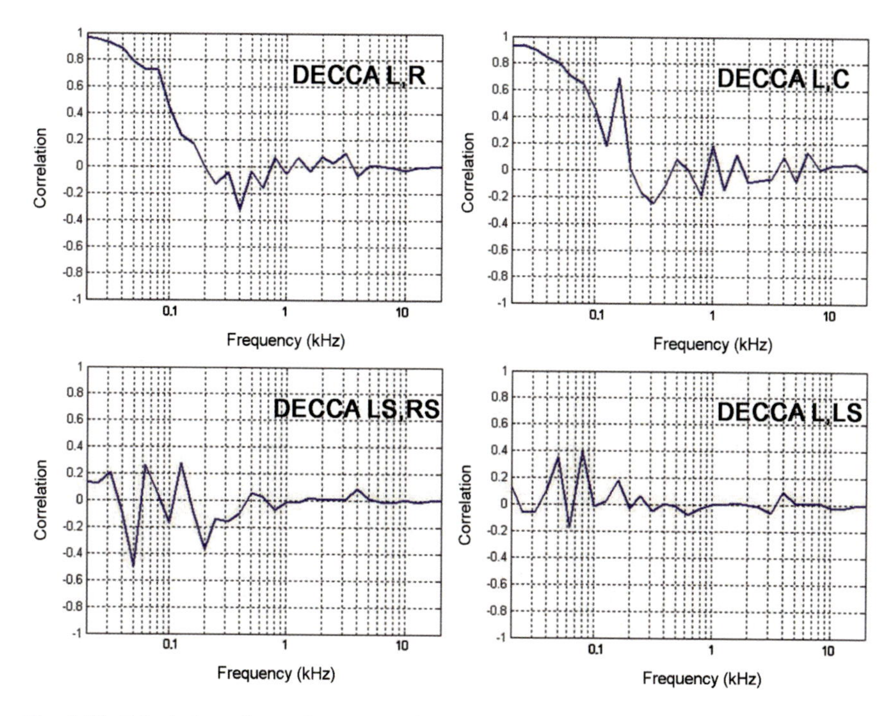

Fig. 7.16 Paired channel *correlation* over frequency for DECCA-surround microphone system signals (2048-point DFT grouped to 31 frequency bands, 1/3rd octave; center frequencies according to ISO; music, 60 s, hall 1) (from Pfanzagl-Cardone 2011)

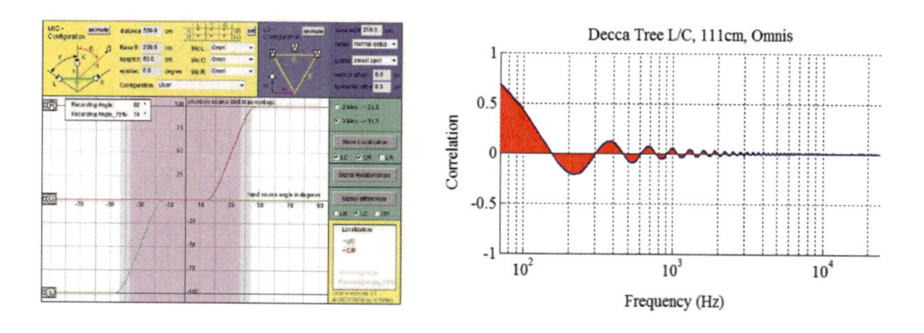

Fig. 7.17 Signal *correlation* over frequency for DECCA-surround L/C or C/R microphone system signals; for spacing L/R = 200 cm; $h = 50$ cm, omni microphones (from Wittek 2012)

channel is rather low: It reaches a high with about 0.5 around 400–500 Hz, but then dies away again toward higher frequencies.

The signals of the rear channels and between the rear and front channels are almost incoherent (apart from a few 'spikes'). The almost complete lack of coherence (i.e. high de-correlation) at lowest frequencies up to about 200 Hz is also the reason for the good spatial impression this microphone system provides.

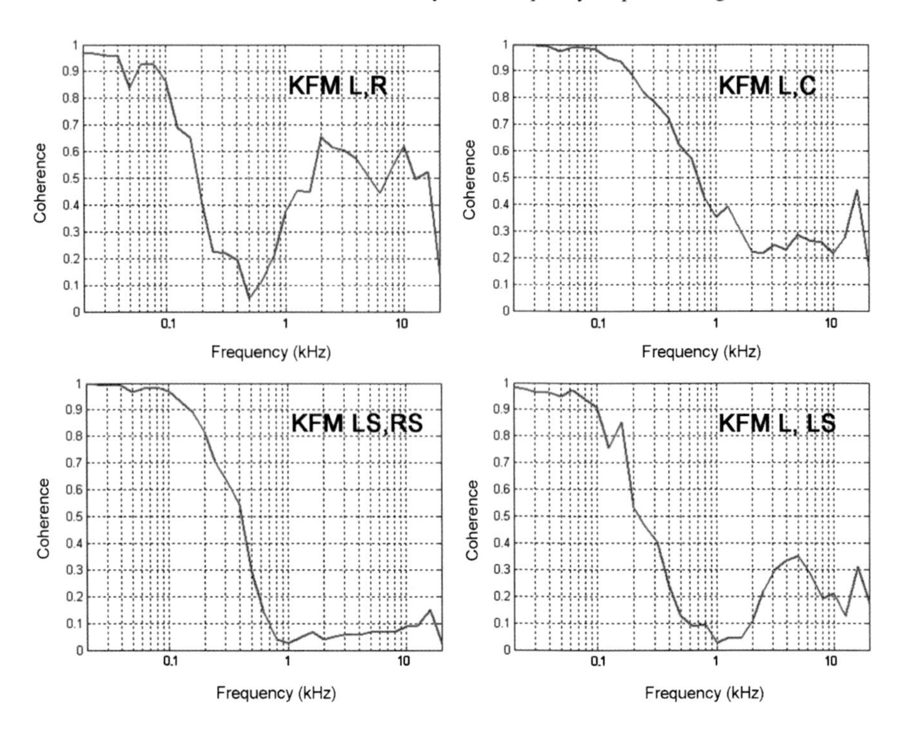

Fig. 7.18 Paired channel *coherence* over frequency for KFM surround microphone system signals (grouped to 31 frequency bands, 1/3rd octave; center frequencies according to ISO; music, 60 s, hall 1) (from Pfanzagl-Cardone 2011)

For completeness, the numerical values for the coherence between the surround channels are listed in Table 7.2.

Also, the analysis of paired signal correlation of the AB-PC system in Fig. 7.21 provides similar results: Below 200 Hz L,R correlation converges toward 0, and above this frequency, it meanders between 0.4 and 0.6. For the channel combination L,C, there is a pronounced area of high correlation with values above 0.6 in the frequency range from 150 Hz-800 Hz. Above 2 kHz, the L,C signals are again largely de-correlated.

The combinations of the rear signals LS,RS as well as the 'side' combinations L,LS (R,RS, respectively) are largely de-correlated as well, apart from a few narrow-band 'spikes' in the frequency range below 200 Hz.

Also, from the numeric values of coherence for the binaural dummy head (re-)recordings in Table 7.2, it can be seen that the AB-PC system is by far the closest to the ORCHestral reference recording (KU81) from the concert hall (same as with correlation).

As was to be expected, the KFM-microphone's surround signals have the highest paired channel coherence for both the ORCHestral recording, as well as the DUO recording. It is surprising to note however, that the OCT array has a slightly higher

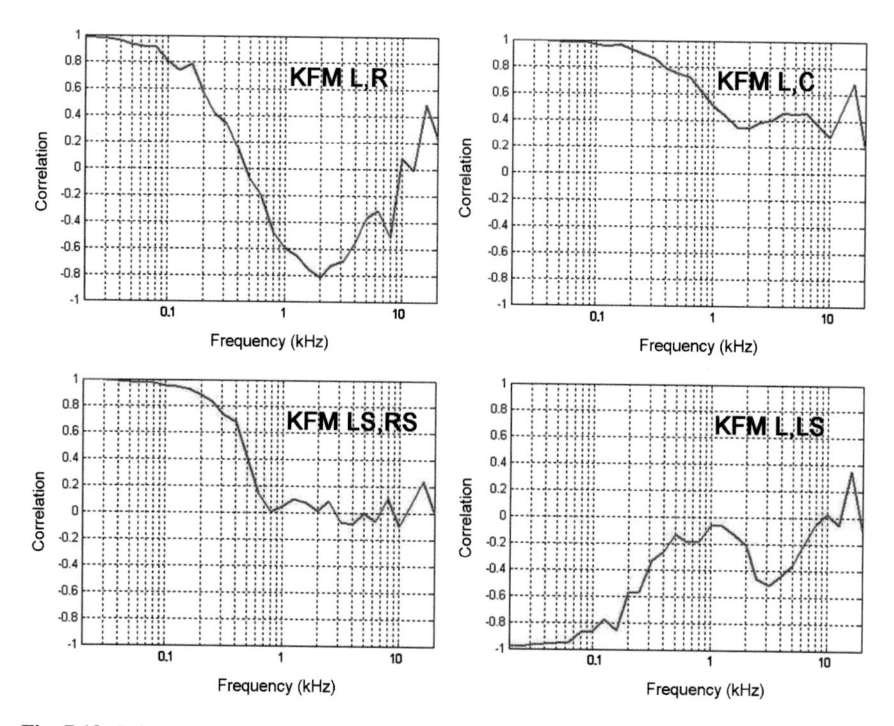

Fig. 7.19 Paired channel *correlation* over frequency (i.e. FCC) for KFM surround microphone system signals (2048-point DFT grouped to 31 frequency bands, 1/3rd octave, center frequencies according to ISO; music, 60 s, hall 1) (from Pfanzagl-Cardone 2011)

value (of 0.74) when it comes to binaural coherence in comparison to the KFM with a value of 0.70, even though the coherence between adjacent channels in the OCT array is always much lower than with the KFM-microphone. A plausible reason for this might lie in the 'dip' in coherence for the KFM mid-frequency region (where out-of-phase—i.e. negative correlation—occurs for the binaurally re-recorded KFM-signal), as well as the high coherence of OCT neighbor-channel signals in the low-frequency region.

As mentioned above, with a coherence value of 0.44, the binaural ABPC-signal is closest to the binaural reference signal of the KU81 with a value of 0.37.

For further comparison, in Table 7.2, the values for the ORCHestral recording, as well as the corresponding values are listed for the DUO recording. Even though the sound source and acoustics of the recording venue are very different, the coherence values found for the different microphone systems are astonishingly similar, which means that they are really typical for each microphone technique and could be considered something like a sonic 'fingerprint.'

As already noted previously, there is a noticeable asymmetry in the front system between L and R with the OCT system (compare L/C, C/R), as has been pointed out and very visible with the graphical analysis of coherence over frequency (see

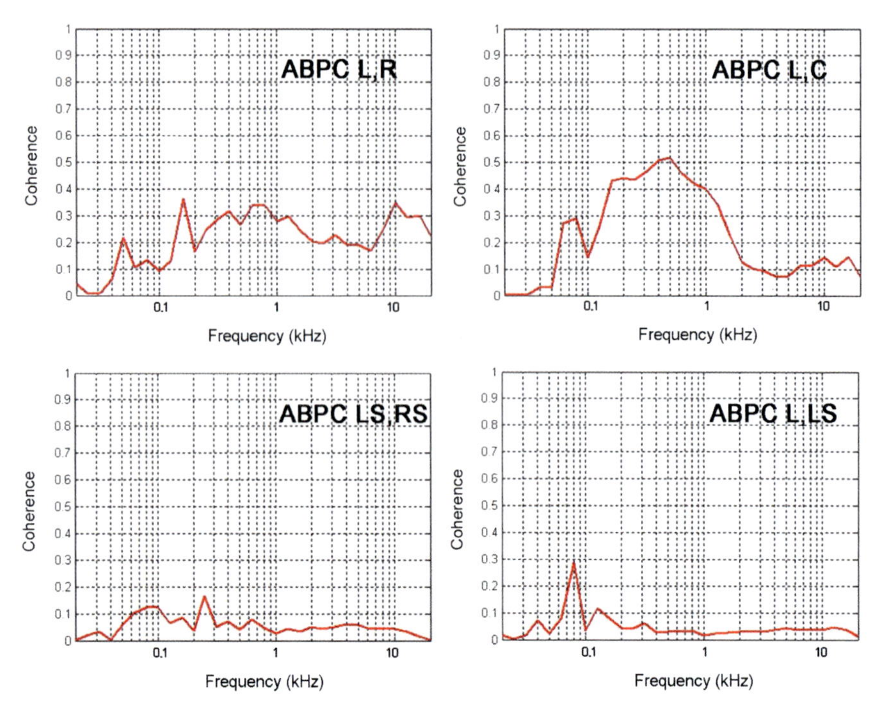

Fig. 7.20 Paired channel *coherence* over frequency (i.e. FCC) for ABPC-surround microphone system signals (2048-point DFT grouped to 31 frequency bands, 1/3rd octave, center frequencies according to ISO; music, 60 s, hall 1) (from Pfanzagl-Cardone 2011)

Table 7.2 Coherence between var. channels of the four surround microphones under test for both ORCHestral, as well as DUO recording; (sample length 60 s); (adapted from (Pfanzagl-Cardone and Höldrich 2008))

ORCH-Rec.	Binaural (Dummy)	Front L-R	L-C	C-R	L-LS	R-RS	Rear LS-RS
OCT	0.74	0.09	0.24	0.31	0.13	0.13	0.20
DECCA	0.68	0.19	0.24	0.24	0.03	0.02	0.05
KFM	0.70	0.49	0.55	0.45	0.37	0.32	0.38
AB-PC	0.44	0.21	0.24	0.25	0.03	0.03	0.04
KU81	**0.37**						
Duo- Rec.	Binaural (Dummy)	Front L-R	L-C	C-R	L-LS	R-RS	Rear LS-RS
OCT	0.56	0.16	0.24	0.36	0.21	0.19	0.22
DECCA	0.50	0.18	0.28	0.29	0.08	0.10	0.10
KFM	0.62	0.54	0.60	0.53	0.36	0.37	0.38
BPT	0.54	0.24	0.42	0.44	0.08	0.09	0.10
KU81	**0.48**						

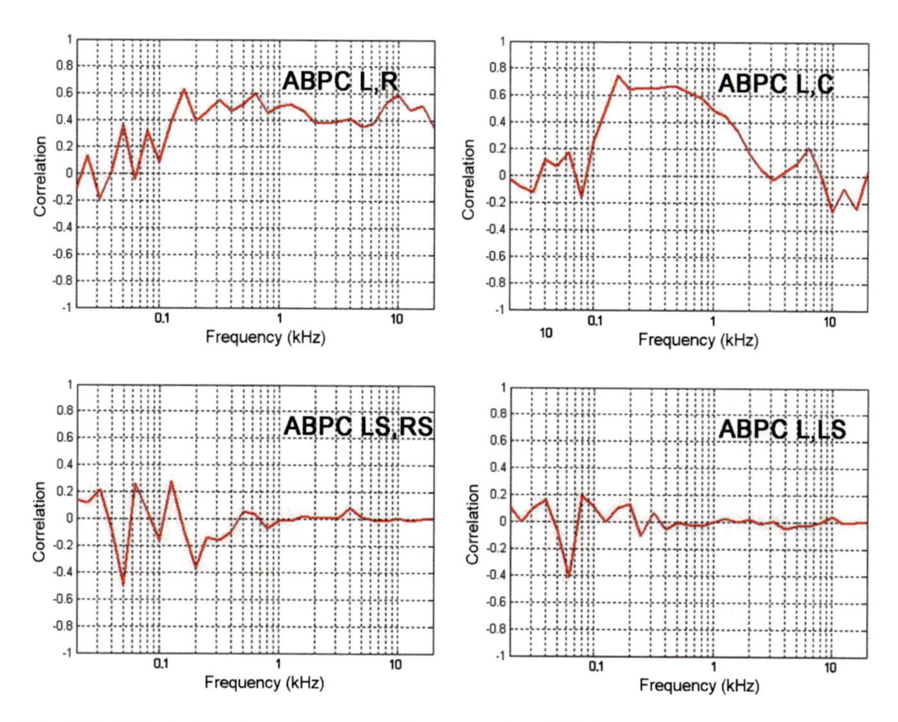

Fig. 7.21 Paired channel *correlation* over frequency (i.e. FCC) for ABPC-surround microphone system signals (grouped to 31 frequency bands, 1/3rd octave; center frequencies according to ISO; music, 60 s, hall 1) (from Pfanzagl-Cardone 2011)

Fig. 7.11). This may be due to a deviation of the R (or L) supercardioid capsule in respect to correct frequency response (maybe due to aged microphone membrane or electronics).

The coherence values found for the DECCA and KFM array in the duo recording of Table 7.2 are very similar to those of the ORCHestral recordings, again a strong argument in favor of the 'fingerprint' qualification of this measurement.

In Fig. 7.22, the measurements of coherence concerning the BPT-surround microphone are documented (DUO recording, music signal, 60 s, hall 2): Up to about 100 Hz signal coherence is very low for the L,R signal combination, then it rises, and from about 200 Hz, upwards coherence varies between approximately 0.25 (min) to 0.5 (max) and after that ($f > 2$ kHz) lowers again with rising frequency, never getting above 0.3.

Due to the strong directional characteristics of the fig.-of-eight capsules coherence between the L and C channel is quite high: At low frequencies up to about 550 Hz, coherence is high with peak values around 0.8, after that there is a decline with rising frequency.

Due to the widely spaced rear cardioids—which are pointing toward the back of the room and are at a distance of more than the reverberation radius from the front

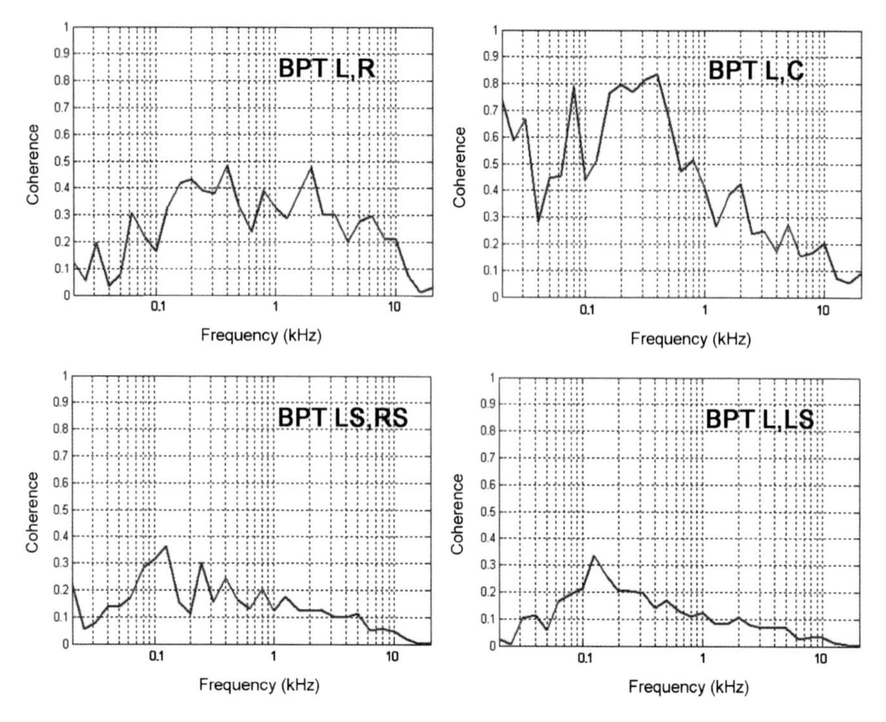

Fig. 7.22 Paired channel *coherence* over frequency (i.e. FCC) for the BPT-surround microphone system signals (grouped to 31 frequency bands, 1/3rd octave; center frequencies according to ISO; music, 60 s, hall 2) (from Pfanzagl-Cardone 2011)

system—signal coherence between the rear channels LS,RS and to the front channel (see the combination L,LS) is very low.

Even if the relatively high coherence at specific frequencies in the mid- and high-frequency bands is not necessarily (psychoacoustically) desirable, the overall low coherence in the LF-region below 200 Hz ensures that the BPT microphone is well suited for good spatial reproduction.

Signal correlation of the BPT array as pictured in Fig. 7.23 shows very similar results: Below about 150 Hz signal correlation is generally low (with the exception of L,C and C,R), which is of advantage for spatial impression. In this respect, it is interesting to relate this to the measurement of the FIACC of the BPT system as documented in Fig. 7.7 (which is for a shorter sample of 16 s duration only).

In this context, it seems important to point out that the high correlation of the BPT signal combinations L,C and C,R in the low-frequency region below 500 Hz can be altered by intentional filtering (i.e. LF-level compensation of the figure-of-8 specific LF-'roll off' in order to 'linearize' the frequency response for the L and R capsule signal, as well as high-pass filtering of the BPT's center-mic signal for an overall enhanced de-correlation at low frequencies) which has been recommended when first describing the BPT microphone in the chapter on surround microphone

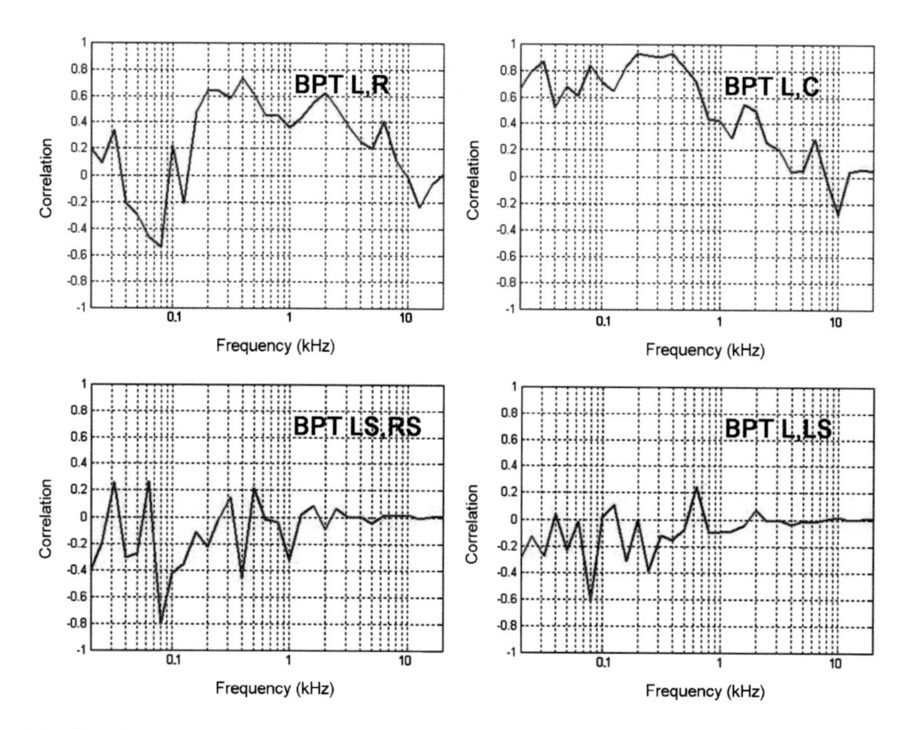

Fig. 7.23 Paired channel *correlation* over frequency (i.e. FCC) for the BPT-surround microphone system signals (grouped to 31 frequency bands, 1/3rd octave; center frequencies according to ISO; music, 60 s, hall 2) (from Pfanzagl-Cardone 2011)

techniques, as such filtering will also alter the phase relationship between the L,C and C,R pairs in a positive way.

As part of the experimental recording session from which the data above have been collected (see Pfanzagl-Cardone and Höldrich 2008) also the signals of an Ambisonic System (SoundField-Microphone) have been recorded. Even though these signals have not been used in the subjective listening test together with the other four systems above, a signal analysis concerning frequency-dependent signal correlation has been conducted, the results of which are presented below for the sake of completeness and also to provide the possibility of a comparison with the graphs of the other systems above.

Looking at the graphs in Fig. 7.24, it is amazing to see that signal correlation between the L and R channel signals is very high, basically over the whole frequency range: Only above about 4 kHz, correlation slowly decreases (after meandering around a factor of 0.8) toward 0.6, which is still considered 'high' correlation. Between the L and C (likewise for R and C; not shown), signal correlation is even higher, essentially around 0.9—therefore almost monophonic.

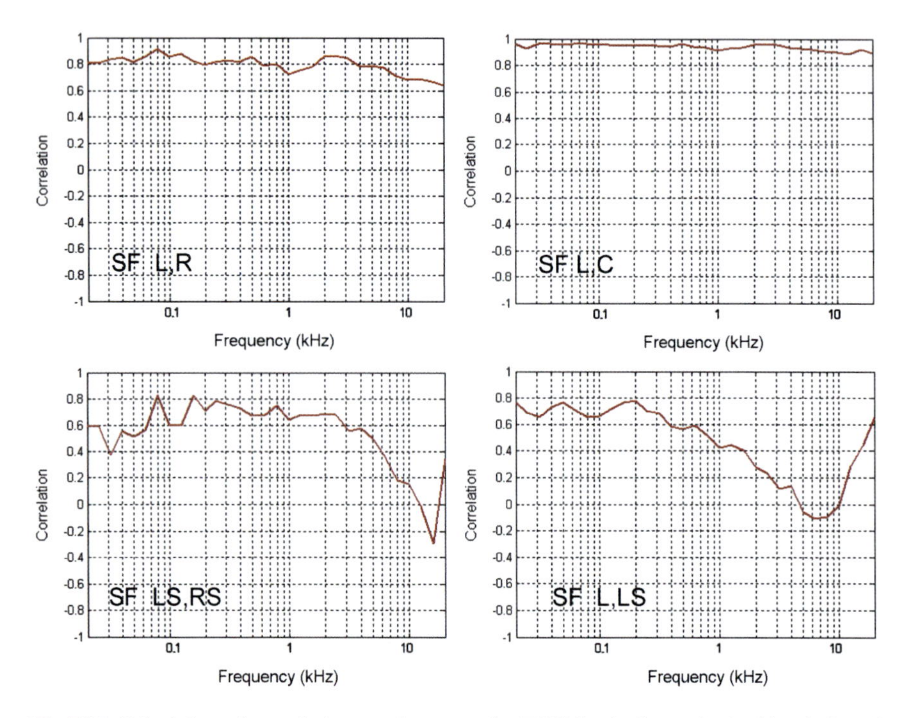

Fig. 7.24 Paired channel *correlation* over frequency (i.e. FCC) for the first-order Ambisonic SoundField MK-V surround microphone system signals (grouped to 31 frequency bands, 1/3rd octave; center frequencies according to ISO; music, 60 s, hall 2); decoded via SoundField MK-V Surround Processor

Even for the rear-channel signals LS, RS correlation over frequency remains very high, usually around or above 0.6 in the frequency range from 20 Hz to 3 kHz, dropping off toward 0 only above 10 kHz. Also, for the channel-pairing L,LS correlation is very high with values above or around 0.6 from 20 Hz to about 600 Hz.

As signal correlation is very high for basically all channel combinations for frequencies below 500 Hz, it is a clear consequence that spatial impression is very poor for the SoundField-Microphone (FOA) First-Order-Ambisonics array, as spatial impression is strongly related to high signal de-correlation at low frequencies (see various papers by Griesinger 1986, 1987, 1988, 1996, 1997, 1999, 2000, 2002 as well as Hidaka et al. 1997 on this subject).

Above, the author has analyzed five surround microphone techniques in detail in respect to frequency-dependent channel coherence, as well as channel correlation. In combination with the analysis of the binaurally re-recorded surround soundfields (see Figs. 7.1, 7.3, 7.4 and 7.7)—to his knowledge—this seems to be the most in-depth analysis for surround microphone systems so far, which was able to unveil characteristics unique to each technique under examination.

In Chap. 3, we have already presented a graph from Muraoka et al. (2007), which shall be reproduced in Fig. 7.25 for reasons of reader's convenience:

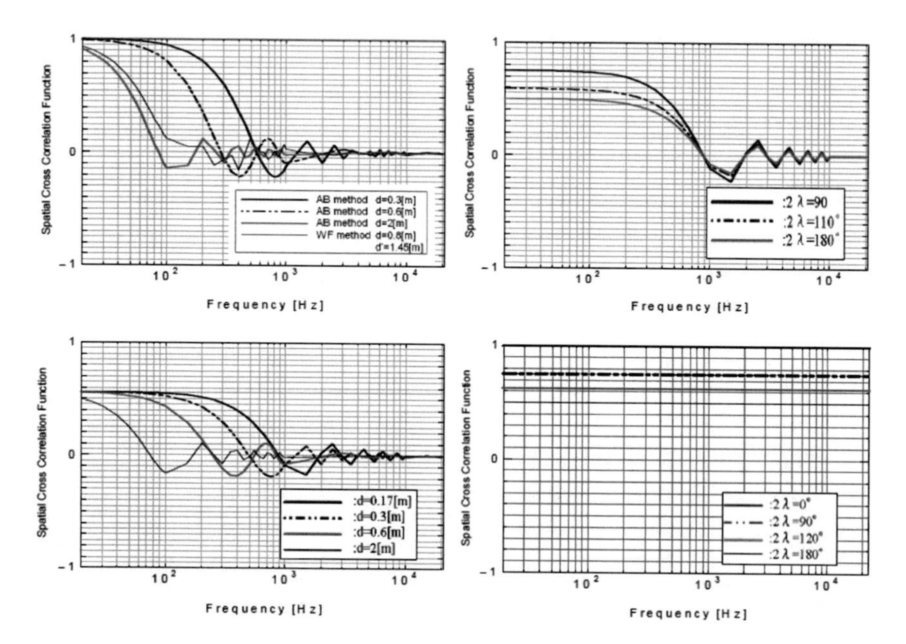

Fig. 7.25 Calculation of the cross-correlation for various stereo microphone techniques; FSCC = Frequency-dependent Spatial Cross-Correlation (from Muraoka et al. 2007). Left up: FSCC-patterns of AB-microphones and WaveFront-microphone; left down: FSCC of ORTF-microphones (capsule spacing dependent); right up: FSCC of ORTF-microphones (directional azimuth dependent); right down: FSCC of XY-microphone (cardioid or supercardioid)

In Wittek (2012), a few stereo and surround microphone techniques have also been analyzed in respect to signal correlation over frequency, but only for two channels each.

For completeness, the data which have been derived from theoretic calculations (not measurements, as in the figures above) are being presented below in a short, alphabetic form [descriptions of the microphone systems are translated from Wittek (2012)]:

(Rem.: Also, it should be noted that the graphs below concern only the 'diffuse field' part of the soundfield, while the graphs of the author of this book show the sum of direct + diffuse sound of acoustically measured signals.)

Hamasaki Square

From Wittek (2012): '4 fig-8 microphones, pointing laterally, with capsule spacings larger than 200 cm

Practical disadvantage: Rather large capsule spacings, array is 'clumsy'

Sonic advantages: 'open' sounding, extremely low DFC (diffuse-field correlation); optimal attenuation for sound from 0°; optimal reproduction of lateral room reflections; calculated recording angle: 18° (according to 'Image Assistant' by Wittek 2002) (Fig. 7.26).

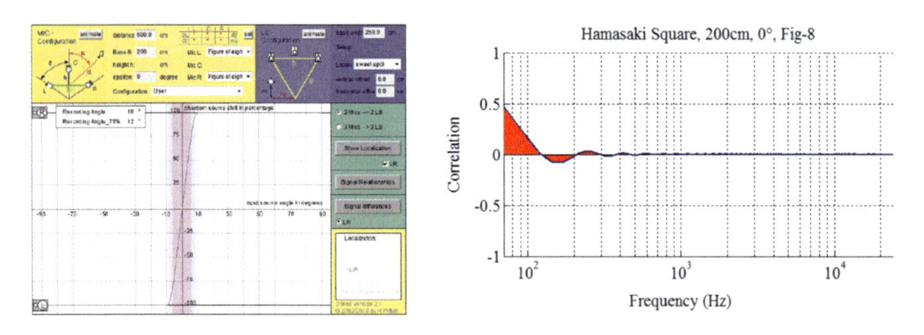

Fig. 7.26 Hamasaki Square; localization curve (left), signal correlation for L/R with spacing 200 cm (right) (from Wittek 2012)

IRT-Cross

Made up of 4 cardioid microphones, there are three possibilities in terms of spacing:

cardioids: capsule spacing/base width 25 cm
supercardioids: capsule spacing/base width 18 cm
wide-cardioids: capsule spacing/base width 31 cm

According to Wittek, the systems is characterized by an 'open' spatial sound and very good 360° imaging. Calculated recording angle: 92° (according to 'Image Assistant' by Wittek 2002) (Fig. 7.27).

ORTF-Surround

4 supercardioids, configuration: 10 cm/100° + 20 cm/80°
Practical advantages: rather compact
Sonic advantages: 'open' sounding, like the IRT-Cross
calculated recording angle: 94° (according to 'Image Assistant' by Wittek 2002) (Fig. 7.28).

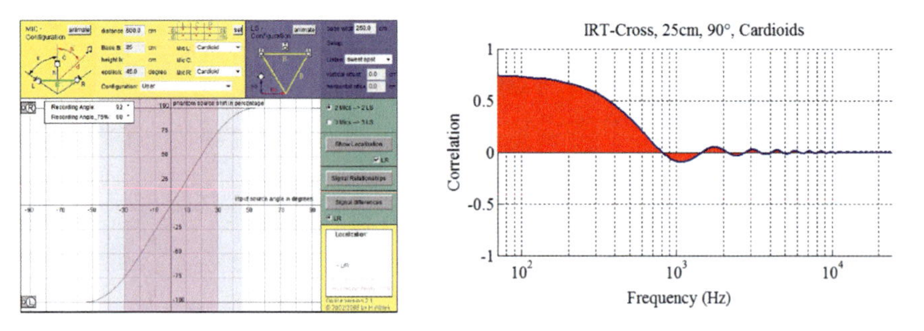

Fig. 7.27 IRT-Cross with cardioids; localization curve (left), signal correlation for L/R with spacing 25 cm (right) (from Wittek 2012)

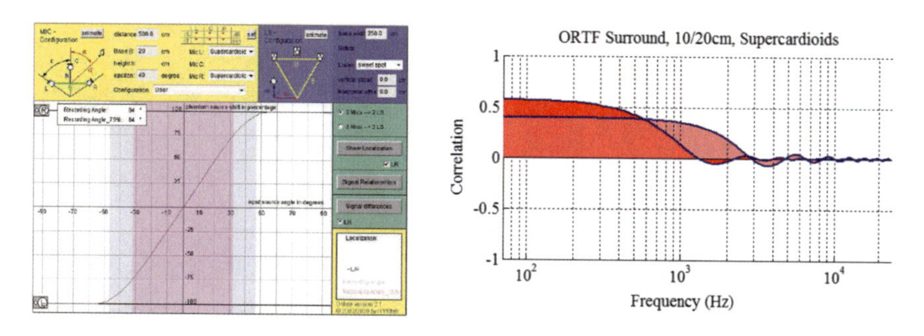

Fig. 7.28 ORTF-surround with supercardioids; localization curve (left), signal correlation (right) (from Wittek 2012)

Theile-Trapezoid

Intended mainly as an array for diffuse sound pickup from the rear (for example in the context of a 9.1 recording setup), it is composed of four cardioids, oriented toward the back, with capsule spacings of 60 cm.

> Sonic advantages: Wittek claims optimal attenuation for sound arriving from 0°
> Calculated recording angle: 60° (according to 'Image Assistant' by Wittek (2002)

L/R capsule spacing = 100 cm; LS/RS capsule spacing = 60 cm (Figs. 7.29 and 7.30).

Equal Segment Microphone Array (ESMA) after Williams

5 cardioid microphones with equivalent angles of 60°, capsule spacing 53 cm, according to Williams (2012a, b); uses cardioid or semi-cardioid microphones;

> Practical disadvantages: not very compact, medium size capsule spacings
> Sonic advantages: very nice sound color; very good spatial impression and imaging capabilities

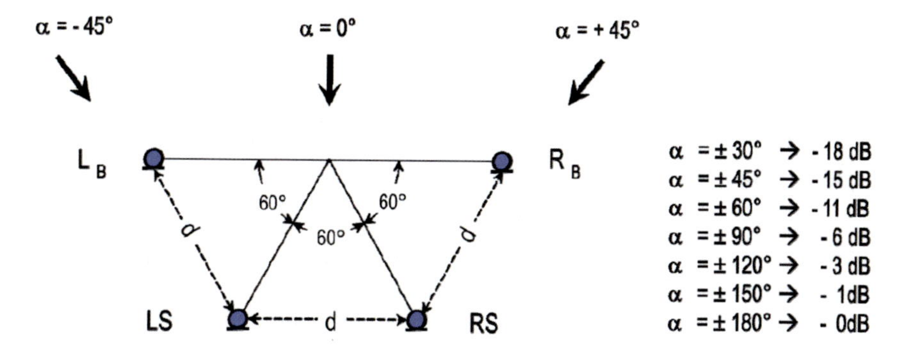

Fig. 7.29 Theile-Trapezoid with cardioids; mic schematic (from Wittek 2012)

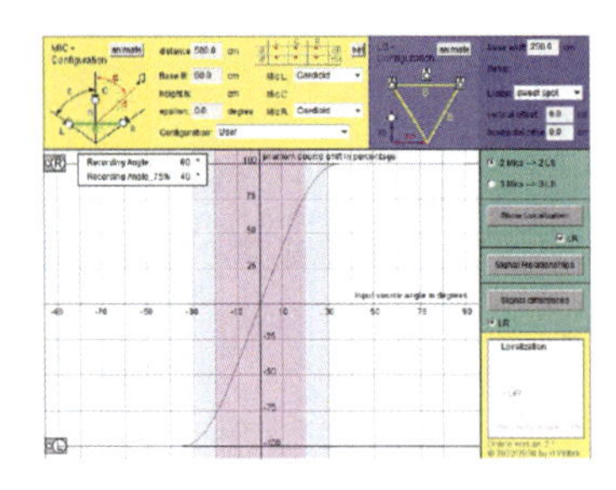

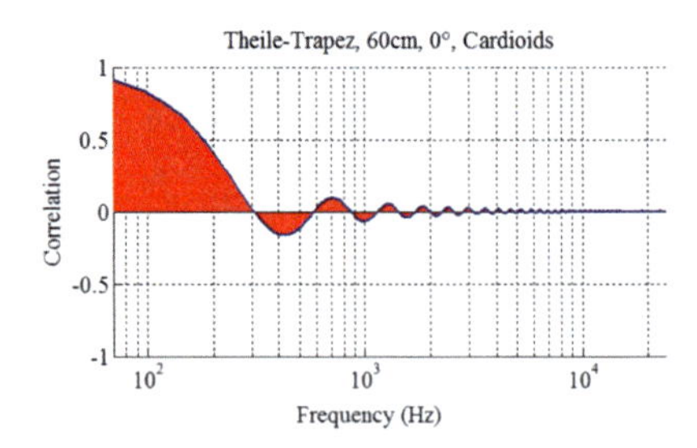

Fig. 7.30 Theile-Trapezoid with cardioids; localization curve (left), signal correlation for a spacing of 60 cm (right) (from Wittek 2012)

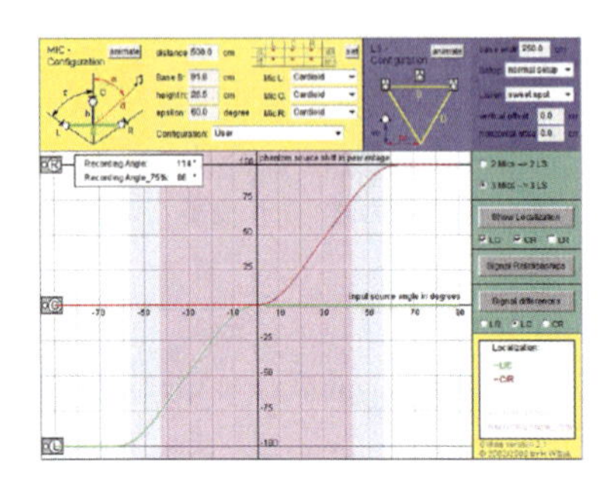

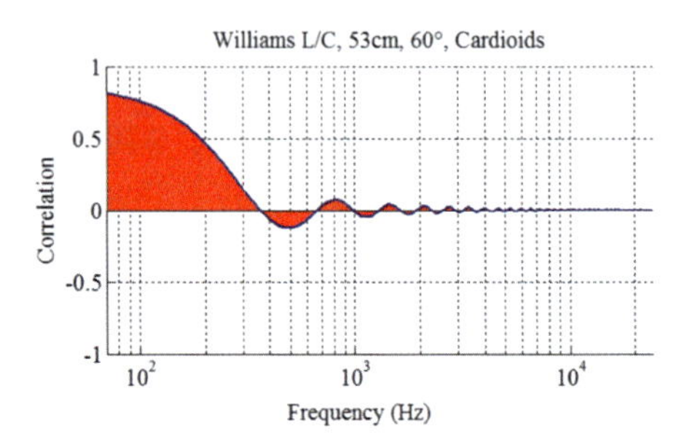

Fig. 7.31 ESMA—after Williams/Theile/Wittek; localization and correlation displayed for L/C (C/R); localization curve (left), signal correlation (right); with $d = 92$ cm, $h = 26.5$ cm, opening angle (L/C) = 60°, 2 cardioids (from Wittek 2012)

Calculated recording angle: 114° (according to 'Image Assistant' by Wittek 2002) (Fig. 7.31).

Coincident Microphones (=Double-MS or first-order Ambisonics)

Practical considerations: compact, but needs proper decoding; Double-MS needs only three capsules (front cardioid, fig-8, rear cardioid);
Sonic considerations: if decoded to more than three output channels, both techniques will be characterized by a high degree of diffuse-field correlation; [Rem. of the author: according to Wittek, a maximum of 4 output-channels makes sense.]' (from Wittek 2012) (Fig. 7.32).

It is interesting to note that the diffuse-field correlation (DFC) of the Double-MS or first-order Ambisonic microphones is at an intermediate value of 0.5 over the whole audio frequency range. This certainly is too high to provide a convincing spatial impression which requires very low correlation, especially in the low-frequency range, as has already been pointed out.

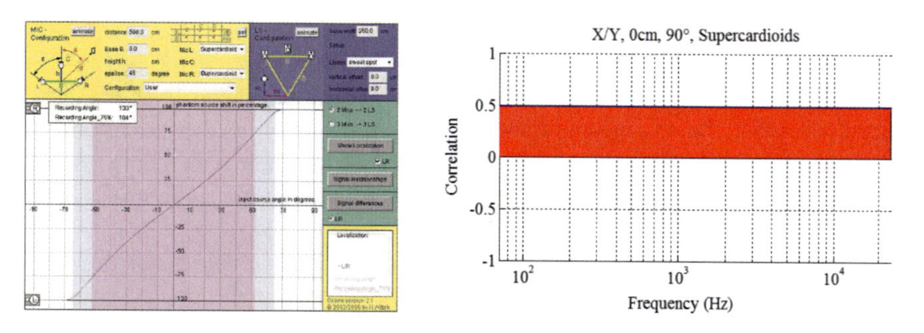

Fig. 7.32 Coincident surround-mic techniques: localization (left) and correlation graphs (right); L/R channel relation, after decoding signal information of coincident mic technique to 4 virtual supercardioids with an orientation of 45°, 135°, 225° and 315° (from Wittek 2012)

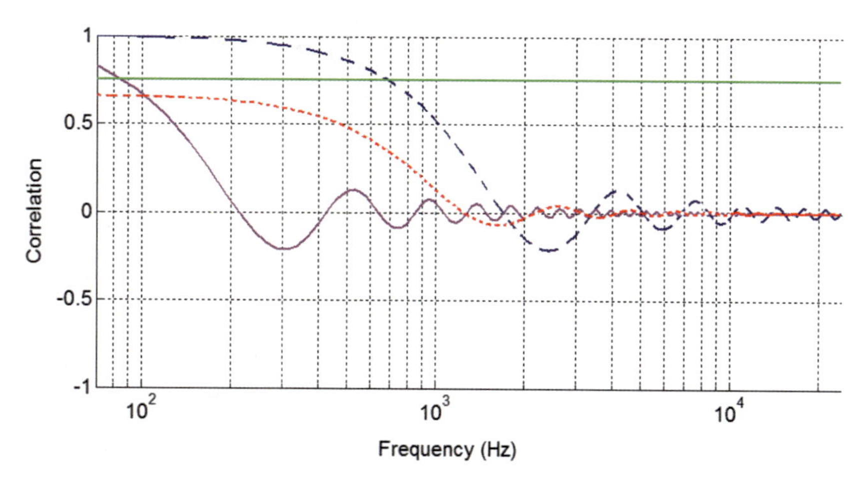

Fig. 7.33 Diffusefield correlation (DFC) for various stereo microphone techniques: (listed from top to bottom) blue dashed line: small AB omnis $d = 10$ cm; green straight line: XY-Cardioids 90°; red dotted line: ORTF Cardioids, 17 cm, 110°; violet: small AB, omnis 80 cm (modified from Wittek 2012)

7.4 Conclusions

Four surround microphone techniques (OCT, DECCA, KFM, AB-PC) have been measured and analyzed in detail by the author in terms of frequency-dependent correlation, as well as frequency-dependent coherence. It was shown that analyzing signal coherence is not able to unveil potential weaknesses of a microphone system (e.g. in respect to phase response anomalies; see KFM), as clearly as when analyzing signal correlation. Taking a look at the signal properties of the KFM surround microphone system, it becomes evident that the out-of-phase signal components, which are clearly recognizable when measuring binaural signal correlation (FIACC) in the 1–3 kHz region (see Figs. 7.1 and 7.3), finds some kind of equivalent—when estimating signal coherence—only in the lowering of signal coherence in the same frequency region (see Fig. 7.10).

As measuring signal correlation seems to have a clear advantage over estimating signal coherence, for reasons of efficiency, we shall restrict ourselves to an analysis of signal correlation over frequency in the following Chap. 8, in which both the FCC and FIACC for some of the most common *stereo* microphone techniques will be analyzed.

In respect to the differences between signal correlation and coherence, as well as the optimum value for signal correlation (or coherence) for microphone systems, it seems interesting to remind the reader of the value for 'Optimized Signal Coherence with Surround Microphone Systems' of $\gamma \geq 0.35$, as proposed by Theile (2000) and reviewed in Sect. 2.2.

In this context, it seems appropriate to note that—as can be seen in Table 7.2—inter-channel coherence values for Theile's OCT surround system stay well below his proposed value of $\gamma \geq 0.35$, with an actual maximum value of 0.22 for one of the outer-speaker combinations (L/R, LS/RS or L/LS, R/RS) for both recording situations ORCHestral, as well as duo (compare with Fig. 2.14).

In fact, the only surround microphone system which does achieve a value higher than 0.35—namely the KFM system—has clearly been downgraded by two groups of test listeners in the double blind listening tests (Rem.: it has to be said though, that this rating seems to have to do mainly with the out-of-phase components inherent to the KFM surround signals and the sound coloration due to phase cancellation and comb-filtering effects which subsequently became evident in the associated binaurally re-recorded KFM-microphone signal).

Apparently, Helmut Wittek, one of the closest collaborators of Günther Theile, has later on revised his view on the necessary minimum value of signal coherence or correlation in order to achieve a 'healthy' surround signal, a matter that is very closely related as to how a microphone technique picks up the diffuse part of a sound field. This matter is also addressed in Chap. 8, regarding the different degree of correlation in which coincident microphone techniques pick up sound (see Fig. 8.9).

7.4.1 Thoughts on Diffuse-Field Correlation (DFC)

In Wittek (2012), he revisits the question how high a value of DFC should be allowed in order not to degrade spatial impression. In previous research, (Riekehof-Böhmer et al. 2010) listening tests have been performed in relation to this question, for which coincident, equivalence and runtime (i.e. AB) microphone techniques have been used. Artificial reverb was reproduced, applying these techniques and test listeners had to rate its 'Apparent Width' (or Stereophonic Width), the results of which are displayed in Fig. 7.34.

Translated and partly summarized from Wittek (2012):

'…It can be assumed that a large apparent width is optimal for diffuse sound/reverb. For AB-based (runtime) microphone techniques the DFC is not

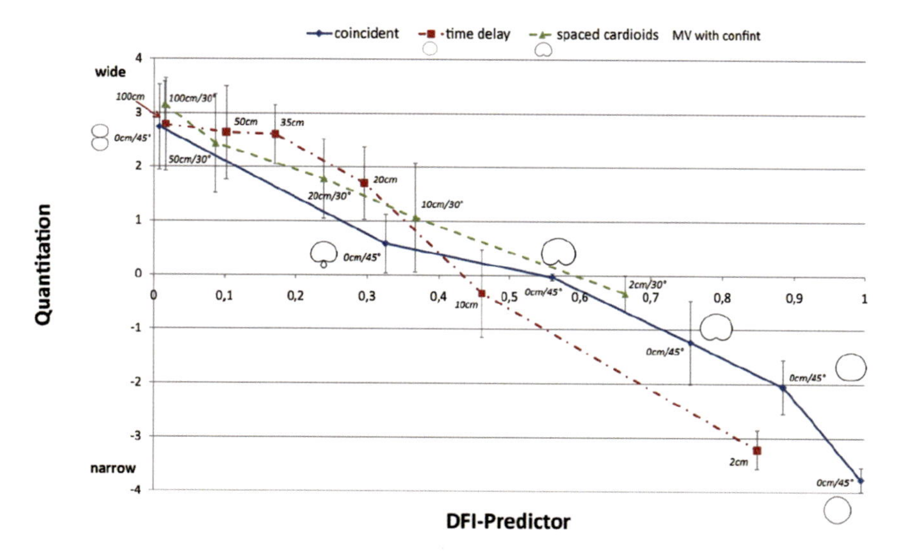

Fig. 7.34 Listener evaluation of apparent width of diffuse sound (reverb), recorded by various microphone techniques (from Riekehof-Böhmer et al. 2010); mean value with confidence intervals

uniform over frequency (see Fig. 7.33) but it can be assumed that it is preferable if the area below the function is small, especially for low frequencies (see also Griesinger 1998). This is why in Riekehof-Böhmer et al. (2010) an asymptotic value for the DFC has been determined, which—in essence—is the integral of the squared, weighted coherence function (low frequencies being more important). With this—the DFI ('Diffuse-Field Image') Predictor—a value is obtained, by which different mic techniques can be compared and the 'apparent width' can be determined (Fig. 7.34).

If we deem a minimum quantitation value of +2 sufficient, the following conclusions can be drawn:

- The Blumlein-Pair was the only 'good' coincident technique, according to the listening test. Its DFC is 0.
- With equivalence and pure runtime techniques the DFI-predictor needs to be smaller than 0.5.

For pure runtime techniques, this means the capsule spacing has to be ≥35 cm.

For equivalence techniques, the spacing can be smaller, depending on the polar pattern and orientation of the microphones involved.

- An optimum, 'wide' reproduction of diffuse sound can be achieved both via coincident, as well as spaced techniques.
- As pure runtime techniques need to have a capsule spacing of at least 35 cm in order to achieve optimum results, multichannel setups using baffles with a smaller effective diameter will not work to satisfaction, as the baffle-effect vanishes with lowering frequency (due to λ).'

The statement above singles out the Blumlein-Pair-based techniques in a positive way in what concerns signal de-correlation, but also makes it clear that small AB-based techniques (also with baffles) with an effective acoustic capsule spacing of less than 35 cm will not produce satisfactory results, due to the lack of sufficient signal de-correlation at low frequencies. Among the surround microphone techniques, this criteria concerns—for example—Jecklin's OSIS technique.

Wittek concludes: '…The diffuse soundfield plays an enormously important role both for spatial impression as well as sound color. Therefore it has to be made sure that it is being reproduced with low correlation. Many good or bad sonic properties of a stereophonic technique are not related to their localization characteristics, but to their ability to reproduce a beautiful, open 'spatial impression'. XY-cardioids are a good example: good localization, but poor spatial impression. In the future much more attention should be paid to diffuse-field correlation. Very often only the desired recording angle is considered as a decisive factor for choosing a specific microphone technique. The DFC is at least as important.' (from Wittek 2012).

Later on, he concedes that the very importance of DFC in relation to the quality of a main microphone has been neglected in many research projects, even of his own. In respect to surround or main microphones for SFX/atmo-recordings, he even regards the DFC to be of higher importance than localization accuracy.

The findings by Wittek, cited above, are in essence what the author of this book has tried to point out in previous publications (see Pfanzagl 2002, Pfanzagl-Cardone 2002, 2011 and Pfanzagl-Cardone and Höldrich 2008), while drawing conclusions and proposing new microphone techniques (e.g. AB-PC, see Chap. 3 on stereophonic microphone techniques).

As has already been outlined in Sect. 2.3, the optimal inter-aural correlation for music perception in concert halls is 0.23, as found by Gottlob (1973).

All this seems to indicate that microphone systems with high signal correlation should probably better be abandoned for the sake of systems that are largely de-correlated. As was found in research by Nakahara (2005) and can be seen in Fig. 2.15, low correlation of the playback signals (L/R vs. C vs. LS/RS) ensures a better compatibility between different listening environments and also leads to an enlarged sweet spot (see also Prokofieva 2007).

Therefore, microphone techniques should be used which are able to provide low signal correlation over the entire frequency range, if possible. Among the four surround microphone systems (OCT, DECCA, KFM, ABPC) which were evaluated via subjective listening tests, as well as objective measurements (FCC and FIACC), the low-correlation AB-PC system (using an ORTF-Triple as a centerfill microphone) turned out most successful: in terms of listener preference and because it has achieved the highest number of 'best mean value' ratings for various acoustic attributes. In addition, it was the microphone system, which managed to physically approximate the original sound-field of the concert hall better than any of the other systems under test.

In contrast, the SoundField-Microphone (Model MK V) has been shown to have a very high overall signal correlation between its channels if decoded to 5.1 surround (especially also in the low-frequency region), which accounts for the poor spatial impression this system provides (Rem.: the MATLAB code, which has been used to measure the FCC and FIACC, as presented in the various figures of this chapter, can be found in the appendix of this book).

References

Griesinger D (1986) Spaciousness and localization in listening rooms and their effects on the recording technique. J Audio Eng Soc 34(4):255–268

Griesinger D (1987) New perspectives on coincident and semi-coincident microphone arrays. Paper 2464 presented at the 82nd audio engineering society convention, May 1987

Griesinger D (1988) Equalization and spatial equalisation of dummy head recordings for loudspeaker reproduction. Paper 2704 presented at the 85th audio engineering society convention, Los Angeles

Griesinger D (1996) Spaciousness and envelopment in musical acoustics. In: Proceedings to the 19. Tonmeistertagung des VDT, pp 375–391

Griesinger D (1997) Spatial impression and envelopment in small rooms. Paper 4638 presented at the 103rd audio engineering society convention

Griesinger D (1998) General overview of spatial impression, envelopment, localization and externalization. In: Proceedings to the audio engineering society 15th int conference on small room acoustics, Denmark, Oct/Nov 1998, pp 136–149

Griesinger D (1999) Objective measures of spaciousness and envelopment. Paper 16-003 presented at the audio engineering society 16th international conference on spatial sound reproduction

Griesinger D (2000) The theory and practice of perceptual modeling—how to use electronic reverberation to add depth and envelopment without reducing clarity. In: Proceedings to the 21. Tonmeistertagung des VDT, Hannover, pp 66–795

Griesinger D (2002) Stereo and surround panning in practice. Paper 5564 presented at the 112th audio engineering society convention, Munich, May 2002

Gottlob D (1973) Vergleich objektiver akustischer Parameter mit Ergebnissen subjektiver Untersuchungen an Konzertsälen. Dissertation, Universität Göttingen, 1973

Hidaka T, Beranek L, Okano T (1995) Interaural cross-correlation, lateral fraction, and low- and high-frequency sound levels as measures of acoustical quality in concert halls. J Acoust Soc Am 98(2)

Hidaka T, Beranek L, Okano T (1997) Some considerations of interaural cross correlation and lateral fraction as measures of spaciousness in concert halls. In: Ando Y, Noson D (eds) Music and concert hall acoustics. Academic Press, London

Hiyama K, Komiyama S, Hamasaki K (2002) The minimum number of loudspeakers and its arrangement for reproducing the spatial impression of diffuse-sound field. Paper 5674 presented at the 113th audio engineering society convention, Los Angeles, Oct 2002

Jacobsen F, Roisain T (2000) The coherence of reverberant sound fields. J Acoust Soc Am 108(1)

Kamekawa T, Marui A, Irimajiri H (2007) Correspondence relationship between physical factors and psychological impressions of microphone arrays for orchestra recording. Paper 7233 presented at the 123rd audio engineering society convention, New York, Oct 2007

Martin G (2005) A new microphone technique for five-channel recording. Paper 6427 presented at the 118th audio engineering society convention, Barcelona

Morimoto M, Meakawa Z (1988) Effects of low frequency components on auditory spaciousness. Acustica 66

Muraoka T, Miura T, Ifukuba T (2007) Ambience sound recording utilizing dual MS (mid-side) microphone systems based upon frequency dependent spatial cross correlation (FSCC). Paper 6997 to the 122nd audio engineering society convention, Vienna

Nakahara M (2005) Multichannel monitoring tutorial booklet (M2TB) rev. 3.5.2. Yamaha Corp 2005, SONA Corp, p 41

Pfanzagl E (2002a) Über die Wichtigkeit ausreichender Dekorrelation bei 5.1 Surround-Mikrofonsignalen zur Erzielung besserer Räumlichkeit. In: Proceedings to the 21. Tonmeistertagung des VDT, Hannover

Pfanzagl-Cardone E (2002b) In the light of 5.1 surround: why AB-PC is superior for symphony-orchestra recording. Paper 5565 presented at the 112th audio engineering society convention, Munich

Pfanzagl-Cardone E (2011) Signal-correlation and spatial impression with stereo- and 5.1 surround-recordings. Dissertation, University of Music and Performing Arts, Graz, Austria. https://iem.kug.ac.at/fileadmin/media/iem/altdaten/projekte/dsp/pfanzagl/pfanzagl_diss.pdf. Accessed Oct 2018

Pfanzagl-Cardone E, Höldrich R (2008) Frequency-dependent signal-correlation in surround- and stereo-microphone systems and the Blumlein-Pfanzagl-Triple (BPT). Paper 7476 presented at the 124th audio engineering society convention, Amsterdam

Prokofieva E (2007) Relation between correlation characteristics of sound field and width of listening zone. Paper 7089 presented at the 122nd audio engineering society convention, Vienna

Riekehof-Böhmer H, Wittek H, Mores R (2010) Voraussage der wahrgenommenen räumlichen Breite einer beliebigen stereofonen Mikrofonanordnung. In: Proceedings to the 26. Tonmeistertagung des VDT, pp 481–492

Shirley B, Kendrick P, Churchill C (2007) The effect of stereo crosstalk on intelligibility: comparison of a phantom stereo image and central loudspeaker source. J Audio Eng Soc 55(10):825

Theile G (2000) Mikrofon- und Mischungskonzepte für 5.1 Mehrkanal-Musikaufnahmen. In: Proceedings to the 21. Tonmeistertagung des VDT, Hannover, 2000, pp 348

Williams M (2012a) Microphone array design for localization with elevation cues. Paper 8601 presented at the 132nd audio engineering society convention, Budapest, April 2012

Williams M (2012b) 3D and multiformat microphone array design for the GOArt project. In: Proceedings to the 27. Tonmeistertagung des VDT, Cologne, Nov 2012, p 739

Wittek H (2002) Image assistant V2.0. http://www.hauptmikrofon.de. Accessed 24 June 2008

Wittek H (2012) Mikrofontechniken für Atmoaufnahmen in 2.0 und 5.1 und deren Eigenschaften. In: Proceedings to the 27. Tonmeistertagung des VDT, Köln

Wittek H, Rumsey F, Theile G (2007) Perceptual enhancement of wavefield synthesis by stereophonic means. J Audio Eng Soc 55:723

Chapter 8
Analysis of Frequency-Dependent Signal Cross-correlation (FCC) in Stereo Microphone Systems

Abstract An enhanced form of audio signal cross-correlation analysis is applied by measuring the frequency-dependent cross-correlation (FCC) as well as the frequency-dependent inter-aural cross-correlation (FIACC) by use of a Neumann KU81 artificial head (introduced as a 'human reference') for the most important stereo main microphone system configurations (large and small AB, XY, ORTF, Blumlein-Pair, Faulkner phased Array, KFM, Jecklin disk) as well as a few 'industry standard' and legacy recordings (TELDEC, RCA 'Living Stereo'). Frequency-dependent signal coherence is not analyzed for these samples, as it has been shown in the previous chapter that frequency-dependent signal correlation is superior in terms of detailed analysis, especially in respect of signal phase. It was possible to show that determining the FCC or FIACC provides a characteristic 'fingerprint' for each stereo microphone technique. In respect of the case studies, it can be seen that the FCC analysis is able to unveil the relative level balance between main microphone signal and the spot microphone signals (TELDEC recording). The measured low degree of signal correlation with the RCA 'Living Stereo' recordings, based on large-AB microphone techniques, turns out to be the reason for the high listener appreciation of this series of legacy recordings.

Keywords Signal correlation · Signal coherence · FCC (frequency-dependent cross-correlation) · FIACC (frequency-dependent inter-aural cross-correlation) · Microphone technique · Living Stereo

A concise description of the most important (two- and multi-channel) stereo microphone systems can be found in Streicher and Dooley (1985) as well as in Chap. 3. The original motivation of undertaking an investigation into researching frequency-dependent signal correlation in microphone systems was in order to compare the properties of 'small AB' versus 'large AB' and to gain knowledge about how 'stereophonic'—or not—the most common stereo microphone techniques are over their entire frequency range.

As we will see in this chapter, there are vast differences in this respect between various techniques.

E. Pfanzagl-Cardone, *The Art and Science of Surround and Stereo Recording*,
https://doi.org/10.1007/978-3-7091-4891-4_8

A study on the 'Phase Relationship in Stereo Signals from Dual Microphone Set-Ups' has been published in Brixen (1994), but the aim of this study was mainly to find out what happens if stereo signals are processed through surround decoders in respect of the spatial distribution and phase relationship of the resulting signals. As has already been reviewed in Chap. 2, in Julstrom (1991) a detailed analysis of signal correlation for XY and MS systems in respect of direct and diffuse sounds has been undertaken. The research by Muraoka et al. (2007)—which will be referred to below—has analyzed spatial cross-correlation for signals of AB, WF ('wavefield'), XY, MS and ORTF microphone systems.

The first group of recordings used for the following analysis was of a chamber music ensemble, which had been recorded simultaneously with 5 different main microphone systems (small AB, ORTF, XY with cardioids, MS and a Schoeps 'Kugelflächenmikrofon'). This material was taken from Jecklin (1999) (see discography), and the recording venue will be referred to as 'hall 3.'

The second group of test recordings has been made at the 'Grosse Festspielhaus' Salzburg concert hall (referred to as 'hall 1') during the years 2001–2002. The following microphone techniques have been used: ORTF, AB (0.2, 0.4, 0.8, 1.3, 3.2, 7.2 and 12.0 m) as well as AB-PC (12 m). Various symphony orchestras served as sound-sources unless noted otherwise. Microphones used were Schoeps MSTC64 (ORTF cardioids) and Schoeps CMC5 + MK3 (diffuse-field compensated omnis).

Additional single recordings have been done during the year 2000 in London in the following locations: a small church with rather 'live' acoustics and a TV recording studio (very 'dry' acoustics). The sound sources involved were: a small orchestral ensemble (church) and an instrumental ensemble of 8 musicians (TV studio). Microphones used were Neumann U-87 in fig-8 and cardioid mode (ORTF, Blumlein-Pair; church), Neumann U-87 in fig-8 and cardioid mode (MS; TV studio).

Another group of recordings comes from a commercially available reference CD (see discography Chesky 1994). On this CD, short pieces, performed by a woodwind trio, can be found which have been recorded in a relatively dry studio. From these recordings, the following techniques were used for this research: XY cardioids with 90° opening angle and Blumlein-Pair (rem.: no information available on the microphones used). One artificial head recording of an organ at the 'Altes Gewandhaus' in Leipzig is also analyzed in this chapter (from Jecklin 1999).

In the following figures, the cross-correlation of the electrical signals of the techniques under test is displayed with a solid black line (FCC), while the frequency-dependent inter-aural cross-correlation is represented by a dotted, colored line (FIACC). The underlying simulation of binaural signals for the measurement of the FIACC has been generated by use of a VST software plugin with the name 'Pano 5' programmed by software company 'Wave Arts.' This plugin is based on the HRTFs of the 'KEMAR' dummy head (manufacturer: Knowles Electronics) which has been measured at the MIT in the USA. For the simulations presented in this research, the plugin parameters were set to simulate 'ear signals' for a loudspeaker setup with speakers at ±30° and a listener distance of 2 m.

If available, two different recordings with the same microphone technique (but in different locations) were analyzed in order to get a more balanced impression about

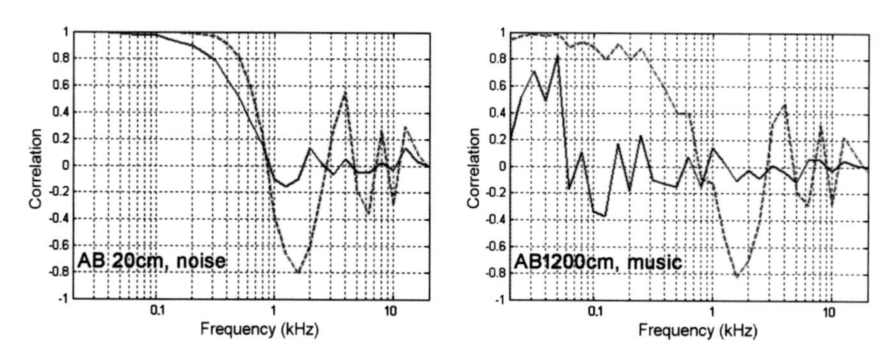

Fig. 8.1 Left:Small AB 20 cm (noise, hall 1) and right: large AB 1200 cm (music, hall 1); FCC (solid line) and FIACC (dotted line)

their 'real-world' sonic properties. Many of the microphone techniques which have been tested are characterized by rather high signal correlation at low frequencies (below approximately 500 Hz), which impedes good spatial impression. Among these are: small AB, XY, ORTF, Faulkner phased array, KFM and Jecklin disk, as will be shown below.

For the FCC of the small AB recording on the left side in Fig. 8.1, an out-phase maximum (of value approximately −0.2) can be seen at a frequency of around 1200 Hz. As has already been shown in Chap. 2, Fig. 2.4 (from Cook et al. 1955), this out-of-phase part in signal correlation is typical for AB microphone systems and the frequency range in which it occurs is directly related to the capsule spacing. A spacing of 40 cm, for example, leads to negative correlation at around 600 Hz, and below this frequency signal correlation increases steeply.

8.1 Calculation of the 'Critical Frequency' in Small AB Microphone Systems

Once the microphone spacing is in the order of half the acoustic wavelength at a given frequency, lateral sound can still be picked up 180° out of phase. Below that frequency, sound will—by law of physics—be picked up with less than 180° phase offset, which means essentially that the correlation coefficient can only get more positive. In other words: as frequency drops below this 'critical frequency' (f_{crit}), the signal becomes more and more monophonic.

As also proposed by Hecker (see Hecker 2000), the critical frequency below which the correlation increases from 0 toward 1 can be calculated approximately as:

$$f_{\text{crit}} = c/(2r) \tag{8.1}$$

with

f frequency in (Hz)
c speed of sound in (m) (i.e. approximately 340 m/s at 20 °C)
r microphone spacing in (cm).

As an example, a 'small AB' microphone spacing of 40 cm will deliver accurate spatial reproduction (for replay via loudspeakers) only down to approximately 430 Hz. This is not suited to ensure sufficient signal de-correlation below 500 Hz, which has been identified of being essential for delivering solid spatial impression to human listeners (see Chap. 1 on spatial hearing). Therefore, in order to achieve spatial fidelity down to 40 Hz, two omni-directional capsules need to be spaced by at least 4.3 m:

$$r = c/(2f) \quad \text{hence} \quad 34{,}000\ \text{cm/s}/(2 \cdot 40\ \text{Hz}) = 425\ \text{cm}$$

This spacing is still narrow enough to work well with small to medium size orchestras. The already very high degree of correlation of a small AB microphone arrangement will be made even higher by the low-frequency 'spatial deterioration' effect of a standard stereo loudspeaker setup [see (Hirata 1983) and Chap. 1, Sect. 1.6.3].

8.2 Calculation of the 'Out-of-Phase Maximum Frequency'

With an AB pair of omni-directional microphones, the maximum negative correlation of the two-channel signal is reached at a frequency f, where the capsule spacing r is essentially two-thirds of wavelength λ:

$$r = 0.66\lambda \tag{8.2}$$

(Rem.: compare to FCC functions in Figs. 8.1, 8.2 and 8.3)

$$\lambda = c/f \tag{8.3}$$

with

λ wavelength

therefore

$$f_{\text{opm}} = 0.66c/r \tag{8.4}$$

Calculation of frequency of the most negative correlation for Figs. 8.2 and 8.3:

Figure 8.2: $\lambda = 34{,}000$ cm/620 Hz s $= 54.84$ cm, therefore $r = 0.66\,\lambda \sim 36.5$ cm

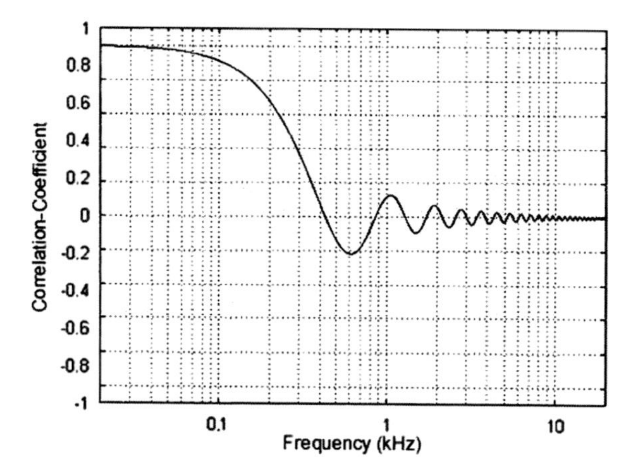

Fig. 8.2 Calculation of the cross-correlation coefficient versus frequency for a pair of omni-directional microphones with a separation of 40 cm in a diffuse sound field (from Martin 2005)

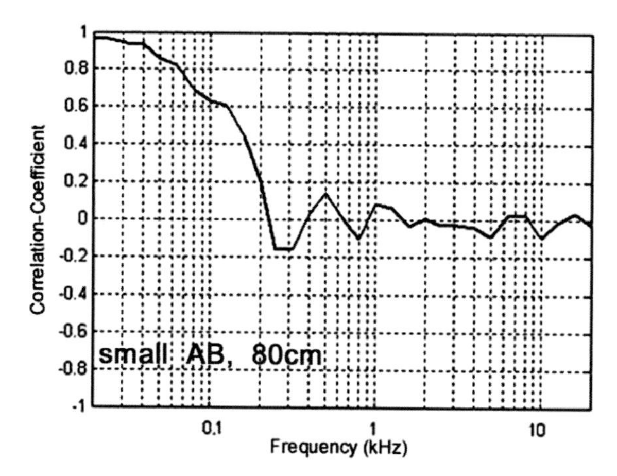

Fig. 8.3 Measured FCC for a 'small AB' pair of omni-directional microphones with 80 cm capsule spacing (f_{opm} = 290 Hz)

Figure 8.3: $\lambda = 34{,}000$ cm/290 Hz s = 117 cm, therefore $r = 0.66\ \lambda \sim 78$ cm

The relationship between capsule spacing (r) and the resulting correlation coefficient R (in a reverberant sound field) in respect of frequency (f) was for the first time calculated and measured in the mid-fifties (see Cook et al. 1955 as well as Sect. 2.1 in Chap. 2), which is about around the same time when the recording teams of various record companies experimented with the challenges of stereo signals and their correlation on a practical level.

For an enlarged capsule spacing of 1200 cm (large AB on the right of Fig. 8.1), both the 'critical frequency' f_{crit} and the 'out-of-phase maximum frequency' f_{opm} shift—at least based on a calculation according to formulas 8.1 and 8.2—to a frequency below 20 Hz:

'Critical frequency':

$$f_{\text{crit}} = c/(2r); \quad \text{or} \quad f = 34{,}000\ \text{cm/s}/(2 \cdot 1200\ \text{cm}) = 14.6\ \text{Hz}$$

'Out-phase maximum frequency' (with cross-correlation coefficient of -0.2):

$$f = c/\lambda \quad \text{and} \quad \lambda = r/0.66 \quad \text{hence} \quad f_{\text{opm}} = 0.66c/r; \text{ or}$$
$$f = (0.66 \cdot 34{,}000 \text{ cm/s})/1200 \text{ cm} = 18.7 \text{ Hz}$$

Looking at Fig. 8.1 (right) in the measured FCC of AB1200, neither f_{crit} nor f_{opm} is clearly visible as—most likely—there are either low-frequency room modes which are relevantly influencing the acoustic properties of the room at the positions of the A and B microphone, or some low-frequency structural noise (air-conditioning system, etc.) was altering the result.

8.3 Comparison of the FCC and FIACC of Various Stereophonic Microphone Techniques

In comparison with Fig. 8.1, the XY stereo microphone arrangements in Fig. 8.4 generally display a higher degree of correlation which diminishes only toward high frequencies. For the XY 90° arrangement, correlation is high (i.e. >0.6) up to about 2 kHz; a similar characteristic can be seen with the XY 120° arrangement, but correlation remains lower also at low frequencies (below 1 kHz) and never reaches a value of +1. This result is very plausible since the larger physical opening angle of the capsules at 120° leads to a better channel separation.

In regard to the calculated FCC of coincident XY microphones with varying opening angles in Fig. 8.5, it is interesting to note that the calculated simulation shows a frequency-independent cross-correlation coefficient for all opening angles of 0°, 90°, 120° and 180° (of +1, +0.75, +0.625 and +0.5, respectively), while the 'real-world' measurements of XY 90° and XY 120° of Fig. 8.4 show a clearly frequency-dependent behavior. This apparent discrepancy can easily be explained by the fact that all calculations of the FCC from Muraoka et al. (2007) take into account

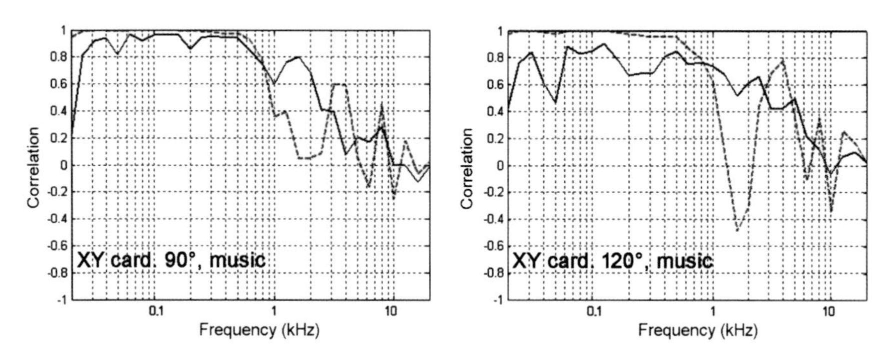

Fig. 8.4 Left: XY cardioids at 90° (music, studio) and right: XY cardioids at 120° (music, hall 3); FCC (solid line) and FIACC (dotted line)

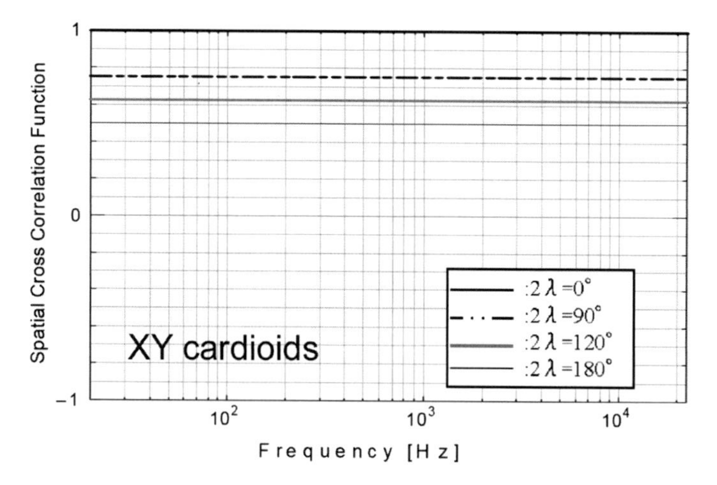

Fig. 8.5 Calculated FCC of coincident XY microphone systems (for diffuse sound only) with 0°, 90°, 120° and 180° opening angles (after Muraoka et al. 2007)

only the diffuse part of the sound field, while for the 'real-world' measurements of Fig. 8.4 direct and diffuse sounds were 'acoustically summed' at the microphone inputs of the XY system.

In respect of the term 'critical frequency', which has been introduced in the context of AB microphone systems, the question arises, whether it may make sense to try to find or define such a frequency also for the other common stereo microphone techniques. As the coincident XY system—at least in theory—does not even touch on a de-correlation of ±0 (while the real-world XY signals do), the 'critical frequency' of an XY microphone system should most likely not be associated with the signal correlation function crossing 0. In fact, as long as signal cross-correlation is low (i.e. below approximately +0.3) this is not 'critical' or negatively affecting the 'spatial impression' aspect of a recorded signal. As it gets critical only once the cross-correlation rises to +0.6 or above, it seems to make more sense to define that the 'critical frequency' has been reached once the cross-correlation coefficient equals—or exceeds—a compromise value in the middle between 0 and +1 of—say—+0.5.

For the measured XY systems with 90° and 120° opening angle, this would be the case in the frequency range between 2 and 3 kHz.

Looking back on our small AB 20 cm spaced microphone pair (see Fig. 8.1), applying the same proposal of 'critical frequency' at a correlation coefficient of + 0.5, it would result in f_{crit} now dropping down to approximately 500 Hz (instead of the previous 900 Hz).

Examining the frequency-dependent cross-correlation coefficient of the two MS systems in Fig. 8.6, it looks like the underlying function may have an almost linear characteristic (along the logarithmic frequency scaling) from +1 (at 10 Hz) declining to 0 (at 20 kHz), with the 'real-world' curve 'meandering' around the 'ideal' value. (Rem: With certainty the concrete value or 'slope' of that curve will depend on

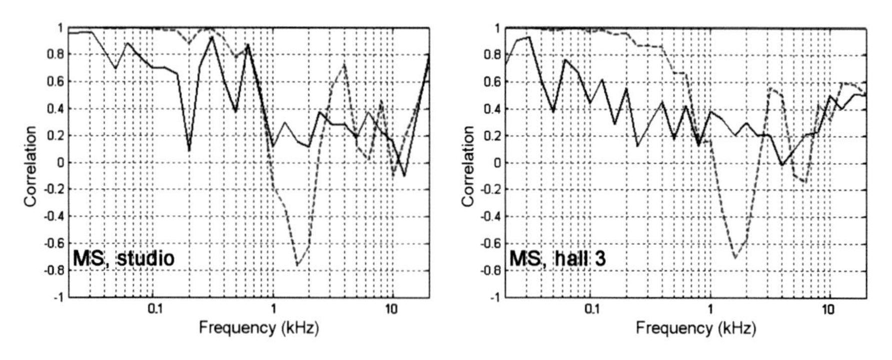

Fig. 8.6 FCC (solid line) and FIACC (dotted line) of two MS systems; left: MS (music, studio); right: MS 110° (music, hall 3)

the recording angle the MS system is set to. Also, it can be noted that with both real-world measurements there seems to be a tendency of a rising FCC from 10 to 20 kHz. The author assumes that this has to do with capsule geometry and inherent deviation from 'ideal' directional pattern with increasing frequency, which is the case for most—even high quality—microphones).

In contrast to the real-world measurements (i.e. direct + diffuse sound), the calculated (diffuse-field correlation) characteristics of MS systems in Fig. 8.7 (Muraoka et al. 2007) show a frequency-independent, 'horizontal' linear cross-correlation coefficient, the value of which depends on the intended opening angle. For an opening angle of 0° the FCC equals +1 (monophonic, or identical information in both channels), for 90° FCC = 0.66, for 120° FCC = 0.25 and for a recording angle of 132° the FCC becomes 0 (complete de-correlation). The authors in Muraoka et al. (2007)

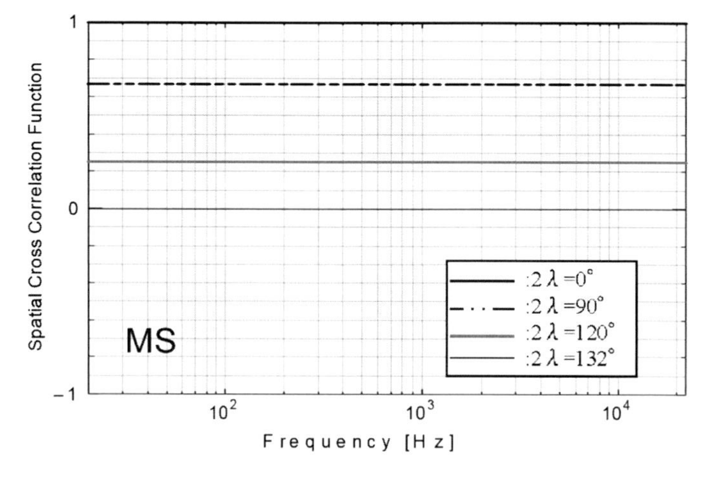

Fig. 8.7 Calculated FCC of MS microphone system (for diffuse sound only) with 0°, 90°, 120° and 132° recording angle (after Muraoka et al. 2007)

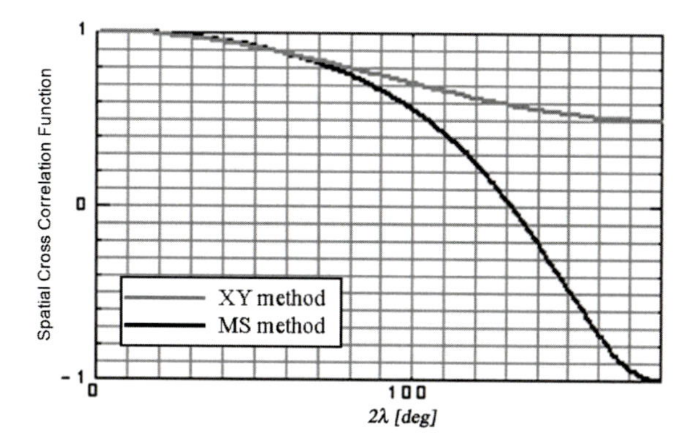

Fig. 8.8 Signal cross-correlation in relation to microphone opening angle for XY and MS microphone systems (from Muraoka et al. 2007)

claim that with the recording angle setting of 132° best results in respect also of diffuse sound pickup were achieved and therefore they favor this arrangement for orchestra concert recording (Fig. 8.8).

In Fig. 8.9, the relation between correlation coefficient, directionality index a and the included angle between two coincident microphone capsules of various characteristics is displayed. It is interesting to note that only with supercardioid, hyper-cardioid and figure-8 microphones, it is possible to achieve complete de-correlation for the signals they deliver, when choosing appropriate opening angles, larger than 90°. The coincident fig-8 arrangement with an opening angle of 90° (Blumlein-Pair) is of course a case of special interest, as the two capsules arranged in this manner capture the whole sound-field within 360° while delivering completely de-correlated signals.

In Fig. 8.10, the FCC functions for two ORTF recordings are displayed: Both exhibit rather high correlation (of >0.5) below 500 Hz, while being pretty much de-correlated above 1 kHz.

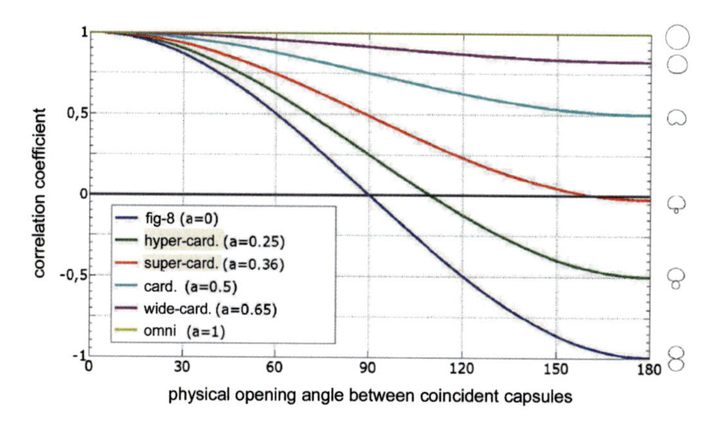

Fig. 8.9 Relation between correlation coefficient, directionality index a and the included angle between two coincident microphone capsules of varied characteristics (adapted from Wittek et al. 2006)

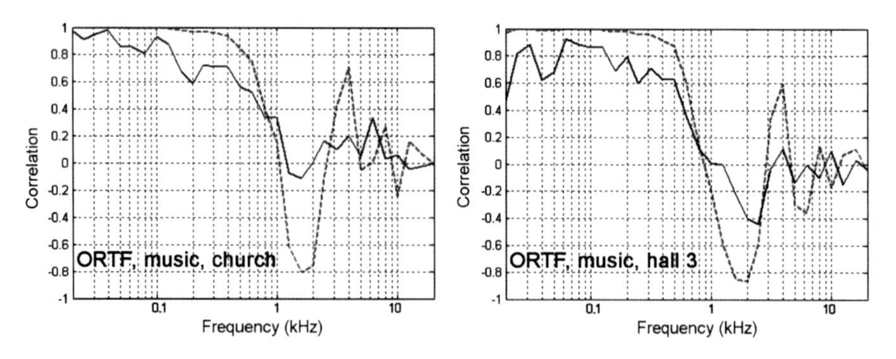

Fig. 8.10 FCC (solid) and FIACC (dotted) of two ORTF systems; left: music, church; right: music, hall 3

In relation to the XY 120° recording from Fig. 8.4, the two ORTF recordings differ technically mainly in respect of the capsule spacing of 17.5 cm, as their physical opening angle is 110°.

When comparing the cross-correlation functions of the XY 120° recording with the right ORTF recording of Fig. 8.10, it is noticeable that both display a dip in the 40–50 Hz region, which is most likely linked to room acoustic reasons, as almost all recordings which have been made in 'hall 3' are characterized by this trait (the only exception: KFM).

The capsule spacing of 17.5 cm, as applied with the ORTF method, finds its visual counterpart in the characteristic dip in the frequency region between 1 and 2.5 kHz, which corresponds to the typical ear-spacing of humans (or small AB systems with similar capsule spacing). While above 1 kHz correlation rolls off only gradually with XY systems, with ORTF systems the signals are already well de-correlated due to the capsule spacing.

For the ORTF calculations in Fig. 8.11, the capsule spacing is held constant at 17.5 cm, while the opening angle is varied to both sides of the 110° specification. According to the calculation for the 90° angle, signal correlation for diffuse sound converges toward a value of +0.75 below the critical frequency. For 110°, it reaches +0.6 with lowering frequency, and for 180° it converges toward +0.5, which was also the case for the calculation of the XY system at 180°.

The signals of the Blumlein-Pair of two crossed figure-of-eight microphones at an included angle of 90° are completely de-correlated in respect of diffuse sound (see Chap. 2, Fig. 2.5, from Elko 2001). The functions of the Blumlein-Pair systems in Fig. 8.12 result from the combination of direct and diffuse sounds in two acoustically very different rooms. A tendency toward de-correlation is noticeable for the whole frequency range displayed.

The following analysis concerns a few less common stereo microphone techniques.

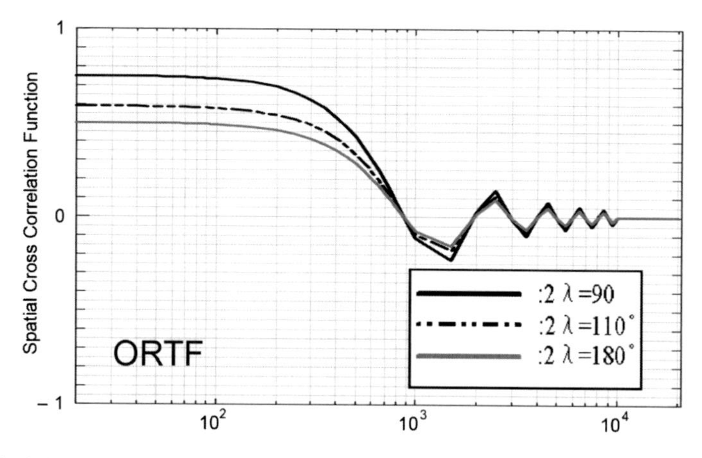

Fig. 8.11 Calculated FCC of ORTF microphone system (for diffuse sound only) with 90°, 110° and 180° recording angle (after Muraoka et al. 2007)

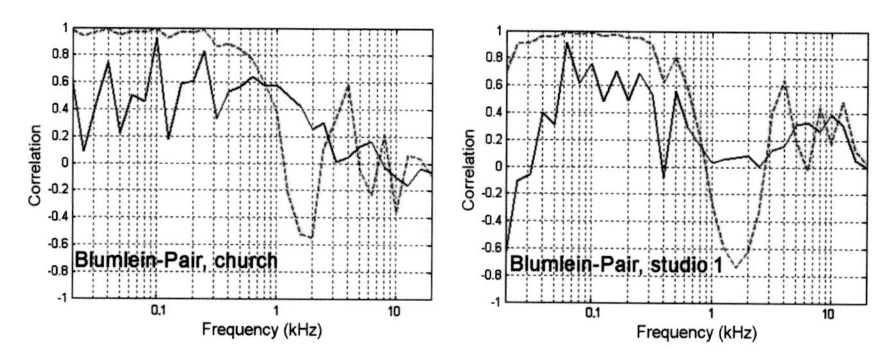

Fig. 8.12 FCC (solid) and FIACC (dotted) of Blumlein-Pair; left: music, church; right: music, studio 1

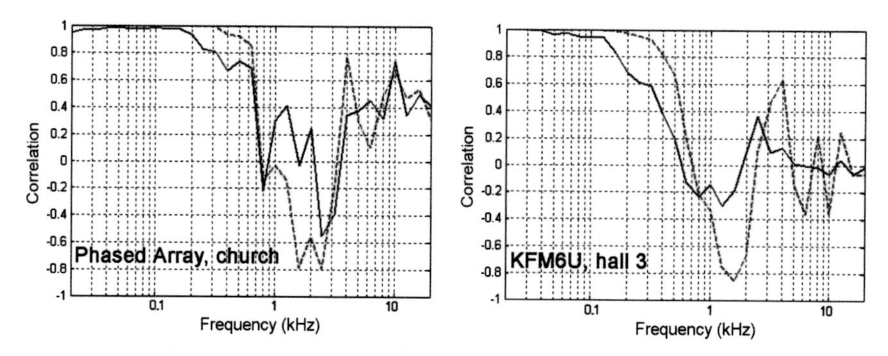

Fig. 8.13 FCC and FIACC; left: phased array (music, church); right: KFM6U (music, hall 3)

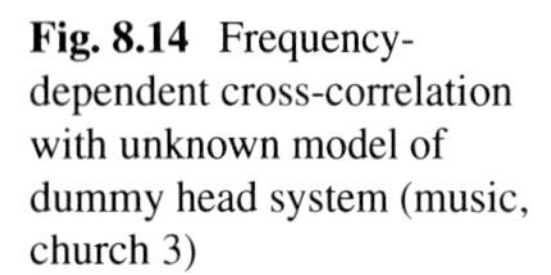

Fig. 8.14 Frequency-dependent cross-correlation with unknown model of dummy head system (music, church 3)

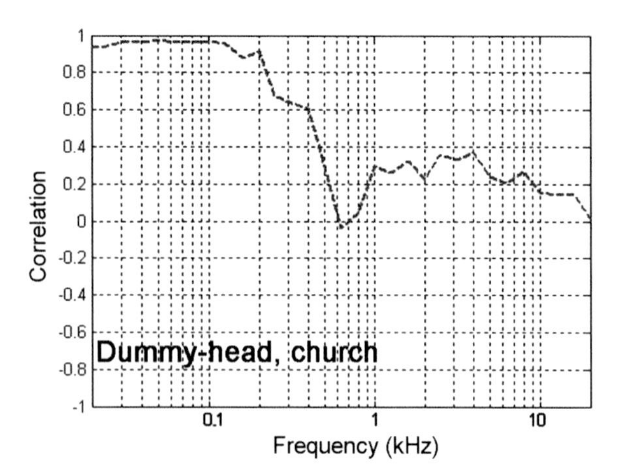

In Fig. 8.13, the cross-correlation function of Faulkner's 'phased array' is displayed (Faulkner 1981), as well as of the Kugelfächenmikrofon KFM6U by company Schoeps.

Both the Faulkner 'phased array' and the KFM6U use a capsule separation of 20 cm: While the phased array uses two parallel front-facing figure-of-eight microphones, the KFM employs two pressure transducers which are integrated into the surface of a sphere.

The FCC measurement of the phased array is characterized by high correlation below 750 Hz and a quite pronounced, narrow out-of-phase region with a value of −0.6 in the frequency band around 2–3 kHz. Above this frequency band, correlation remains in the range from +0.3 to +0.7.

The correlation function of the KFM system on the other hand is very similar to that of an artificial head, which is not very surprising (compare the FCC of Figs. 8.13 right side and 8.14).

As another special case of 'baffle-recording,' the correlation function of the 'OSS system' (Jecklin disk) has been analyzed. As already described in the chapter on stereo microphone techniques, this system—based on two capsules with omnidirectional characteristics (i.e. pressure transducers) with a separating, absorptive disk in between—delivers a basically monophonic signal at very low frequencies (see Jecklin 1981). In Fig. 8.15, it can be seen that correlation is quite high up to 800 Hz with a value of >0.6, after which there is the narrowband, slight out-phase maximum between 1 and 2 kHz which is typical for AB systems, and in the frequency range above, the absorptive disk becomes effective as baffle and alters the normally rather de-correlated signal relation: The measurement of the OSS system in Fig. 8.15 shows a relatively broad out-of-phase band which extends from 9 to 20 kHz.

The ABPC array in Fig. 8.15, right side, consists of a large-AB system with 12 m capsule spacing in combination with an ORTF-Triple as centerfill system (for more details, please see the chapter on stereo microphone techniques) and is characterized by only lightly correlated signals (≤0.4) down to about 300 Hz. Below that frequency,

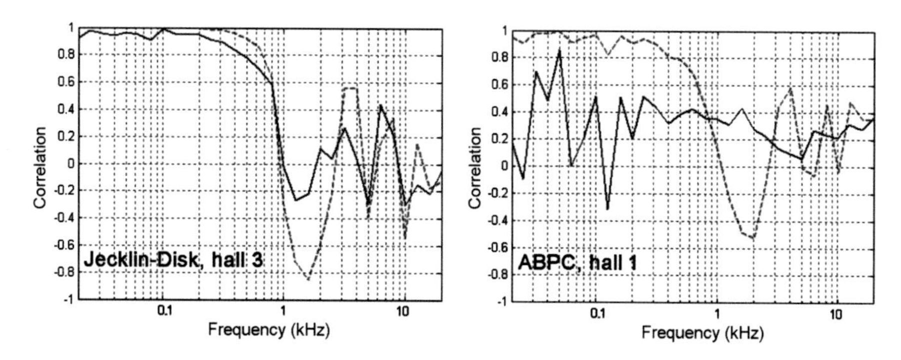

Fig. 8.15 FCC (solid) and FIACC (dotted); left: 'OSS'—Jecklin disk (music, hall); right: ABPC system (music, hall 1)

there are some narrowband peaks with higher correlation which are most likely caused by the room acoustics, as the correlation dips at 200, 130 and 50 Hz are equally noticeable in the ABPC system in Fig. 8.15 as well as in the large-AB system in Fig. 8.1, right side. The same is the case for correlation peaks at approximately 240, 160, 80, 40 and 20 Hz.

Essentially, the FCC of the ABPC system results more or less from a summing of the large-AB (1200 cm) function of Fig. 8.1, right side (identical with Fig. 8.16, left side) with the function of the ORTF system of Fig. 8.16, right side, which—due to the high correlation in the low-frequency range below 500 Hz—pushes up the 'low-correlation' curve of the large-AB system (rem.: it is not a 'full level' and frequency linear summation though, as the signal of the center cardioid of an ORTF-Triple is missing, and also because the ORTF system's signals are high-pass filtered at around 100–150 Hz, according to the proposal which can be found in Pfanzagl-Cardone (2002) in order to preserve good signal de-correlation at low frequencies).

As a final comparison, we want to take a look at two 'industry recordings', which—stylistically—belong to two different 'aesthetic schools' and use main microphone

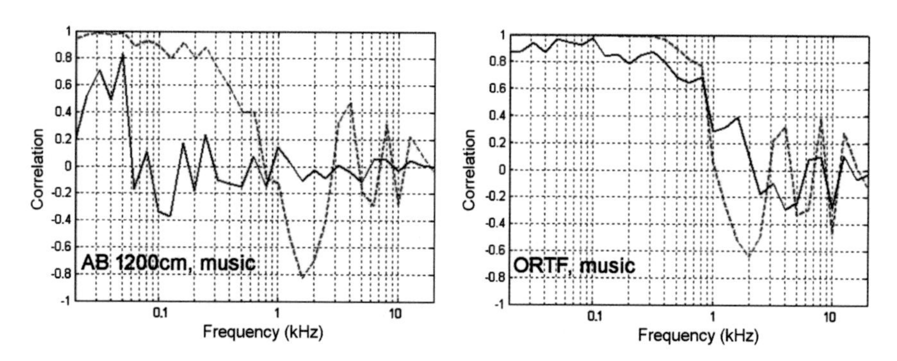

Fig. 8.16 FCC (solid) and FIACC (dotted); left: large-AB 1200 cm (music, hall 1); right: ORTF system (music, hall 1)

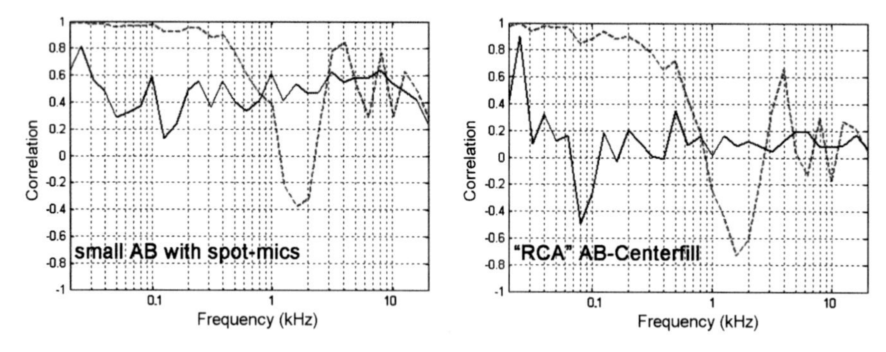

Fig. 8.17 FCC (solid) and FIACC (dotted); left: small AB + spot-mics (TELDEC; music, hall); right: AB-Centerfill system (RCA 'Living Stereo'; music, hall)

systems in combination with spot microphones, as opposed to the previously analyzed recordings, which make use of main microphones only.

In Fig. 8.17, left side, the FCC and FIACC of a TELDEC recording can be seen (see Discography Teldec 1995), made by Tonmeister Eberhard Sengpiel, who has also been lecturer at the UdK (Universität der Künste), Berlin. The microphone technique he has used is small AB in combination with spot microphones.

The second analysis in Fig. 8.17 (right side) has been taken from the 'Red Seal' ('Living Stereo') series of recordings (see Discography RCA 1957), the label under which the US-American company RCA has released their first stereo recordings which they made from the mid-1950s onward. The microphone techniques used can be considered rather purist, as they were employing a minimum of microphones [usually based on a 'large-AB with centerfill' system; see (Valin 1994)].

These recordings are cherished by music lovers worldwide and have been re-released in CD format in the 1990s and from 2003 onward also in the 'multichannel SA-CD' format (stereo and 3-channel).

As can be seen in Fig. 8.17, the 'AB-Centerfill' recording is de-correlated over the whole frequency range down to 20 Hz (meandering around a cross-correlation value of approximately +0.1), while the TELDEC small AB recording alternates between relatively low correlation of +0.4 and relatively high correlation of +0.6 (with a presumed mean correlation value of +0.5).

Analyzing the FCC and FIACC functions in Fig. 8.17, a question arises concerning the level balance between the small AB main microphone and the spot microphones of the TELDEC recording: As the typical high-correlation characteristic (rising up to a value of +1) of small AB spacings for the low-frequency band is not visible in Fig. 8.17 (left side), it is rather obvious that the signals of the spot-microphones must have been added to the mix at such a high level that it seems discussable whether the small AB microphone system can still be considered to have been used as a true 'main microphone' system, which—at least in the opinion of the author—should contribute to at least about two-thirds of what is being heard in the final mix. (Rem.: As a 'rule of thumb', the signal of spot-microphones usually gets mixed in at a relative level

−10 dB in respect of signal of the main microphone. Therefore, it would need a lot of spot-microphone signals (or: mixing them in at rather high levels) in order to ‘overrule’ the signal of the main microphone).

What regards the measured high degree of signal de-correlation with the RCA ‘Living Stereo’ recordings, based on large-AB microphone techniques, it can be assumed that this is (apart from the musical quality of course) most likely the main technical reason for the high listener appreciation this series of legacy recordings has received from the public. As has already been pointed out in Chap. 2 on the topic of ‘spatial hearing,’ an appropriate degree of signal de-correlation on replay leads to a higher apparent source width and—in the low-frequency range—is also essential for achieving spatial impression, which again is highly related to listener preference as was pointed out in the analysis of the listening tests in Chap. 6, as well as the in the findings of other researchers (see Chaps. 2 and 6 for references) (rem.: the MATLAB code, which has been applied to measure the FCC and FIACC, as presented in the various figures of this chapter, can be found in the appendix of this book).

8.4 Conclusions, Further Analysis and Literature

So what characterizes a ‘healthy’ stereo signal? In order to enable good spatial impression, it needs to be de-correlated at low frequencies. In addition, it should preferably also display low correlation at frequencies above 1 kHz, as higher correlation in the upper frequency band in the electrical (replay) signal will also lead to high correlation for the FIACC, which is not desirable, as we have already seen when examining the surround recordings.

Other detailed analyses of frequency-dependent cross-correlation in stereo microphone systems can be found, e.g., in Muraoka et al. (2007), the results of which are displayed in Fig. 8.18.

Also, Wittek has made a very interesting analysis on the importance of diffuse-field correlation (DFC), as being conveyed through stereo and surround microphone signals in Wittek (2012) (Fig. 8.19).

In the context of the research conducted by Riekehof-Böhmer et al. (2010), listening tests have been performed in relation to the question of how much the diffuse-field component influences the ‘apparent width’ (or stereophonic width) of a sound event. In their research, an asymptotic value for the DFC has been determined, which—in essence—is the integral of the squared, weighted coherence function (low frequencies being more important). With this—the DFI (‘diffuse-field image’) predictor—a value is obtained, by which different mic techniques can be compared and the ‘apparent width’ can be determined.

In reference to Fig. 8.20, the following evaluation can be found in Wittek (2012), as has already been pointed out in the final Sect. 7.4.1. of Chap. 7: ‘If we deem a minimum quantitation value of +2 sufficient, the following conclusions can be drawn:

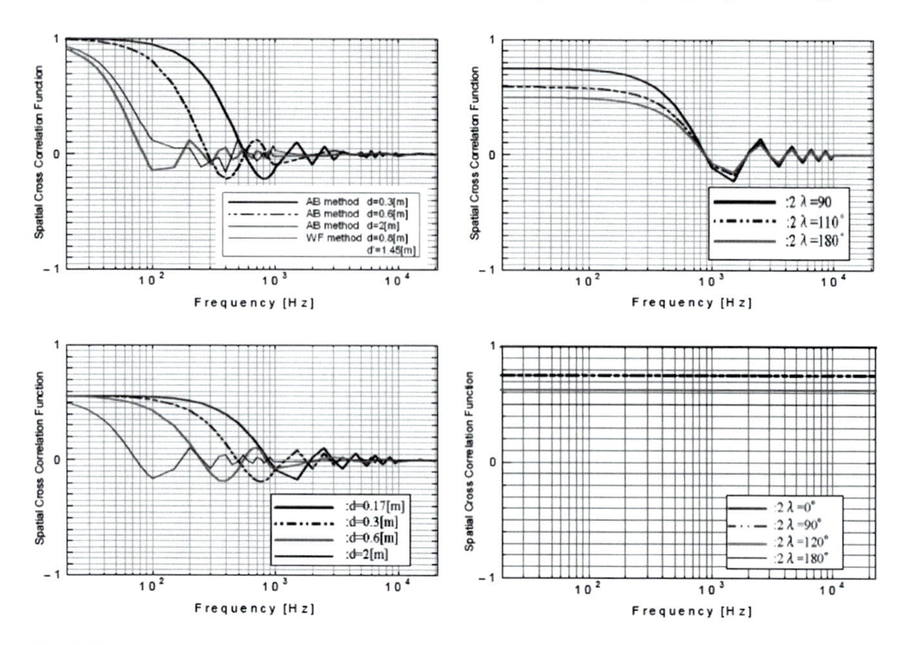

Fig. 8.18 Calculation of the cross-correlation for various stereo microphone techniques; FSCC = frequency-dependent spatial cross-correlation (from Muraoka et al. 2007). Left up: FSCC patterns of AB microphones and wavefront microphone; left down: FSCC of ORTF microphones (capsule spacing-dependent); right up: FSCC of ORTF microphones (directional azimuth-dependent); right down: FSCC of XY microphone (cardioid or supercardioid)

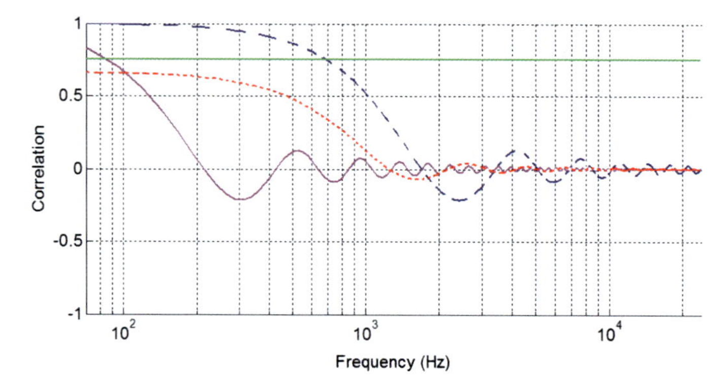

Fig. 8.19 Diffuse-field correlation (DFC) for various stereo microphone techniques: (listed from top to bottom) blue dashed line: small AB omnis $d = 10$ cm; green straight line: XY cardioids 90°; red dotted line: ORTF cardioids, 17 cm, 110°; violet: small AB, omnis 80 cm (modified from Wittek 2012)

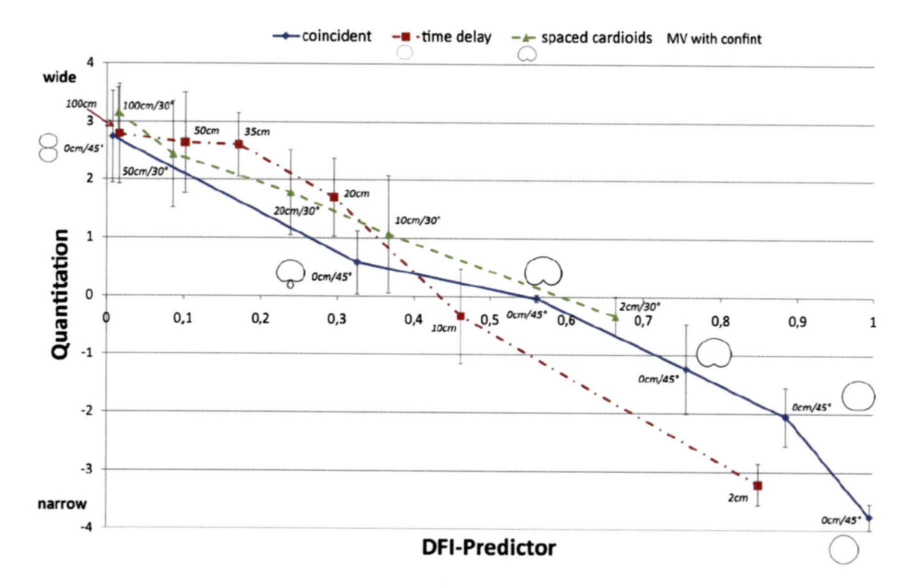

Fig. 8.20 Listener evaluation of apparent width of diffuse sound (reverb), recorded by various microphone techniques (from Riekehof-Böhmer et al. 2010); mean value with confidence intervals

- The Blumlein-Pair was the only 'good' coincident technique, according to the listening test. Its DFC is 0.
- With equivalence and pure runtime techniques, the DFI predictor needs to be smaller than 0.5.
  For pure runtime techniques, this means the capsule spacing has to be ≥35 cm. For equivalence techniques, the spacing can be smaller, depending on the polar pattern and orientation of the microphones involved.
- An optimum, 'wide' reproduction of diffuse sound can be achieved both via coincident and spaced techniques.
- As pure runtime techniques need to have a capsule spacing of at least 35 cm in order to achieve optimum results, multichannel setups using baffles with a smaller effective diameter will not work to satisfaction, as the baffle effect vanishes with lowering frequency (due to λ).'

The statement above singles out the Blumlein-Pair of crossed figure-of-eights in a positive way in what concerns signal de-correlation, but also makes it clear that small AB techniques (also with baffles) with an effective acoustic capsule spacing of less than 35 cm will not produce satisfactory results, due to the lack of sufficient signal de-correlation at low frequencies. Among the stereo microphone techniques, this criterion concerns—for example—Jecklin's OSS technique (Jecklin disk) and the 'KFM' Kugelflächenmikrofon by company Schoeps.

Wittek further concludes: '… The diffuse sound field plays an enormously important role both for spatial impression as well as sound color. Therefore it has to be made sure that it is being reproduced with low correlation. Many good or bad sonic properties of a stereophonic technique are not related to their localization characteristics, but to their ability to reproduce a beautiful, open 'spatial impression'. XY- cardioids are a good example: good localization, but poor spatial impression. In the future much more attention should be paid to Diffuse Field Correlation. Very often only the desired recording angle is considered as a decisive factor for choosing a specific microphone technique. The DFC is at least as important' (from Wittek 2012).

Later on, he concedes that the very importance of DFC in relation to the quality of a main microphone has been neglected in many research projects, even of his own. In respect of surround or main microphones for SFX/atmo-recordings, he even regards the DFC to be of higher importance than localization accuracy.

The findings by Wittek, cited above, are in essence what the author of this book has tried to point out in previous publications (see Pfanzagl 2002; Pfanzagl-Cardone 2002, 2011; Pfanzagl-Cardone and Höldrich 2008), while drawing conclusions and proposing new microphone techniques (e.g. AB-PC; see Chap. 3 on stereophonic microphone techniques).

As has already been outlined in Sect. 2.3, the optimal inter-aural correlation for music perception in concert halls is 0.23, as found by Gottlob (1973). All this seems to indicate that microphone systems with high signal correlation should probably better be abandoned for the sake of systems that are largely de-correlated. As was found in research by Nakahara (2005) and can be seen in Fig. 2.15, low correlation of the playback signals (L/R vs. C vs. LS/RS) ensures a better compatibility between different listening environments and also leads to an enlarged sweet spot (see also Prokofieva 2007).

Further notice

As a closing note to this chapter, I would like to direct the interested reader toward some documentary videos which concern signal cross-correlation over frequency, which are related to the information provided in Figs. 8.1, 8.2, 8.3, 8.4, 8.5, 8.6, 8.7, 8.8, 8.9, 8.10, 8.11, 8.12, 8.13, 8.14, 8.15, 8.16 and 8.17. These videos can be found on my YouTube channel by the name futuresonic100: Watch out for videos containing the description 'mic tech analysis' in their title.

References

Brixen EB (1994) Phase relation in stereo signals from dual microphone set-ups. Paper 3825 presented at the 96th Audio Eng Soc Convention, Amsterdam 1994

Cook RK, Waterhouse RV, Berendt RD, Edelman S, Thompson MC (1955) Measurement of correlation coefficients in reverberant sound fields. J Acoust Soc Am 27(6):1072

Elko GW (2001): Spatial coherence functions for differential microphones in isotropic noise fields. In: Brandstein M, Ward D (eds) Microphone arrays. Springer, pp 61

Faulkner T (1981) A phased array. Hi Fi News Record Rev, July 1981

Gottlob D (1973) Vergleich objektiver akustischer Parameter mit Ergebnissen subjektiver Untersuchungen an Konzertsälen. Dissertation, Universität Göttingen, 1973

Hecker P (2000) The decision of the microphone spacing and its creative benefit (German). In: Proceedings to the 21. Tonmeistertagung des VDT, Hannover, 2000, pp 796–804
Hirata Y (1983) Improving stereo at L.F. Wireless World, Oct 1983, p 60
Jecklin J (1981) A different way to record classical music. J Audio Eng Soc 29(5):329–332
Julstrom S (1991) An intuitive view of coincident stereo microphones. J Audio Eng Soc 39(9)
Martin G (2005) A new microphone technique for five-channel recording. Paper 6427 presented at the 118th Audio Eng Soc Convention, Barcelona 2005
Muraoka T, Miura T, Ifukuba T (2007) Ambience sound recording utilizing dual MS (Mid-Side) microphone systems based upon Frequency dependent Spatial Cross Correlation (FSCC). Paper 6997 to the 122nd Audio Eng Soc Convention, Vienna, 2007
Nakahara M (2005) Multichannel monitoring tutorial booklet (M2TB) rev. 3.5.2. Yamaha Corp 2005, SONA Corp, p 41
Pfanzagl E (2002) Über die Wichtigkeit ausreichender Dekorrelation bei 5.1 Surround-Mikrofonsignalen zur Erzielung besserer Räumlichkeit. In: Proceedings to the 21. Tonmeistertagung des VDT, Hannover, 2002
Pfanzagl-Cardone E (2002) In the light of 5.1 surround: why AB-PC is superior for symphony-orchestra recording. Paper 5565 presented at the 112th Audio Eng Soc Convention, Munich, 2002
Pfanzagl-Cardone E (2011) Signal-correlation and spatial impression with stereo- and 5.1 surround-recordings. Dissertation, University of Music and Performing Arts, Graz, Austria. https://iem.kug.ac.at/fileadmin/media/iem/altdaten/projekte/dsp/pfanzagl/pfanzagl_diss.pdf. Accessed Oct 2018
Pfanzagl-Cardone E, Höldrich R (2008) Frequency-dependent Signal-Correlation in Surround- and Stereo-Microphone Systems and the Blumlein-Pfanzagl-Triple (BPT). Paper 7476 presented at the 124th Audio Eng Soc Convention, Amsterdam, 2008
Prokofieva E (2007) Relation between correlation characteristics of sound field and width of listening zone. Paper 7089 presented at the 122nd Audio Eng Soc Convention, Vienna 2007
Riekehof-Böhmer H, Wittek H, Mores R (2010) Voraussage der wahrgenommenen räumlichen Breite einer beliebigen stereofonen Mikrofonanordnung. In: Proceedings to the 26. Tonmeistertagung des VDT, pp 481–492
Streicher R, Dooley W (1985) Basic microphone perspectives—a review. J Audio Eng Soc 33(7/8)
Valin J (1994) The RCA bible—a compendium of opinion on RCA living stereo records, 2nd edn. The Music Lovers Press, Cincinatti, Ohio
Wittek H, Haut C, Keinath D (2006) Doppel-MS – eine Surround-Aufnahmetechnik unter der Lupe. In: Proceedings to the 24. Tonmeistertagung des VDT, Leipzig, 2006
Wittek H (2012) Mikrofontechniken für Atmoaufnahmen in 2.0 und 5.1 und deren Eigenschaften. In: Proceedings to the 27. Tonmeistertagung des VDT, Köln, 2012

Discography (Compact Disks)

Chesky (1994) Best of Chesky Classics & Jazz and Audiophile test disc—volume 3. Tracks 15, 18. Chesky Records, JD 111
Jecklin J (1999) The low art of recording. Personal Audio Signal Reference-CD
RCA (1957) Antonin Dvorak, Symphony No 9 in E minor, op95, 1st movement, RCA-Victor (BMG Classics) Nr 09026 62587 2, 1957 (CD Re-Release: 1995)
Teldec (1995) Gustav Mahler, Symphony No 9 in D major, 3rd movement. TELDEC Nr 4509-90882-2, 1995

Chapter 9
Discussion and Conclusion: An Attempt at a Qualitative Ranking of Stereo and Surround Microphone Techniques—An Examination of Various Studies in the Field

Abstract First, we take a look at the newly defined 'critical frequency,' below which the FCC (frequency-dependent cross-correlation) of a microphone technique consistently exceeds a value of ≥+0.5. The most important stereo microphone techniques are analyzed in this respect, based on the measurements of Chap. 8. Coincident techniques with various microphone patterns are analyzed in relation to various aspects of their sonic character, including also their ability to reproduce the diffuse sound field in an appropriately de-correlated manner. A critical look at 'small AB' versus 'large AB' is taken, before some data on the frequency-dependent radiation characteristics of string and brass instruments and their impact on sound image stability are presented. The frequency-dependent reverberation radius (or 'critical distance' CD) of wind instruments is analyzed. Acoustic 'zoom effects,' resulting from the use of close-up mic techniques and their impact on ASW (apparent source width) are discussed. A ranking of important stereo mic techniques, based on the results of a listening test by Ceoen, is presented. In respect of surround mic techniques, the advantages and disadvantages of various surround speaker layouts are analyzed by use of FIACC (frequency-dependent inter-aural cross-correlation) measurements. Afterward, a qualitative ranking of 12 surround microphone techniques, derived from five extensive listener-based comparative tests, is presented.

Keywords Critical frequency · Diffuse sound · Reverberation radius · Critical distance · Zoom effect · Frequency-dependent cross-correlation

Before undertaking the daring attempt to arrive at a qualitative ranking of stereo (and further down also 5.1 surround) microphone techniques, the author would like to state his conviction that—of course—there is certainly not one particular microphone technique, which could be considered 'best' for all possible recording situations. As most recording engineers will agree, the variables of sound source size and complex (i.e. frequency-dependent) radiation characteristics, room acoustics, external noise sources, the necessity to be able to rebalance individual sound sources in an ensemble, etc., do all contribute to the fact that the appropriate microphone technique for capturing a sonic event needs to be chosen very carefully on an individual basis. Therefore, it can happen that a microphone technique does not render satisfactory results in one situation, but may have an advantage over other techniques

E. Pfanzagl-Cardone, *The Art and Science of Surround and Stereo Recording*,
https://doi.org/10.1007/978-3-7091-4891-4_9

under different conditions. In listening tests, sometimes the program selection itself (e.g. solo instrument instead of small ensemble as sound source) may already be a determining factor, without even changing the room acoustics.

Nevertheless—with knowledge gained through experience, experiments and well-founded scientific understanding of underlying acoustic principles—it is possible to arrive at 'informed decisions' which should be very helpful in choosing an appropriate technique for a particular event beforehand. In Sect. 9.1 below, we will try to give an outline by describing the general aesthetic attributes of the most common stereo microphone techniques and by comparing some measured acoustic parameters.

When analyzing the frequency-dependent signal cross-correlation in surround microphone systems in a previous chapter, it was shown that estimating the signal coherence of a microphone system is not able to unveil potential weaknesses (e.g. in respect of phase anomalies; see KFM) as clearly, as by measuring signal correlation (see var. figures in Chaps. 7 and 8).

Measuring signal correlation seems to have a clear advantage over estimating signal coherence, and therefore—and also for reasons of efficiency—we have restricted ourselves to an analysis of signal correlation over frequency in the chapter of stereo signal analysis, in which both the FCC and FIACC for some of the most common stereo microphone techniques have been documented.

9.1 Stereo Microphone Techniques

Whenever available, two different recordings with the same microphone technique (but in different locations) have been analyzed in Chap. 8 in order to get a more balanced impression of their 'real-world' performance. Many of the microphone techniques tested are characterized by rather high signal correlation at low frequencies (below approximately 500 Hz), which impedes good spatial impression. Among these are:

Small AB,
XY (with cardioids),
ORTF,
Faulkner phased array,
KFM and
Jecklin disk (OSS).

Based on the results of the data analyzed in this chapter, only the microphone techniques MS, Blumlein-Pair, large AB, ABPC and AB-Centerfill provide sufficiently de-correlated signals also at low frequencies to ensure good spatial impression.

The 'critical frequency' has been defined in Chap. 8, below which the FCC (frequency-dependent cross-correlation) of a microphone technique reaches (for the first time and then successively surpasses) a high value of $\geq$+0.5 and tendency-wise (depending on the technique under test) converges toward +1 (Rem.: the previous criterion of only surpassing a correlation value of zero and then converging toward

+1, as in the case of the spaced AB microphone technique, can be abandoned for the sake of a more widely applicable—and probably psycho-acoustically more relevant—higher correlation value of +0.5 which should be the 'anchor' for determining the 'critical frequency' of any microphone system in a specific room).

If we make a ranking of microphone techniques according to $f_{crit\ 0.5}$ based on the measurements of the FCC presented in the last chapter, we arrive at the following:

The order of microphones in Table 9.1 shows quite clearly that techniques which are coincident or based on small capsule spacings usually have a much higher value for their $f_{crit\ 0.5}$ than techniques which are either based on 'large AB' capsule spacings or well de-correlated coincident signals, due to high microphone directivity combined with an appropriate recording angle, which results in enhanced channel separation. Just one additional remark to Table 9.1: The fact that for XY 90°, a lower $f_{crit\ 0.5}$ value of 2200 Hz is displayed than for XY 120° (with $f_{crit\ 0.5} = 2800$ Hz) may be a bit misleading. Looking at the rather complex function of the FCC in Fig. 8.4, it can be seen that XY120 has a lower overall cross-correlation value than XY90, as can also be expected by the undoubtedly higher channel separation due to the greater physical opening angle. The higher value of this readout is due to the individual variation (deviation) in the FCC function from an 'ideal' curve, due to individual sound source and room acoustic properties in that specific recording situation.

As also pointed out in Muraoka et al. (2007) in general, spaced-pair microphone setups have an advantage in terms of ambience pickup (i.e. diffuse sound) but sometimes unstable image localization, while coincident microphone systems can be excellent in localization but unsatisfactory in terms of ambience, with a rather narrow spatial impression. This was confirmed—among others—through the

Table 9.1 'Critical frequency' (at correlation value of +0.5) for stereo main microphone systems (based on the measurements presented in Chap. 8)

Microphone technique	$f_{crit\ 0.5}$ (Hz)
Small AB with spot mics (TELDEC)	12,000
XY 120°	2800
XY 90°	2200
Jecklin disk	800
ORTF	600–900
MS	200–800
Faulkner phased array	700
Small AB 20 cm	500
Blumlein-Pair	500–1200
Kugelflächen microphone	350
AB-PC with ORTF-Triple	250
MS 110° (MS 'hall 3')	210
Large AB 1200 cm	45
AB-Centerfill (RCA 'Living Stereo')	18

following research projects Ceoen (1972), Wohr and Nellseen (1986) and Takahashi and Anazawa (1987).

Already in Dickreiter (2011), a graphical representation of the main microphone technique 'classes' in general (i.e. coincident, equivalent, spaced-pair, artificial head) and their 'behavior' in relation to localization, spatial impression, stage depth and presence can be found.

More recently, Olabe has proposed a stereo technique selection scheme (Olabe 2014), based on direct point evaluation, which takes into account objective qualities (such as mic separation and the measurement of the FCC), as well as subjective qualities (such as the grading, which a specific microphone technique has received through listening tests according to the Berg and Rumsey psychoacoustic scale (see Berg and Rumsey 2003)), and is meant to be a help in respect of the selection of the best suited microphone technique for a specific purpose (Fig.9.1).

In Streicher and Dooley (1985) 'Basic Microphone Perspectives—A Review,' a concise description of the most common stereo microphone techniques can be found. Below I would like to cite from this source, regarding the overall characteristics of each technique.

Stereo technique selection using DPE method (direct point evaluation)

Stereo mic technique	**Objective qualities**		**Subjective qualities**		Mono compatibility
	Mic separation	FCC	Scoring in all of B&R except for LOC and SDIST	Scoring for LOC & SDIST	
XY coincident	+	+ + +	+	+ +	+ + +
ORTF equivalence	+ +	+ +	+ +	+ +	+ +
AB Δt–stereoph.	+ + +	+	+ + +	+	+
Decca tree	+ + +	+	+ + +	+	+ +

FCC = Frequency dependent Cross Correlation
LOC = Localisation attribute
SDIST = Source distance attribute
B&R = Berg & Rumsey psychoacoustic scale

Fig. 9.1 Stereo technique selection by direct point evaluation (modified after Olabe 2014)

9.1.1 Coincident XY with Cardioids and Hyper-cardioids

From Streicher and Dooley (1985): ' …The microphone pair is typically set at an included angle of between 60 and 120°. The specific angle chosen determines the 'apparent width' of the stereo image, and the choice of this angle is subjective, with consideration given to the distance of the microphone pair from the sound source, the actual width of that source, and the polar pattern of the microphones. A critical factor to consider when using this technique is polar response. As the individual microphones are oriented at an angle to most of the sound source, considerable off-axis sound coloration is possible. As with any stereo technique, the microphones comprising the pair should have as good a polar response as possible. Further, they should be closely matched with regard to polar and frequency response, since any differences will cause the image to wander with changes in pitch.

1. **Cardioid microphone**. Use of cardioid microphones is common in coincident techniques, typically with an included angle of 90–120°, and placed fairly close to the sound source. Often the axes of the microphones are aimed at the extremes of the sound source. As the direct-to-reverberant-sound ratio of this approach is high, this can offer some rejection of unwanted sound from the rear of the pair. Sometimes a distant pickup with a large reverberation component is desired. In such circumstances, included angles as large as 180° may be employed.
2. **Hypercardioid microphone**. Using a hypercardioid pair is similar to using cardioids, except that the included angle is typically narrower to preserve a solid center image. The increased reach of the hypercardioid allows a more distant placement for a given direct-to-reverberant-sound ratio. With their small reverse-polarity lobes, using hypercardioids is a good compromise between implementing XY with cardioids and the Blumlein technique. …'

David Griesinger also cherishes the use of hyper-cardoids at an included angle of 133° as the signals derived from such a pair are largely de-correlated, which he claims important in order to achieve good spatial impression in a recording (Griesinger 1987).

Due to the relatively high signal correlation of the XY90 technique (see Fig. 8.4, and the values of $f_{crit\,0.5}$ in Table 9.1, as well as the calculations in Muraoka et al. (2007), it is characterized by a high degree of 'mono-compatibility,' which is probably one of the reasons for its frequent use by radio sound-engineers, at least as long as this was an issue for broadcast. As this technique is also bare of the risk of out-of-phase components in the low-frequency range, it was a 'safe choice' for recordings intended for vinyl record release, for which (strong) out-of-phase LF signal information represents a 'no-go' due to mechanical reasons connected to the process of master disk cutting.

Having said all this, it is also clear that XY90 is not the most 'stereophonic' of all microphone techniques, as—looking at Fig. 8.4, left—below approximately 1 kHz (where $\rho = 0.6$) the FCC converges toward a value of $+1$, so essentially the signal becomes almost monophonic!

9.1.2 Coincident XY with Figure-of-Eights (Blumlein-Pair)

From Streicher and Dooley (1985): '…The *crossed pair of figure of eights* is the earliest of the XY techniques, and is configured with two bidirectional microphones, oriented at an included angle of 90°. It was developed in the early 1930s by British scientist Alan Blumlein and was presented in his seminal patent (1931).

One attribute of this technique is that the rear lobes of these microphones record the rear 90° quadrant in phase but out of polarity, and place this into the stereo image (cross channeled) together with the front quadrant. Signals from the two side quadrants are picked up out of phase. Placement is therefore critical in order to maintain a proper direct-to-reverberant-sound ratio and to avoid strong out-of-phase components. Typically, this technique works very well in a wide room or one with minimal side-wall reflections, where strong signals are not presented to the side quadrants of the stereo pair. It is often commented that this configuration produces a very natural sound …'

This listener appreciation may have very much to do also with the ideal property of the Blumlein-Pair signal pickup characteristics in relation to diffuse sound (i.e. completely de-correlated, which is essential for good spatial reproduction), which has been outlined in Chap. 2 (see also Elko 2001 and Julstrom 1991).

However, despite the above appraisal of the 'very natural sound,' the Blumlein-Pair apparently also has one weak point, as Tomlinson Holman describes it in (2000, p. 95): '… The system aims the microphones to the left and right of center; for practical microphones, the frequency response at 45° off the axis may not be as flat as on axis, so centered sound may not be as well recorded as sound on the axis of each of the microphones.' In practice this means that—depending on the acoustics of the recording room, the size and distance of the sound source, etc.—the listener might experience something like a 'hole-in-the-middle' effect, similar to the one that can be found with large AB microphone arrangements (if no additional 'centerfill' system is used).

This is of course one of the main reasons why it makes very much sonic sense to add—for example—a third figure-of-eight capsule to the Blumlein-Pair arrangement (see proposal for the BPT Microphone in Sect. 3.1.5), as this helps to stabilize the center image of a sound source and essentially can be used to fill the hole in the middle. In addition, this gives the sound-engineer a high degree of freedom, of how broad (i.e. 'stereophonic') or narrow ('monophonic') he would like the sound image to be, by simply varying the level of the center capsule in relation to the L and R capsules. For normal recording applications, the level of the center microphone will usually be set between −6 and −3 dB in relation to the L and R microphones. Deviations of this are, of course, possible, and for some applications it might make sense to set it at −10 dB or—in case it was necessary to emphasize the center of the sound source—it may be set to 0 dB or even higher. In order to shield off unwanted signals from the rear, it is preferable to place an acoustic barrier (e.g. in the form of a sound-absorptive acoustic panel) behind the BPT microphone. This helps to achieve a better direct/diffuse sound (i.e. direct signal/reverb) ratio and enables the sound-engineer to move the system further away from the sound source, thereby usually achieving also a better balance between instruments (or instrumental groups) in case of larger ensembles.

As pointed out by Rumsey in (1994), in Clark et al. (1958) it has been shown "... that the Blumlein-Pair arrangement delivers output-signals which differ in amplitude with varying angle over the frontal quadrant by an amount which gives a very close correlation between the true angle of offset of the original sound source from the center-line and the apparent angle of reproduction *on loudspeakers which subtend an angle of 120°* (sic) to the listening position. At lesser angles the change in apparent angle is roughly proportionate as a fraction of total loudspeaker spacing, maintaining a correctly proportioned 'sound stage'. It was also shown that the difference in level between channels should be smaller at high, than low frequencies in order to preserve constant correlation between actual and apparent angle, and this may be achieved by using an equalizer in the difference channel which attenuates the difference channel by a few decibels for frequencies above 700 Hz [this being subject of a British Patent by Vanderlyn (1957)] ...

[See also Sect. 1.6.1 on 'Frequency dependent localization-distortion in the horizontal plane' and the research by Knothe and Plenge (1978), as well as Griesinger (2002)].

Gerzon (1986) has suggested a figure between 4 and 8 dB, depending on program material and spectral content, calling this 'spatial equalization'. The ability of a system based only on level differences between channels to reproduce the correct timing of *transient* information at the ears has been questioned, not least in the discussion following the original paper presentation by Clark et al. (1958), but these questions were tackled to some extent by Vanderlyn (1979) a much later paper, in which he attempted to show how such a system can indeed result in timing differences between the neural discharges from the ears to the brain, taking into account the integrating effect of the hearing mechanism in the case of transient sounds. He quoted experimental evidence to support his hypothesis, which is convincing. If a system based only on level differences did not cope accurately with transients, then one would expect transients to be poorly localized in subjective tests, and yet this is not the case, with transients being very clearly located in coincident-pair recordings.

Recordings which having been made by use of coincident microphone techniques in terms of psychoacoustic perception are making use of what is called 'Summation localization theories' by Theile (1991), who dismisses such theories as not allowing for the correct natural inter-aural attributes of sound signals required for spatial reproduction. Theile states that the correct way to derive signals suitable for natural stereo reproduction is to generate 'head referred' signals, i.e. signals similar to those used for binaural systems, since these contain the necessary information to reproduce a sound image in the 'simulation plane' between two loudspeakers. Subjective test results have shown that binaural signals equalized for a flat frequency response produce convincing stereo from loudspeakers. There is strong disagreement between those who adhere to this theory and those who adhere more closely to traditional summation theories of stereo. …" (from Rumsey 1994). The reader is referred to Rumsey (1991) for a broader discussion of the subject.

Coming back to the point that the Blumlein-Pair arrangement apparently '… gives a very close correlation between the true angle of offset of the original sound source from the center-line and the apparent angle of reproduction on *loudspeakers which subtend an angle of 120° (sic)* to the listening position …' the author would like to point out that this property renders a 'back-to-back' setup of two 'Blumlein-Pfanzagl-Triples' (with an absorptive panel in between; see Fig. 4.13 in Chap. 4) ideal for use with a 6-channel surround loudspeaker replay setup with equally included angles of 60° between the speakers (see Fig. 9.2, speaker setup '6a' or '6b').

In case of a limit of 6 reproduction speakers, this 6-channel speaker setup is ideal also for replaying moving sound sources, and also very well suited for replaying de-correlated diffuse sound in the horizontal plane. The first BPT microphone needs to be positioned to capture signals for the front channels L, C, R, while the BPT behind the panel captures the rear-channel signals LS, CS (center surround) and RS (rem.: obviously, this loudspeaker setup differs in terms of loudspeaker positions from the regular 6.1 surround specification).

With a larger loudspeaker base setup of +/−60° from center (which is at 0°) for the L and R speakers—as is the case for the speaker setup '6a' in Fig. 9.2—it becomes even more apparent that there is the need for an additional center signal, as provided by the BPT's center capsule, in order to fill a potential acoustic 'hole in the middle.'

9.1.3 Small Versus Large AB—Revisited

Now we are going to examine an orchestra recording setup:

Recordings in the form of coincident, near coincident or small AB techniques have one major disadvantage in comparison with large AB, as they almost always provide a perspective 'from the inside' of the orchestra, meaning from somewhere along the center line that splits the orchestra in left and right of the conductor's position. Usually, the microphones are also quite close to the orchestra, especially to the string instruments next to the conductor's podium. Only, XY techniques with crossed cardioids or figure-of-eights (Blumein-Pair) might allow to move the microphones half the stage width—or even more—out in the hall (see Faulkner 1981). For most of the other arrangements, the microphones stay inside the critical distance (reverberation radius), usually slightly back from the conductor's position and between 3 and 5 m above the stage.

Large AB on the other hand captures the orchestra from a more 'balanced' position (see Fig. 9.3). Since the recording often gets played back on home systems with loudspeaker spacings of much smaller dimension, it also seems to be the right approach to try to capture the orchestra in its full width, which will suffer 'down-scaling' on playback anyway. It is the author's experience (with orchestra as well as opera recording) that widely spaced main microphone techniques, which make

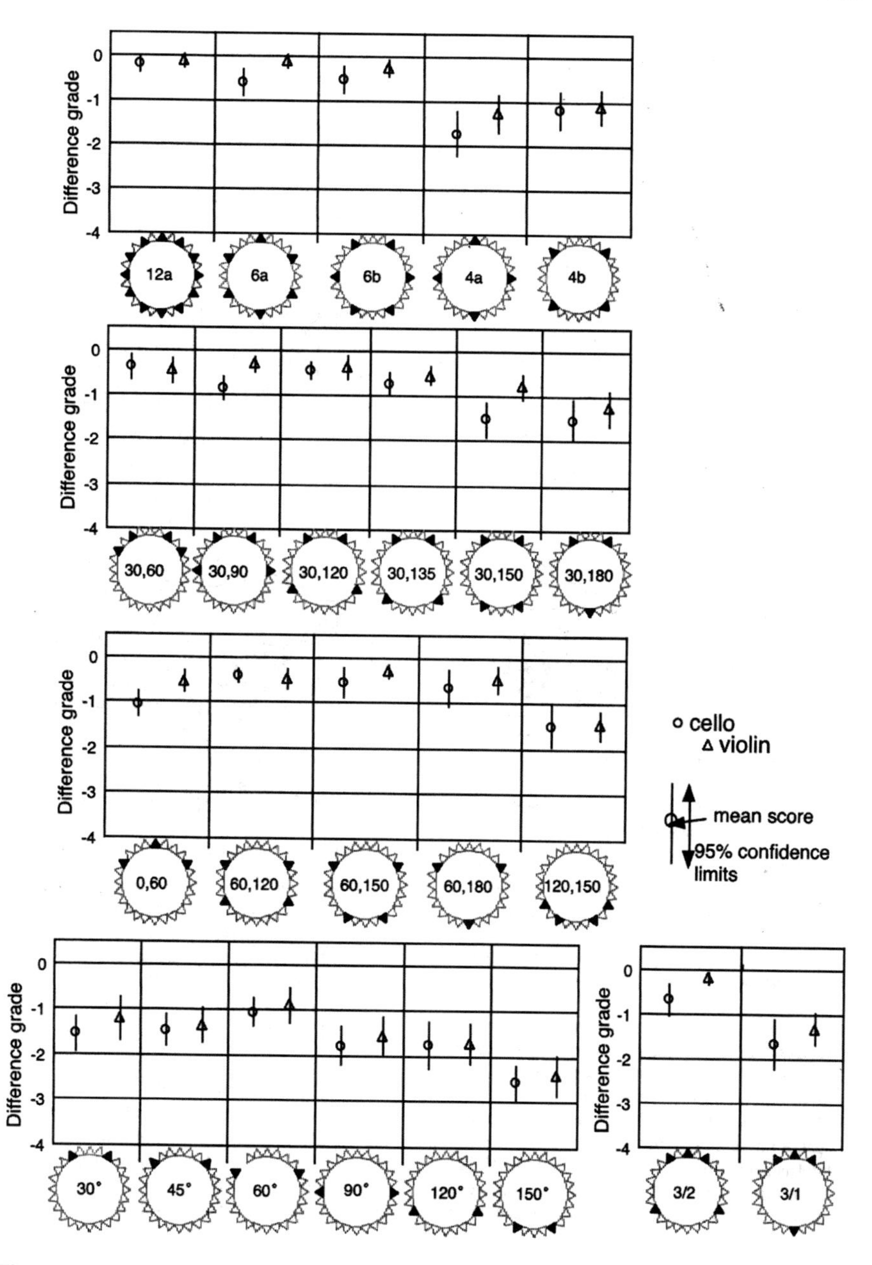

Fig. 9.2 Multichannel loudspeaker setups from the test of Hiyama et al. (2002)

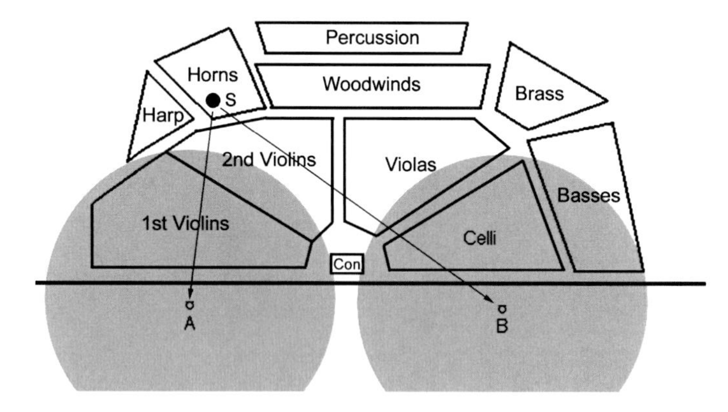

Fig. 9.3 Large AB, capsule spacing approximately 12 m; reverberation radius approximately 5.8 m (grey area)

use of only very few spot-microphones, preserve the sense of 'space' much better when played back on low-fi sound systems with very small speaker spacings (TV monitors, for example) than most other techniques.

Looking at Fig. 9.3, we can see that with a large AB system, about half of the orchestra instruments are within the critical distance (or reverberation radius) of the two microphones A and B.

In addition, the majority of the rear instrumental sections of the orchestra are quite loud and have highly directional radiation characteristics, which helps in terms of both localization and direct-to-diffuse sound ratio.

In addition to better L/R amplitude separation (due to the level attenuation with distance, according to the inverse square law applied for sound propagating from point sources), time-of-arrival differences in a 'large AB' system are much bigger in comparison with a 'small AB' system.

With 'small AB' technique, sound sources outside the critical distance from the main system will not be localized properly since the diffuse field dominates the direct sound. As a consequence, the need for the use of spot-microphones arises. In case of a 'large AB' system, amplitude differences, as well as the much larger time-of-arrival differences, allow localization even for sound sources outside of the reverberation radius.

The 'hole-in-the-middle' effect, inherent to the large AB technique, can easily be compensated for with an appropriate centerfill system, as will be shown later. With large AB, there may be a certain degree of localization distortion, as some critics are claiming. Most sources refer to theoretical calculations based solely on loudness differences, as published in the original paper on the experiments at the Bell Laboratories from 1934 (Steinberg and Snow 1934). One of the authors of this paper has revised these statements in a later publication (Snow 1953), in which also time-of-arrival differences and quality differences (referring to the amplitude–frequency spectrum) have been taken into consideration.

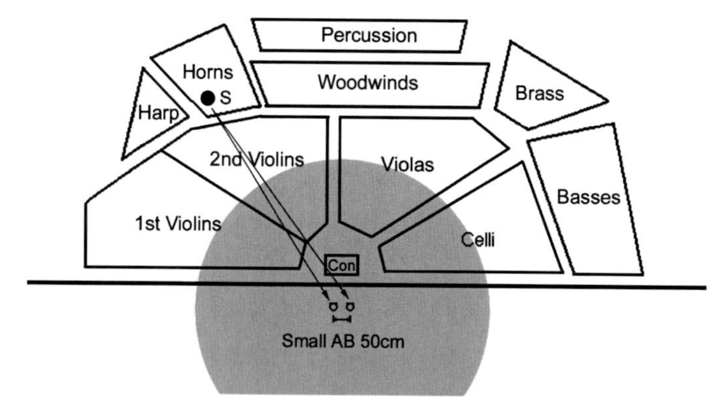

Fig. 9.4 Small AB, capsule spacing 50 cm; reverberation radius approximately 5.8 m (grey area)

Looking at Figs. 9.3 and 9.4, it should become clear that a large AB-based recording system immediately delivers a higher amount of balanced information on a large percentage of the orchestra, which reduces the need for extra spot-microphones to a minimum, while the small AB technique delivers unsatisfactory spatial impression due to high signal correlation at low frequencies and needs to be supported by a lot of spot-microphones in order to achieve a proper balance and direct-to-diffuse sound ratio for all the instruments within the orchestra.

The capsule spacing in large AB systems has even one more advantage:

It is also common practice among sound-engineers to set up so-called ambience or room microphones at a certain distance to the sound source, which are usually omni microphones that are meant to capture de-correlated diffuse sound. Of course, integrating their signal into the mix later on needs to be done with an appropriate level, as essentially they represent a recording of the sound source from 'far away' so there is a relevant time delay of this signal in relation to the signals of the main microphone (e.g. small AB, XY cardioids, etc.) which could lead to unwanted acoustical effects, among them 'echo'.

When using a 'large AB'-style capsule spacing which exceeds the reverberation radius (as is the case in Fig. 9.3, for example), the 'large AB' pair of microphones themselves already fulfill the purpose of ambience microphones: The A microphone on the left side will function as an ambience microphone for the right side of the orchestra, and the B microphone will do the same for the left orchestra side!

In the case of good acoustic properties of the concert hall, the amount of diffuse sound arriving at the A and B microphone will be just fine and the overall sonic picture obtained by the large AB pair will certainly be much clearer than a mix of—for example—a 'small AB' main microphone and an additional pair of 'room microphones.'

This aesthetic-acoustic fundamental possibility has already been evident to the great Harry F. Olson and Herbert Belar from RCA Laboratories in New Jersey as they stated in their AES Paper from 1960, titled 'Acoustics of Sound Reproduction in the Home':

"… In stereophonic reproduction it is possible to restore reverberation to the amount optimum for direct listening. For instance, instruments whose direct sound is reproduced from one speaker can have the reverberant sound reproduced from the other and vice versa. This suggests that for a maximum spacious room-effect the instruments be made to favor either one or the other microphone or groups of microphones for direct pickup and that the other microphones are angled so as not to pick up direct sound but mostly reverberant sound for those instruments unless the difference in the path to the two microphones or groups of microphones is large enough to be counted as a first reflection. In this case the direct sound can also be picked up by the second group of microphones. The ability to carry the room sound will be least when the source is to appear in the center between the two speakers during reproduction. Therefore, depending upon the program material, with stereophonic reproduction it is possible to vary from intimate 'highly-damped-room' sound reproduction to the effect of a great hall just adjoining and opening into the listener's living room. … "
(see Olson and Belar 1960, p. 10)

9.1.4 Microphone Technique, Reverberation Radius and Frequency-Dependent Radiation Patterns of Musical Instruments

The 'reverberation radius' (also called 'critical distance,' often abbreviated as D_{crit} or CD) is located at the distance from a sound source where direct sound and diffuse sound (i.e. reverb) have the same level. When recording, normally a sound-engineer will want to stay well within the reverb radius with his microphone, at least as long as spot microphones are concerned. In the context of the recording of a large sound source, a main microphone will usually be outside the reverb radius at least for part of the orchestra, which is one of the reasons why the use of spot microphones is almost inevitable. However, for 'purist style' recordings, which aim to use only very few or even just one pair of microphones, the decision of where to put these microphones in relation to the reverberation radius of the room becomes crucial.

$$D_{crit}[\mathrm{m}] = 0.057\ \mathrm{sqrt}\left(V\left[\mathrm{m}^3\right] / \mathrm{RT}_{60}[\mathrm{s}]\right) \tag{9.1}$$

with V volume of the room in cubic meters and RT_{60} reverb time in seconds.

With formula (9.1), the reverberation radius or critical distance of an omni-directionally radiating sound source can be calculated, based on the volume of the room and the (medium) reverberation time. If the reverb time in several octave bands is known, the calculation for these bands can be performed individually to determine the critical distance over frequency (or 'frequency-dependent reverberation radius,' an example of which can be found in Chap. 6 in Tables 6.1 and 6.2) (rem.: to be precise—for sound sources with a more directional radiation pattern, a directivity factor Q (in the case of musical instruments and loudspeakers, also frequency dependent) would need to be added to the equation) (Fig. 9.5).

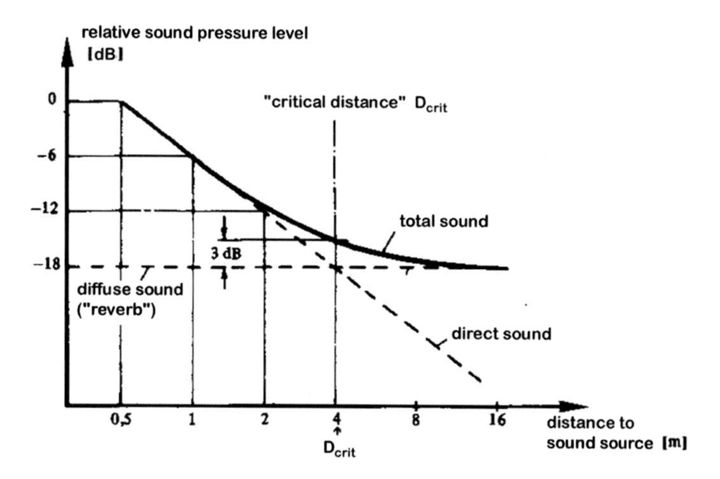

Fig. 9.5 'Reverberation radius' or 'critical distance' (D_{crit})

Table 9.2 Reverb time and resulting critical distance for performance spaces of different sizes

Hall	Volume (m^3)	Volume (ft^3)	Seats	RevTime (s)	D_{crit} (m)	D_{crit} (ft)
Ottobeuren (church, Germany)	130,000	4,590,300	NA	7.00	7.77	25.48
Royal Albert Hall, London	86,650	3,059,612	5080	2.40	10.83	35.53
Amsterdam's Concertgebouw	18,700	660,297	2206	2	5.51	18.08
Boston Symphony Hall	18,740	661,709	2631	1.80	5.82	19.08
Vienna Musikverein	14,600	515,526	1680	2.05	4.81	15.78
Teatro della Scala (opera house, Italy)	11,252	397,308	2289	1.24	5.43	17.81
Eisenstadt Castle—concert hall	6800	240,108	NA	1.70	3.60	11.83
King's Theatre, London	4550	160,661	NA	1.55	3.09	10.13
Brahmssaal, Vienna Musikverein	3390	119,701	604	1.63	2.60	8.53
Esterháza Castle—concert hall	1530	54,024	NA	1.20	2.04	6.68
Control room (typ. rec. studio)	118	4167	NA	0.24	1.26	4.15

In Table 9.2, it can be seen that the reverberation radius of performance spaces can vary in a wide range from more than 10 m (Royal Albert Hall, London), to little more than 1 m (control room of a typical, damped recording studio).

Also, when we are talking about the reverberation radius of a room, the reader should keep in mind that this refers only to the 'medium' reverberation radius, i.e. reverberation radius measured and calculated for the 1 kHz octave-band, as the reverb-radius is usually also frequency dependent (unless the reverb time was really the same for all frequencies, which is a rather unlikely event, as the higher frequencies usually having a shorter reverb time).

In Table 6.2 for example, the reverberation radius has been calculated for a large rehearsal stage with a medium reverb time of 1.5 s, for which the reverberation radius was approx. 2 m. In the same room the reverb time at 8 kHz was only 0.58 s, resulting in a reverberation radius of almost 3.20 m, while at 63 Hz the reverb time was almost 2 s, with an associated critical distance of 1.73 m (which is almost only half of the reverb radius at 8 kHz …).

Therefore it might happen that a sound source in that room, placed at a distance of 2 m from a recording microphone (with omnidirectional characteristics) will be within the reverberation radius for high frequencies, but outside of it (and therefore embedded in 'diffuse sound') for the very low frequencies …

But the interaction of sound source, microphone and room and—hence—the ratio of direct versus diffuse sound at the microphone gets even more complex, as the 'typical' sound source itself is usually not an 'omni-directional radiator' (or point source), but exhibits frequency-dependent radiation characteristics.

As an example, we can take a look at the typical, pitch-dependent radiation characteristics of a violin:

For violin recording, it is very common to position a spot microphone at a distance of 0.5–1 m (1.5–3 ft.) looking down on the instrument: Examining the radiation characteristics as documented in Fig. 9.6, it can be seen that while the largest part of the audio spectrum is well represented at that point, and for the frequency bands around 500 and 1500 Hz the actual sound output of the instrument will not be picked up in an accurate proportion, which may result in (audible) dips in the frequency spectrum.

On the other hand, it is also a commonly known effect that picking up the sound of violins only from above, in a frontal position to the stage or orchestra (like with large AB 'outriggers') there is a tendency that the overall sound might be too sharp, which is not a big surprise, as the main portion of energy in the octave band from 2500 to 5000 Hz is radiated in this direction by the instrument (see Fig. 9.6).

Looking at Fig. 9.7, it can be seen that a member of the audience seated on the balcony of the Musikverein in Vienna will be far outside of the reverb radius for the majority of the orchestra, but well within that radius for an instrument with an extremely pronounced directional characteristic like the trumpet, for which the 'critical distance' at frequencies higher than 10 kHz even reaches the balcony area (Fig. 9.8).

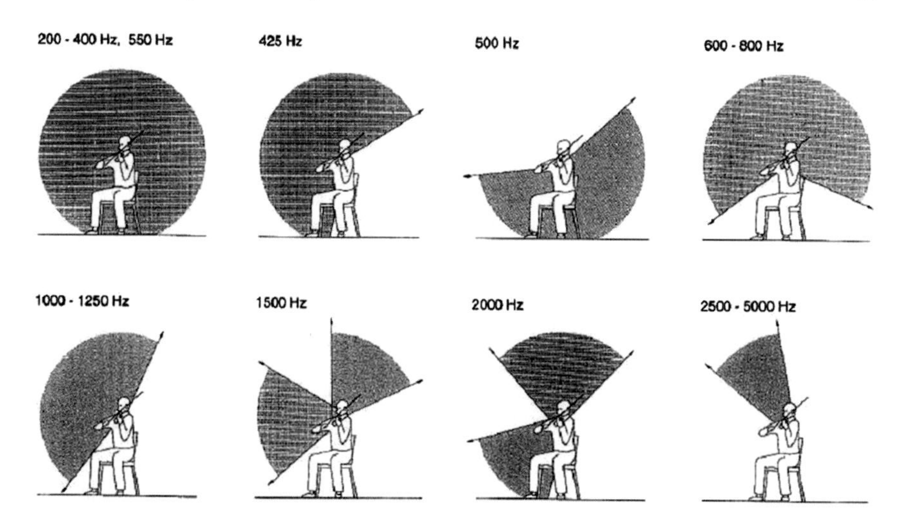

Fig. 9.6 Frequency-dependent radiation characteristics (0 to −3 dB) of a violin in the vertical plane [(Fig. 137 from Meyer (1999, p. 211)]

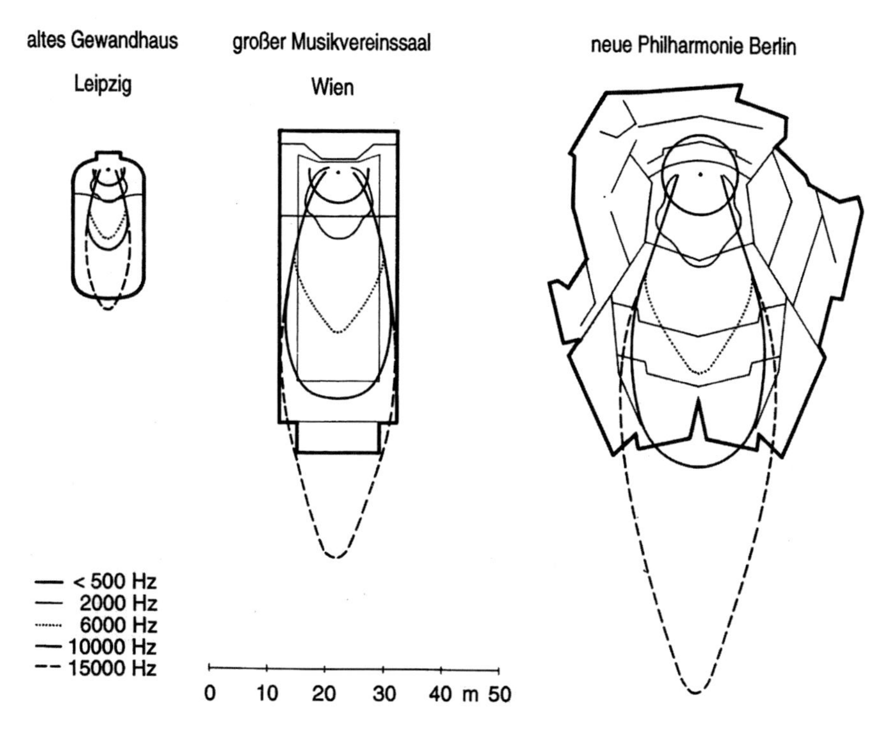

Fig. 9.7 Frequency-dependent 'reverberation radius' (i.e. critical distance) for a trumpet in three different concert halls [Fig. 109 from Meyer (1999, p. 168)]

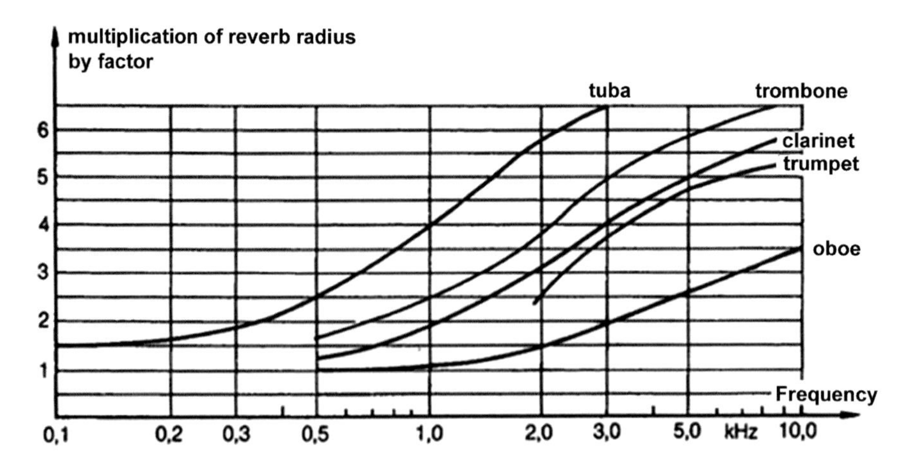

Fig. 9.8 Frequency-dependent reverberation radius of wind instruments [modified from Fig. 1.8E (Dickreiter 2011, p. 32)]

The importance of considering the complex (i.e. frequency-dependent) radiation characteristics of musical instruments when trying to choose an appropriate microphone technique cannot be over-emphasized:

9.1.4.1 Sound Image Stability

The image stability of a recorded sound source depends very much on the microphone technique used. Apart from 'natural movement of the sound source' (e.g. a soloist moving his body—and therefore also the instrument—while playing), it is again the frequency-dependent radiation characteristic of musical instruments which is responsible if audible 'jumps' of localization occur upon playback of the recording. This so-called localization distortion (music instruments which seem to 'jump' between the loudspeakers) can happen quite easily if stereo (or surround) main microphones are being used relatively close to instruments with strongly non-uniform (non-omni-directional), frequency-dependent radiation characteristics.

A very good example is the solo-violin, for which the orientation of the 'sound beam' of maximum radiation may differ from note to note. If such an instrument is being picked up with a small AB type of microphone setup, for which localization relies also on level differences between the two capsules (apart from the temporal information contained in the form of time-of-arrival differences of sound, of course), it may happen that for the listener sound source localization 'quavers' of 'jumps' between the loudspeakers, when the instrumentalist plays a scale.

With the use of one-point microphone techniques (like XY with cardioids or Blumlein figure-of-eights), this effect may be minimized, but even there remains the danger that the strong directional characteristics of the microphones involved (in combination with the frequency-dependent radiation characteristics of the instrument) might lead to localization distortion, even though—more likely—only for the higher frequencies which have a wavelength in the order of the capsule diameter [rem.: for low frequencies the level differences at the capsules will become more and more negligible, as the sound waves 'bend' around these relatively small acoustic 'obstacles'; in addition—with many (cardioid) microphones—the pickup pattern tends to become less 'correct' with lowering frequency, meaning that they behave more and more omni-directional; figure-of-eight microphones might be considered an exception in that respect, depending on the specific type of manufacture (single or double membrane)].

A solution to the problem of unwanted sound source 'wandering' or 'jumping' upon playback may be found in narrowing the stereo panorama, in case a stereo spot-microphone was being used. However, this is of course no proper solution if the unstable sound source is the solo instrument, which is being picked up by the main microphone together with the rest of the musical ensemble or orchestra.

Therefore, it seems advisable to use main microphone techniques which actually try to keep the sound information of a soloist as 'separate' as possible from the main 'body of sound' (orchestra), or use a microphone technique which tries to capture the performance of orchestra and soloist 'as a whole.' Usually, the second option will mean to either move the main microphone further out into the hall to achieve a better balance between soloist and orchestra, or move the capsules of the main microphone system far away from the soloist (which can be achieved using a large AB technique for example) and—for example—use an additional 'centerfill' microphone in a double function as spot-microphone on the soloist in an large A-B-C style of recording arrangement. The center microphone in this setup can have omni or cardioid characteristics, depending on the chosen mic pattern and orientation of the microphone, but also on the intended overall sonic effect (e.g. a cardioid, suspended right above and looking straight down on the soloist for maximum separation from the rest of the orchestra; or an omni microphone at a reasonable distance from the soloist in order to represent a real 'centerfill' also for the orchestra, but also capturing the soloist and the orchestra behind with a proper volume balance) (Fig. 9.9).

The author has used this technique (with one omni-directional microphone as a centerfill) even on occasions when two soloists (harp and violin, placed to the left of the conductor) and orchestra had to be recorded with great success: This had been done for the sake of sonic clarity and in order to avoid unwanted comb-filtering effects which may have resulted in case of the use of two spot microphones (one for each solo instrument) due to cross talk. Of course, great care had to be taken in terms of proper placement of this centerfill microphone in order to create

(a) the proper direct/diffuse sound ratio for the solo instruments, as well as
(b) the right level balance between the solo instruments and the orchestra in this 'C' microphone.

Coming back to the topic of apparent 'wandering' or 'jumping' of solo instruments when being picked up by a (probably too close) main microphone, or also the potential

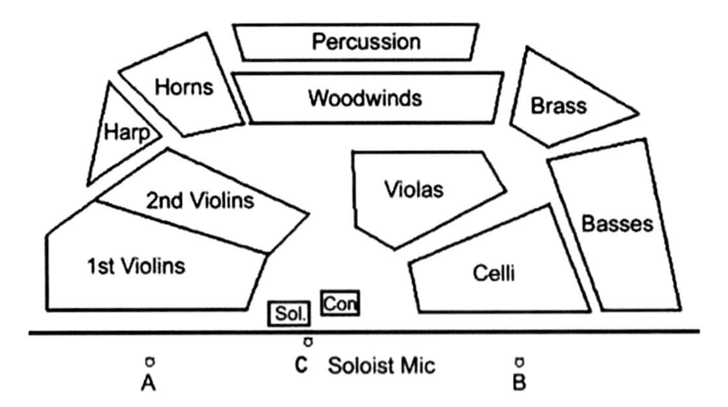

Fig. 9.9 Large AB-Centerfill system similar in style to the early mercury 'Living Presence' recordings (see Pfanzagl-Cardone and Höldrich 2008, 2011 for further details)

problem that this solo instrument is being picked up too close (and too loud) by this main microphone:

In the opinion of the author, this is one of the most pronounced problems also with a very popular stereo microphone technique, the DECCA-triangle. While also the results of listening tests analyzed further down in this chapter show that the DECCA technique provides very pleasing results also when used for surround recording (see Camerer and Sodl 2001), the problem starts as soon as a solo instrument (in close proximity to the conductor) is added. This soloist is usually being picked up not only too loud, but also too close, resulting in an unnatural apparent source width (ASW) of the instrument.

This is due to a so-called zoom effect which can happen with certain microphone techniques in close proximity to an acoustic instrument.

In order to avoid this problem, the author proposes to use a technique which tries to capture the orchestra and soloist in a 'separate' manner (see Chap. 3 on stereo techniques, Sect. 3.4.1 and Figs. 3.35, 3.36, and 3.37).

9.1.4.2 Stereo Width and Apparent Source Width (ASW)

This aspect describes how 'wide,' 'large' or 'broad' ('voluminous' in a spatial sense) a sound source appears to be. Apparent source width is a psychoacoustic 'impression,' which applies to both sound sources in a real acoustic space (where a broadening of the sound source can be caused by reflections on nearby surfaces, as well as standing wave phenomena), as well as the re-presentation of a sound source through a recording. In the second case, it depends entirely on the choice of the applied microphone technique how large (if any) ASW will be perceived by the listener. Therefore, the sound-engineer is well advised to think carefully before making a final decision on how to combine signals of a main microphone system with those of several (usually mono but sometimes stereo) spot- or soloist-microphones.

One of the temptations or dangers in this respect is to portray (a soloist, for example) 'larger than life' by use of a non-appropriate microphone technique or fader/level setup on the mixing desk.

Very often, it is close-up microphone techniques (small AB or XY) which lead to an exaggerated stereo-panning of the sound source/instrument, which may sound 'nice' and 'impressive' for a solo-recording of that instrument, but inappropriate if that signal has to be combined with other spot- or main microphone signals. Those kind of techniques are frequently (and appropriately) used in the field of pop music, but are not recommendable for classical music, unless the instrument (or singer) is really a soloist (and not a soloist accompanied by an orchestra) in the recording concerned. Alternatives to the typical small AB approach for soloist recording may be the 'coincident omnis' technique as proposed by Ron Streicher (see Streicher and Dooley (1985) and also Sect. 3.1.4 in this book), a technique which has also been applied by veteran sound-engineer Burce Swedien (see Swedien 1997), which gives a more subtle 'stereo' representation of a solo sound source.

Common examples of non-proportionate portrayal of solo instruments can be found in quite a number of recordings of concerts for piano and orchestra, in which the piano often fills about two-thirds of the orchestral base width, which—in turn—leads to an acoustic impression that the piano would have to be at least 6–8 m (approximately 18–24 ft.) long.

The sound-engineer's intentions may be the best, trying to give the listener a detailed and 'in-depth' (acoustic) view on the solo instrument by—sort of—looking at it with a magnifying lens (or looking glass), but the effect obtained is certainly detrimental to the perceived naturalness of the recording as a whole.

It is a bit like trying to integrate a 'close-up zoom' of a video image of an instrument in a 'picture-in-picture' style manner into the 'total stage view' of a camera which is way back in the hall and trying to capture the overall perspective. It may be interesting to look at, but it will certainly not look (or in this case: sound) 'natural.'

9.1.5 Ranking of a Few of the Most Common Stereo Main Microphone Techniques

In respect of the attempt of a proposed 'ranking' of stereo microphone techniques, one listener-evaluation-based test from the past sticks out: the one conducted by Ceoen in (1972), including a total of 64 test listeners, out of which 34 were seated in ' … a favorable stereophonic listening area (sufficient symmetrical emplacement, listening angle varying from 40 to 90°), while 30 were seated in a comparatively bad listening area. … The listening test signals were supplied by six different stereophonic microphone systems, placed in front of an orchestra, and were simultaneously and separately recorded. Neither reverberation nor other effects were added. The audience, listening to unidentified stereophonic excerpts, was invited to classify different quality parameters (liveness, intimacy, perspective, stage continuity, extra width, dynamic range, warmth, brilliance and preference). The listening tests were carried out at the first Central Europe Convention of the Audio Engineering Society in Cologne, Germany, March 1971 (Fig. 9.10).

When listening sequences were finished, the audience was informed about the results of some tests carried out on the six discussed stereophonic systems.

Also, a mobile point source has been moved on a graduated circumference around each system. People listening to the recordings have noted the apparent incident

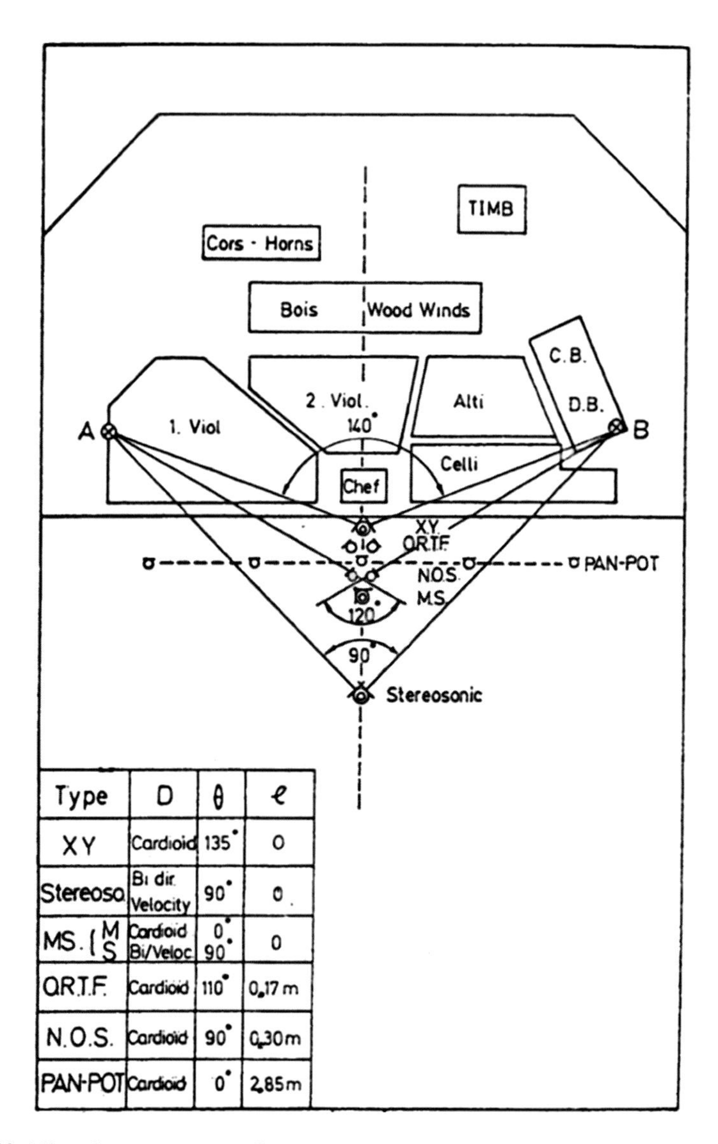

Type	D	θ	ℓ
XY	Cardioid	135°	0
Stereoso.	Bi dir. Velocity	90°	0
MS. (M / S	Cardioid / Bi/Veloc	0° / 90°	0
O.R.T.F.	Cardioid	110°	0,17 m
N.O.S.	Cardioid	90°	0,30m
PAN-POT	Cardioid	0°	2,85m

Fig. 9.10 Microphone arrangement from the test by Carl Ceoen [Fig. 7 from Ceoen (1972)] (l (i.e. length): distance between capsules of the respective microphone technique)

angle of the reproduced sound spot. Figures 9.11 and 9.12 show the results of these investigations.

1. The apparent perceived angle φ versus the true incident angle γ (Fig. 9.11). Ideally, this relation should be represented by a straight line that can be perfectly matched by the 'pan-pot' system, of course. However, the MS, ORTF and stereosonic systems (the last in a restricted opening angle of $+/-45°$) meet this requirement very well.
2. The angular standard deviation σ versus the perceived angle φ (Fig. 9.12, left column). This expression, a function of the spreading among individual estimates, depicts in fact the lack of resolution of a given system. In spite of the fact that the XY, MS and stereosonic systems are all three based upon pure 'intensity stereophony,' we note the impaired resolution of the MS system. This should be attributed to the strong influence of accidental phase deviations in the M and S channels. The stereosonic principle enhances the ratio of the diffuse sound field to direct sound information and is also responsible for the dissolution of the reproduced virtual sound spot.
3. The alteration of the perceived stereophonic loudness Wst versus the perceived angle φ (Fig. 9.12, right column). The six involved systems all avoid the appearance of the so-called hole or hump in the middle.
4. Finally, a notion about the correlation between left and right signals. The method of an oscilloscope display of Lissajous pattern can give valuable information about compatibility chances. The oscillograms of Fig. 9.13 are taken again and again on the first beat of the tenth bar of the tenth Slavonic dance of A. Dvorak.

The XY system ensures the highest congruence between left and right signals and, as a result, guarantees good compatibility. The less elliptical MS pattern indicates once more the vulnerability of this system to undesired phase errors. The well separated 'pan-pot' microphones, as well as the figure-of-eight pattern of the stereosonic microphones, display right and left signals in an almost random phase relation. The spaced microphone pairs of the ORTF and NOS systems can also impair the quality of the sum-derived monophonic signal by producing possible interference phenomena.

The results are shown separately in Fig. 9.14 for the favorably and badly seated listeners as well as for the entire audience in general. As it could be expected, the listening position has very little influence on some parameters like intimacy, dynamic range and brilliance. On the other hand, correct impressions of liveness, warmth and extra width phenomena require a very critical listening area.

Among other logical results, we quote the following:

1. The correct stage continuity produced by the 'pan-pot' distributed system,
2. The impression of good perspective ensured by the stereosonic system (i.e. Blumlein-Pair) and resulting from the more distant microphone emplacement of this very discriminative sound-collecting system,
3. The expected consequences of the phase-reversal phenomena associated with the stereosonic system: an excellent reproduction of the diffuse sound field (liveness) and a partial cancelation of the bass information (lack of warmth),

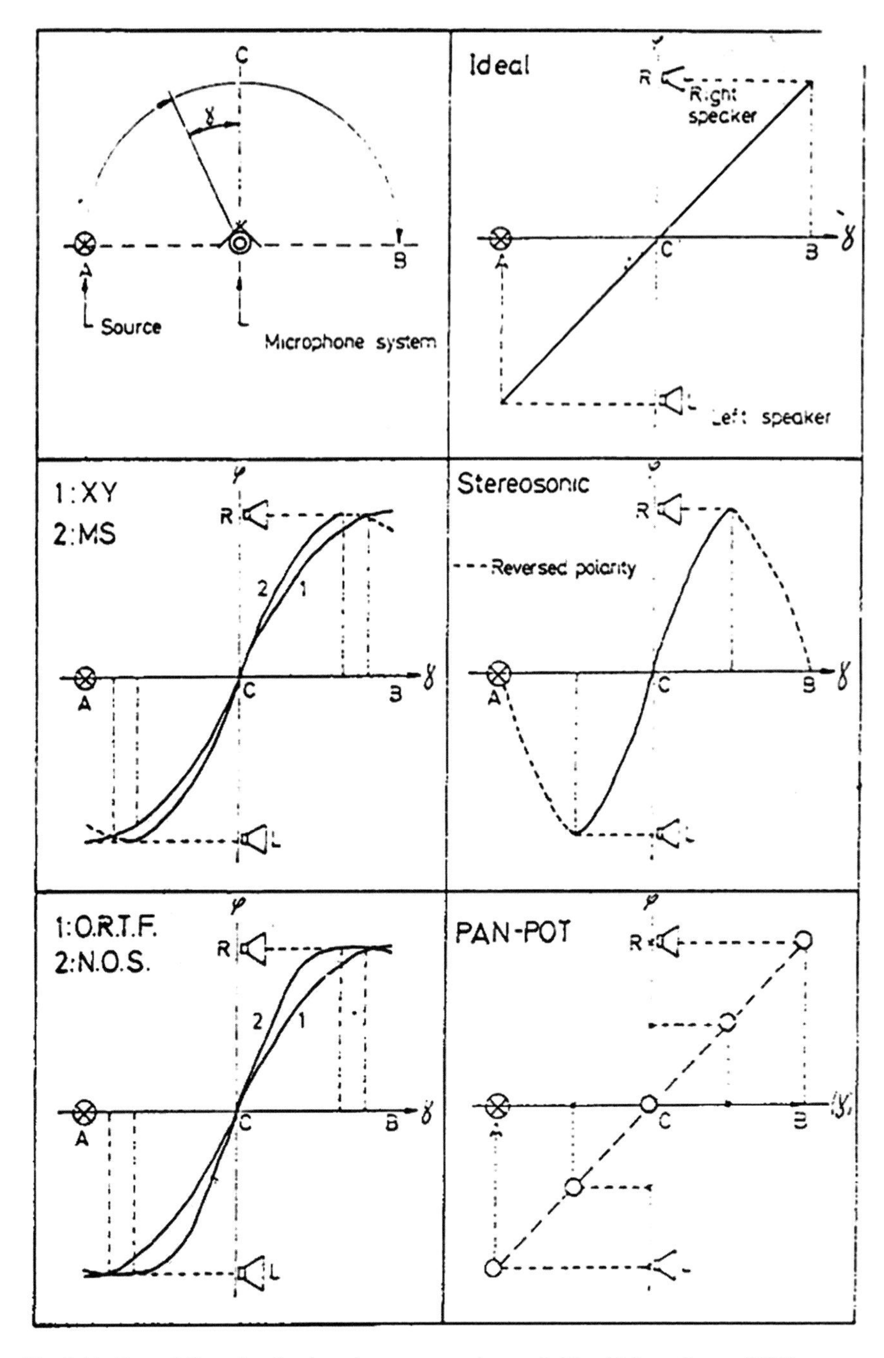

Fig. 9.11 Frontal distortion for the mic systems under test in Fig. 11 from Ceoen (1972)

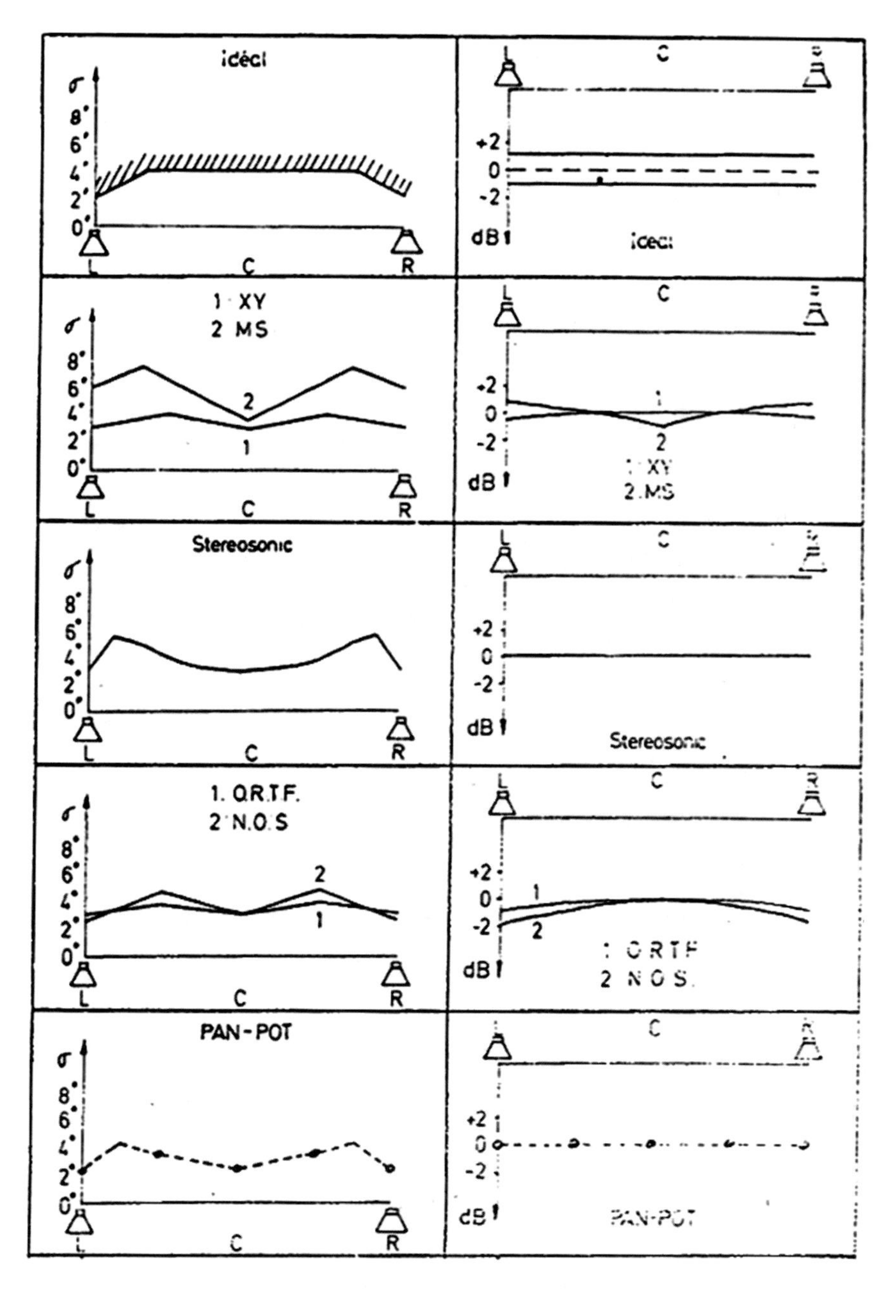

Fig. 9.12 Left column: resolution distortion and right column: presence distortion for the mic systems under test [Fig. 12 from Ceoen (1972)]

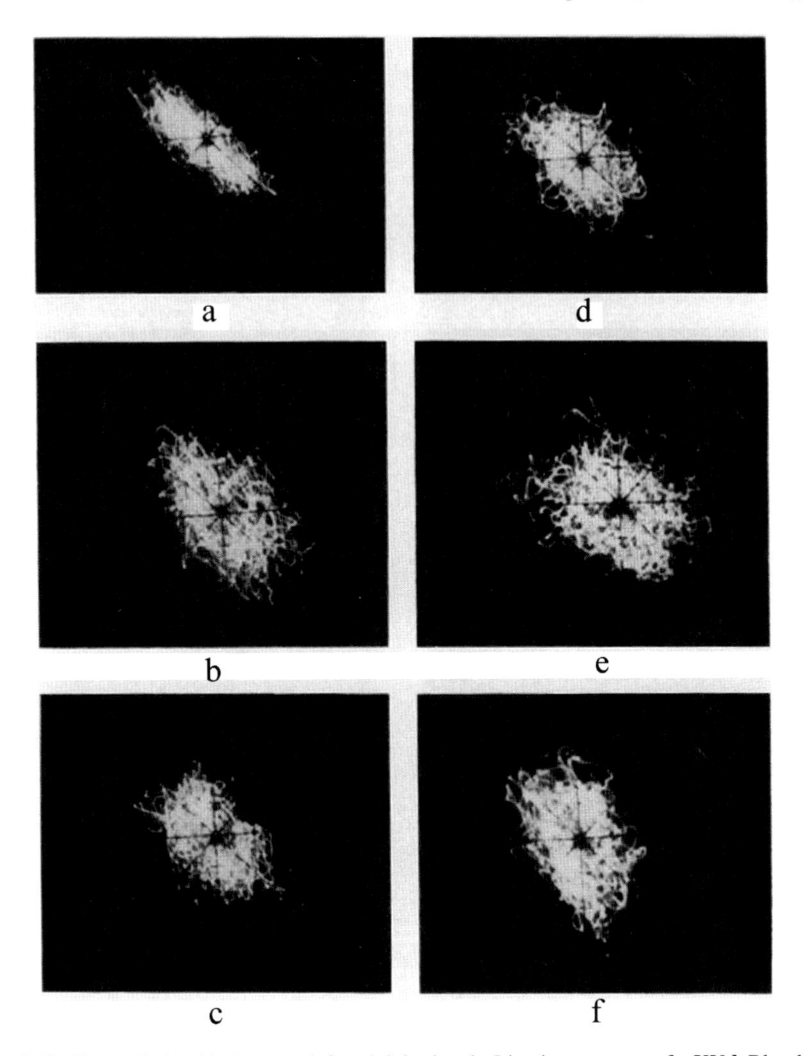

Fig. 9.13 Phase relationship between left and right signals: Lissajous patterns of **a** XY, **b** Blumlein-Pair, **c** MS, **d** ORTF, **e** NOS and **f** Pan-pot system (Fig. 13 from Ceoen 1972)

4. Finally, we note that the general agreement of the audience to distinguish the lack of intimacy of the MS system and to select the ORTF system as a best overall compromise is surprisingly high. …' (from Ceoen 1972).

The results presented in Ceoen (1972) are very interesting and clearly in favor of the—back then—newly developed ORTF technique. They also seem to coincide well with the sonic properties pointed out in Chap. 3 and the current chapter. However, given the fact that recording veteran John Eargle proposes an orchestra recording technique which uses a combination of ORTF with large AB omnis (see Fig. 9.37) and also Tomlinson Holman has a similar technique which he employs for 5.1 surround

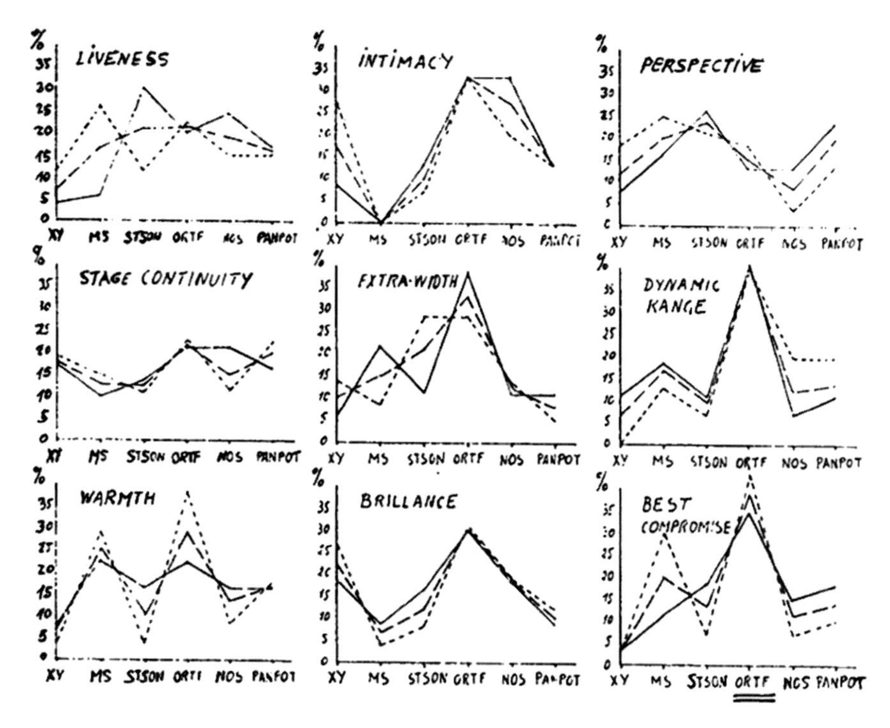

Fig. 9.14 Audience classification of quality parameters; solid line (__): well-seated listeners; dotted line (…): badly seated listeners; dashed line (- - -): entire audience [Fig. 14 from Ceoen (1972)]

recordings (see Holman 2001) seems to indicate that a combination of ORTF with large AB-style flanking microphones (or 'outriggers') produces even more favorable results.

Therefore, the author would like to point out again the possibility to use 'combined microphone techniques,' as described in Sect. 3.4, as they are able to provide results that are superior to any of the single techniques.

9.2 Surround Microphone Techniques

As has already been pointed out in Sect. 9.1 for the reproduction of sound sources laid out (or moving) around the listener in a 360° horizontal plane, a 6-channel surround speaker layout according to versions '6a' and '6b' of Fig. 9.2 would be favorable.

However, this requirement has apparently not been the main objective when the ITU has recommended another 6-channel format, which was to become the de-facto standard in surround sound reproduction, namely the 5.1 surround sound standard, as defined in the Recommendation BS.775-1, which includes 5 (preferably full-range)

Specifications of concert halls.

	Site B	Site C	Site D
Seats	1204	1301	1100
Frontage	18 m	18 m	18 m
Depth	15 m	14 m	15 m
Height	8 m	9 m	8 m
Reverberation time	1.9 seconds	1.6 seonds	1.8 seconds

Side view of microphone arrangement.

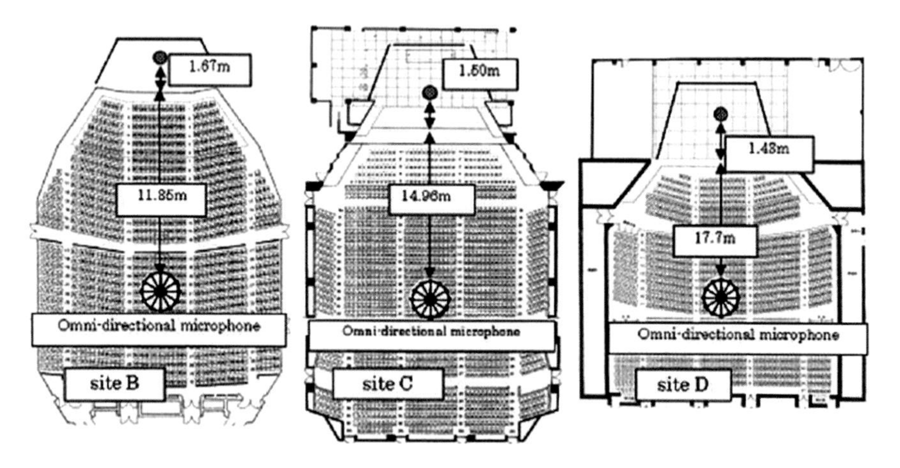

Fig. 9.15 Concert hall layout, data and configuration details of field test (after Muraoka and Nakazato 2007)

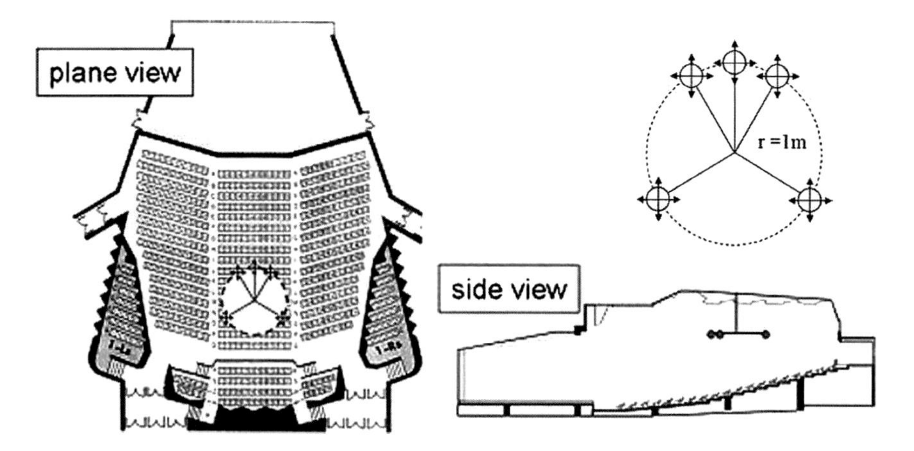

Fig. 9.16 Diffuse-field microphone array in a concert hall as proposed by Muraoka and Nakazato (adapted from Muraoka and Nakazato 2007)

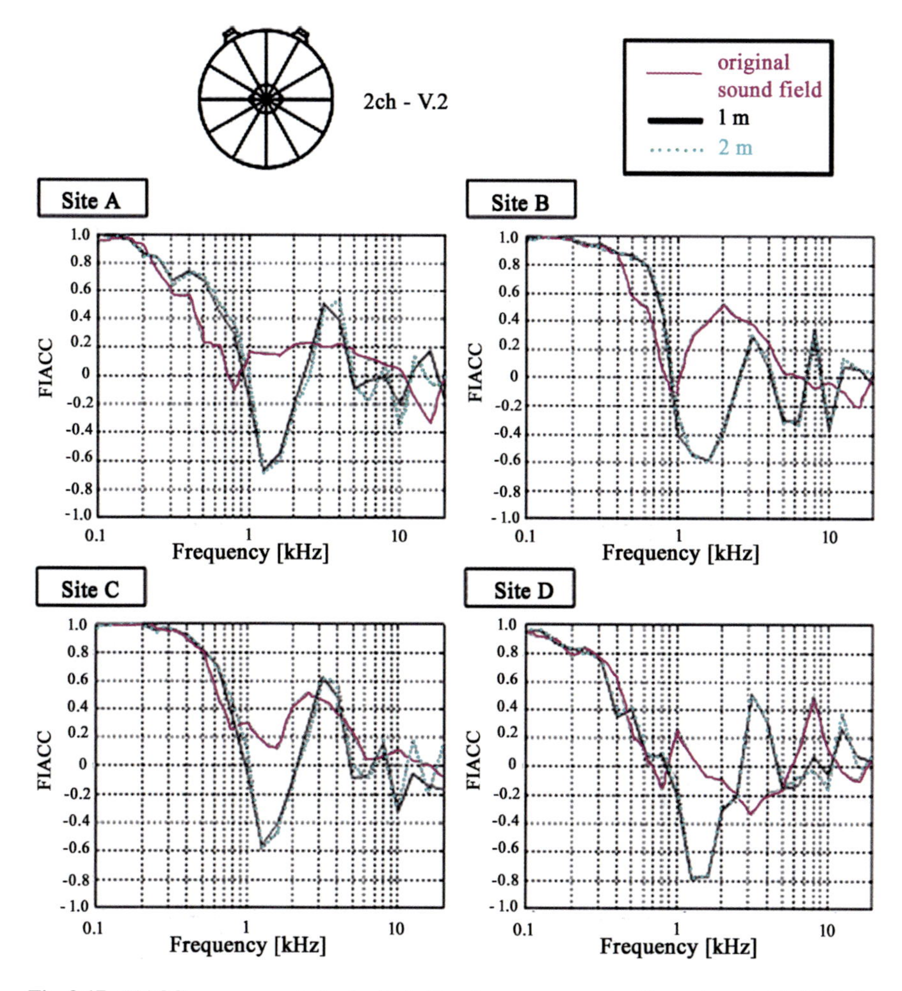

Fig. 9.17 FIACC measurements for the '2ch-2' replay arrangement with speakers at +/−30° (after Muraoka and Nakazato 2007)

channels and one additional LFE channel (low-frequency extension), the layout of which can be seen in Fig. 9.2, version '3/2' (see lower right corner of figure).

In a paper by Muraoka and Nakazato (2007), an examination of multi-channel sound-field re-composition utilizing frequency-dependent interaural cross-correlation (FIACC) has been made, with the outcome that the—in respect of the front-back distribution of loudspeakers—asymmetrical layout of the 3/2 arrangement according to ITU-R BS.775-1 is actually better suited for the replay of surround sound signals which also contain de-correlated signal components, than the 6- or even 12-channel loudspeaker reproduction layouts which have also been examined.

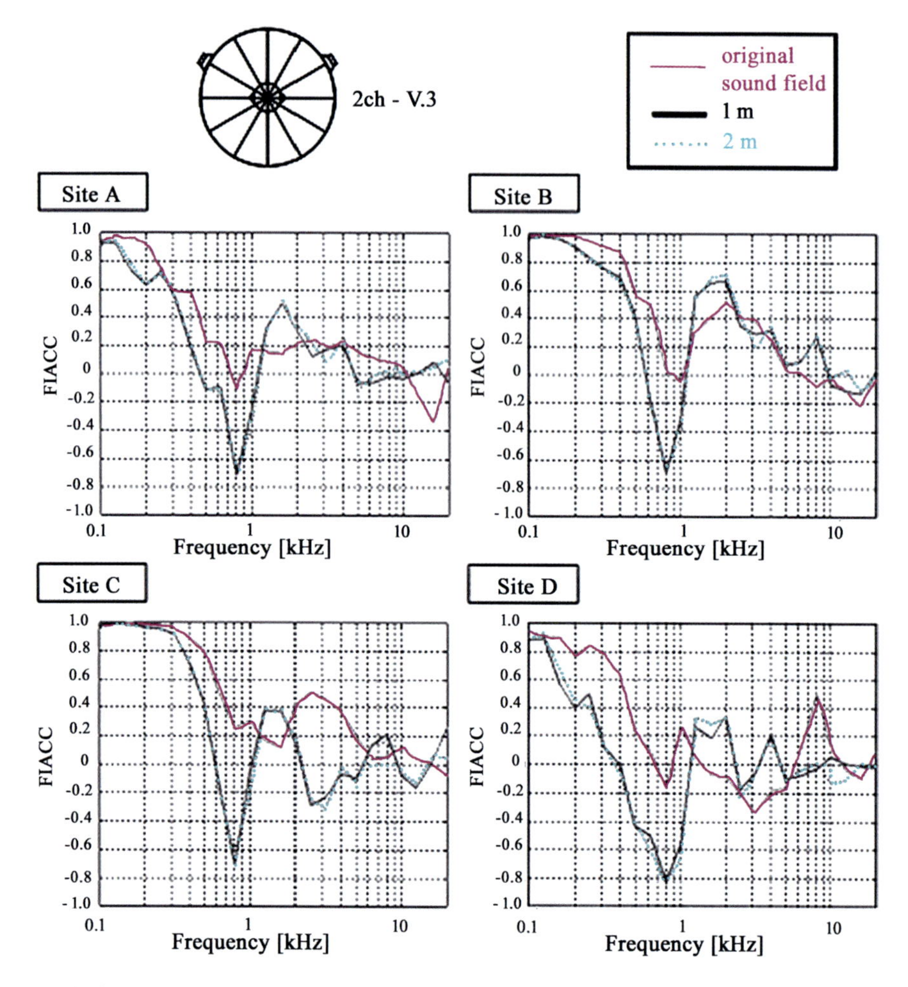

Fig. 9.18 FIACC measurements for the '2ch-3' replay arrangement with speakers at +/−60° (after Muraoka and Nakazato 2007)

For their experimental analysis, they had set up twelve microphones around a dummy head and recorded pink noise replayed through an omni-directional loudspeaker (Victor GB-10) placed on the stage of three different concert halls and a podium of one auditorium (lecture room) (Fig. 9.15).

As one of the aims of the researchers was to determine the ideal (up to 12-channel) speaker layout for the reproduction of de-correlated sound signals, these 12 signals of the microphones were then replayed to the same dummy head in an anechoic chamber. Subsequently, the resulting binaural, frequency-dependent interaural cross-correlation of the re-recording was numerically compared with the original FIACC of the dummy head recording from the concert halls and the resulting error E was calculated (E being the 'squared error' calculated from the FIACC difference between

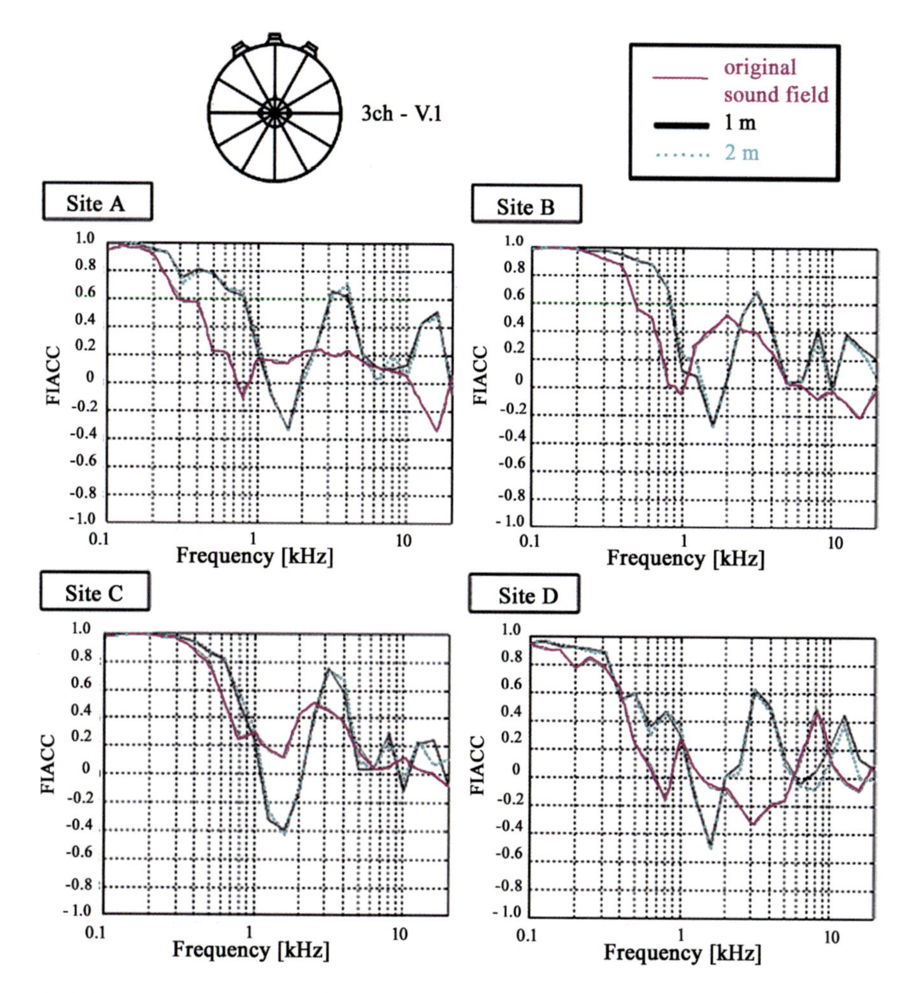

Fig. 9.19 FIACC measurements for the '3ch-1' replay arrangement with speakers at +/−30° (after Muraoka and Nakazato 2007)

the original site and the reproduction site; calculated both for the 'full' band, summed up from measurements in 23 individual bands with center frequencies between 100 Hz and 20 kHz, as well as the 'fundamental' band, composed of 10 individual bands for frequencies from 100 Hz to 1 kHz, the frequency range of which was determined with reference to studies by Hiyama et al. (see his publications from 2000, 2002, 2003), who proved experimentally that the FIACC in this range is related to the impression of spaciousness).

As an outcome of their research, Muraoka and Nakazato propose a circular 5-channel microphone array (radius = 1 m) with omni microphones, the layout and spacing of which is oriented at the speaker layout of ITU-R BS.775-1. It should be noted though that this microphone array is not intended to serve as a traditional

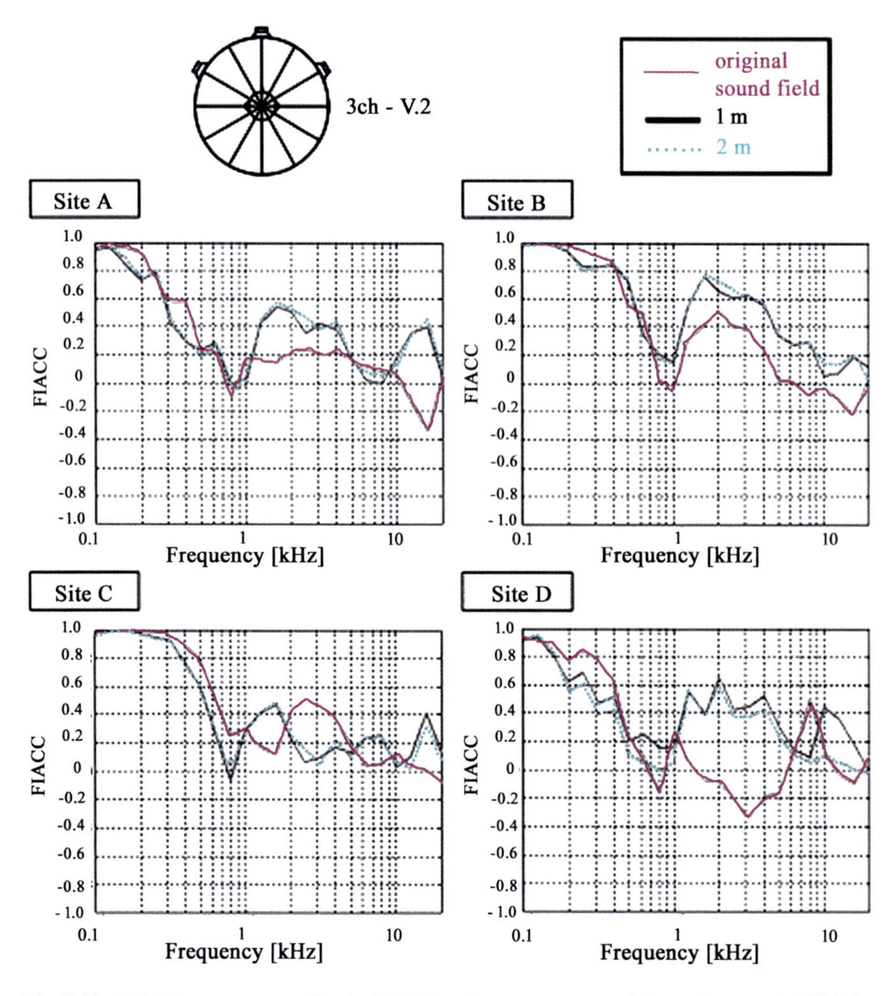

Fig. 9.20 FIACC measurements for the '3ch-2' replay arrangement with speakers at +/−60° (after Muraoka and Nakazato 2007)

(surround) 'main microphone' system in relative vicinity to the sound source, but is instead intended to serve as a diffuse sound-field pickup (similar to a 'Hamasaki Square') which is suspended in the rear of the concert hall. To arrive at a satisfactory direct/diffuse sound ratio for a final recording, the authors suggest the use of additional spot microphones on stage, apparently with the complete omission of a traditional 'main' microphone.

Even though this is—without any doubt—a feasible technique, it should also be said that their proposal is a bit unusual for sound-engineers, as many of them try to find an appropriate surround main microphone system and choose an ideal place for it in order to achieve an optimized direct-to-diffuse sound ratio that is being captured

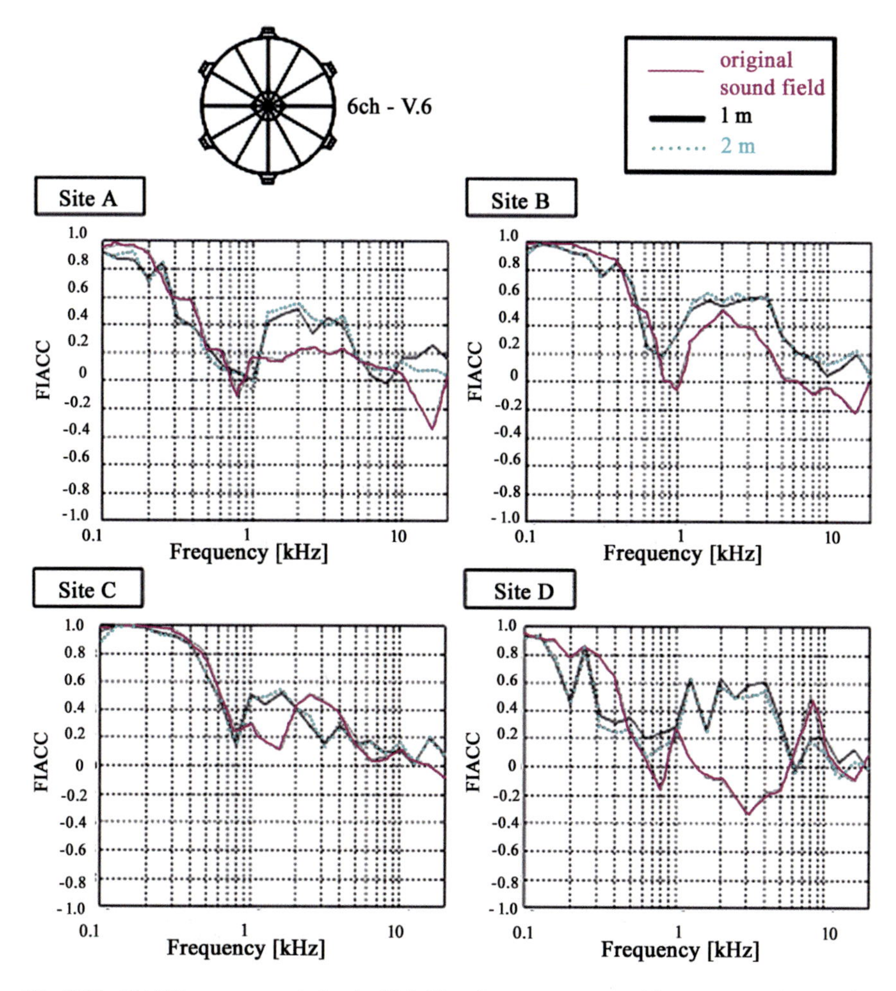

Fig. 9.21 FIACC measurements for the '6ch-6' replay arrangement with an even speaker spacing of 60° (after Muraoka and Nakazato 2007)

simultaneously by just one microphone array (and not resulting from the 'overlay' or mix of a diffuse-field mic array and direct sound (or spot-) microphones) (Fig. 9.16).

As an interesting side result of their research, it can be noted that the replay of 2-channel diffuse sound in their experimental setup (see the configurations '2ch-2' and '2ch-3') resulted in strong out-of-phase signal information in frequency bands with center frequencies around 1200 Hz and 800 Hz, respectively, according to the combination of microphone and loudspeaker spacing. The relatively pronounced FIACC with a value of down to −0.8 at these frequencies also seems to be a result of the anechoic replay conditions, while in a listening room with more 'normal' reverb times due to at least partially reflecting surfaces (as is the case in control and living rooms) this rather extreme out-phase FIACC value would have been lowered

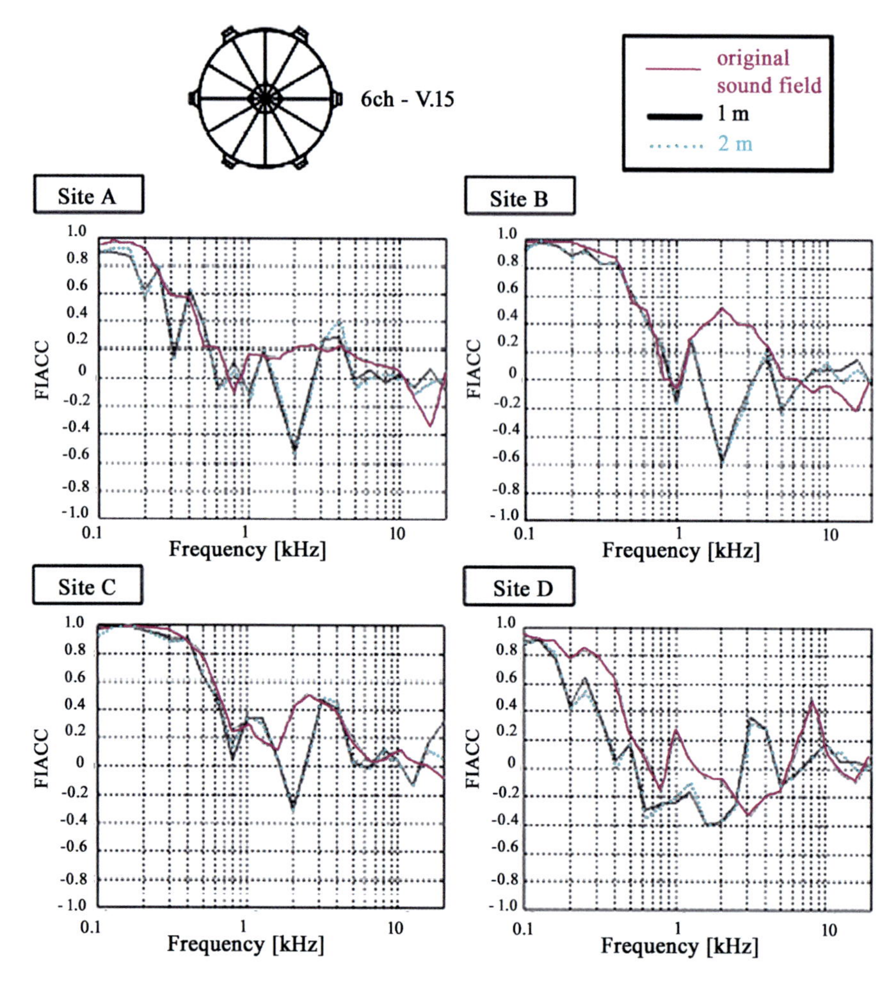

Fig. 9.22 FIACC measurements for the '6ch-15' replay arrangement with an even speaker spacing of 60°, but offset by 30° in comparison with the layout '6ch-6' (after Muraoka and Nakazato 2007)

due to overlapping diffuse sound resulting from the acoustics of the replay room. However, this result from the study—i.e. the creation of an unnatural out-phase signal present in the FIACC at the reproduction site, which was not present in the FIACC of the dummy head recordings at the original site—is one more argument against the use of small AB recording techniques whenever the main goal is a faithful capture and reproduction of sound events. Because this is essentially what we are seeing in Figs. 9.17: the resulting cross-correlation of the (binaurally re-recorded) signals of two closely spaced omni capsules (i.e. small AB), when being replayed through the standard replay loudspeaker layout of +/−30°, which differs quite drastically from the original sound field, as picked up by the dummy head in the concert hall (Fig. 9.17).

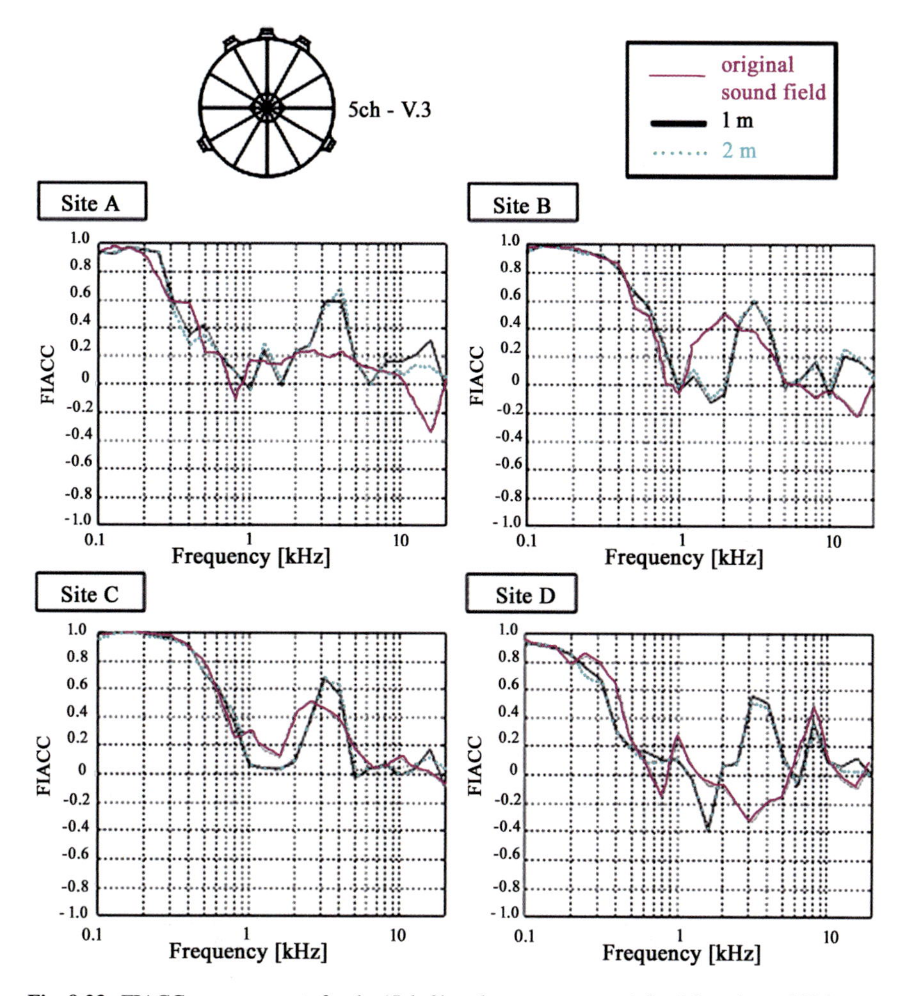

Fig. 9.23 FIACC measurements for the '5ch-3' replay arrangement (after Muraoka and Nakazato 2007)

For the FIACC measurements of the '2ch-3' arrangement, it is interesting to note that the out-phase frequency has moved to a lower frequency band around 800 Hz, as was to be expected due to the larger capsule spacing of the recording microphones as well as the replay loudspeakers (Fig. 9.18).

For the '3ch-1' arrangement, a center speaker has been added in comparison with the arrangement of version '2ch-2.'. The result for the FIACC is that the very strong out-phase peak of −0.8 has been reduced to a maximum of −0.5 (in the case of hall D), so the addition of the center speaker was certainly of benefit and a step in the direction of a more faithful reproduction (Fig. 9.19).

For the '3ch-2' arrangement, a center speaker has been added in comparison with the arrangement of version '2ch-2' with the L and R speaker placed at +/−60°. The

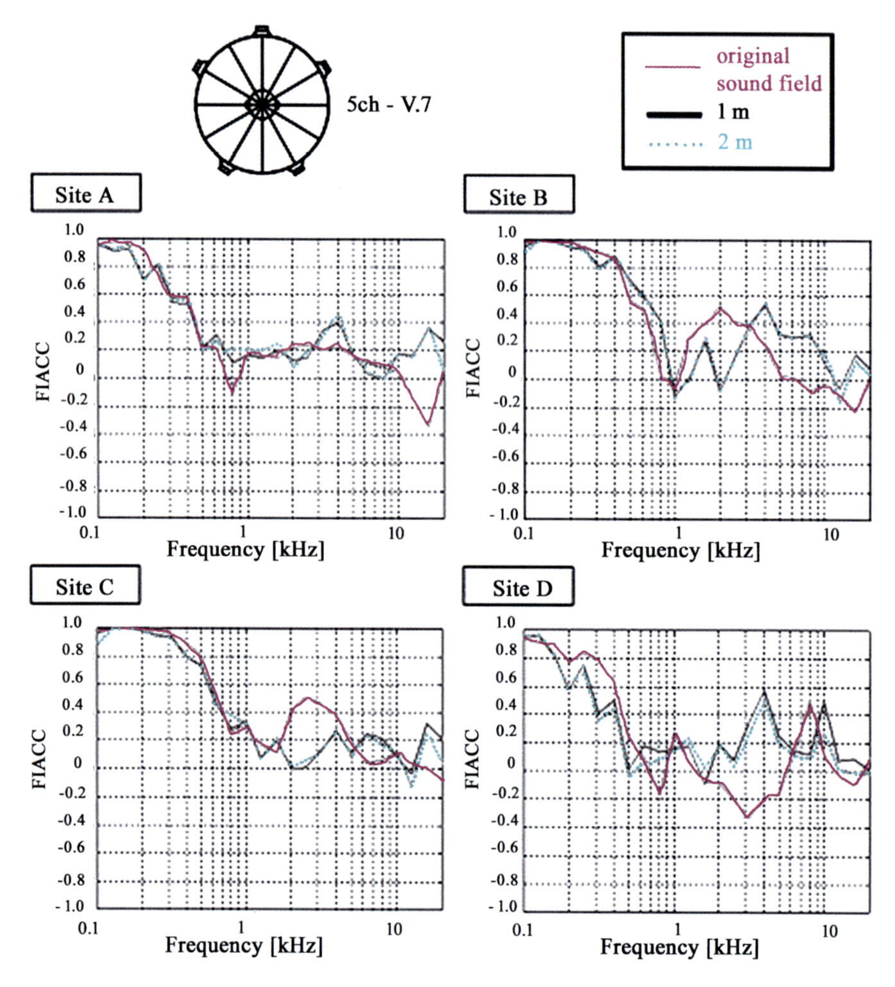

Fig. 9.24 FIACC measurements for the '5ch-7' replay arrangement (after Muraoka and Nakazato 2007)

result for the FIACC is that the resulting FIACC function seems to follow the FIACC of the original recording in the concert hall even more closely, especially considering the absence of strong out-phase peak components which for the arrangement '3ch-2' reach not even −0.1 on the occasion of site C in configuration '3ch-2' (Fig. 9.20).

When looking at the 6-ch arrangements, which have been examined by Murakoa and Nakazato, the '6ch-6' arrangement seems favorable over the '6ch-15' arrangement, as the latter again exhibits relevantly strong out-of-phase components (especially for sites A and B), which are not present in the original dummy head FIACC (see Figs. 9.21 and 9.22).

If we abandon the concept of front/rear symmetrical layout and even spacing of loudspeakers, which is better suited for the replay of direct (and rotating) sound

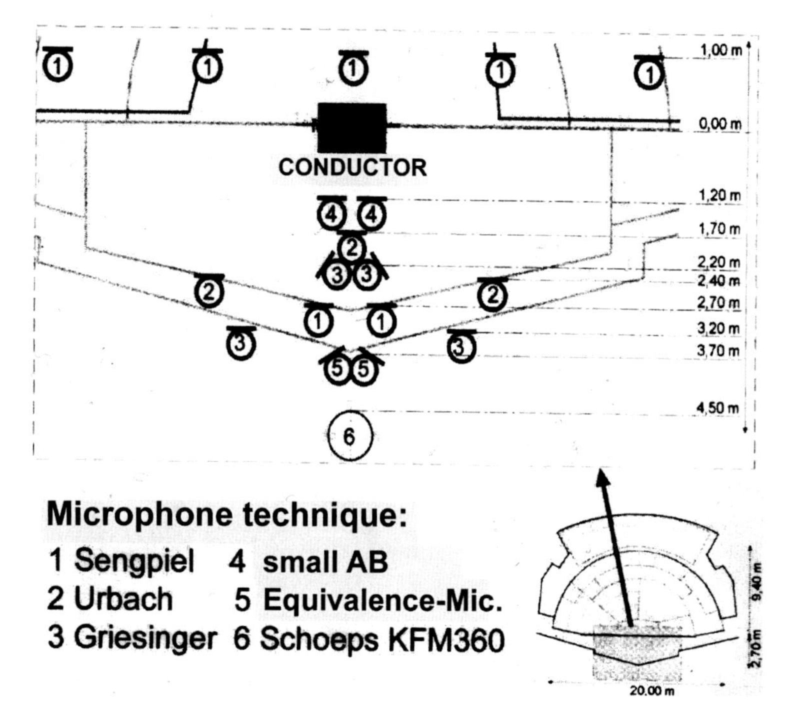

Fig. 9.25 Microphone positions for the 6 surround microphone systems tested at the 'Neue Philharmonie Berlin' (adapted from Hildebrandt and Braun 2000)

sources from around 360° for the sake of optimized (authentic) diffuse sound reproduction as well as the convenience of a better compatibility of the loudspeaker layout to the preexisting (surround) 5-channel film sound format, we end up with 5-channel speaker layouts, as presented—for example—in the arrangements '5ch-3' and '5ch-7.' Of these two, the '5ch-3' arrangement seems slightly more faithful in terms of diffuse sound reproduction, but the difference (squared error) is rather small and seems to depend on both the room acoustics of the specific recording site and the diameter of the circle in which the replay speakers have been set up (1 or 2 m radius) (Figs. 9.23 and 9.24).

As already pointed out above, one of the very interesting outcomes of the research by Muraoka and Nakazato (2007) was that the asymmetrical layout of the 3/2 arrangement according to ITU-R BS.775-1 is actually better suited for the replay of surround sound signals which also contain de-correlated signal components (i.e. diffuse sound), than the 6- or even 12-channel loudspeaker reproduction layouts which have also been examined.

Fig. 9.26 Surround microphone setup for the 'Vienna test' at the 'Grosse Sendesaal' of ORF (Austrian National Radio) (from Camerer and Sodl 2001)

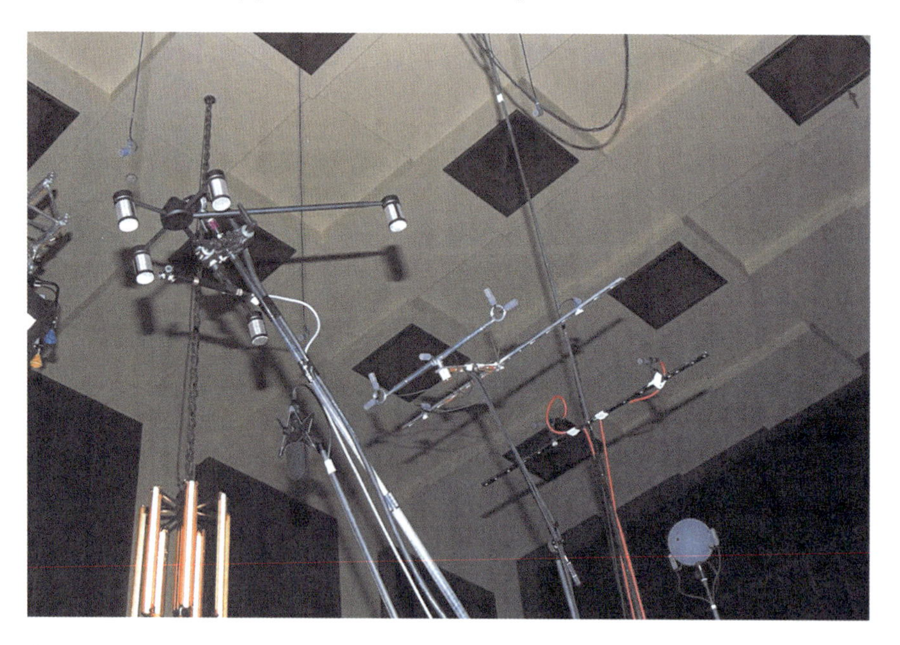

Fig. 9.27 Placement of the 6 surround microphone systems of the 'Vienna test' (from Camerer and Sodl 2001)

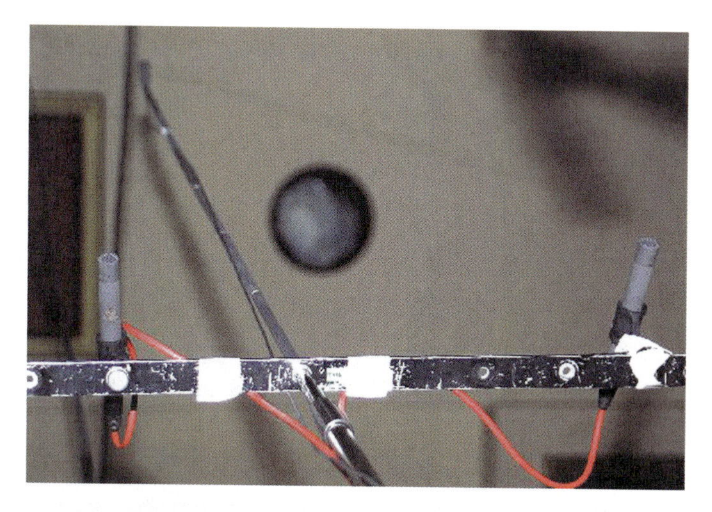

Fig. 9.28 Stereo+C (after Gernemann) (from Camerer and Sodl 2001)

Fig. 9.29 Hamasaki Square (with Atmo-Cross in the middle) (from Camerer and Sodl 2001)

9.2.1 Comparative Surround Microphone Tests

Quite a number of surround microphone comparisons have happened since the 1990s, many of which are based on listening tests. Some of them will be analyzed below in

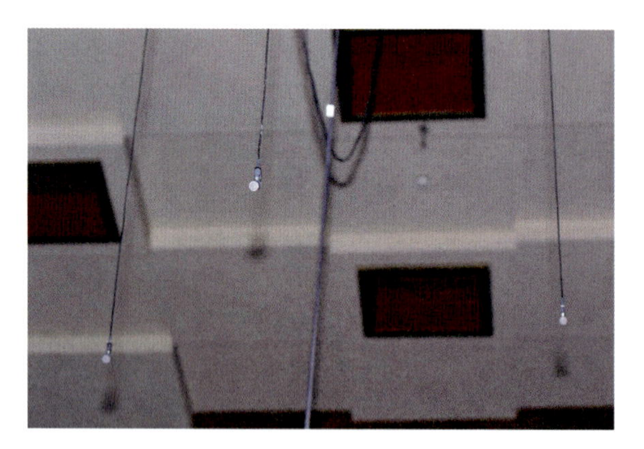

Fig. 9.30 DECCA-tree (from Camerer and Sodl 2001)

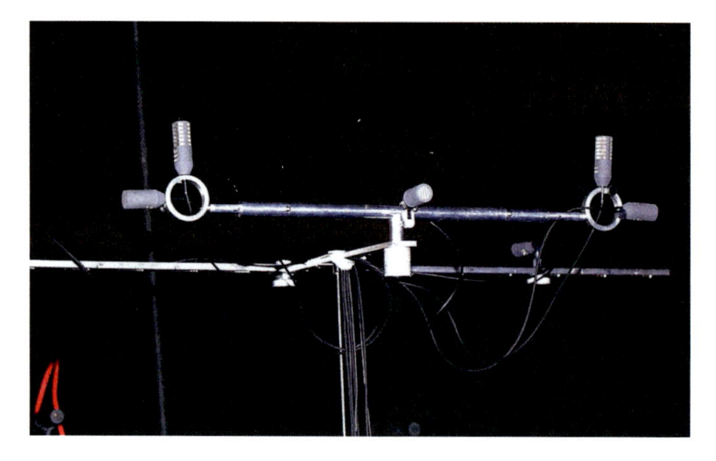

Fig. 9.31 OCT (from Camerer and Sodl 2001)

order to take their results into consideration for our attempt to arrive at a qualitative ranking of 5.1 surround techniques.

9.2.1.1 The 'Neue Philharmonie Berlin Test'

In Hildebrandt and Braun (2000), the results of a listening test, based on a comparative surround microphone recording which took place at the 'Neue Philharmonie Berlin,' have been analyzed (Fig. 9.25).

A short description of the microphone setups:

1. **Microphone setup 'S' by E. Sengpiel** (TELDEC): 'small AB' (capsule spacing =110 cm, height = 5 m), combined with 'microphone curtain' parallel to the rim of the stage, made up of 3–5 microphones above the strings. The signal of the

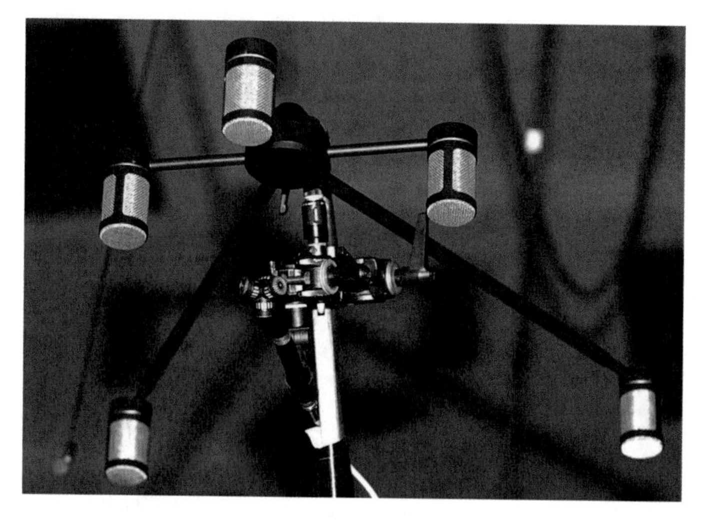

Fig. 9.32 INA 5 (from Camerer and Sodl 2001)

Fig. 9.33 KFM 360 sphere microphone with 2 fig-of-eight microphones (from Camerer and Sodl 2001)

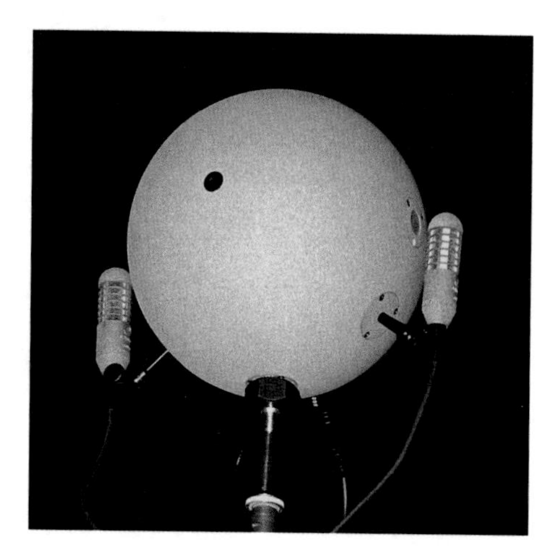

microphone in the middle is routed to the center channel only, while the signals of the other ones are being panned hard L and R (and mid-L, mid-R in case of the use of 5 microphones) and the L/R summing bus. The balance between the 'small AB' main system and the 'microphone curtain' should be adjusted dynamically, according to the music (and resulting psychoacoustic needs).

2. **Microphone setup 'U' by H. Urbach** (DENON): 3× omnis, the exact positions of which need to be determined 'by ear'; mic signals are being routed directly to L, C, R. This technique is claimed to deliver a very 'open sounding' spatial

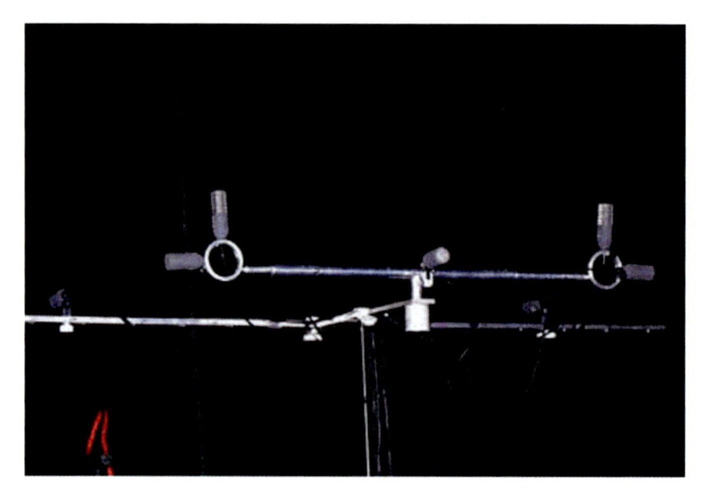

Fig. 9.34 OCT surround (with rear cardioids) (from Camerer and Sodl 2001)

Fig. 9.35 Soundfield MKV microphone (from Camerer and Sodl 2001)

impression; should preferably be used without filters or artificial reverb. Microphone spacing L-R: usually 2.80–5 m, in case of the 'Neue Philharmonie Berlin' 4.80 m.

3. **Microphone setup 'G' by D. Griesinger** (LEXICON): 2× cardioids with a spacing in the order of the ‚critical distance'(i.e. reverb radius of the hall) combined with 2 'centerfill' hyper-cardioids (capsule spacing 80 cm, physical opening angle 100°) close to the conductor. The weaker bass reproduction of the cardioids and hyper-cardioids is compensated for by EQ-ing. The balance between these microphones should be adjusted dynamically.
4. **'small AB'**: 2× omnis, capsule spacing 52 cm; resulting recording angle 180° (after Sengpiel).
5. **Equivalence mic setup 'E'**: 2 cardioids, spaced by 25 cm, opening angle 50°; resulting recording angle 120° (according to Williams-Curves).

Fragebogen / *Questionnaire*

Name / *Name*: .... **Gruppe:** ....

Beispiel 1A / *Item 1A*:

Question-Nr	Räumliche Darstellung des Orchesters / *Spatial presentation of the orchestra*						
		1	2	3	4	5	
1	breit / *wide*						schmal / *narrow*
2	nahe / *close*						entfernt / *distant*
3	tief / *depth*						flach / *flat*
4	stabil / *stable*						instabil / *unstable*
5	präzise / *precise*						verschwommen / *blurred*

Question-Nr	Klangfarbe des Orchesters / *Sound colour of the orchestra*						
		1	2	3	4	5	
6	befriedigend / *satisfactory*						unbefriedigend / *unsatisfactory*

Question-Nr	Raumabbildung / *Spatial imaging*						
		1	2	3	4	5	
7	vollkommener Raumeindruck / *perfect spatial impression*						unvollkommener Raumeindruck / *imperfect spatial* impression
8	zu starker Raumanteil / *too much indirect sound*						zu schwacher Raumanteil / *too little indirect sound*
9	Surround Signale lokalisierbar / *surround channels identifiable*						Surround Signale nicht lokalisierbar / *surround channels not identifiable*

Fig. 9.36 Questionnaire for the 'Vienna' surround microphone listening test (from Camerer et al. 2001)

6. **Schoeps KFM 360 setup 'K'**: 'Kugelflächenmikrofon' (sphere microphone) 2× pressure transducers (flush-mounted in sphere) in combination with 2 figure-of-eight microphones: Via MS-matrixing, a total of 4 surround signals is derived (L, R, LS, RS). A center-channel signal can be derived via Gerzon-Matrix (rem.: the KFM360 on its own has an effective recording angle of 110° according to the manufacturer's specifications).

The results of the test-recording at Neue Philharmonie Berlin were published in the paper by Hildebrandt and Braun (2000).

Sengpiel=S, Urbach=U, Griesinger=G, Equivalence-Setup=E and sphere microphone Schoeps KFM360=K.

rating: A>B means "A is clearly better than B", A≥B "A is somewhat better than B", A=B "A equal to B"

Spaciousness:

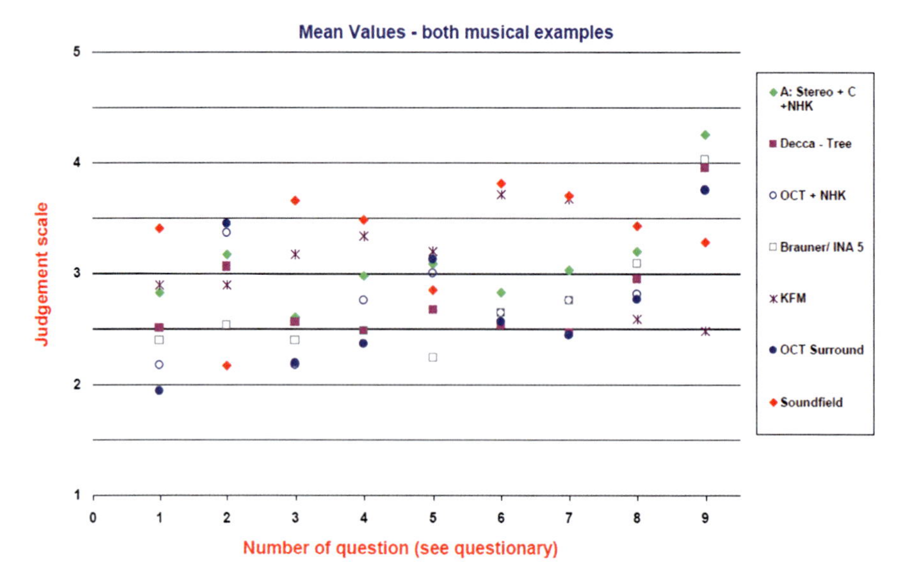

Fig. 9.37 Mean values of the ratings for the different acoustic attributes of the 7 surround microphone techniques compared in the 'Vienna test' (from Camerer and Sodl 2001)

Sound example - string quartett: U=G≥S>E=K, therefore the 3-channel microphone techniques (U, S) were rated 'better' in general.

Sound example - orchestra recording: S≥G=E≥U=K; for this recording there was a clear preference for the Sengpiel method (qualitative differences between the other methods were negligible)

Localization:

– string-quartet S≥G≥U>E≥K; from this rating it can be deducted that microphone techniques with 3 front-channels enable better localization than mere 2-channel techniques (such as E, K)

Orchestra:

S≥G≥U≥E=K, tendency-wise also for the orchestra-recording the 3-ch techniques have won. A more detailed analysis of the listening tests, taking into account also the listener's position in the room, came to the result that listeners more to the sides preferred the Sengpiel method, while listeners in the sweet spot had a significant preference for the 'Kugelflächenmikrofon" recording.

Sound color:

String quartett: G≥U=S≥E>K

– The only clear result was the preference of Griesinger's method over the sound of the Kugelflächen microphone. The other perceived differences were insignificant.

Sound color:

Orchestra: K=U≥G=S=E

– No apparent preference. In an A/B comparison method Sengpiel was rated inferior to the KFM360, but only for listeners in an 'ideal' listening position.

The last point in the questionnaire was related to the quality of replay of the woodwinds via the center loudspeaker in terms of overall reproduction quality and sound color. With a rating of S=G≥U=K≥E also in this case, the 3-channel microphone techniques performed better. Especially at off-center listening positions, listener preference was clearly in favor of the 3-channel techniques.

Summary: The main result of this research study was a preference of 3-channel microphone techniques in respect of spaciousness and localization especially for off-center listening positions.

This result coincides well with the findings of the Americans Snow and Steinberg, who had already researched into the quality of 2- and 3-channel loudspeaker reproduction techniques at Bell Laboratories in the early 1930s (see Steinberg and Snow 1934).

9.2.1.2 The 'Vienna Test'

In the year 2001, a comparison of seven main microphone setups for 5.1 surround sound was effectuated by means of a simultaneous recording with consecutive subjective listening tests at ORF ('Österreichischer Rund-Funk', i.e. the Austrian National Broadcasting Agency). For the surround recording, the 'RSO' (Radio Symphony Orchestra, Vienna) played excerpts from two different pieces of music: W.A. Mozart's 'Maurerische Trauermusik' in c-minor, KV 477, and Luciano Berio's 'Concert for Trombone and Orchestra' [see Camerer and Sodl (2001) and Camerer et al. (2001) from where most of the information below, including photographs and the results of the listening test—displayed in graphical form—has been reproduced, with kind permission of Florian Camerer, ORF)].

9.2.1.3 Details of the Recording

Recording hall: Grosser Sendesaal (large radio hall) of Austrian National Radio

The recording console was an AMS/Neve Capricorn; the monitoring setup consisted of 2× Genelec 1034 for L and R; 3× Genelec 1037 for C, LS + RS; and no subwoofer (no LFE signal had been recorded). Recording machines were 2× Sony PCM 3324 DASH Multitrack recorders; staff-engineers Florian Camerer and Christian Sodl were the sound-engineers on this project (Fig.9.26).

9.2.1.4 The Microphone Setups

'A Decca-tree and a Brauner ASM5 (an INA5 setup) were suspended right above the conductor at a height of 3.2 m (the Decca-tree a bit higher at about 4 m height with the center mic down by 40 cm). The Soundfield MKV was recessed 1 m behind at the same height as the ASM5. Next came OCT at a height of 3.5 and 1.5 m behind the conductor. Following in line were the Stereo+C arrangement (2.5 m behind and 4 m above ground; the center-omni 6 m above ground) and the Schoeps KFM360 Surround as the furthermost system with a distance of 3.5 m and a height of 3.6 m. The Hamasaki Square was added to some configurations to reproduce the room information. The individual recordings were mixed and optimized according to taste by the relevant protagonists if present (Theile/Wittek for OCT and Gernemann for Stereo + C). The other systems were adjusted by qualified participants of the recording session, who already had specific experience regarding one or the other surround microphone system (Fig. 9.27).

9.2.1.5 Stereo + C

This system, developed by Gernemann (see Gernemann 2000), is based on a regular main microphone for 2-channel stereophony like AB or MS with an additional omnidirectional microphone for the center channel high above the main pair (>2 m); As a result of the big distance, the center signal is assumed to be relatively de-correlated with the main pair and can be used from barely noticeable to being mixed in at the same level as the L and R signals. The main reason for this solution is compatibility with non-ideal 'real-world' listening room conditions in private homes, where the center speaker is often located either beneath or above the TV set and/or often of inferior audio quality. A second point is the inherent backward compatibility with 2-channel stereo, which is very much of interest in times of tighter budgets and—hence—shorter time for post-production and for mixing program material in both formats for 2-channel stereo and 5.1 surround. The microphones used were two Schoeps MK 2S (omnis) for L and R (AB) and one Schoeps MK 2H (omnis) for the center channel (Fig. 9.28). As surround setup, the 'Hamasaki Square' was used (Fig. 9.29).

9.2.1.6 Hamasaki Square

Or—as it was internally re-named—just 'NHK,' after its inventor Kimio Hamasaki of the NHK Science and Research Laboratory. The Hamasaki Square was made up of 4 figure-of-eight microphones (Schoeps CCM 8) arranged in a square of about 2 m length, placed not too far behind the main microphone (about 5 m), which [according to the authors Camerer and Sodl] has the advantage of a greatly reduced delay or echo problem with respect to the signal timing of the frontal pickup. The 4 figure-of-eight mics face the sidewalls of the hall and have therefore minimum direct sound pickup. The panning was as follows: The rear figure-of-eights were panned directly to left surround (LS) and right surround (RS), and the front figure-of-eights were panned half-way between front and surround on each side, respectively (Fig. 9.29).

9.2.1.7 Decca-Tree + Hamasaki Square

This well-known setup with three omnis (here: Schoeps CCM 2S) is used as the standard configuration for the numerous recording sessions taking place in the 'Grosser Sendesaal' at the ORF (base length: 1.5 m). The system was reinforced with two additional Schoeps CCM 2S (omnis) on the sides (3.5 m), the so-called outriggers (Fig. 9.30).

9.2.1.8 OCT + Hamasaki Square

This system was developed by Günther Theile (for details, see Theile 2001) and relies on unequivocal localization between the two stereo bases L-C and R-C. Based on Theile's 'localization theory,' OCT uses supercardioids for L and R, which are pointing 90° away from the orchestra (base distance $b = 70$ cm) and a forward facing cardioid for C, which is also a bit closer to the orchestra ($h = 8$ cm; h being the 'height' of the triangle which is formed between the L, R and C microphones). For example, a half-left signal is then mainly picked up by the left supercardioid and the central cardioid and has sufficiently little cross talk into the right supercardioid due to the fact that the signal arrives at it in the minimum sensitivity part of its polar pattern. Therefore, only L and C are relevant to the panorama position, resulting in a distinct localization. Special side address supercardioids are used, as their main pickup is from 90°, ensuring that the frequency response is equal to 0° frontal pickup, albeit with lower level. The somewhat poorer bass response due to construction principles is compensated with the addition of two omnis adjacent to the supercardioids which are low-pass-filtered at 100 Hz to provide low end down to 20 Hz. The microphones used were from the Schoeps CCM series (two CCM 41 V, one CCM 4 and two CCM 2S) (Fig. 9.31).

9.2.1.9 INA 5

(Ideale Nieren-Anordnung = Ideal Cardioid Arrangement) after Herrmann/Henkels (see Hermann et al. 1998): This system is based on William's localization curves, so that the recording angles of adjacent microphone pairs just touch each other. The German company SPL together with microphone manufacturer Brauner developed a special crate with five arms with the respective microphones and a corresponding processor for remote powering and control of polar patterns, level ganging, LFE split-up, 2-channel downmix, etc. (ASM5-Microphone + SPL Atmos-5.1-Processor) (Fig.9.32).

9.2.1.10 Schoeps KFM 360 Surround

After Bruck (1996), a sphere microphone (Schoeps KFM360, after G. Theile (see Theile 1986)) with a smaller diameter than the original sphere is combined with two figure-8 microphones aiming forward and closely mounted to the pressure transducers. By matrixing (sum and difference like MS stereo), the four channels are derived. An optional center signal is derived through a special 2/3 matrix. All matrixing is done in a digital processor box, the Schoeps DSP4 KFM360 (Fig. 9.33).

9.2.1.11 OCT Surround

Instead of the Hamasaki Square, for the surround signals two cardioids (Schoeps CCM 4) were used right behind (50 cm) the frontal OCT array and with a slightly extended base (100 cm). The cardioids are oriented straight away from the orchestra (180°), so that their direct signal pickup is minimal (Fig. 9.34).

9.2.1.12 Soundfield MKV + Processor SP451

The Soundfield microphone captures the sound-field through a tetrahedral array of 4 (sub-)cardioid capsules. The primary signals of those cardioids (also called 'A-format') are then matrixed so that 4 new signals are created—individually named W, X, Y and Z; the so-called B-format—representing 4 new characteristics: omni (W) and three figure-of-eights (X, Y, Z) perpendicular to each other. This B-format signal is recorded and further down the signal chain played back on a desired speaker arrangement through a suitable decoder (rem.: concerning the speaker layout, there are many possibilities, theoretically limitless). The Soundfield microphone forms the practical basis of the Ambisonics system (see Gerzon 1973) that can provide complete periphonic sound reproduction. For 5.1 loudspeaker setups, the special decoder (B-format to 5.1) required is called 'Vienna' decoder and the resulting speaker feeds are called G-format. The Soundfield company produces their own B-format-to-5.1 converter, the SP451. It shall be noted that for horizontal reproduction the Z-channel is omitted (Fig. 9.35).

9.2.1.13 The Listening Test

During the ORF Surround seminar and at another, later occasion, the samples were compared in A/B listening tests. The 18 participants in Vienna and 14 participants at Fachhochschule Düsseldorf had to assess different characteristics of these sound samples in a blind test.

On the mixing desk for replay, every setup was saved as a snapshot and then recalled according to a prepared order only known to the 'master of ceremony,' Christian Sodl from ORF Radio. One excerpt from the rehearsed works (Mozart +

Berio), each approximately two minutes long, was presented to the listeners with all seven systems. One complete test procedure therefore lasted about 40 min. At the national Austria broadcasting agency ORF in Vienna, 18 subjects (sound-engineers) were divided into 6 test groups (3 persons each). The test listeners were allowed to move around freely during the listening test' (from Camerer and Sodl 2001) (Fig. 9.36).

A detailed description of the results of the listening test can be found in Camerer et al. (2001). The results of the assessments are reproduced as mean values in Figs. 9.37, 9.38, 9.39, 9.40, 9.41, 9.42, 9.43, 9.44, 9.45 and 9.46.

As can be seen in Fig. 9.38, the sound image of the Soundfield Microphone is rated significantly narrower than almost all other techniques (with the exception of Stereo+C and KFM, but only for the ORF group of listeners).

In the recording of the Soundfield microphone, the orchestra sounds are significantly closer than all other techniques. In terms of mean value, the INA5 configuration, which uses only cardioids, seems to provide the second 'closest' sounding results.

Due to the results regarding the two previous acoustic attributes, it may not come as a surprise that again the Soundfield microphone provides statistically significant results: It sounds much flatter in respect of the depth of the orchestra image, than all other microphone techniques, with the possible exception of the KFM (which is second, at least in terms of mean values) and Stereo+C for which the error bars (of the 95% confidence intervals) overlap.

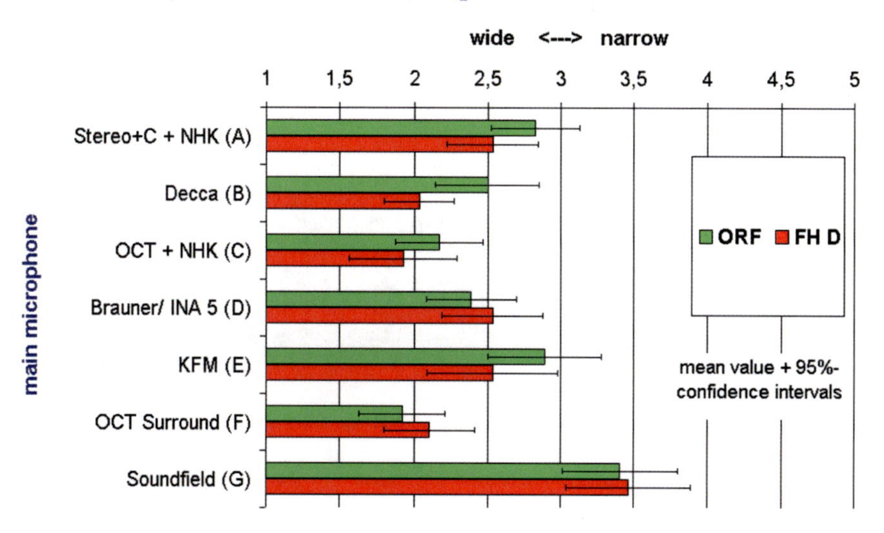

Fig. 9.38 ASW ('apparent source width') of the recorded orchestra in relation to microphone technique (from Camerer et al. 2001)

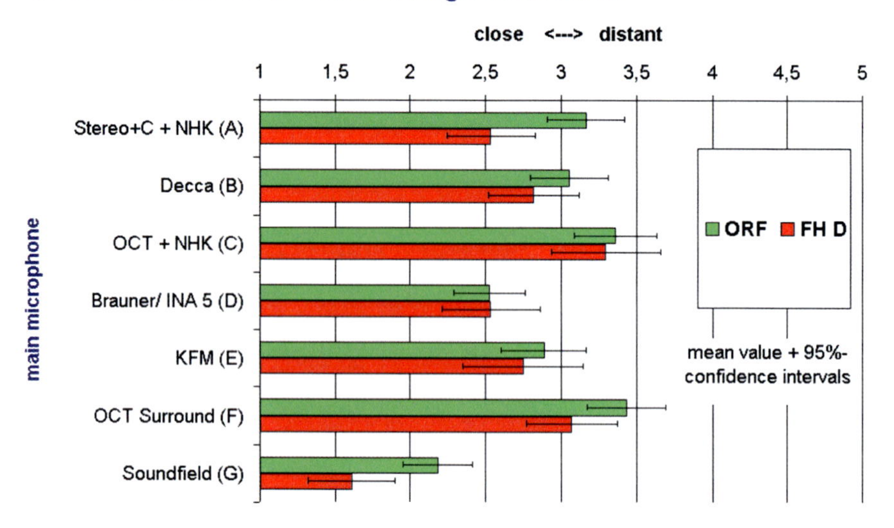

Fig. 9.39 Perceived distance of the recorded sound source (orchestra) (from Camerer et al. 2001)

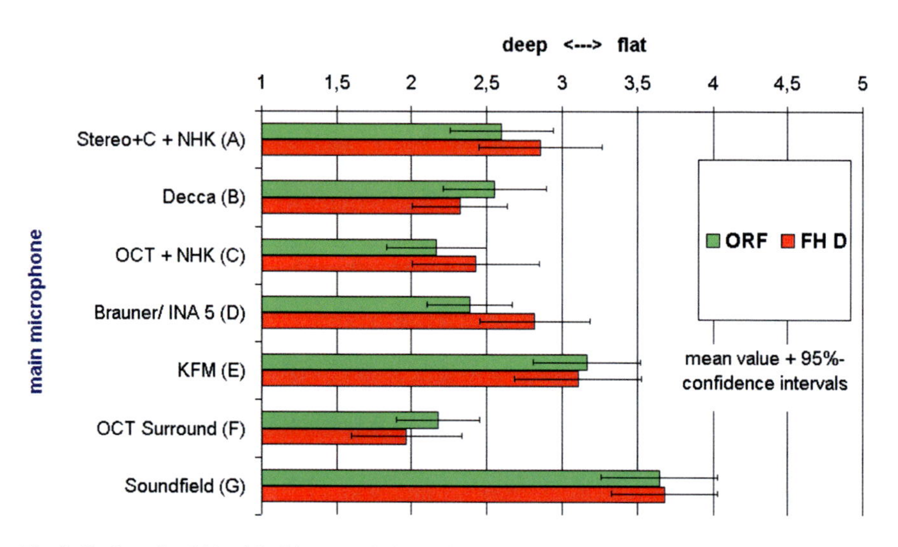

Fig. 9.40 Perceived 'depth' of the recorded sound source (i.e. orchestra) (from Camerer et al. 2001)

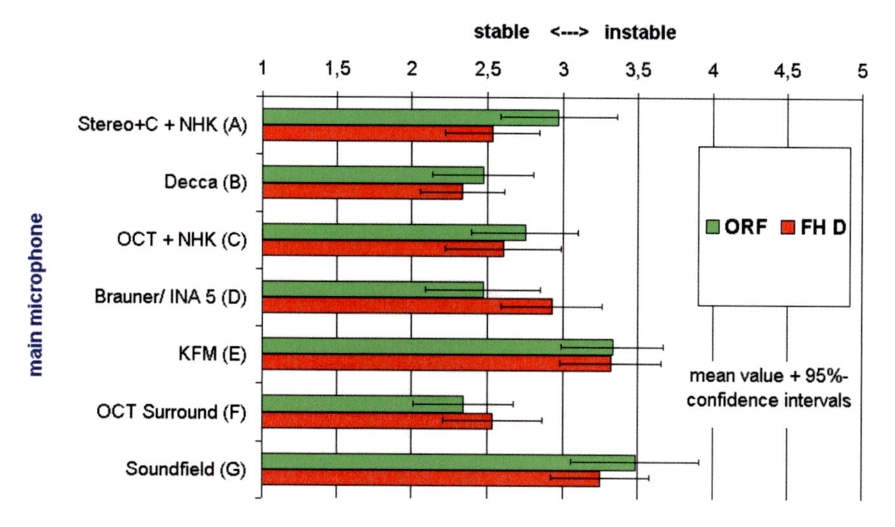

Fig. 9.41 Perceived stability of the sound image (from Camerer et al. 2001)

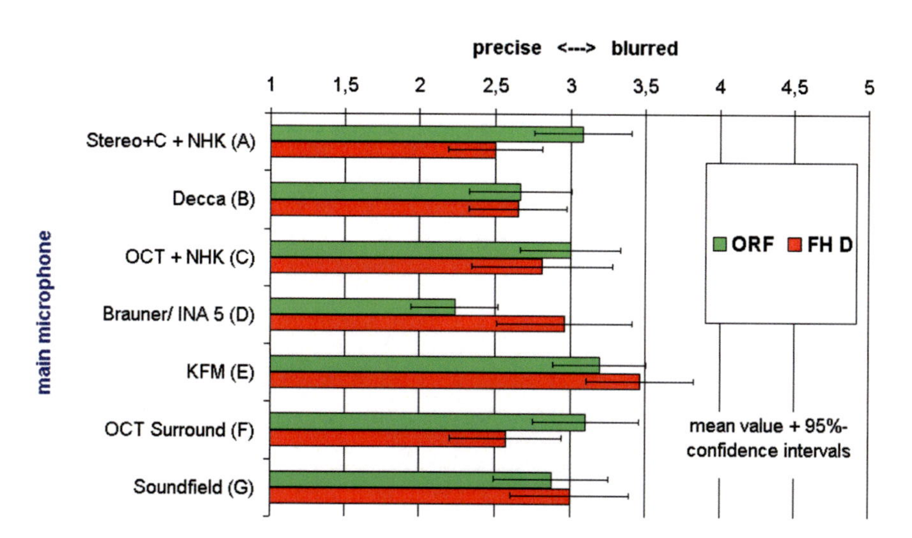

Fig. 9.42 Perceived precision of the sound image (from Camerer et al. 2001)

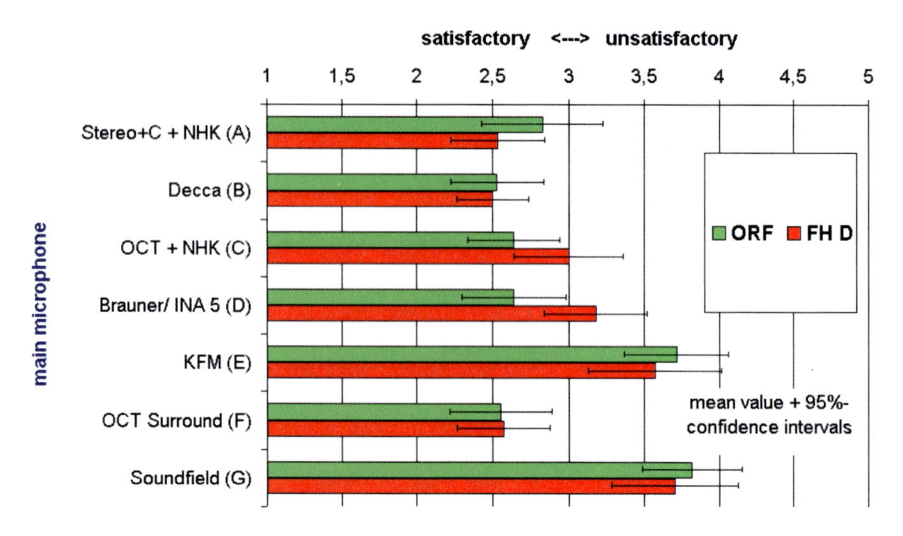

Fig. 9.43 Perceived sound color of the sound source (from Camerer et al. 2001)

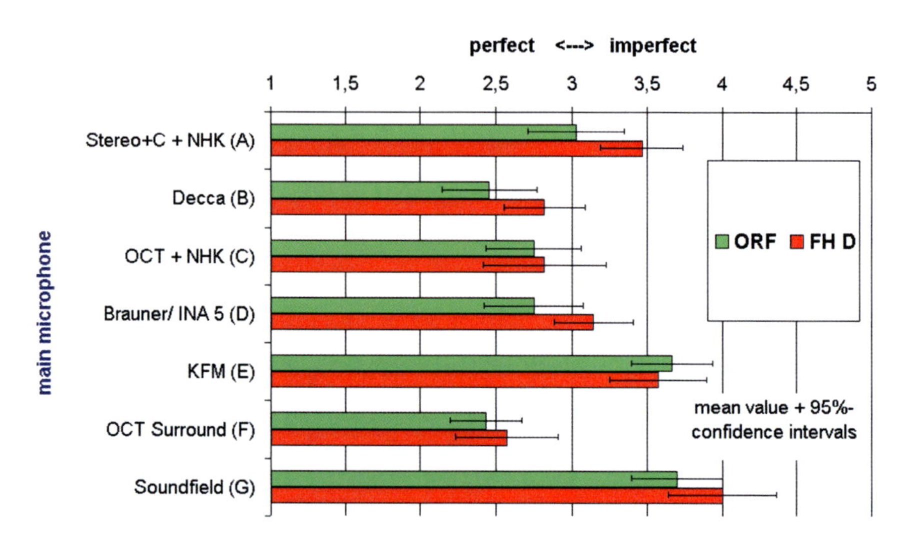

Fig. 9.44 Perceived room impression of the orchestra recording (from Camerer et al. 2001)

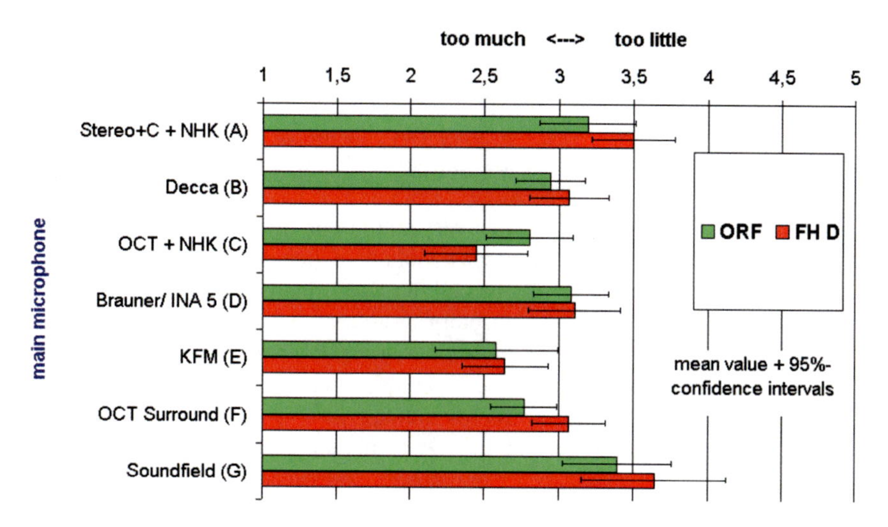

Fig. 9.45 Perceived presence of room information in the orchestra recording (from Camerer et al. 2001)

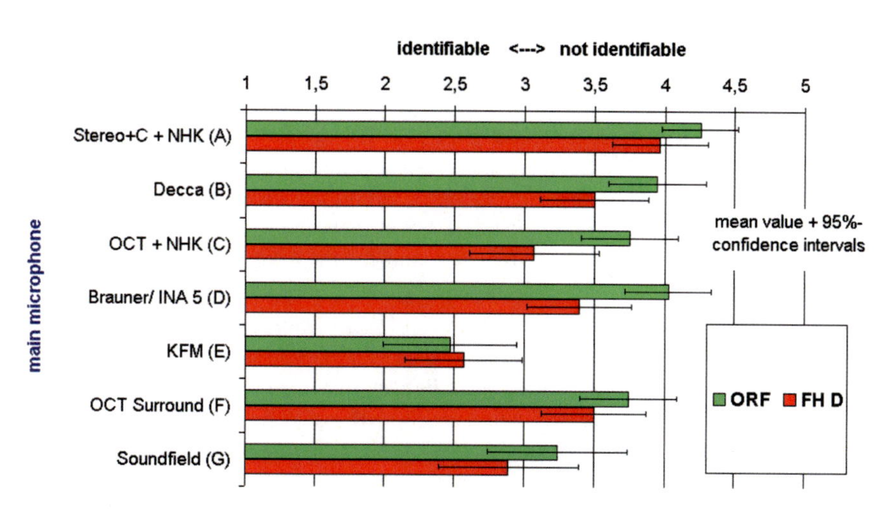

Fig. 9.46 Identifiability of the surround signals (rear signals) of the orchestra recording (from Camerer et al. 2001)

The mic systems providing the greatest impression of 'depth' seem to be OCT surround and OCT+NHK, followed by Decca.

In respect of their mean values, both Soundfield and KFM seem to be slightly on the 'instable' side, while all other techniques are clearly toward the 'stable' side of sound source image reproduction. However, the error bars of these two techniques partly overlap with the error bars of three other techniques (INA5, OCT+NHK and Stereo+C). OCT surround and Decca seem to be slightly more stable than the other systems (mean value).

In Fig. 9.42, the KFM seems to be very slightly on the 'blurred' side, while all other systems seem more precise, at least looking at their mean values.

Again, the Soundfield microphone and the KFM stand out from the crowd, this time in terms of being more 'unsatisfactory' in respect of the sound color. Decca and OCT surround seem to score best (at least in terms of mean value), but statistically there is no significant difference to the other systems (Fig. 9.43).

The room impression (spatial impression) seems to be the least perfect for the Soundfield microphone followed by the KFM and Stereo+C microphone techniques (Fig. 9.44).

Decca and OCT surround seem to score best.

The signals of the Soundfield and Stereo+C microphones tendency-wise contain too little room information (i.e. too dry), while in relation to those two, the KFM and OCT+NHK systems carry—also statistically—significantly more room information.

What remains a bit unclear is the question whether it is considered an advantage or a disadvantage if the rear (surround) signals in the orchestra recording can be identified. Apart from this, it seems that with the KFM system these signals seem somewhat identifiable; for the Soundfield MKV microphone, the situation seems rather neutral, while for all other systems the signals tend to be not identifiable, which seems more desirable.

Conclusion: As can be seen in Fig. 9.37, according to the mean value ratings (smaller value = better) given by test listeners from both groups (Vienna + Düsseldorf groups), the Soundfield microphone—and for many of the questions also the KFM—is on the non-favorable side (sounding flatter, reproducing the sound source with less 'width', etc.), while the OCT surround seems to be the clear overall favorite among the surround microphone systems. INA5 and Decca seem to be second best, while the 'Stereo+C +NHK' appears to be the system, which is the most 'in between.'

The results of the listening evaluation expressed by use of mathematical operands:

OCT-Surround $>$ INA5 $\approx$ Decca $\geq$ OCT+NHK $>$ Stereo-C+NHK $>$ KFM $\geq$ Soundfield

9.2.1.14 The 'Düsseldorf-Test'

In Hermann et al. (1998), a surround microphone test is described, in which four techniques were tested via subjective evaluation by 88 test listeners. The microphone systems under test were: Decca-Triangle, INA, KFM and SF, which is an Ambisonics-based technique, similar to the Soundfield-Microphone. The orchestral sound example used for the test was from 'Thus spoke Zarathustra' (R. Strauss), performed by Düsseldorf Symphonic Orchestra.

The ratings for spatial impression were:
Decca-Triangle > INA $\geq$ KFM $\geq$ Soundfield
The ratings for localization:
INA > Decca-Triangle > KFM > Soundfield
The ratings for 'overall sound' (sound-color/preference):
INA $\approx$ Decca-Triangle > KFM >> Soundfield

While there seems to be an inversion in respect of the order of the Decca-Triangle and INA system in respect to the attributes 'spatial impression' and 'localization', for the other two techniques, the ranking seems clear: KFM is better than the Ambisonic technique.

9.2.1.15 The 'Osaka Test'

In this context, a study with the title 'Correspondence Relationship between Physical Factors and Psychological Impressions of Microphone Arrays for Orchestra Recording' by Kamekawa et al. (2007) should be mentioned. Citing (with slight changes) from the paper:

' … Eight types of well known surround microphone-array techniques were recorded in a concert hall and subsequently compared in subjective listening test in respect to seven attributes such as *spaciousness*, *powerfulness* and *localization* using a method inspired by MUSHRA (MUltiple Stimuli with Hidden Reference and Anchor). The results of the experiment show similarity and dissimilarity between each microphone array. It is estimated that the directivity of a microphone and distance between each microphone are related to the character of the microphone array and that these similarities are changed by the character of the music (e.g. opera, or concert recording). The relations of the physical factors of each array were compared as well, such as SC (Spectral Centroid), LFC (Lateral Fraction Coefficient), and IACC (Inter Aural Cross-correlation Coefficient) from the impulse response of each array or recordings by a dummy head [Rem.: the 'Spectral Centroid' is a measure used to characterize a frequency spectrum. It indicates where the 'Center of mass' of the spectrum is. Perceptually, it has a robust connection with the impression of 'brightness' of a sound (see Grey and Gordon 1978). It is calculated as the weighted mean of the frequencies present in the signal, determined using a Fourier transform, with their magnitudes as weights (see IRCAM 2003; Schubert et al. 2004)].

The correlation of the physical factors SC, LFC, IACC and the attribute scores show that the contribution of these physical factors depends on the music and its temporal change.

As we have already shown in Chap. 7, in Fig. 7.5 the IACC of the 8 different surround techniques under test is displayed for a time-span of 60 s, through which the temporal dependence of inter-aural signal cross-correlation becomes evident. It can be noted that at the dummy head, the resulting IACC can have very differing

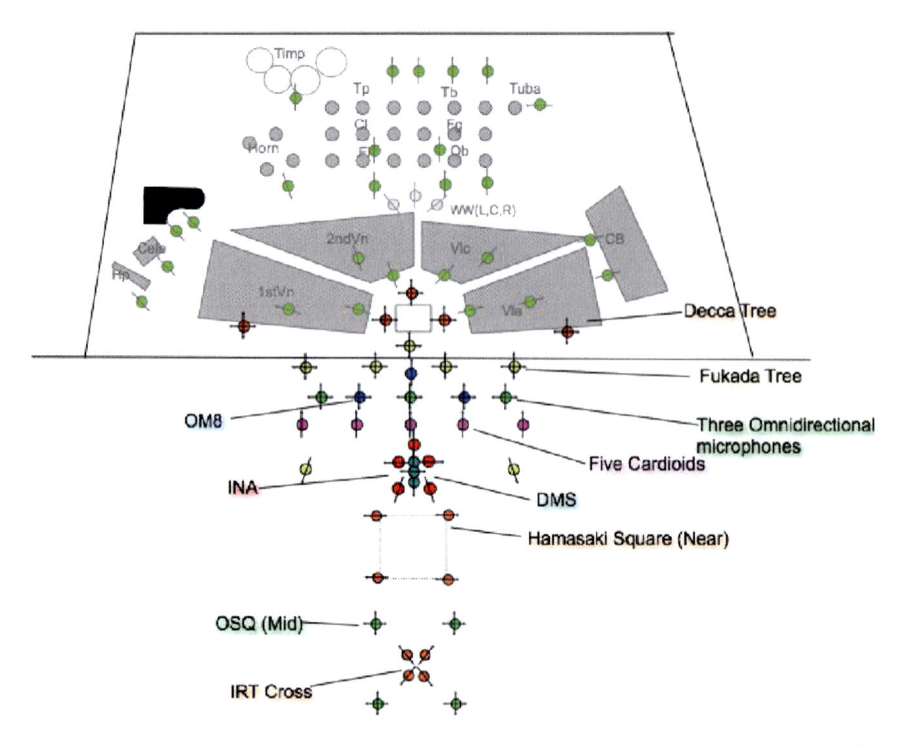

Fig. 9.47 Layout of main and ambient microphones for the experiment of Kamekawa et al. (2007) (abbreviations: *DMS* double-MS, *OM8* omni+8, *DTO* Decca-tree + omni square, *DTH* Decca-tree+Hamasaki Square, *5CH* 5 cardioids + Hamasaki Square, *3OI* 3 Omnis + IRT cross)

values, depending on the surround microphone technique used. In Fig. 9.47 the layouts of the main and ambient microphone arrays are displayed:

9.2.1.16 The Recording Session

Was held at the Symphony Hall in Osaka from September 25th to 27th, 2006, performed by Osaka Philharmonic Orchestra. There were eight types of main microphone arrays, seven types of ambience microphone arrays, and 40 spot microphones, placing a total of 98 microphones in the hall. These microphones were connected to microphone amplifier (Studer 962) and recorded with 96 kHz/24 bit resolution.

To provide the optimum sound of each microphone array, the microphones were placed at optimum locations found through discussion among several recording experts.

9.2.1.17 Acoustic attributes and panel of listeners

Each microphone array was evaluated by the following seven types of attributes

1. *Spaciousness* (Spc): The width of frontal image
2. *Envelopment* (Env): The enveloped feeling; surrounded lateral and backward
3. *Depth* (Dep): The apparent spatial distance of the sound source from the listener
4. *Localization* (Loc): The apparent location of the sound source.
5. *Powerfulness* (Pow): Strong or heavy impression (Opposite meaning is 'weak' or 'feeble')
6. *Softness* (Sof): 'Mild' or 'silky' impression (Opposite meaning is 'hard' or 'harsh')
7. *Preference* (Pref) Listener preference for a recording over the other recordings.

The rating of eight microphone arrays was done for one attribute each, chosen in random order from the seven attributes. The exception was *preference*, which was presented as the final attribute of the session for all subjects.

Group of listeners: Twenty-two subjects, including 13 students and 9 recording experts.

Selection of musical examples:

1. Overture from the opera 'The marriage of Figaro' (Mozart), smaller orchestra;
2. The beginning of 'The Pines of Rome' (Respighi)*)*, large orchestra;
3. an excerpt from the middle part of 'Wellington's Victory' (Beethoven), brass section;

Lengths of each piece were approximately one minute for the listening test.

Subjects were allowed to listen to each piece repeatedly until they were satisfied with evaluating all eight microphone arrays.'

The two results which stand out clearly in Fig. 9.48 are:

1. that the combination DT+OSQ (Decca-tree with omni square) has received the highest ratings for all attributes except *spaciousness* and *localization*, and
2. that the DMS technique (double-MS) has clearly scored lowest for all attributes (with an overlap with INA for the attribute *envelopment*).

The brackets in the figures show groupings of stimuli based on 95% confidence intervals. There were some groups of microphone arrays distinguished significantly within some of the attributes such as *spaciousness* and *powerfulness*, *softness* and *preference* at 'Figaro.' But there were different results at the other two pieces. For all three music excerpts, there were significant differences between each microphone array regarding *spaciousness* and *powerfulness*, but there was no difference on *depth* and *localization*. Also, the differences on *envelopment*, *softness* and *preference* were dependent on the musical character of the piece (Fig. 9.49).

However, it can be said that in terms of overall tendency for all three musical excerpts the Decca-tree-based surround techniques (in both variations: with omni square and with Hamasaki Square) usually occupied the first and second place in terms of rating, while the double-MS technique was usually last. Also, for the musical

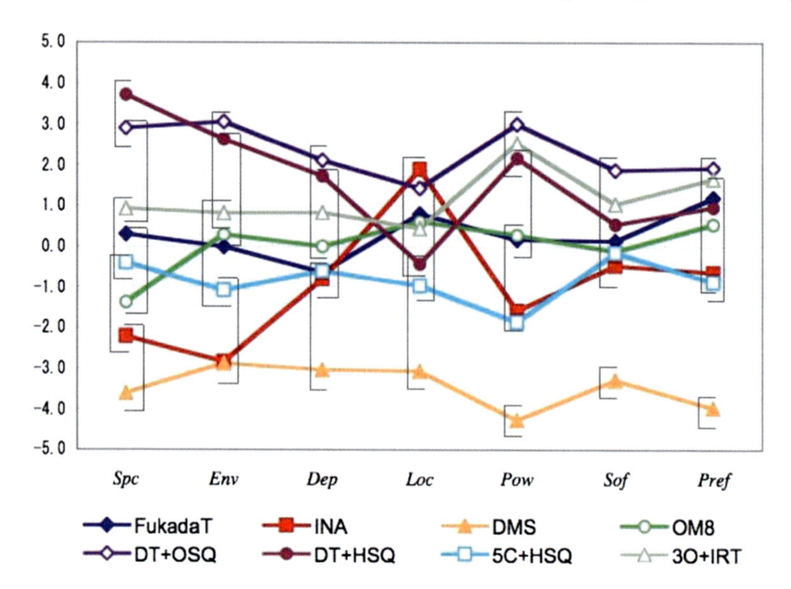

Fig. 9.48 Average rating of seven attributes and listener preference for 8 surround microphone techniques for the opera-ouverture 'Figaro' [see Fig. 3 from Kamekawa et al. (2007)]

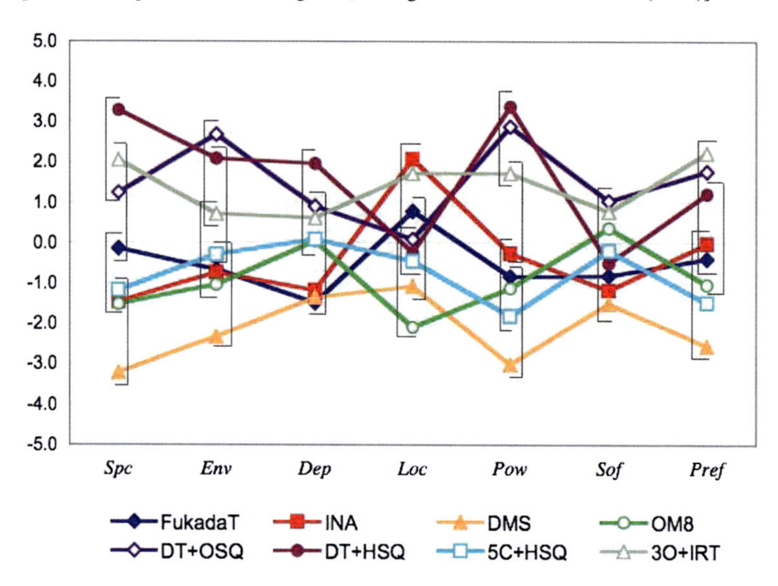

Fig. 9.49 Average rating of seven attributes and listener preference for 8 surround microphone techniques for 'Wellington's Victory' [see Fig. 5 from Kamekawa et al. (2007)]

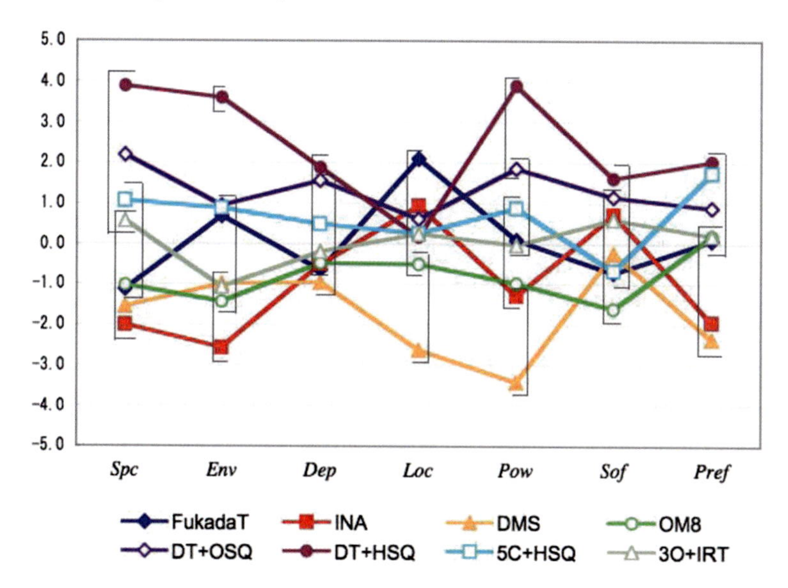

Fig. 9.50 Average rating of seven attributes and listener preference for 8 surround microphone techniques for 'Pines of Rome' [see Fig. 4 from Kamekawa et al. (2007)]

excerpts of 'Figaro' and 'Wellington's Victory' the 3O+IRT (three omnis combined with the IRT cross) seems to occupy a relatively stable third place in terms of listener rating.

In addition to the above analysis of the average score, an INDSICAL (individual difference scaling) analysis was conducted on correlation coefficients computed from the subjects for the score of each microphone array. Typical use of INDSCAL analysis in psychoacoustics is to derive a map of stimuli from psychological distances obtained from global dissimilarity ratings. The assumption behind the use of correlation coefficients in this case was that correlation values computed from *all* subjective attribute ratings are highly related to global dissimilarities in the psychological domain.

From the stepwise multiple regression analysis with each of the INDSCAL dimensions as dependent variable and seven attribute scores from all subjects as independent variables, the relation between each dimension and attributes was calculated.

The results were that for two of the three musical excerpts (i.e. for 'Figaro' and 'Pines of Rome') *envelopment* and *spaciousness* resulted as the first dimension.

For all three excerpts, the second and third dimension was represented by *powerfulness* in combination with either *spaciousness*, *localization* or *softness*, with the exception of 'Figaro,' for which the second dimension was represented by the combination *spaciousness/preference*.

In order to be able to make a comparison of physical factors and psychological factors, IR (impulse response) measurements were made for each microphone array, from which SC (spectral centroid) and LFC (lateral fraction coefficient) of each array were calculated (see Hanyu 2002). While SC was expected to be related with the timbral attributes such as *powerfulness* and *softness*, LFC was expected to be related to the attributes of spatial impression such as *spaciousness*, *envelopment* and *depth*.

To compare the physical factor of each piece, IACC (inter-aural cross-correlation coefficient) and SC were calculated for the recordings done using a dummy head microphone (from company 'HEAD Acoustics') in the same position as for the listeners in the microphone comparison.

Comparing Figs. 9.51 and 9.52, it is interesting to note that in both through the analysis of the listener ratings (INDISCAL), and the physical measurements and subsequent calculations the DMS technique is somehow separate from the rest. Not

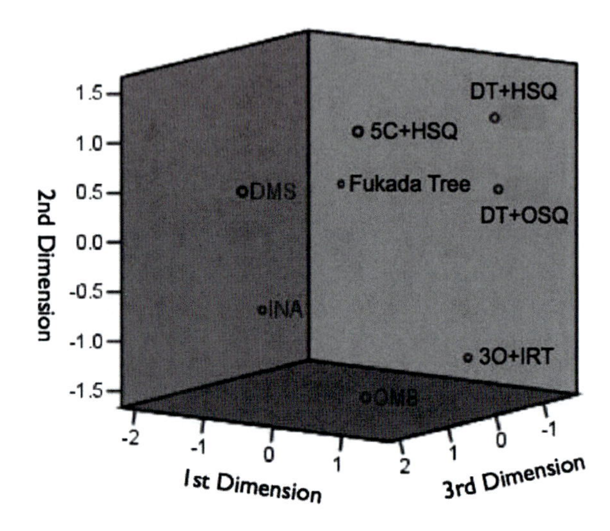

Fig. 9.51 Spatial configuration of eight microphone arrays on three dimensions according to INDISCAL analysis for 'Figaro' [Fig. 6 from Kamekawa et al. (2007)]

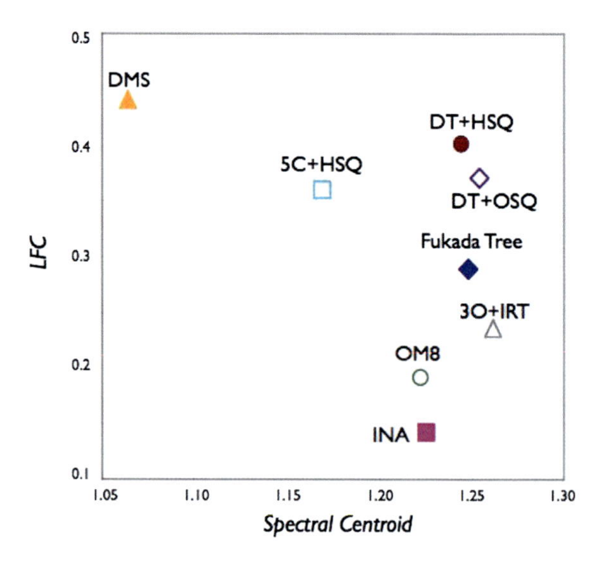

Fig. 9.52 Relation of SC (spectral centroid) and LFC (lateral fraction coefficient) for eight microphone techniques, calculated from IR measurement [Fig. 10 from Kamekawa et al. (2007)]

surprisingly, the two Decca-based techniques DT+HSQ and DT+OSQ (which had been the 'winners' in the averaged ratings for two of the three musical examples; see Figs. 9.48, 9.49 and 9.50) are displayed as closely related.

As an attempt to arrive at a 'final verdict' in terms of quality of the eight surround microphone techniques—mainly based on listener preference—it seems that the Decca techniques DT+OSQ and DT+HSQ have scored highest, with 3O+IRT in a third position, while DMS has scored lowest for almost all attributes in the three musical excerpts. It is a pity that Theile's OCT+Surround technique had not been included in the test, as it had been part of several other comparative studies and therefore could have served as common reference. In this context, it is of interest to note that the INA technique, which has been included in the test by Kawekama et al., has received relatively low scores for almost all attributes (except *localization*, in which it has performed very well), which somehow seems to put it in the second-to-last position.

The other microphone techniques are somewhere in between the abovementioned 'top-quality' group and the two microphone techniques, which have performed worst.

If we try to put the qualitative ranking, derived from the test by Kawekama et al. in an order which uses mathematical operands we arrive at the following:

$$\mathrm{DT+HSQ} \geq \mathrm{DT+OSQ} \geq \mathrm{3O+IRT} \approx \mathrm{FT} \approx \mathrm{5C+HSQ} \approx \mathrm{OM8} \geq \mathrm{INA} \geq \mathrm{DMS}$$

The study by Kamekawa et al. from 2007 stands out among works in the field, as it tries to arrive at a qualitative ranking of surround microphone techniques not only by listener comparison, but in addition through signal analysis of some acoustic parameters (LF and SC). In this respect, there is a similarity in approach to the research method as applied by the author of this book (see Pfanzagl-Cardone 2002, 2011; Pfanzagl-Cardone and Höldrich 2008).

Not based on a mere numeric evaluation, but taking into account the underlying physical principle of each surround microphone technique, it seems interesting to note that microphone techniques—the front arrays of which are using omni microphones and are based on time-of-arrival differences (i.e. 'DT' = DECCA-tree including 'large AB' style 'outriggers')—seem to have delivered the most convincing sonic results for most listeners. They seem to have a clear advantage over the surround microphone systems which use mainly directional microphones and are—therefore—also placed further away (5C+HSQ, INA), as well as the 'one-point' coincident technique of DMS.

9.2.2 A Final Ranking for Several Common 5.1 Surround Microphone Techniques

The results from the Vienna Test as described in a similar form:

OCT Surround > INA5 $\approx$ Decca$\geq$ OCT+HamSqr > Stereo-C+ HamSqr > KFM $\geq$ Soundfield

The immediate results from the Berlin Test (in which only the acoustic attributes of spaciousness, localization and sound color were analyzed) are:

- Three-channel techniques seem to have an advantage over 2-channel techniques.
- The Schoeps KFM performed worst.

Taking a look at the results from the surround microphone test conducted in Salzburg by the author of this book (see Chap. 4, microphone technique 'AB-PC' and Chap. 6, results presented in Fig. 6.8), in which 4 surround microphone techniques have been evaluated: Rather few statistically significant differences between the microphone techniques under test (ABPC, DECCA, OCT, KFM) became evident for the various acoustic attributes. The KFM system was significantly drier than the others, resulting in a much poorer spatial impression and perceived depth of the sound source. Also, the volume balance within the sound source (i.e. balance of the different instrumental groups within the orchestra) was judged clearly inferior to the other three techniques. In terms of source width, the two techniques DECCA and ABPC, both using (also) omni-directional microphones and therefore relying on time-of-arrival differences, were rated better than OCT and the KFM surround system, which was rated worst.

In terms of balance and localization accuracy, as well as sound color ABPC performed best, which also resulted in a tendency-wise 'best mark' in terms of overall listener preference, while the KFM was clearly the least appreciated. Not surprisingly, the same was also the case for the acoustic attribute of naturalness.

Therefore, based on listener preference, expressed by the use of mathematical operands (with overlapping 95% confidence intervals, therefore statistically not significant, except for the KFM technique):

$$\text{ABPC} \geq \text{OCT} \approx \text{DECCA} >> \text{KFM}$$

Now, before undertaking the brave attempt to 'stitch together' all results from the above tests, a few remarks seem due: Of course—strictly speaking—it is impossible to arrive at a clear and completely 'fair' or correct ranking among microphone techniques, as also some of the tests cited above have shown that the perceived 'quality' of a microphone technique will depend on varied parameters such as the room acoustics (large or small hall, diffusion, etc.), the nature of musical piece (orchestral, chamber music, opera) to name just the two most important criteria. Therefore, the apparent lack or pitfall of one microphone technique under a given set of circumstances may turn out to be a positive asset at another place and time (see, e.g., the much 'drier' characteristics of the KFM and Soundfield microphones, which may become very useful in case of 'too reverberant' recording conditions).

However—having said all this—it is nevertheless true that some surround microphone techniques seem to have a consistently better 'overall performance' than others; otherwise writing this book would have been in vain.

Also, in the above tests, it can clearly be seen that some microphone techniques repeatedly perform worse than others, which is also a rather clear indicator in respect of their quality.

For the sake of correctness and clarity, it needs to be pointed out that the DECCA technique—as used in the Vienna and Salzburg test—is essentially the DECCA-triangle, while the DT technique in the Osaka test is a complete DECCA-tree including (large AB) outriggers, which may be of advantage in respect of better spatial impression (rem.: also the Fukada tree in the Osaka test uses omni-outriggers).

Systems like the IRT cross have initially been intended as a (quadraphonic) surround microphone system in their own right, but seem to be put to use—at least in the context of the Osaka test—mainly for capturing the diffuse sound part of a 5.1 surround signal. The same is true for the 'omni square' (OSQ) and the Hamaski square (HSQ) in respect of their main use as 'diffuse sound pickup' system.

Let us quickly review the main results of the above tests in 'telegram' style (rem.: some of the abbreviations have partly been written out to make things more easily understandable. Also, the NHK system of the Vienna test is in fact the Hamaski square and therefore both have below been re-named as HamSqr, for easier comparison and understanding).

Berlin-Test:

– 3-channel techniques performed better that 2-channel techniques

Since most of the surround microphone techniques in the Berlin-Test are rather 'individualistic' (some are a bit unusual and all of them defined in detail by their inventors), only a few qualify for being regarded as a 'standard' technique.

The one result, which clearly stood out was:

KFM = worst

Düsseldorf-Test:

Decca-Triangle $\geq$ INA > KFM > Soundfield

Osaka-Test:

DeccaTree+HamSqr $\geq$ DeccaTree+OSQ $\geq$ 3Omni+IRT $\approx$ FukadaTree $\approx$ 5Card+HamSqr $\approx$ OM8 $\geq$ INA $\geq$ DMS

Salzburg-Test:

ABPC $\geq$ OCT-Surr $\approx$ Decca-Triangle >> KFM

Vienna-Test:

OCT-Surr > INA $\approx$ Decca-Triangle $\geq$ OCT+HamSqr > Stereo-C+HamSqr > KFM $\geq$ Soundfield

Looking at the results of the Salzburg and Vienna test, it seems that the OCT system may be slightly superior to the Decca-Triangle. However, the fact that the OCT combined with the Hamasaki Square is inferior to the Decca-Triangle shows that the presumed superiority of the OCT (front system) depends very much on which rear system it is combined with.

If we try to combine the results from above in a summarized manner (and of course without claiming completeness):

ABPC $\approx$ OCT-Surr $\approx$ Decca > 3Omni+IRT $\approx$ FukadaTree $\approx$ 5Card+HamSqr $\approx$ OM8 $\approx$ Stereo-C+HamSqr $\geq$ INA $\geq$ DMS $\geq$ KFM $\geq$ Soundfield

ABPC, OCT surround and Decca (triangle and/or tree) seem to belong to the 'best group,' while KFM and the Soundfield-Microphone (which is based on a first-order Ambisonic principle) are clearly worst. All other surround microphones and combinations of techniques seem to be somewhere in between, while their exact order will likely depend on the specific room acoustics and music example, as well as other parameters of a given situation. The above-described relation between INA, DMS and KFM can only be estimated, as there is no test which includes all three systems. Therefore, the ranking described above is a 'best guess' made from a combination of the Berlin, Osaka, Salzburg and Vienna tests.

From the above summary, derived from the results of extensive subjective listening tests, it also seems clear that surround microphone techniques, which use large capsule spacings (i.e. ABPC and Decca-tree techniques with large A/B outriggers), or highly directional microphone techniques (OCT)—which both result in high channel separation and therefore low correlation between the signals—are superior to other techniques, which do not seem to provide as much channel separation (or signal de-correlation). INA, DMS, KFM and Soundfield/Ambisonic seem to belong to the latter group.

Therefore measuring the FCC (Frequency dependent Cross-Correlation) or the binaural equivalent of it, the FIACC (frequency-dependent inter aural cross-correlation) can provide a strong first indicator for the perceived 'quality' of a microphone system.

In the following chapter, a new measure, the BQI_{rep} ('Binaural Quality Index of reproduced music'), is proposed, which may also be helpful in determining the perceived quality of a recording.

References

Berg J, Rumsey F (2003) Systematic evaluation of perceived spatial quality. In: Proceedings to the 24th International Conference of the Audio Eng Soc. Available at: Surrey Research Insight Open Access, 2012. http://epubs.surrey.ac.uk/567/. Accessed 20 Sept 2018

Blumlein AD (1931) Improvements in and relating to sound-transmission, sound-recording and sound-reproducing Systems. British Patent 394,325, 14 Dec 1931 (reprinted in: Anthology of stereophonic techniques. Audio Eng Soc, 1986, p 32–40)

Bruck J (1996) Solving the surround dilemma. In: Proceedings to the 19. Tonmeistertagung des VDT, pp 117–124

Camerer F, Sodl C (2001) Classical music in radio and tv—a multichannel challenge. The IRT/ORF Surround Listening Test. http://www.hauptmikrofon.de/stereo-surround/orf-surround-techniques. Accessed 31 May 2019

Camerer F, Sodl C, Wittek H (2001) Results from the Vienna listening test. http://www.hauptmikrofon.de/ORF/ORF_und_FHD.htm. Accessed 31 May 2019

Ceoen C (1972) Comparative stereophonic listening tests. J Audio Eng Soc 20(1)

Clark H, Dutton G, Vanderlyn P (1958) The 'Stereosonic' recording and reproduction system: a two-channel system for domestic tape records. J Audio Eng Soc 6:102–117
Dickreiter M (2011) Mikrofonaufnahme, 4th edn. S. Hirzel Verlag Stuttgart
Elko GW (2001) Spatial coherence functions for differential microphones in isotropic noise fields. In: Brandstein M, Ward D (eds) Microphone arrays. Springer, pp 61
Faulkner T (1981) A phased array. Hi Fi News Record Rev, July 1981
Gernemann A (2000) Stereo+C - an all purpose arrangement of microphones using three frontal channels. Paper to the 110th audio eng soc convention, Amsterdam 2001
Gerzon M (1973) Periphony: with-height sound reproduction. J Audio Eng Soc 21(1)
Gerzon M (1986) Stereo shuffling: new approach, old technique. Studio Sound, July 1986, pp 122–130
Grey JM, Gordon JW (1978) Perceptual effects of spectral modifications on musical timbres. J Acoust Soc Am 63 (5):1493–1500
Griesinger D (1987) New perspectives on coincident and semi-coincident microphone arrays. Paper 2464 presented at the 82nd audio eng soc convention, May 1987
Griesinger D (2002) Stereo and surround panning in practice. Paper 5564 presented at the 112th audio eng soc convention, Munich, May 2002
Hanyu (2002) Room acoustical parameters. J Acoust Soc Japan 60(2):72–77
Hermann U, Henkels V, Braun D (1998) Comparison of 5 surround microphone methods. In: Proceedings to the 20. Tonmeistertagung des VDT, Karlsruhe, 1998, p 508
Hildebrandt A, Braun D (2000) Untersuchungen zum Centerkanal im 3/2 Stereo-Format. In: Proceedings to the 21. Tonmeistertagung des VDT, p 455
Hiyama K, Ohkubo H, Komiyama S (2000) Examination of optimum speaker layout in multispeaker sound field reproduction. Paper resented at the Spring Conference of AST, March 2000
Hiyama K, Komiyama S, Hamasaki K (2002) The minimum number of loudspeakers and its arrangement for reproducing the spatial impression of diffuse sound field. Paper 5674 presented at the 113th Audio Eng Soc Convention, Los Angeles, Oct 2002
Hiyama K, Ohkubo H, Komiyama S, Hamasaki K (2003) Examination of optimum speaker layout for restoration of diffusive sound field. Paper presented at the Fall Conference of AST, Sept 2003
Holman T (2000) 5.1 Surround sound – up and running. Focal Press (Elsevier)
Holman T (2001) Mixing the sound (Part 2): Perspective – where do the sounds go? Surround Professional, May/June 2001, p 35
IRCAM (2003) A large set of audio features for sound description. Section 6.1.1, Technical Report IRCAM
Julstrom S (1991) An intuitive view of coincident stereo microphones. J Audio Eng Soc 39(9)
Kamekawa T, Marui A, Irimajiri H (2007) Correspondence relationship between physical factors and psychological impressions of microphone arrays for orchestra recording. Paper 7233 presented at the 123rd Audio Eng Soc Convention, New York, Oct 2007
Knothe J, Plenge G (1978) Panoramaregler mit Berücksichtigung der frequenzabhängigen Pegeldifferenzbewertung durch das Gehör. In: Proceedings to the 11th Convention of the VDT, Berlin, p 136–143
Meyer J (1999) Akustik und musikalische Aufführungspraxis, 4th edn. Verlag Erwin Bochinsky (PPV Medien)
Muraoka T, Nakazato T (2007) Examination of multichannel sound-field recomposition utilizing frequency-dependent interaural cross correlation (FIACC). J Audio Eng Soc 55(4):236–256
Muraoka T, Miura T, Ifukuba T (2007) Ambience sound recording utilizing dual MS (Mid-Side) microphone systems based upon frequency dependent spatial cross correlation (FSCC). Paper 6997 to the 122nd Audio Eng Soc Convention, Vienna, 2007
Olabe I (2014) Técnicas de grabación de música clásica. Evolución histórica y propuesta de nuevo modelo de grabación. Dissertation, Universitat de les Illes Balears. http://hdl.handle.net/10803/362938. Accessed 12 Oct 2015
Olson HF, Belar H (1960) Acoustics of sound reproduction in the home. J Audio Eng Soc 8(1):7–11

Pfanzagl-Cardone E (2002) In the light of 5.1 surround: why AB-PC is superior for symphony-orchestra recording. Paper 5565 presented at the 112th Audio Eng Soc Convention, Munich, 2002
Pfanzagl-Cardone E, Höldrich R (2008) Frequency-dependent signal-correlation in surround- and stereo-microphone systems and the Blumlein-Pfanzagl-Triple (BPT). Paper 7476 presented at the 124th Audio Eng Soc Convention, Amsterdam, 2008
Pfanzagl-Cardone E (2011) Signal-correlation and spatial impression with stereo- and 5.1 surround-recordings. Dissertation, University of Music and Performing Arts, Graz, Austria. https://iem.kug.ac.at/fileadmin/media/iem/altdaten/projekte/dsp/pfanzagl/pfanzagl_diss.pdf. Accessed Oct 2018
Rumsey F (1991) Creating new images. Studio Sound
Rumsey F (1994) Acoustics and acoustic devices - stereo. In: Talbot-Smith M (ed) Audio engineer's reference book. Focal Press
Schubert E, Wolfe J, Tarnopolsky A (2004) Spectral centroid and timbre in complex, multiple instrumental textures. In: Proceedings to the 8th international conference on music perception & cognition, North Western University, Illinois
Snow WB (1953) Basic principles of stereophonic sound. SMPTE (Soc Motion Picture and Television Eng) 61:567–589
Steinberg JC, Snow WB (1934) Auditory perspective – physical factors. Electrical Engineering 53(1):12–15
Streicher R, Dooley W (1985) Basic microphone perspectives – a review. J Audio Eng Soc 33(7/8)
Takahashi and Anazawa (1987) Recent research on recording technology in Europe. J Acoust Soc of Japan 6
Theile G (1986) Das Kugelflächenmikrofon. In: Proceedings to the 14. Tonmeistertagung des VDT, Munich, 1986, p 277
Theile G (1991) On the naturalness of two-channel stereo sound. Paper presented at Audio Eng Soc 9th Int Conference, Detroit, 1–2 February 1991
Theile G (2001) Multichannel natural music recording based on psychoacoustic principles. In: Proceedings to the 19th Int Conf of the Audio Eng Soc, 2001, pp 201–229
Vanderlyn PB (1957) British Patent 781,186 14 Aug 1957
Vanderlyn P (1979) Auditory cues in stereophony. Wireless World, Sept 1979, pp 55–60
Wohr M, Nellseen B (1986) Untersuchung zur Wahl des Hauptmikrofonverfahrens. In: Preprints to the 14. Tonmeistertagung des VDT, Munich, 1986

AV-Sources

Swedien B (1997) Recording with Bruce Swedien. VHS-Videotape, Palefish Enterprises (Inc/AcuNet Corporation), White Salmon, WA

Chapter 10
Binaural Quality Index of Reproduced Music (BQIrep)

Abstract The 'Binaural Quality Index' BQI, defined by Keet in 1968, has been identified as 'one of the most effective indicators of the acoustic quality of concert halls' by Beranek in 2004. The acoustics of concert halls—being ideal locations for musical performances—should remain the basic reference point on how music should sound, also when being replayed to the listener in the living room. Therefore, a newly defined BQI_{rep}—'Binaural Quality Index of reproduced music' is proposed as a measure for the 'naturalness' of recorded music.

Keywords BQI—Binaural Quality Index · Concert hall acoustics · BQIrep · IACC—Interaural cross-correlation coefficient · Home listening · Naturalness

The 'Binaural Quality Index' (BQI) as defined by de Keet (1968) has already been described in the chapter covering spatial impression and has been identified by Beranek (2004) as a measure for determining the acoustical quality of concert halls.

It has been defined by Keet as (see also Chap. 1):

$$BQI = (1 - IACC_{E3}) \tag{10.1}$$

The subindex E3 denotes the early sound energy in the time window from 0 to 80 ms in the octave bands with center frequencies at 500 Hz, 1 kHz and 2 kHz.

As a result of the research by Pfanzagl-Cardone (2011), it turned out that one of the microphone techniques under test ('ABPC'—AB-Polycardioid Centerfill; see chapter on surround microphone techniques) was able to reproduce the BQI—measured in the concert hall—much better than the other microphone techniques under test (see Table 10.1). This was verified using the new BQI_{rep}—'Binaural Quality Index of reproduced music,' which is defined as follows:

$$BQI_{rep} = (1 - IACC_3) \tag{10.2}$$

with $IACC_3$ being the mean value of the cross-correlation coefficients of the octave bands 500 Hz, 1 kHz and 2 kHz measured between the L and R binaural signals.

E. Pfanzagl-Cardone, *The Art and Science of Surround and Stereo Recording*,
https://doi.org/10.1007/978-3-7091-4891-4_10

Table 10.1 BQI_{rep} and rho (500, 1000, 2000), based on acoustic measurement of the re-recordings of ORCH-Sample #4 (60 s duration) as captured through the surround microphone techniques OCT, DECCA, KFM and ABPC by means of the Neumann KU81-dummy head; also the original value of the KU81 dummy head recording from the concert hall is being displayed (i.e. reference)

Mic array	rho_{500}	rho_{1000}	rho_{2000}	BQI_{rep}
OCT	0.83	0.64	0.66	0.29
DECCA	0.77	0.61	0.57	0.35
KFM	0.90	0.61	0.39	0.37
AB-PC	0.43	0.26	0.11	0.73
KU 81	**0.41**	**0.19**	**0.20**	**0.73**

Determining the BQI_{rep} is related to the BQI, used for concert hall evaluation, as also in the case of measuring the BQI the cross-correlation of the ear signals of a dummy head is being evaluated for the octave bands 0.5, 1 and 2 kHz.

In contrast to the concert hall related BQI measurement, for the BQI_{rep} 'early' and 'late' reflected sound energy cannot be distinguished, as the measurement cannot make use of an impulse response, but instead the final versions of music recordings (usually CDs or final masters) are being analyzed. This is also why the subindex '$_{E3}$' is missing, in which the E stands for 'early' and takes into account the early sound energy in the time range from 0 to 80 ms. For the sake of a clearer distinction from already existing acoustic measures like $IACC_{E3}$ (i.e. interaural cross-correlation coefficient of the early energy [0–80 ms]) and $IACC_{L3}$ (i.e. Inter-Aural Cross-Correlation Coefficient of the late energy [80–3500 ms]) the new term BQI_{rep} is being proposed, which already indicates in its name that this measure refers to the degree of correlation in (previously recorded) music signals.

The restriction of the BQI_{rep} to be applied only to pre-recorded (music as well as non-music) signals, which are being re-recorded via dummy head in the listening room, can also be seen as an advantage, as therefore any 2-channel recording can be subjected to this evaluation process.

The graphs displayed in Chap. 7, Fig. 7.5 are an indicator for the necessity to have 60 s as a minimum duration for acoustic analysis like above (concerning the BQI_{rep}), since a shorter duration may yield inaccurate (only short term, instead of more 'averaged') results: looking at surround technique '5CH' (brown) it can be seen that signal correlation is strongly out-of-phase with a value of –0.6 for the time-span 24–29 s, while it is in-phase with a value of +0.5 in the time segment 44–48 s.

In order to show that the results displayed for BQIrep in Table 10.1 above are meaningful, for comparison the results of measurements by Hidaka et al. (1996) are listed in Table 10.2, with very similar values.

The position used for the original KU81 dummy head recordings corresponds with 'Position 11' in Fig. 10.1. In the measurements by Hidaka, Position 11 was characterized by unusually high IACCE3—values of 0.90 when sender position S0 was used. When averaged over the measurement positions listed in Table 10.2, $IACC_{E3}$ resulted

Table 10.2 $IACC_E$, $IACC_L$ and $IACC_A$; '3 bands' = mean of the values measured for the octave bands 500 Hz, 1 kHz, 2 kHz (E = 0–80 ms, L = 80–3500 ms, A = 0–3500 ms (from Hidaka et al. 1996)

Festspielhaus, Salzburg [IACC]

		$IACC_E$					$IACC_L$					$IACC_A$				
S0		500	1k	2k	**3 bands**	4k	500	1k	2k	**3 bands**	4k	500	1k	2k	**3 bands**	4k
	101	0.851	0.850	0.861	**0.85**	0.748	0.257	0.185	0.140	**0.19**	0.049	0.444	0.535	0.530	**0.50**	0.541
	102	0.550	0.625	0.475	**0.55**	0.545	0.209	0.203	0.099	**0.17**	0.071	0.322	0.417	0.275	**0.34**	0.335
	11	0.883	0.877	0.945	**0.90**	0.823	0.287	0.126	0.102	**0.17**	0.054	0.509	0.493	0.735	**0.58**	0.688
	14	0.702	0.620	0.624	**0.65**	0.416	0.223	0.105	0.080	**0.14**	0.053	0.405	0.353	0.372	**0.38**	0.236
	17	0.511	0.537	0.669	**0.57**	0.491	0.206	0.115	0.100	**0.14**	0.114	0.325	0.327	0.445	**0.37**	0.362
	18	0.491	0.308	0.211	**0.34**	0.328	0.090	0.139	0.108	**0.11**	0.039	0.258	0.223	0.143	**0.21**	0.245
	21	0.797	0.729	0.714	**0.75**	0.463	0.162	0.089	0.064	**0.11**	0.056	0.542	0.418	0.395	**0.45**	0.291
	23	0.587	0.329	0.405	**0.44**	0.352	0.180	0.067	0.065	**0.10**	0.058	0.478	0.236	0.282	**0.33**	0.253
	24	0.371	0.487	0.534	**0.46**	0.306	0.175	0.063	0.085	**0.11**	0.054	0.273	0.317	0.350	**0.31**	0.242
	25	0.619	0.410	0.239	**0.42**	0.283	0.188	0.145	0.054	**0.13**	0.033	0.471	0.324	0.187	**0.33**	0.207
	Mean	0.636	0.577	0.568	**0.59**	0.476	0.198	0.124	0.090	**0.14**	0.058	0.403	0.364	0.371	**0.38**	0.340
Sp	101	0.844	0.719	0.283	**0.62**	0.550	0.161	0.167	0.075	**0.13**	0.061	0.398	0.410	0.180	**0.33**	0.405
	102	0.487	0.538	0.548	**0.52**	0.504	0.112	0.072	0.078	**0.09**	0.072	0.252	0.246	0.314	**0.27**	0.318
	11	0.938	0.838	0.791	**0.86**	0.842	0.101	0.088	0.094	**0.09**	0.067	0.582	0.520	0.569	**0.56**	0.723
	14	0.400	0.488	0.554	**0.48**	0.459	0.056	0.084	0.056	**0.07**	0.073	0.222	0.293	0.284	**0.27**	0.300
	17	0.630	0.574	0.610	**0.60**	0.517	0.119	0.106	0.144	**0.12**	0.051	0.390	0.346	0.408	**0.38**	0.415
	21	0.18	0.321	0.431	**0.31**	0.366	0.289	0.111	0.084	**0.16**	0.065	0.158	0.154	0.269	**0.19**	0.260
	23	0.551	0.228	0.206	**0.33**	0.328	0.233	0.084	0.136	**0.15**	0.055	0.321	0.168	0.075	**0.19**	0.203

(continued)

Table 10.2 (continued)

Festspielhaus, Salzburg [IACC]

	$IACC_E$					$IACC_L$					$IACC_A$				
24	0.585	0.231	0.247	**0.35**	0.156	0.249	0.089	0.092	**0.14**	0.059	0.370	0.127	0.119	**0.21**	0.102
25	0.488	0.187	0.271	**0.32**	0.312	0.167	0.083	0.118	**0.12**	0.060	0.275	0.091	0.161	**0.18**	0.205
Mean	0.567	0.458	0.438	**0.49**	0.448	0.165	0.098	0.097	**0.12**	0.063	0.330	0.262	0.264	**0.29**	0.326

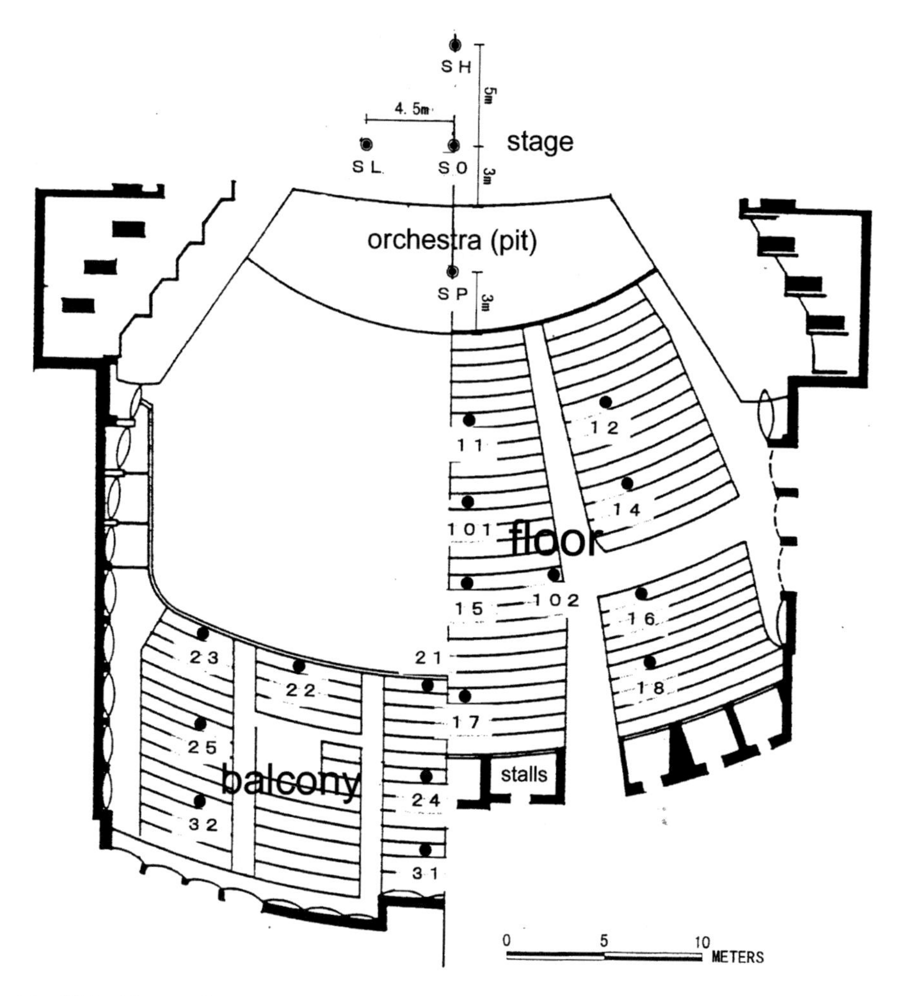

Fig. 10.1 Scheme of sender and receiver positions at Grosses Festspielhaus, Salzburg (modified from Hidaka et al. 1996)

in 0.59. If we also take a look at the values of $IACC_{A3}$ (A = 'all'; time window 0–3500 ms), averaged over all receiver positions ('mean'), the resulting value for S0 is 0.38. For sender position Sp, for example, an even smaller mean value (averaged over all receiver positions) for $IACC_{A3}$ was found with a value of 0.29. Therefore, the averaged $IACC_{A3}$ mean values for S0 and Sp results in 0.34 which means that 1-$IACC_{A3}$ equals 0.66.

If we take into consideration that the original dummy head recording with the KU81—of course—did not have the omni-directionally radiating dodecahedron loudspeaker as sound source (used by Hidaka and his team), but a much larger and more complex radiating sound source (i.e. symphony orchestra) the 1-$IACC_3$ measured for the KU81 with a value of 0.73 (see Table 10.1) and the above-calculated

1-$IACC_{A3}$ with a relatively close value of 0.66, based on the measurements by Hidaka and his team (see Hidaka et al. 1996) seem very plausible.

Measuring only the BQI_{rep} is—most likely—not yet a criterion strong enough for a final evaluation of the quality of any given (3D/2D) surround- or 2-channel stereo recording. Otherwise, this would have had to show up more clearly in the results of the subjective evaluation of the 4 surround recordings DECCA, OCT, KFM and ABPC, which have been analyzed in the chapters on surround microphone techniques (see Chaps. 6 and 7); also, the ABPC technique would have had to assume a much clearer leading position, given that its BQI_{rep} was identical to the one measured with the KU81 in the concert hall (and—with a value of 0.73—much higher than the competitive microphone systems under test, which had achieved only values between 0.29 and 0.37 (see Table 10.1). Nevertheless, it seems very plausible that the apparent listener preference for the ABPC technique with respect to the attribute 'spatial impression' (see Fig. 6.9) is closely related to the actual value of the BQI_{rep}. If the high degree of de-correlation of low-frequency components in the concert hall is to have such a strong influence on listener preference ($IACC_{E3}$ and BQI, respectively; see Beranek (2004), then it seems plausible that similar things should apply to the sonic impression in the control or listening/living room (even though it needs to be noted that the degree of low-frequency de-correlation captured through a microphone technique may be partially masked or 'overruled' by the individual acoustics of small rooms, due to strong room modes, etc. (see, e.g., Griesinger 1997 on this matter).

I would also like to point the interested reader to a visual documentation series of measurements which are related to the BQI_{rep}, which can be found on my Youtube channel 'futuresonic100,' when looking for clips with the term 'mic tech analysis' in the title. These video clips have been produced utilizing MAAT's '2BC multiCORR' VST plugin, which displays the cross-correlation of two audio channels *over frequency* in realtime. Several audio samples, based on recordings with well-documented microphone techniques, are analyzed with respect to their signal correlation, which unveils interesting details concerning their overall sonic character (Rem.: more information on the '2BC multiCORR' plug-in can be found in Appendix C).

10.1 Conclusion

As can be deducted from the results presented in Table 10.1, the newly defined BQI_{rep} is likely to be a measure worth to be subject to a more detailed research with respect to the perceived 'naturalness' of high-quality recording and reproduction of music, or—to put it more general—acoustic signals in quality performance spaces.

References

Beranek L (2004) Concert halls and opera houses: music, acoustics and architecture, 2nd edn. Springer, New York

de Keet VW (1968) The influence of early lateral reflections on spatial impression. 6th Int Congress on Acoustics, Tokyo

Griesinger D (1997) Spatial impression and envelopment in small rooms. Paper 4638 presented at the 103rd Audio Eng Soc Convention

Hidaka T, Masuda S, Arai T, Nakajima H (1996) Acoustical Measurement of Grosses Festspielhaus, Salzburg. Measurement report by Takenaka Research & Development Institute, Japan, 1 July 1996

Pfanzagl-Cardone E (2011) Signal-correlation and spatial impression with stereo- and 5.1 surround-recordings. Dissertation, University of Music and Performing Arts, Graz, Austria. https://iem.kug.ac.at/fileadmin/media/iem/altdaten/projekte/dsp/pfanzagl/pfanzagl_diss.pdf. Accessed Oct 2018

Chapter 11
A Short History of Microphone Techniques and a Few Case Studies

Abstract The discovery of 'stereophonic sound' happened about 140 years ago, and around the turn of the 1930s, both in the UK and the USA, specialists undertook relevant research in that field: while Alan Dower Blumlein's 'Blumlein-Pair' of two crossed figure-of-eight microphones is a coincident system, Arthur Keller at Bell Laboratories went for a 'broader' approach with his wavefront-oriented ABC technique. In the 1950s many record companies entered the field of stereophonic recording and names like 'DECCA-tree,' RCA 'Living Stereo' and Mercury's 'Living Presence' are still present today due to the legacy of highly valued recordings of those labels. The techniques used back then are analyzed in detail through about 40 schematics or photographs and testimonies of their inventors. A hypothesis regarding the listener preference for some of these recordings is made with hints toward content of Chap. 8 and the 'Stethoscope' qualities of Frequency-dependent signal cross-correlation analysis (FCC and FIACC). In the second part, we are giving a look to a few more recent case studies: live- and studio recordings of orchestra with and without solo instruments, ensemble recording with acoustic instruments and vocalists, down to solo-recording of the piano. At the end, some work recommendations—based on the author's 'natural perspective' aesthetic approach—are given.

Keywords Microphone technique · Legacy recordings · Decca · Living stereo · Blumlein · Sound recording

11.1 The Early Beginnings of 'Stereophonic Sound'

The first documented case of a stereophonic microphone technique can be found already before the end of the nineteenth century on the occasion of the 'International Electricity Exhibition' in Paris in 1881: Clement Ader had fixed a total of 10 microphones along the rim of the stage at the Paris Opera, which was 3 km from the exhibition site, which were each wired to (monaural) ear-phones in a room at the 'Palais de l'Industrie.' Visitors noted that '… Every one who has been fortunate enough to hear the telephones at the Palais de l'Industrie has remarked that, in

E. Pfanzagl-Cardone, *The Art and Science of Surround and Stereo Recording*,
https://doi.org/10.1007/978-3-7091-4891-4_11

Fig. 11.1 Ader microphones installed at the Opera House stage (from 'Die Elektricität im Dienste der Menschheit' reproduced in Eargle 1986)

listening with both ears at two telephones, the sound takes a special relief and localization which a single receiver cannot produce …' so—essentially a more spacious sound image. (from 'La Lumière Electrique' and 'Scientific American') (Figs. 11.1, 11.2, 11.3, and 11.4).

The first recording of stereo signals apparently took place by means of the 'Multiplex Graphophone Grand', which used three separate funnels and pickup units (Fig. 11.5). In the price of US$1000 were included:

3 recording and/or playback styluses,
1 stand for the funnel,
1 metal-funnel length 135 cm,
3 metal-funnels with a length of 135 cm,
12 pre-recorded 'Multiplex Grand' waltzes,
6 blank 'Multiplex Grand' waltzes.

This machine is from 1889 and uses—as opposed to the electrical microphones of Clemens Ader—the old mechanical principle for recording and reproduction of sound events.

A next, well documented step in terms of 'stereo recording' happened 1928 at Bell Laboratories in the USA, where Arthur C. Keller made first 2-channel recordings of speech, piano, as well as the 'Roxy' Orchestra (see Keller 1981).

This research at Bell Laboratories was undertaken under the title of 'Telephony' in order to gain knowledge on human speech and acoustic sounds in general and on their transmission. As a side product of this research about 6000 gold-plated wax-lacquers with various recordings were produced, among them 128 recordings with

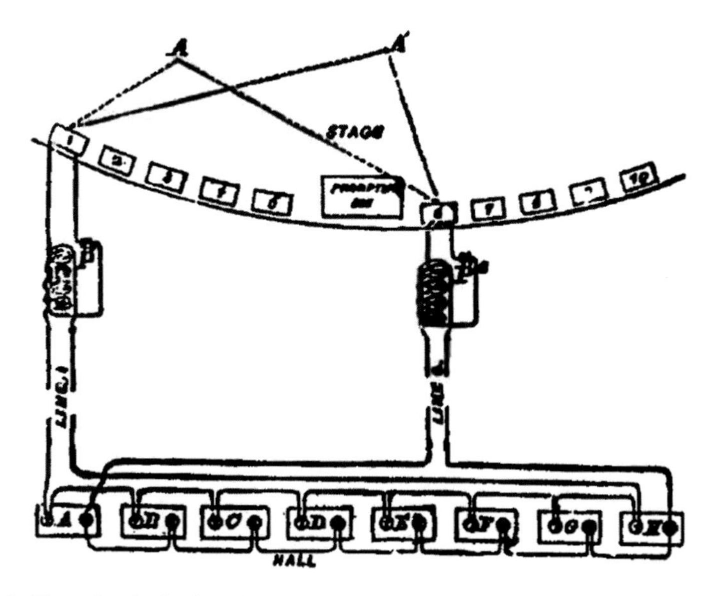

Fig. 11.2 Plan of paired microphone connection to receivers (from ‘Scientific American,’ reproduced in Eargle 1986)

the ‘Philadelphia Orchestra’ and Leopold Stokowski conducting. These recordings were made between 1931 and 1932, with a frequency response from 30 Hz to 10 kHz, making use of a ‘large ABC’ microphone technique (Fig. 11.6).

The added center microphone C helps compensate the latent ‘hole-in-the-middle’ effect, which can occur with ‘large AB’ microphone recordings, by effectively closing the sound gap between the A and B microphone. Depending on the spacing of the 3 microphones in the ABC system, there may be the danger of sound coloration due to phase cancellation when the signals of the three microphones are mixed to a stereo bus, but if the spacing is large enough (e.g. larger than the ‘reverberation radius’ or ‘critical distance’), this should not represent a problem usually.

Especially with the newer replay formats of 5.1 surround, for example, in which the channels L, C, R are kept separate and replayed through individual loudspeakers, this problem—created due to electrical signal summing on a bus—does not even occur.

At the same time, when the engineers at Bell Laboratories were experimenting with two and three-channel stereo, the chief engineer at EMI’s Abbey road studios, Alan Dower Blumlein was researching into stereophonic sound as well. This lead to a number of very significant inventions in the field of stereo sound transmission, recording and reproduction. (see, among others, Blumlein 1931).

In this patent, Blumlein describes several microphone techniques for stereo recording, as well as the 45° side writing for disk cutting. Due to the interruption caused by World War II, it took almost a quarter of a century until the latter invention was commercially used for vinyl disk cutting.

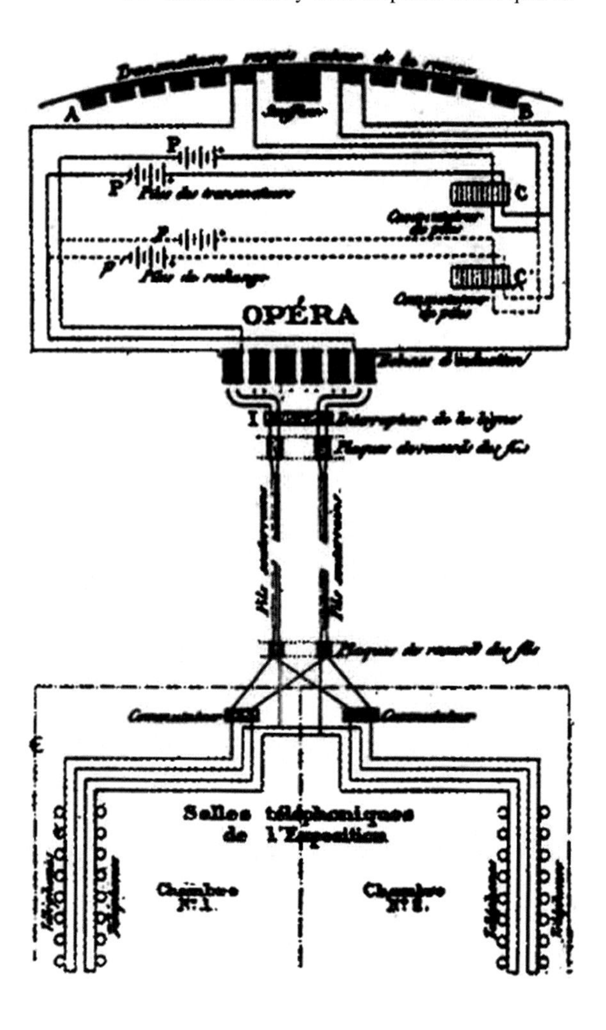

Fig. 11.3 Diagram of installation (from 'La Lumière Électrique,' 1881 reproduced in Eargle 1986)

The well-known 'Blumlein-Pair' microphone technique consists of two figure-of-eight microphones crossed at an angle of 90°, usually with a symmetrical layout of ±45° relative to the center of the sound source. In his patent, Blumlein also describes the same arrangement rotated by 45° which is essentially an MS arrangement with two fig-8 microphones, which allows the sound-engineer to modify the width and localization of the sound-source. However, the first arrangement described is the 'classical' Blumlein-Pair (Fig. 11.7).

It is also interesting to note that in his patent, the inventions of Blumlein address not only the needs for sound recording, but are also directed toward the use in film sound and also live sound reinforcement (PA-systems) (see Blumlein 1931, pg. 8).

One vulnerable point in the use of the 45° side writing for disk cutting is its sensitivity in respect to strong low-frequency out-of-phase signal components, which can cause relevant problems and render the master disk unusable. Taking this into account,

Fig. 11.4 Exhibition telephone room where visitors heard binaural transmission of Paris Opera (from 'La Lumière Électrique,' 1881 reproduced in Eargle 1986)

it is not surprising that Blumlein—even though he also describes a 'small AB' microphone system which uses omni-directional microphones—has his attention focused on XY type figure-of-eight arrangements, as they deliver a mono-compatible signal.

Another interesting point, which can be understood when studying the patent carefully, is that an additional motivation for undertaking research in the field of stereo sound was the fact that designing recording studios for monophonic recordings required a much higher financial investment in order to modify the room acoustics. In his patent from 1931, Blumlein describes the potential acoustic problems which may occur with mono recording: '… When recording music considerable trouble is experienced with the unpleasant effects produced by echoes, which in the normal way would not be noticed by anyone listening in the room in which the performance is taking place. An observer in the room is listening with two ears, so that echoes reach him with the directional significance, which he associates with the music performed in such room. He therefore discounts these echoes and psychologically focuses his attention on the source of sound. When the music is reproduced through a single channel the echoes arrive from the same direction as the direct sound so that confusion results. It is a subsidiary object of this invention so to give directional significance to the sounds that when reproduced the echoes are perceived as such. … The herein described acoustic system while being especially applicable to talking pictures is not limited to such use. It may be employed in recording sound quite independently of any picture effects and in this connection (as well as when used in cinematograph work) it seems probable that the binaural effect introduced will be found to improve the acoustic properties of recording studios and to save any drastic acoustic treatment

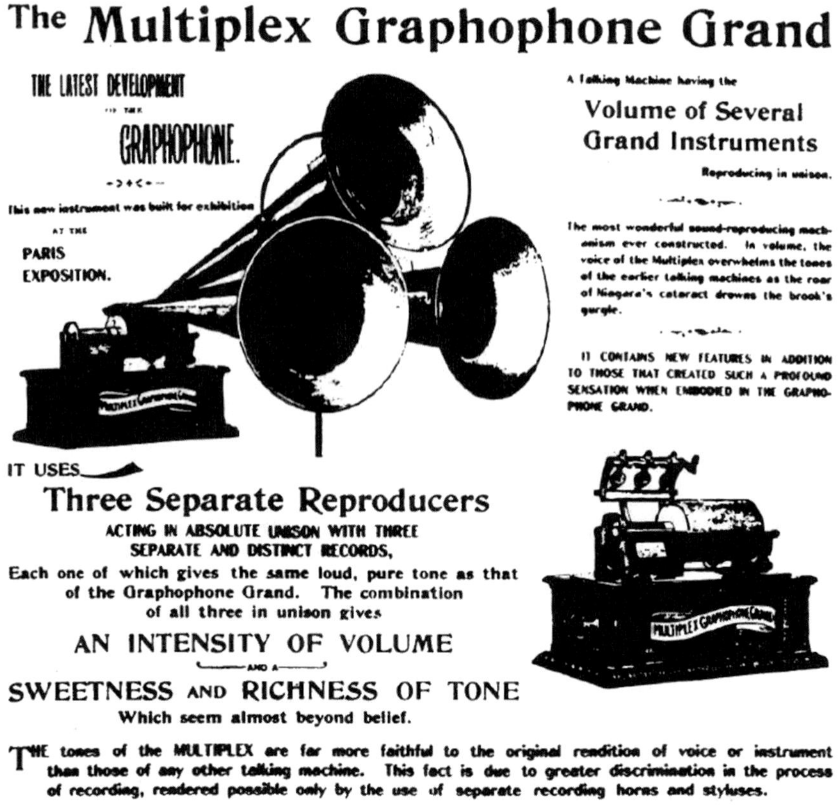

The Multiplex Graphophone Grand

THE LATEST DEVELOPMENT OF THE GRAPHOPHONE.

This new instrument was built for exhibition AT THE PARIS EXPOSITION.

A Talking Machine having the Volume of Several Grand Instruments Reproducing in unison.

The most wonderful sound-reproducing mechanism ever constructed. In volume, the voice of the Multiplex overwhelms the tones of the earlier talking machines as the roar of Niagara's cataract drowns the brook's gurgle.

IT CONTAINS NEW FEATURES IN ADDITION TO THOSE THAT CREATED SUCH A PROFOUND SENSATION WHEN EMBODIED IN THE GRAPHOPHONE GRAND.

IT USES Three Separate Reproducers

ACTING IN ABSOLUTE UNISON WITH THREE SEPARATE AND DISTINCT RECORDS,

Each one of which gives the same loud, pure tone as that of the Graphophone Grand. The combination of all three in unison gives

AN INTENSITY OF VOLUME

AND A

SWEETNESS AND RICHNESS OF TONE

Which seem almost beyond belief.

THE tones of the MULTIPLEX are far more faithful to the original rendition of voice or instrument than those of any other talking machine. This fact is due to greater discrimination in the process of recording, rendered possible only by the use of separate recording horns and styluses.

Fig. 11.5 Multipex Graphophone Grand (reproduced in Eargle 1986)

thereof while providing much more realistic and satisfactory reproduction.' (from Blumlein 1931).

As a consequence, the advancements of recording techniques from mono to stereo were reflected in the room acoustic treatment of Abbey Road's 'Studio One,' which up to date is mainly used for orchestral and large ensemble recording. As famous violinist Yehudi Menuhin—who had recorded with Sir Elgar Elgar at Abbey Road as early as 1931—recalled in an interview: '…Studio One has seen more transformations than any other. At one time, they felt that sound had to be just pure sound with no reverberation whatsoever, that was a false conception. Then they went the other way and the room was dotted with loudspeakers all the way round so they would create echo effects (rem.: known as Ambiophony) that could be heard for miles around. Now I think it has a very fine acoustics.' (from Southall et al. 1997).

In essence, from the information given above it can be understood that rooms designed for mono-recording need to be much 'drier' (i.e. have a much shorter reverberation time) than recording venues designed or intended for stereo recording. At Abbey Road Studios the consideration of this fact had led to the installment

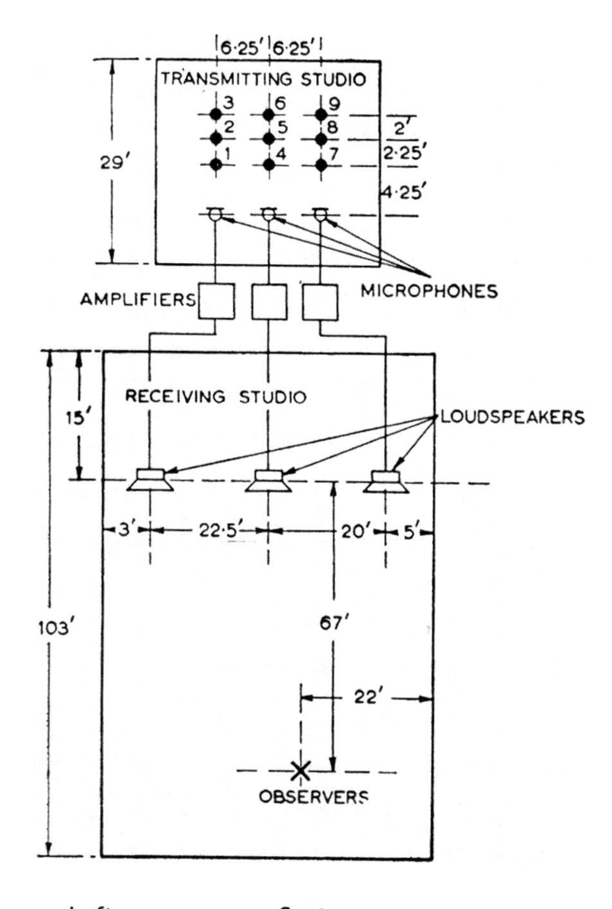

Fig. 11.6 'Wavefront'-Arrangement used by Bell Laboratories in 1934 (from Keller 1981)

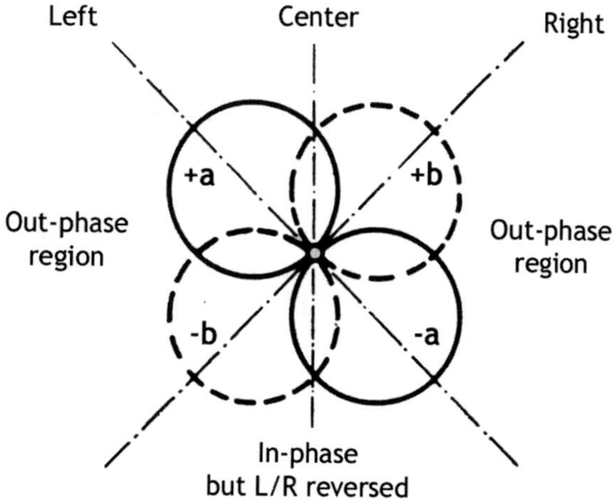

Fig. 11.7 Blumlein-Pair of two crossed figure-eight microphones (after Rumsey 2001)

of a large number of absorbers in 'Studio 1' during the 1930s. As over time, it became clear that many recordings required longer reverberation times, in 1958 an 'Ambiophony' system with a total of 100 loudspeakers, situated on the side walls and ceiling was installed. As back then it was already clear to scientists that reverb was essentially a phenomenon of many reflections (or echoes) in a room, these speakers replayed a room response signal that was delayed by means of 'magnetic delay drums', depending on their position in the room. This system was rather complex, but allowed for variable room acoustics. Unfortunately, as it happens occasionally, there was also a relevant discrepancy between theory and practice in respect to the acoustic quality of the reverb that could be achieved from this method, and the shortcoming thereof finally led to the abandonment of the technique in this form and the whole system at Abbey Road. (As a side remark: at almost the same time an 'Ambiophony' system also got installed at the—back then newly constructed—concert venue 'Grosses Festspielhaus' in Salzburg, obviously intended for use with live concerts and opera) (Fig. 11.8).

11.2 DECCA-tree

The triangular configuration, known to us today as 'DECCA-tree' has been invented in 1953 by DECCA engineer Roy Wallace, who also had to develop a proper mixing console, capable of handling these signals, as the underlying approach is fundamentally 3-channel. The first recording with a DECCA-tree took place in spring 1953 at the 'Victoria Saal' in Geneva with the 'Orchestre de la Suisse Romande.' Three Neumann M-49 (switched to cardioid) were used, placed above the conductor at a height of about 3.30 m (10 feet). The L and R microphone had an included angle of 60° in relation to the Centre mic, pointing diagonally across the orchestra, vertically angled by 30° (Figs. 11.9, 11.10 and 11.11).

During the course of the following years, the DECCA team around Roy Wallace and Kenneth Wilkinson experimented with various microphones from German microphone manufacturer Neumann, also with cardioid and figure-of-eight characteristics (M-49 [omni, cardioid, fig-8], M-50 [omni], as well as KM-53 [omni] and KM-56 [omni, cardioid, fig-8]), sometimes even using absorptive baffles between (and behind) the microphones for the sake of better acoustic separation of the signals. (Rem.: while the M-50, M-49 and U-47 microphones have a large membrane, the abbreviation KM stands for 'Kleinmembran Mikrofon' which means 'small membrane microphone') (Figs. 11.12, 11.13 and 11.14).

According to (Gray 1986) '… From his initial trio of M-49's Wallace soon shifted to sets of baffled Neumann M-50's, and then in the spring of 1957 to openly mounted KM-56's. All of these changes in main tree microphones were preceded by months of testing, retesting and evaluation, comparing not just the M-49, M-50 and KM-56 with each other, but with the omni-directional KM-53 in a variety of tree and baffle configurations. Adding mikes to the tree's mix to broaden the stereo image, something that Wallace's mixer permitted but which Wallace himself opposed, was, however,

Fig. 11.8 An aerial shot of Abbey Road Studio One, during an opera recording. Apart from the orchestra in the foreground, there is also a small ensemble on the rear of the stage. The soloists can be seen standing on a clearly numbered floor area, which facilitated accurate continuity in stereo positioning for any subsequent retakes. Note too, the numerous little black Ambiophony speakers around the walls ((c) by Godfrey MacDomnic)

another story, for it not only contravened his determination to use only the tree for the stereo pickup but would also have made it much harder to tell if the tree's own mikes were working properly.' M-50s flanking a K-56 tree, according to Brown, produced more 'air' in the stereo picture than adding 50's to a 50 tree, which already had plenty of ambient information in its pick-up. Flanking a 56 tree with figure-eight patterned KM-56 outriggers (which was tried in London when outriggers first appeared there) captured the strings and produced a broader stereo image, but still failed to convey the sense of space that materialized when the 56 outriggers were replaced by M-50s, and the 56-tree thereafter by an M-50 tree (Figs. 11.15 and 11.16).

Wallace was followed by Kenneth Wilkinson, for who '... microphones were always the means to an end; for his objective, in his own words, was to achieve 'as natural a sound as possible' with the 'quality' that was 'more important than anything

Fig. 11.9 DECCA-tree with three Neumann condenser microphones set up by Engineer Jimmy Lock (from Decca 2003); reprinted by kind permission of Decca Music Group Limited

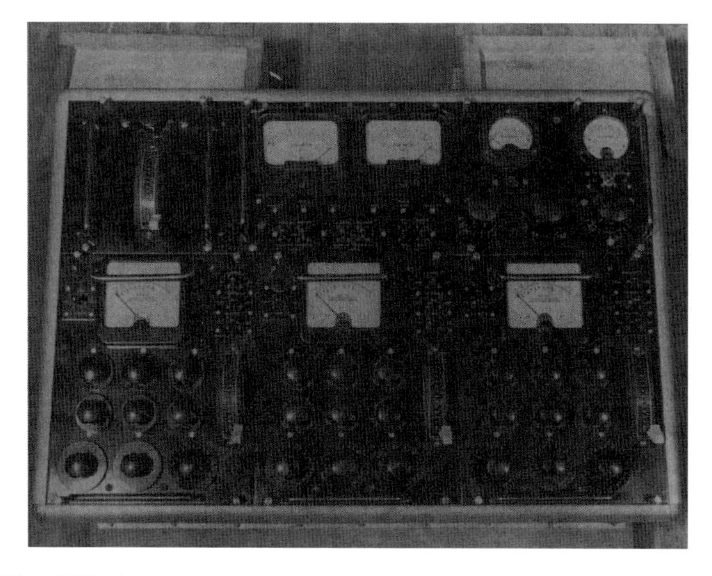

Fig. 11.10 DECCA 9-channel Mixer (from Decca 2013, pg. 130); reprinted by kind permission of Decca Music Group Limited

on a recording ....' This was especially true after the arrival of stereo, for Wilkinson, sooner, perhaps, than other Decca balance engineers, concluded that cardioid mikes, such as KM-56s, were not the way to get the best possible sound. 'The 56 naturally gave a good stereo effect,' he says, 'but I was never happy with the quality of the sound. I don't think a directional mike gives you a good sound from an orchestra

Fig. 11.11 Decca-Recording of Berlioz's Requiem at Masonic Auditorium (choir placed up on the audience balcony); conductor: Lorin Maazel [from (Decca 2013), pg. 185, © by DECCA/Peter Hastings]; reprinted by kind permission of Decca Music Group Limited

anyway. It's very good on solos and solo work, but I don't think it is for orchestral.' (from Gray 1986).

From this, it can clearly be seen that the original 'DECCA-Tree' (or 'DECCA Triangle,' as it is also called) did not strictly consist of omni-directional microphones, with the intention to mainly make use of the time-of-arrival differences of sound in order to achieve localization and spatial impression in a recording. Instead, the main objective was a satisfactory channel separation of signals and a fundamental 3-channel recording approach, which made use of all acoustic means available, such as the use of directional microphones and acoustic absorptive panels to minimize 'crosstalk' between the L, C and R channel-signals (see Figs. 11.17 and 11.18).

Wilkinson describes a typical DECCA setup of the early days as follows: 'You set up the tree just slightly in front of the orchestra. The two outriggers, again, in front of the first violins, that's facing the whole orchestra, and one over the cellos. We used to have two mikes on the woodwind section—they were directional mikes, 56s in the early days. You'd see a mike on the tympani, just to give it that little bit of clarity, and one behind the horns. If we had a harp, we'd have a mike trained on the harp. Basically, we never used too many microphones. I think they're using too many these days.' (from Gray 1986).

Wilkinson described how finding the right balances worked, starting with a rough equality in the mix between the tree and the outriggers. 'The spot mikes you don't use so much, because you do get the balance between the tree and the outriggers

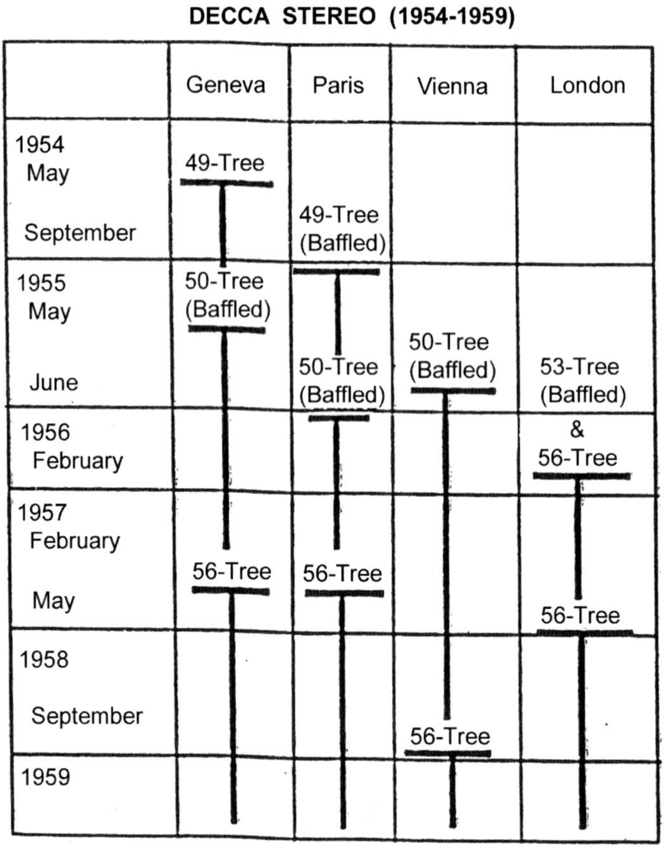

DECCA STEREO (1954-1959)

	Geneva	Paris	Vienna	London
1954 May	49-Tree			
September		49-Tree (Baffled)		
1955 May	50-Tree (Baffled)			
June		50-Tree (Baffled)	50-Tree (Baffled)	53-Tree (Baffled)
1956 February				& 56-Tree
1957 February				
May	56-Tree	56-Tree		56-Tree
1958				
September			56-Tree	
1959				

Fig. 11.12 Timeline of DECCA recordings (after Valin 1994)

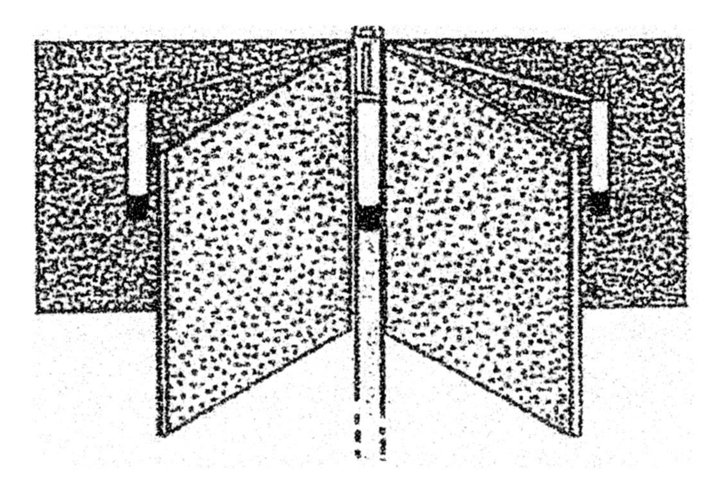

Fig. 11.13 Schematic of baffled DECCA-tree with microphones (after Valin 1994)

Fig. 11.14 DECCA recording session with baffled tree at Sofiensäle; reprinted by kind permission of Decca Music Group Limited

Fig. 11.15 Neumann TLM50 condenser omni microphone (successor to the famous M-50)

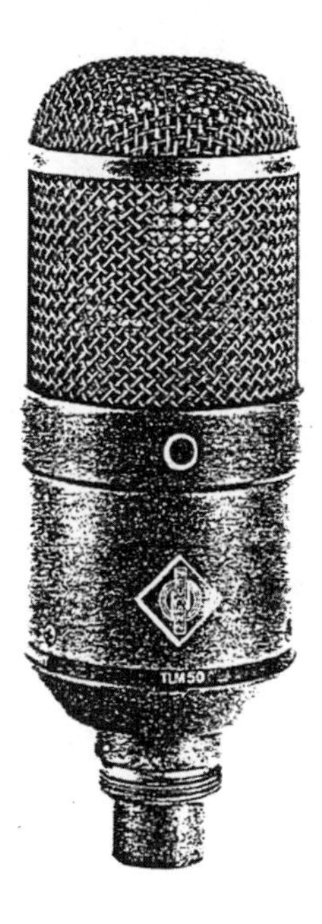

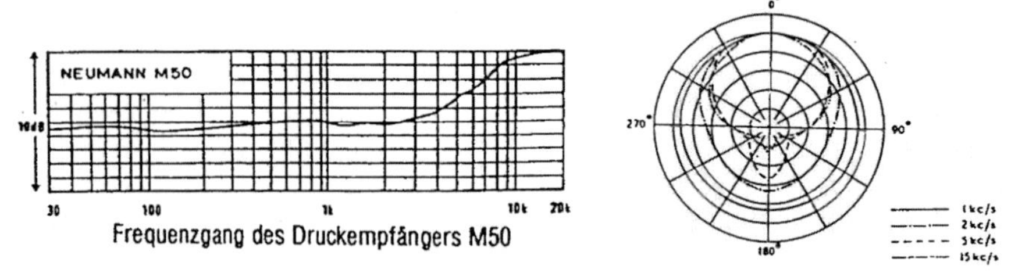

Fig. 11.16 Frequency response graph and polar pattern of the diffuse-field compensated Neumann M50 omni microphone (from Sengpiel 1992)

Fig. 11.17 Decca recording at the Amsterdam Concertgebouw (audience seats taken out) conductor George Solti in 1961 (note the baffle behind the—small dimension—Decca-Triangle and the large AB 'outriggers) [photograph from (Decca 2013), pg. 165; © by DECCA/Carel de Vogel]; reprinted by kind permission of Decca Music Group Limited

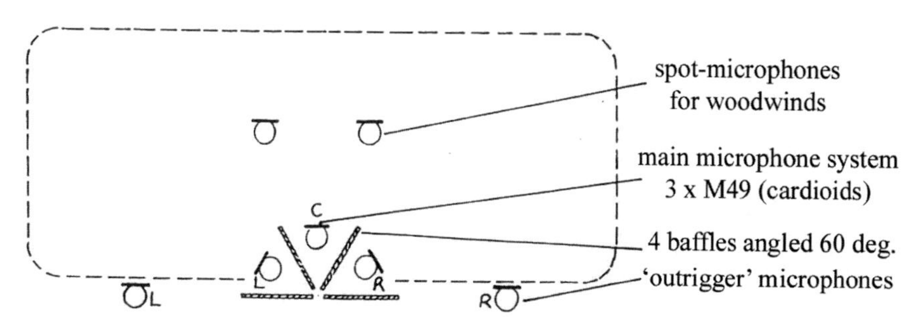

Fig. 11.18 Basic layout of DECCA early baffled tree recording with outriggers and woodwind spot-microphones (after Sengpiel 1992)

from the whole orchestra. It's only to give just that little bit more definition on the tympani, the horns, and the woodwinds that we use these spot mikes. We didn't rely on those for the whole sound. You basically get that through the front mikes. I think woodwinds should not be brought forward because they are behind the orchestra, more or less, so therefore they should sound a little bit distant. You need the clarity, naturally, to hear what they're doing.' Clarity and quality were what it came down to. 'That's the whole reason, to get the sound as natural as you can. You're trying to make a record sound the way you hear the orchestra in the hall.' (from Gray 1986) (Figs. 11.19).

Examining the hand drawing of the Decca 'Le Papillon' recording at Kingsway hall (see Figs. 11.20, 11.21 and 11.22), we can see that the height of the Decca-Triangle was at 10′6″ (roughly 3 m), the height of the 'Left-Hand Outrigger' (LHO) and 'Right-Hand Outrigger' (RHO) at 11′0″ and the two spot-microphones for the woodwinds at 10′6″. While the Decca-Triangle was at its usual position above the conductor, the LHO and RHO microphones were set up slightly in front of the orchestra. It is interesting to note that the harp and its spot microphone were set-up in a soloist position to the right-hand side of the conductor, slightly in front of the orchestra. In the listing of the track-sheet, the RHO is followed by the spot-mic of the double-basses (KM53, set up in the middle of the basses right-hand side), then the Woodwinds (WW) (KM64, cardioid), the spot microphone for the Horns (NB: placed BEHIND them) (KM64), then Tympani (KM64), Percussion (KM56), Harp (KM56) and the room microphones (named 'AIR'), for which Neumann KM53s were used. As can partly be seen on the track-sheet listing, the KM56—in cardioid mode—was frequently used as a spot microphone for woodwinds, harp and also for soloists. On this recording session, for the Decca Triangle—as well as the 'outriggers'—Neumann M50 microphones have been used.

In his aesthetic approach, especially Decca engineer Kenneth Wilkinson, from 1960 onwards, tended toward use of the 'more open sounding' Neumann M50 microphones, characterized by a frequency response meant for diffuse-field compensation. The M50 microphones were used a lot as so-called outriggers—essentially largely spaced AB omnis—especially when the Decca-Triangle was using KM-56 microphones. (see Valin 1994, pg. 124). Under the guidance of Kenneth Wilkins, from

Fig. 11.19 DECCA Studio-style opera recording with choir and soloists at Sofiensaal; note that the harps have been 'singled out' to a position behind the conductor for better sonic separation; reprinted by kind permission of Decca Music Group Limited

approximately 1960 onwards (not included in Fig. 11.12), it seems that the DECCA-tree had found its 'final format' with mainly the Neumann M50 microphone employed for both the DECCA-triangle (without baffles), as well as the outriggers, which is essentially the setup which is still in use today when sound-engineers or 'Tonmeisters' rig their DECCA-trees with omni-directional microphones.

Nowadays, the term 'Decca-tree' is used for both the 'basic' Decca-Triangle, as well as the complete set-up of Decca-Triangle with added 'outriggers,' which can be a bit confusing.

The Decca engineers themselves used to call the Decca-Triangle ''Decca-tree' and—sticking to this name—subsequent triangular arrangements of similar kind have been referred to as 'trees' by other sound-engineers as well. For the sake of clarity, it would be preferable if sound-engineers could agree on distinguishing between the simple 'Triangle' construction and the more advanced 'Tree' setup, which includes also the 'outriggers,' but I am afraid there is little hope for this to happen.

The Decca-Triangle was also used for the recording of chamber music, depending on the size of the ensemble with or without outriggers. In Fig. 11.23, a recording session with the famous Wiener Oktett at Sofiensaal in Vienna is shown. Note the relatively high position of the Decca-Triangle above and slightly in front of the ensemble. Also, the small membrane condenser spot-microphones for the double-bass and woodwinds have been put on high stands. As we have already seen for

Fig. 11.20 Decca-Recording of Benjamin Britten's 'War Requiem' at Kingsway Hall, 1963 (note the baffling between the microphones for better channel separation) (from Decca 2013, pg. 169); reprinted by kind permission of Decca Music Group Limited

the spot-mic of the horns in the 'Le Papillon' orchestra recording, this has been positioned behind the horn. Also, it is interesting to note that a baffle has been set up behind the horn, most likely in order to prevent the instrument from exciting the hall too much and thereby generating excessive reverb.

Taking a look at another Decca orchestra recording, the 'Planets' by Holst with the Los Angeles Philharmonic Orchestra, conducted by Zubin Mehta at the Royce Hall at UCLA in 1971: Due to the large symphonic orchestra, a rather extensive microphone setup was used, which employed not only the standard Decca-Triangle (seemingly with the omission of the center-microphones, see track-sheet …) but two sets of 'outriggers,' namely LO/R ('left outer-rig') and LI/R ('left inner-rig') and of course RI/R and RO/R on the right-hand side of the conductor (Fig. 11.24).

Let's analyze the information provided on the session-sheet, starting with track 1:

Celeste (Neumann KM-64), Piano (KM-64), Organ L (KM-53), Organ R (KM-53), Echo Return L (KM 53) and Echo Return R (KM-53) from the 'Green Room Reverb Chamber,' LO/R, LI/R, L-Tree, R-Tree, RI/R, RO/R all using Neumann M-50 microphones, Basses (M-50), FWW L and R (the forward row of woodwinds) picked up by two Neumann 64 microphones, BWW L and R (the backward row of woodwinds) picked up by two small diaphragm KM-56 microphones, Timpani picked up by two Neumann KM-64, Tuba (KM-56), Trumpets (KM-64), Trombones

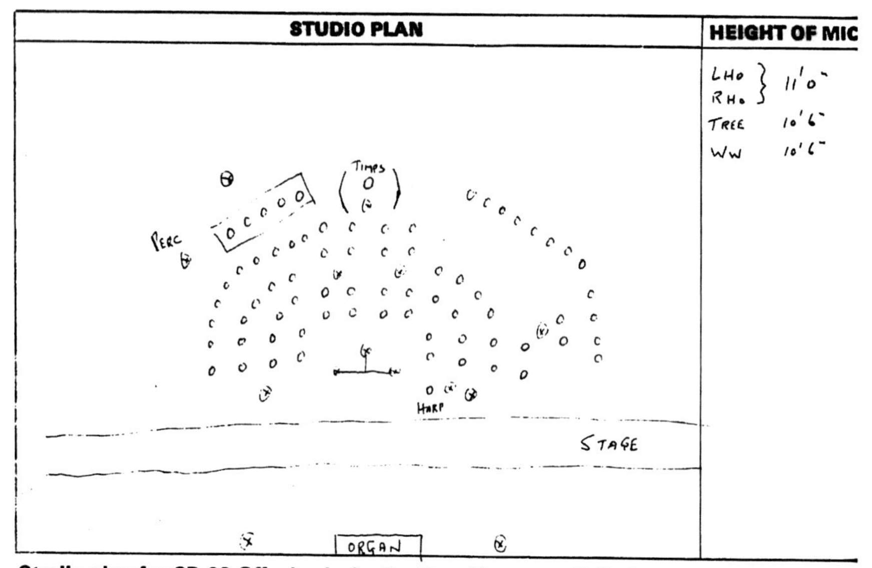

Studio plan for CD 22 Offenbach: *Le Papillon*, Kingsway Hall, January 1972
NB height of microphones:
left-hand orchestra [LHO] /
right-hand orchestra [RHO] 11′0″
tree 10′6″
woodwind [WW] 10′6″

Fig. 11.21 Hand drawing of orchestra layout for recording of Offenbach's 'Le Papillon' at Kingsway Hall with microphone positions (partly including height notes) (from Decca 2013, pg. 143); reprinted by kind permission of Decca Music Group Limited

(KM-56), the Horns picked up by two KM-64, the Harps picked up by two KM-56, the Percussion section splayed out over 6 tracks, miked with 4 pieces of KM-56 and one KM-64 … for a total of 30 channels on the mixing desk! Looking at the large orchestra apparatus that needed to be captured, it seems understandable that so many microphones were put to use, even though some years earlier the same recording session would inevitably have happened with much fewer microphones as there simply would not have been a mixer available, able to handle such a large number of signals … (Figs. 11.25 and 11.26).

Looking at the main microphones involved, the Decca-Triangle (by omission of the center microphone apparently reduced to something like an 'AB pair' of omnis) plus two sets of outriggers reminds us somehow of the 'Microphone Curtain' or 'Multiple Microphone' system we have encountered—for example—on the occasion of the Berlin Surround microphone comparison (see Chap. 9).

Many of the legendary DECCA recordings were re-released on CD in the year 1999, and as several boxed sets of these recordings from 2012 onwards (Figs. 11.27 and 11.28).

STORM II

ELECTRICAL RECORD OF SESSION

DATE: 24-26th JAN 1972 | STUDIO: K. H. | ENGINEERS: K.E.W / J.D | PRODUCER: M. WOOLCOCK

TITLE: OFFENBACH – LE PAPILLON | ARTIST: L.S.O / BONYNGE | LABEL: DECCA

CHANNEL	1	2	3	4	5	6	7	8	9	10	11	12	13	14	15	16	17	18	19	20	21	22	23	24	E.R.1	E.R.2	E.R.3	E.
LOCATION	LHO	←—	TREE	—→	RHO	BASS	← WW	→	HORNS	TYMP?	PERC	HARP?	← A	R →														
TYPE MIC.	50	50	50	50	50	53	64	64	64	64	56	56	53	53														
TREB. FREQ.																												
TREB. DB.	–	+1	+1	+1	–	–	+1	+1	+1	+3	+1		+1	+1														
BASS FREQ.																												
BASS DB.	–	–	–	–	–	–	–	–	–	-2	-1		-1	-1														
PRESENCE																												
ECHO BF/AF.																												
PLATE/CHAMBER																												
CHAN. TO GROUP																												
GROUP TO TRACK																												
ECHO GAIN																												
ECHO PAN																												
PRE-SET GAIN	L	M	J	M	M	N	I	I	J	I	I	I	M	N														
SIG. PAN	L	8L	C	8R	R	8R	6L	6R	7L	C	3L	9R	L	R														
SIG. FADER	9	9	9	9	9	12	12	12	15	12	15	15	12	12														

ECHO SEND	1	2	3	4
REVERB. TIME				
TREB. FREQ.				
TREB. DB.				
BASS FREQ.				
BASS DB.				
FADER				

MASTER FADERS	1	2	3	4	5	6	7	8	ST.	GR.1	GR.2	GR.3	GR.4	L.S
									13½	12	12	21		4

LIMITER	1	2	3	4
THRESHOLD				
GAIN				
RECOV. TIME				
POSITION				

TAPE MACHINES	
TYPE	C.37 x 2
REPLAY	N.A.B. +4 dB
SPEED	15 i.p.s.
TAPE	SCOTCH 206

REMARKS: QUADROPHONIC RECORDED SEPERATELY

Fig. 11.22 Session-sheet for the Decca Recording of Offenbach's 'Le Papillon' at Kingsway Hall in 1972 (from Decca 2013, pg. 142); reprinted by kind permission of Decca Music Group Limited

11.3 RCA 'Living Stereo' Recordings

Around the same time in the second half of the 1950s and early 1960s, the recording team of EMI was experimenting at Abbey Road studios—staying true to the 'house technique'—by using a Blumlein-Pair arrangement of two Neumann M-49s in 'figure-of-eight' mode and in the USA the RCA ('Radio Corporation of America') made their first stereo recording using a 'large AB' setup with Neumann M50 omnis, as well as U-47s (in omni or cardioid mode) (Figs. 11.29 and 11.30).

RCA producer Jack Pfeiffer came up with the microphone positions for this first RCA stereo recording based on a wealth of experience gained during all previous 'mono' recording sessions and judging by 'eye and ear'—and turned out the be completely right with his decision—as this 'A/B time-intensity' or 'spaced mike' stereophony arrangement of microphones should become the RCA standard setup (to which spot-microphones were added, as needed, during later sessions) (Figs. 11.31 and 11.32).

In the microphone setup of the 'Gaieté Parisienne' two ribbon (figure-of-eight) RCA 77-DX spot-microphones have been added for the wind instruments in addition to the large-AB main microphone pair made of Neumann M-50 (or U-47). Since there is still the potential problem of a 'hole-in-the-middle effect' a centerfill microphone (Neumann M-49 or U-47) was added, as can be seen in the mic setup for the recording of 'Pictures at an Exhibition' from 1957 (Figs. 11.33 and 11.34).

Fig. 11.23 Members of the 'Wiener Oktett' at a recording session at Sofiensaal, Vienna (pg. 68 from Decca 2013) © by DECCA/Elfriede Hanak; reprinted by kind permission of Decca Music Group Limited

ELECTRICAL RECORD OF SESSION

DATE: APRIL 1971 | STUDIO: ROYCE HALL UCLA | ENGINEERS: J.L CDM. J.B. | PRODUCER: JHM.

TITLE: PLANETS (HOLST) | ARTIST: MEHTA LAPO | LABEL: DECCA

CHANNEL	1	2	3	4	5	6	7	8	9	10	11	12	13	14	15	16	17	18	19	20	21	22	23	24	25	26	27	28
LOCATION	[illegible]	PIANO	ORGAN	ORGAN	ECHO RES	ECHO RES	L o/R	L i/R	L TREE	R TREE	R i/R	R o/R	BASS	L F.WW	R F.WW	L B.WW	R B.WW	TIMP	TUBA	TPTS	TRBS	HRNS	HRNS	1st HARP	2nd HARP	—PERCS—		
TYPE MIC.	64	64	53	53	53	53	50	50	50	50	50	50	50	64	64	56	56	2×64	56	64	56	64	64	56	56	64 56	56 56	56
TREB. FREQ.	[illegible]												5kHz					2 | 5				2 | 5	2 | 5	-1	-1	-1 -1	-1 -1	-1
TREB. DB.	-2		-1	-1									+2					+2 | +2				+2 | +4	+2 | +4					
BASS FREQ.					100 | 50	100 | 50												50 | 125				50 | 125	50 | 125					
BASS DB.	-2		-1	-2	-6 | -6	-6 | -6	-1	-1	–	–	–	–	–	-1	-1	-1	-1	-4 | -2	-1	-2	-2	-4 | -2	-4 | -2	-1	-1	-1 -1	-1 -1	-1
PRESENCE																												
ECHO BF/AF.																												
PLATE/CHAMBER																												
CHAN. TO GROUP																												
GROUP TO TRACK																												
ECHO GAIN							L	L	L	R	R	R	-	L	R	L	R	-	R	R	R	L	L	R	R	(VARIED)		
ECHO PAN							G	G	H	G	G	H		G	G	F	G		←VARIED→					G	G			
PRE-SET GAIN	G		M	I	K	K	M	M	M	M	M	M	L	K	K	N	N	D	N	K	H	J	K	L	L	VARIED		
SIG. PAN	L		L	R	L	R	L	L	L5	R5	R	R	R	L	R	L	R	R2	R7	R	R8	L	L7	R5	R	VARIED		
SIG. FADER																												

ECHO SEND	1	2	3	4	MASTER FADERS	1	2	3	4	5	6	7	8	ST.	GR.1	GR.2	GR.3	GR.4	L.S	REMARKS:
REVERB. TIME																				PERC MIKES CONTINUALLY
TREB. FREQ.					LIMITER	1	2	3	4	TAPE MACHINES										VARIED from Piece to Piece
TREB. DB.					THRESHOLD					TYPE	A62 ×2									Gen Room Echo.
BASS FREQ.					GAIN					REPLAY	NAB STRETCHED									
BASS DB.					RECOV. TIME					SPEED	15 IPS									
FADER					POSITION					TAPE	202 SCOTCH									

Fig. 11.24 Track-sheet of Decca orchestra recording of the 'Planets' at the Royce Hall at UCLA in 1971 (from Decca 2013 pg. 136); reprinted by kind permission of Decca Music Group Limited

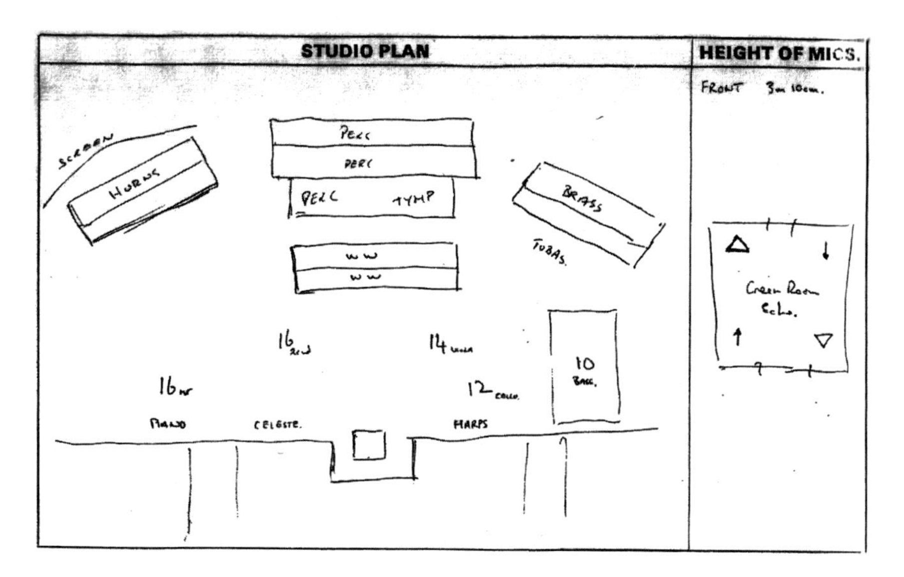

Fig. 11.25 Orchestra setup for Decca orchestra recording of the 'Planets' at the Royce Hall at UCLA in 1971 (from Decca 2013 pg. 137); reprinted by kind permission of Decca Music Group Limited

Fig. 11.26 Playback session for Decca orchestra recording of the 'Planets' at the Royce Hall at UCLA in 1971 with Zubin Mehta and Decca producer Ray Minshull (from Decca 2013 pg. 138); reprinted by kind permission of Decca Music Group Limited

Fig. 11.27 A Cover from the DECCA-Legends CD-reissue of Karajan's recording from Sofiensaal, 1959 (from DECCA-Legends 2000); reprinted by kind permission of Decca Music Group Limited

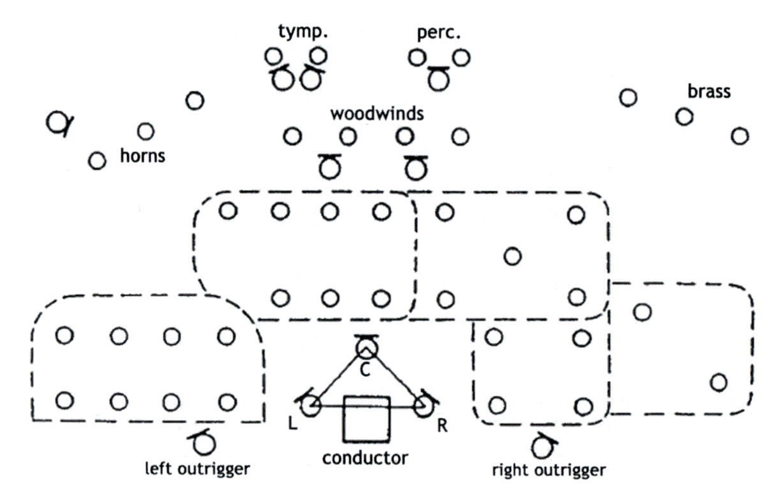

Fig. 11.28 Typical microphone setup for a DECCA orchestral recording with tree, L/R outriggers and spot microphones for the woodwinds, percussion and tympani; note that the horns are miked from behind (after Sengpiel 1992)

Fig. 11.29 Neumann 'M-49' large membrane directional condenser microphone (omni, cardioid, figure-of-eight)

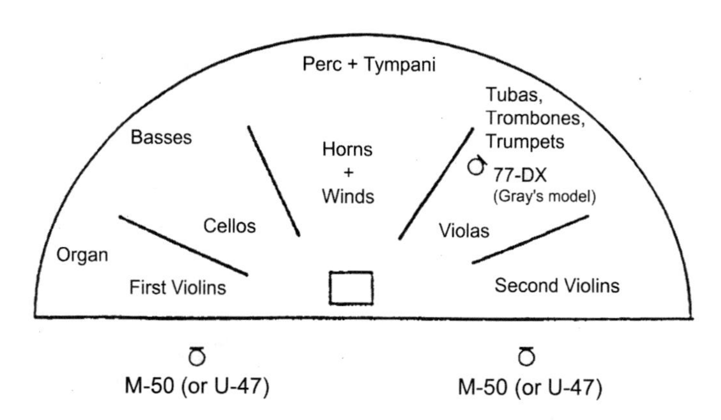

Fig. 11.30 RCA miking setup for 'Thus said Zarathustra,' recorded March 8, 1954 at Orchestra Hall (after Valin 1994)

With the Neumann U-47 polar pattern plots it is interesting to note that toward high frequencies both omni and cardioid patterns get more directional. Also, independent of polar pattern, the U-47 displays a level-boost around the 10 kHz region, which has remained—although maybe not as pronounced—with one of its famous successor models, the U-87 (Fig. 11.35).

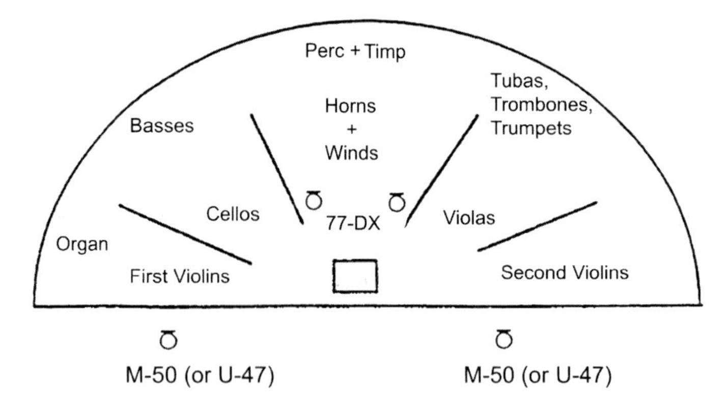

Fig. 11.31 RCA miking setup for 'Gaieté Parisienne,' recorded at Boston Symphony Hall, in June 1954 (after Valin 1994)

Fig. 11.32 RCA Model 77-DX ribbon microphone (figure-eight or 'cardioid' pattern)

Over the years more 'helper-' or spot-microphones have been added, as can be seen with the addition of an M-49 for the second violins and a U-47 for the Cellos on the occasion of the recording of 'Don Quixote' in 1959 (Fig. 11.36).

Of course, there have been variations in the basic ABC main microphone ('AB-centerfill') set-up, like the one which is documented for the recording of 'Pines of Rome' from 1959: the mono center microphone has been replaced with a stereo

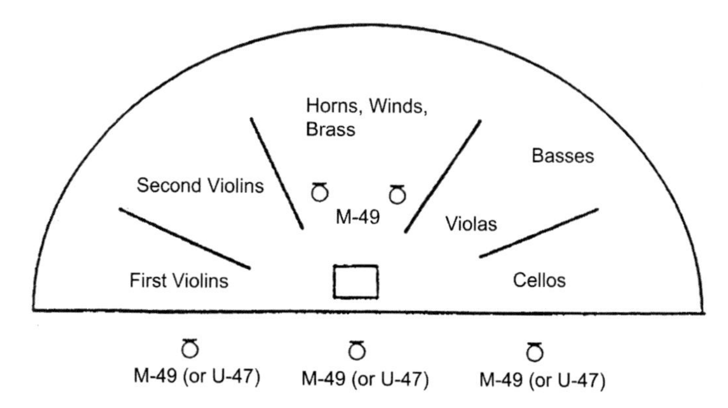

Fig. 11.33 RCA miking scheme for 'Pictures at an Exhibition,' recorded December 1957 [according to Michael Gray, graphic (after Valin 1994)]

Fig. 11.34 Neumann U-47 large membrane condenser microphone (variable pattern: omni or cardioid)

microphone (Neumann condenser SM-2). It is also interesting to note that for this recording the traditional AB pair of spot microphones for the wind section has been replaced by what looks pretty much like a DECCA-triangle (made up of M-49s); in addition a single spot-microphone for the group of harp, horns, celesta and piano (RK-5). This DECCA-tree style 'insert' to the orchestra makes very much sense, as the double-basses have been split into two groups, left and right in the back of the second violins and cellos (Fig. 11.37).

Over the years this 'large AB' setup matured to become a 'large ABC' setup with an added 'centerfill' microphone in the form of an omni microphone or stereo microphone like the Neumann SM-2 (see Figs. 11.37 and 11.38).

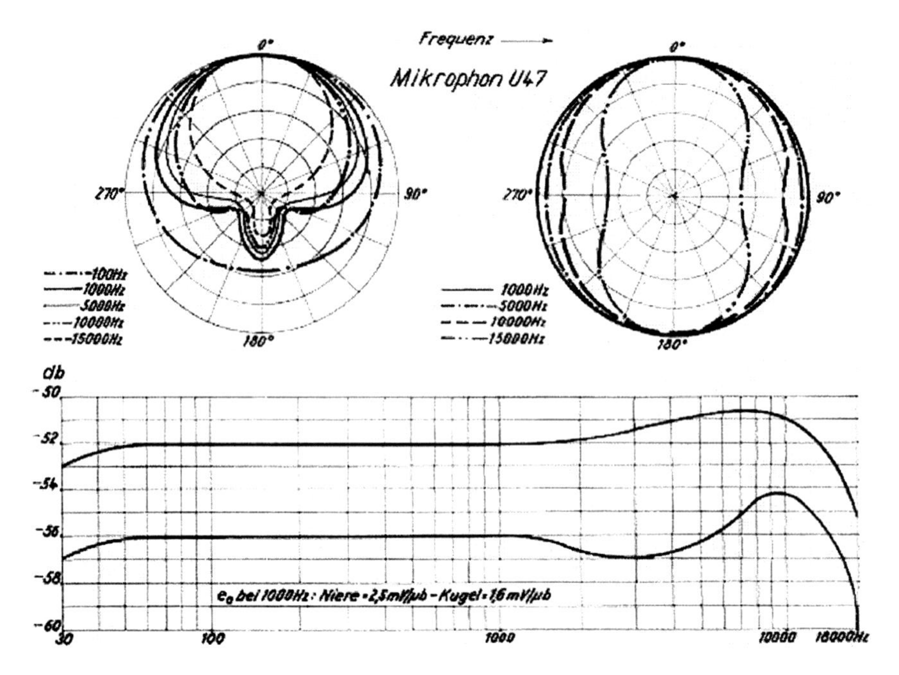

Fig. 11.35 Neumann U-47 large membrane condenser microphone; frequency response and polar pattern (courtesy of Neumann, Germany)

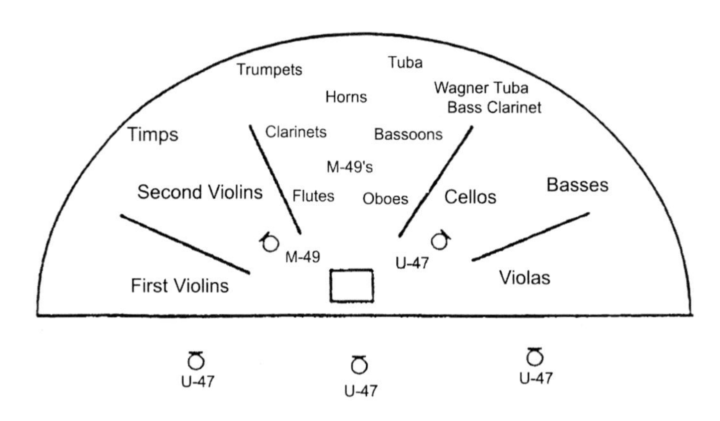

Fig. 11.36 Layton's 'mature' miking technique: a diagram of orchestra and microphone placement in recording Strauss' 'Don Quixote' in April 1959 (after Valin 1994)

Over the years, the number of spot-microphones for the RCA recordings grew, which—according to Jack Pfeiffer—lead to a somehow 'blurred' sound image. Therefore, when Jack Pfeiffer again assumed his role of producer at the RCA a few years later, he insisted on going back to using simpler microphone layouts, similar to the early days of stereo (from Valin 1994, pg. 6). As Jack Pfeiffer put it in

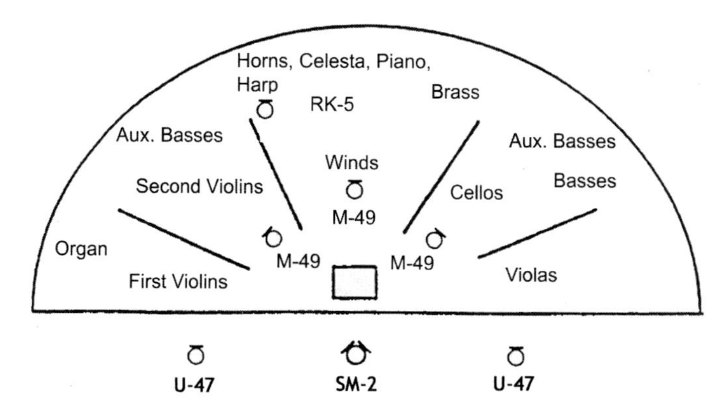

Fig. 11.37 RCA miking scheme for 'The Pines of Rome,' recorded October 1959 (from Valin 1994)

an interview: '... I think the early stereo experiment proved the point that the fewer microphones you have, the more likely you are to get a really first-class recording ... the more microphones you have, the more phase differences you get, plus you pick up all the reflections from the acoustical environment. It [can] all add up to a mess' (from Pfeiffer 1992). This is also the reason why—when put into charge as a producer for the later 'Dynagroove' records of RCA—he insisted that the engineers would throw out the extra mikes and return to the simpler mike setups, with minimal EQ that he and Leslie Chase had started to use during their 1954–55 recording session (see Valin 1994).

Of course depending on the hall, the height of the large-AB microphones, used at RCA sessions was usually between 3.30 m and 3.60 m (roughly between 10″ and 11″) looking down on the orchestra with an angle of about 30° (Figs. 11.39, 11.40 and 11.41).

The RCA 'Living Stereo' recordings, using the 'Large-AB (Centerfill)' microphone approach, are criticized by some for the seemingly 'wrong' approach (meaning: using large-AB instead of small AB capsule spacings, the latter of which adhere to the 'rules of psychoacoustics' in terms of localization by providing 'proper' ITDs (inter-aural time difference), while large-AB capsule spacings don't) but at the same time are cherished by music lovers worldwide for the 'open sound' they provide. In addition to the fact that recording and mixing music is—of course—not just a technical matter, but involves very much also an artistic sensibility from the side of the mixing engineer, I would be tempted to put the argument forward: 'What pleases the listener, can't be totally wrong ...!'.

Apart from this purely aesthetic evaluation, there is of course also a physical/acoustic 'reality': in previous chapters, especially Chaps. 7 and 8, we have seen—by means of analyzing the frequency dependent cross-correlation coefficient FCC and the FIACC (the binaural equivalent of the FCC)—that some microphone techniques are much better able to reproduce the sound-field of the concert hall in the listening room, than others. One typical RCA 'Living Stereo' recording has also been

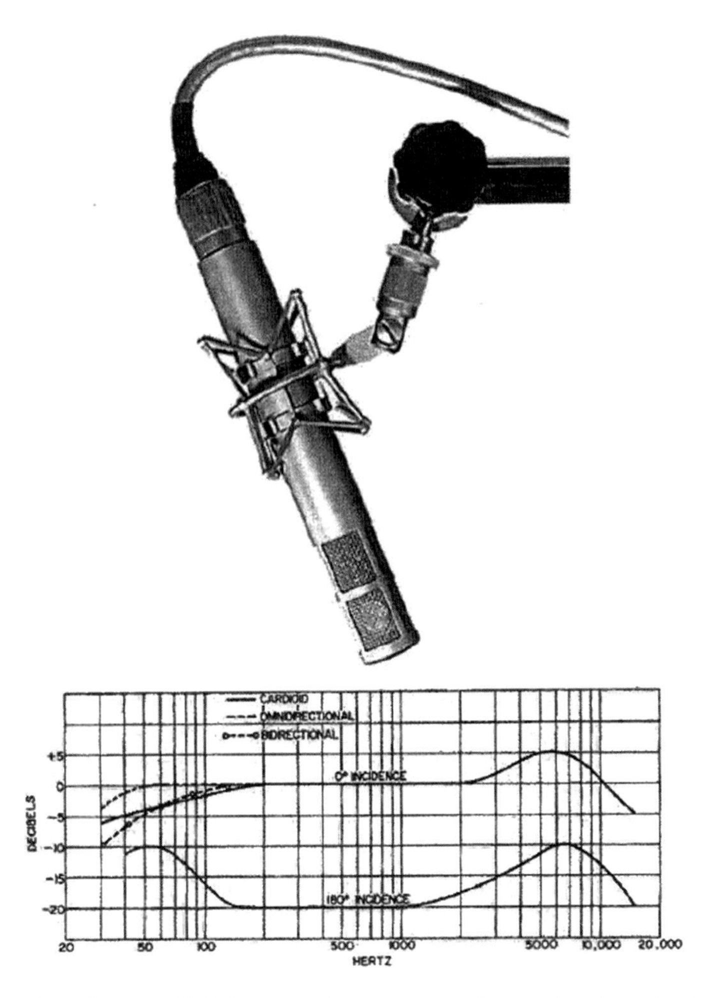

Fig. 11.38 Top: Neumann SM-2 large membrane stereo condenser microphone; Bottom: SM-2 amplitude-frequency plot for omni, cardioid and fig-8 mode (courtesy of Neumann, Germany)

analyzed in this way (see Fig. 8.17), and it has been shown that large-AB techniques with appropriate 'centerfill' belong to the category of microphone techniques, which are most true to the original concert hall sound field. It has been shown in the previous chapters that this has to do mainly also with the much better 'spatial impression' these techniques provide. According to results of subjective listening tests conducted by the author, but also other researchers, good 'spatial impression' itself is highly correlated to 'listener preference' (see the results and conclusion of Sect. 6.10 in Chap. 6). As also found in other studies the evaluation of the subjective listening tests of the author has unveiled a high correlation between the perceived 'naturalness' of a recording and listener preference, which is not surprising (see Pfanzagl-Cardone 2012). This is also where the 'Binaural Quality Index of reproduced music' (BQI_{rep}), as introduced

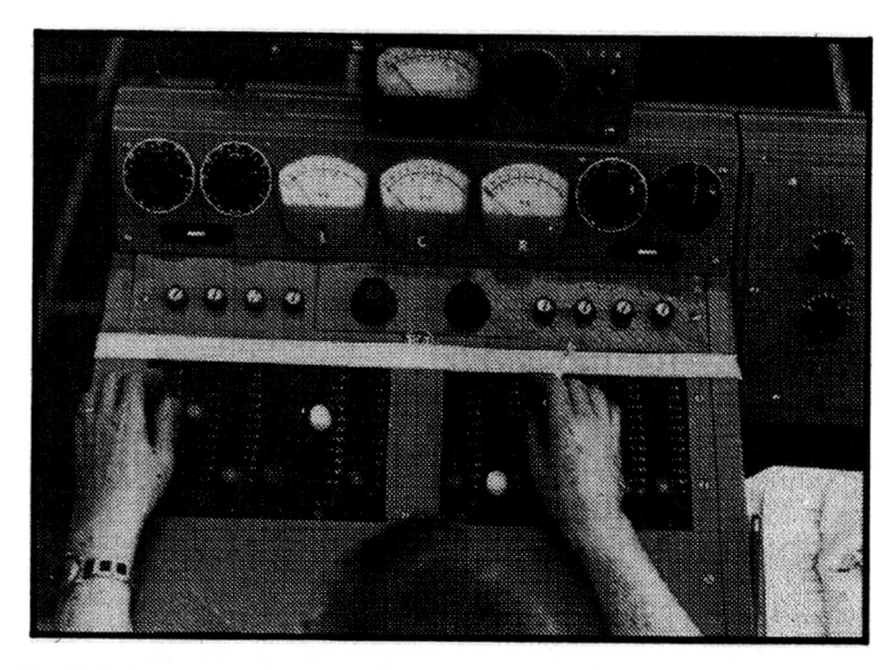

Fig. 11.39 Engineer Lewis Layton at RCA's 3-channel mixing desk (1957) reprinted by kind permission of RCA/Sony Music Entertainment

Fig. 11.40 An early RCA 'Living Stereo' recording session at Webster Hall, NYC (1957) reprinted by kind permission of RCA/Sony Music Entertainment

Fig. 11.41 RCA 'Living Stereo' re-release of Dvorak's 'New World Symphony' in 2- and 3-channel stereo on SACD (Super-Audio CD); reprinted by kind permission of RCA/Sony Music Entertainment

in Chap. 10, comes into play: microphone techniques, which manage to recreate the original sound-field of the concert hall better than the competing techniques will also be characterized by a higher $\mathrm{BQI}_{\mathrm{rep}}$ upon playback in a living or control room. The measurements in the previous chapters indicate that the RCA 'Living Stereo' series belongs to this group of recordings.

The original 'Binaural Quality Index' BQI, as defined in (Keet de 1968), has been identified as 'one of the most effective indicators of the acoustic quality of concert halls' by Beranek, one of the most important experts in respect to concert hall acoustics (see Beranek 2004). In the opinion of the author, the acoustics of concert halls—being ideal locations for musical performances—remain the basic reference point on how music should sound, also when replayed to the listener in the living room. Therefor the newly defined $\mathrm{BQI}_{\mathrm{rep}}$—'Binaural Quality Index of reproduced music' is proposed as a measure for the impression of 'naturalness' of recorded music.

The high 'stereo spread' (or ASW—Apparent Source Width) which the large-AB microphone technique based RCA 'Living Stereo' recordings are characterized by, results in largely de-correlated stereo channel information and also a high degree of 'spatial impression' which are both reasons why these recordings—apart from their musical quality, of course—despite dating back to the 1950s and 60s, are still highly appreciated even today. (For more detailed info on this matter please see Chap. 1, Sect. 1.5 'The influence of loudspeakers and listening room acoustics on listener preference')

A third reason, why the RCA 'Living Stereo' series is appreciated by many music lovers, might be rooted in the way how most of those these recordings have captured the diffuse sound of the concert hall: When using a 'large AB'-style capsule spacing which exceeds the reverberation radius of the hall (as is certainly the case with the majority of the 'Living Stereo' orchestral recordings) the 'large AB' pair of microphones themselves already fulfill the purpose of ambience- (or room-) microphones, which are traditionally used to capture diffuse sound in a recording venue in addition to the main microphone and spot-microphones, which are there to mainly capture direct sound: the 'A' main microphone on the left side will function as an ambience microphone for the right side of the orchestra and the 'B' main microphone will do the same for the left orchestra side!

In the case of good acoustic properties of the concert hall, the amount of diffuse sound arriving at the A and B microphone will be just fine and the overall sonic picture obtained by the large AB pair will certainly be much clearer than a mix of—for example—a 'small AB' main microphone and an additional pair of 'room microphones' …

This aesthetic-acoustic fundamental property of the 'large AB' technique has already been evident to the great Harry F. Olson and Herbert Belar from RCA laboratories in New Jersey as they stated in their AES Paper from 1960, titled 'Acoustics of Sound Reproduction in the Home':

> … In stereophonic reproduction it is possible to restore reverberation to the amount optimum for direct listening. For instance, instruments whose direct sound is reproduced from one speaker can have the reverberant sound reproduced from the other and vice versa. This suggests that for a maximum spacious room-effect the instruments be made to favor either one or the other microphone or groups of microphones for direct pickup and that the other microphones are angled so as not to pick up direct sound but mostly reverberant sound for those instruments unless the difference in the path to the two microphones or groups of microphones is large enough to be counted as a first reflection. In this case the direct sound can also be picked up by the second group of microphones. The ability to carry the room sound will be least when the source is to appear in the center between the two speakers during reproduction. Therefore, depending upon the program material, with stereophonic reproduction it is possible to vary from intimate 'highly-damped-room' sound reproduction to the effect of a great hall just adjoining and opening into the listener's living room. … (see Olson and Belar 1960, pg. 10)

To summarize, it seems fair to assume that it is various sonic characteristics of the acoustical properties inherent in the recordings of the RCA 'Living Stereo' series, which contribute to their success with music lovers: Due to the large-AB capsule spacing—which is the solid basis of all RCA 'Living Stereo' orchestral recordings—not only the direct sound of the orchestra, but also the diffuse sound (i.e. reverb and early reflections) coming back from the hall, have both been captured in a 'quasi-optimum' de-correlated manner, which also ensures a high 'apparent source width,' which has also been identified as an important factor, closely related to listener preference [see—among others—(Berg and Rumsey 2001), as well as (Bech 1998)].

In order to record diffuse sound in an optimized, de-correlated manner, you need to either use two omni directional microphones with a capsule spacing larger than the reverberation radius (or 'critical distance') of the recording venue, or an XY coincident pair of highly directional microphones with figure-of-eight (Blumlein-Pair, 90°) or hyper-cardioid (at 133° included angle) characteristics (for the latter see Griesinger 1987).

While signal de-correlation at frequencies below 500 Hz is majorly important for good 'spatial impression' (for more details and literature see Chap. 1), the psychoacoustic sensation of 'apparent source width' is something that has to do with appropriate signal de-correlation (i.e. signal dissimilarity, as is needed for the L and R channel of a true 'stereo' signal) in the mid-frequency range from a few hundred to a few thousand Hz (see Chap. 1, Fig. 1.15).

In contrast to the 'old style' RCA 'Living Stereo' series, most modern recordings usually make use of a plenitude of spot-microphones in addition to the main microphone system and this results in a very different sound, which is characterized by a higher FCC (Frequency–dependent Cross-Correlation) value, as can also be seen in the left-hand side graphic of Fig. 11.42, which displays the FCC for a small AB recording with many spot-microphones, made by record company TELDEC.

Judging by eye, the left graph is characterized by an average correlation value of around 0.5, while the right graph has an average value of around 0.1, therefore being almost completely 'de-correlated.' In the past, various scientists have already researched the question of 'optimized signal correlation.' In his research, Gottlob arrives at the conclusion that the optimal interaural cross-correlation IACC (for live music in a concert hall) equals 0.23 (see Gottlob 1973; Cremer und Müller 1978, pg. 482). More recent research by Muraoka et al. (2007) seems to indicate that

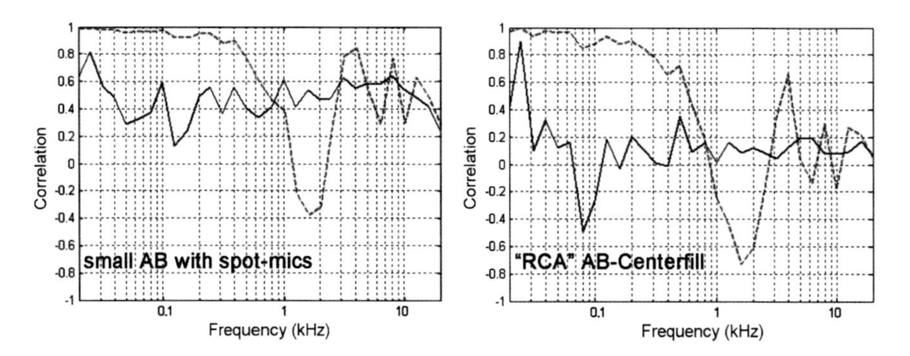

Fig. 11.42 Frequency-dependent Cross-correlation Coefficient (FCC, solid line) and Frequency-dependent Inter-Aural Cross-correlation Coefficient (FIACC, dashed line); Left: small AB + spot-mics (TELDEC; music, hall); Right: AB-Centerfill system (RCA 'Living Stereo,' music, hall) (Rem.: same as Fig. 8.17; the two graphs of Fig. 11.42, as well as the many other ones from Chaps. 7 and 8 have been calculated by means of the MATLAB function 'mcorrfrequ 48 k' listed in Appendix C)

for the reproduction of recorded music the correlation of the recorded stereo signal should optimally be zero. This evaluation seems to be based primarily on an informal evaluation by the team of the authors. In their paper 'Ambience sound recording utilizing dual MS (Mid-Side) microphone systems based upon frequency-dependent spatial cross-correlation (FSCC)' the authors claim that they '… studied the sound pickup characteristics of typically used main microphones to clarify their ambience representation, employing frequency-dependent spatial cross-correlation (FSCC) as a criterion, and concluded that MS microphone set with a directionality azimuth at 132° is best. FSCC of the microphone output becomes uniformly zero (Rem.: see also Chap. 8, especially Figs. 8.7 and 8.8 for more details). This makes [the] listener feel most natural ambience, and the effect was certified through actual orchestral recording.' The fundamental requirements for the main microphone systems under test—as set out by the authors—were:

1. sufficient ambience description, and
2. exact sound source localization.

Looking at the results of the research by Muraoka et al. (2007), it seems very plausible that the almost completely de-correlated RCA recording with an average FCC of 0.1 would be preferred by most listeners over the TELDEC recording with an average FCC of 0.5, at least when it comes to spatial reproduction. (For more details, see Chap. 2, Sect. 2.3)

A more in-depth analysis of what can be seen in the graphics of Fig. 11.42 and the quasi 'Stethoscope'-qualities of the FCC and FIACC (Frequency-dependent inter-aural cross-correlation) analysis can be found in Chap. 8, Sect. 8.3. This section also contains an analysis on related reasons for the apparent high listener appreciation of the RCA 'Living Stereo' legacy recordings. For acoustic examples and further analysis of different microphone techniques, the interested reader may also visit the author's Youtube channel by the name of 'futuresonic100,' looking for the term 'mic tech analysis' in the title of video clips which are related to the topic.

11.4 MERCURY 'Living Presence'

Since November 1955, similar to the RCA, also record label Mercury used an Ampex 300 ½ inch 3-track tape machine for their stereo recordings. Chief sound engineer Robert Fine had a very 'purist' approach for his recordings: 3 Telefunken '201' (or Neumann 56) omni microphones with a rising high end (rem.: 'diffuse-field compensation') were lined up along the front of the orchestra. The right balance between the instrumental groups was achieved by re-arranging the positions of the musicians on stage and not by moving the microphones or introducing spot-microphones. Even for more complex recording situations—like opera, or works with large orchestra and choir—he did not shy away from this 'Spartan' technique. Therefore, it was the full responsibility of the conductor to ensure the 'correct' balance between the instruments or instrumental groups of an orchestra over the whole course of a performance.

Fig. 11.43 Mercury 'Living Presence' recording session: Howard Hanson conducts the Eastman-Rochester Orchestra (with 3 suspended Telefunken 201 condenser microphones); reprinted by kind permission of Decca Music Group Limited

As Robert Fine, chief recording engineer at Mercury described it in his own words: '… three (omni-directional) microphones were … hung in front of the orchestra at fairly widely placed intervals along a single plane … the tape at the session was never electronically altered; microphones were never moved; optimum recording levels were set at the beginning of the session and never changed. The conductor … was the person solely responsible for the musical flow of the performance.' (from (Valin 1994), pg. 128).

Watford Hall was Mercury's preferred location, while DECCA used Walthamstow Town Hall and—as EMI—also Kingsway Hall in London for their recordings (Figs. 11.43 and 11.44).

11.5 A Few Case Studies of Live and Studio Recordings with Orchestra (and Solo-Instruments)

An interesting example of an orchestra recording with solo piano, with a quite typical microphone setup for the period of time in which the recording took place, can be found in Gould (2002). The photograph has been taken on the occasion of a recording session of the Schönberg piano concerto at Massey Hall, Toronto with Glenn Gould in 1961. Apparently only 6 microphones have been used, which look like AKG C12 or Telefunken 201 condenser microphones. There is only one (mono) spot-microphone for the piano (to the right of the conductor), three spot microphones covering the left and right rear sections of the orchestra (one of which seems to be set up mainly for the percussion section, the other two for the wind instruments) and—as a kind of

Fig. 11.44 MERCURY 'Living Presence' re-release of Brahms and Khatchaturian violin concertos, recorded at Watford Town Hall in 1962 and 1964; reprinted by kind permission of Decca Music Group Limited

main microphone—a large-AB pair, pointing diagonally across the string sections left and right of the conductor.

Due to the use of a relatively small amount of microphones, it makes very much sense that they are positioned relatively high above the various instrumental groups of the orchestra in order to achieve a balanced coverage for the whole group (Fig. 11.45).

Another example of a recording for orchestra and piano happened more recently at the Musikverein-Saal in Vienna on the occasion of a Mozart piano concerto recording with pianist Lang Lang and conductor Nikolaus Harnoncourt. According to what can be seen on the photographs, the main microphone setup consists of a mid-size AB pair of omnis, approximately in the 4th row in front of the orchestra, as well as an (alternative) small AB pair of two Neumann M50 diffuse-field compensated omnis. For the piano, there seems to be an equivalence microphone pair of directional microphones on a high stand in the extension of the piano lid, facing down on the high and low strings for a nice 'stereo' imaging of the instrument. In addition to these main microphones, there are 'Orchestra L' and 'Orchestra R' microphones on high stands with gallows, overlooking the respective sides of the orchestra, similar to the miking approach we have seen with the Glenn Gould recording at Massey Hall above (Figs. 11.46 and 11.47).

In addition to the main microphone systems explained above, spot microphones for various instruments may have been used, which can be seen in the photographs. However, not having been present at the recording session, this remains an uncertainty, as the hall is regularly used by the National Austrian Broadcasting Agency ORF, the technicians of which tend to leave a multitude of spot microphones rigged in their standard positions (Fig. 11.48).

Fig. 11.45 Glenn Gould recording the Schönberg piano concerto with Robert Craft at Massey Hall, Toronto in 1961; photographer: Don Hunstein © Sony Music Entertainment; reprinted by kind permission of Sony Music Entertainment

Fig. 11.46 Main microphone setup for a recording of a Mozart piano concerto at Vienna Musikverein with Lang and Harnoncourt [from documentary video (Mission Mozart 2014); see listing of 'A/V-references']

Fig. 11.47 Lang/Harnoncourt recording: detail view of piano microphones, woodwind spot microphone [from documentary video (Mission Mozart 2014); see listing of 'A/V-references']

Another possibility, how to mike a concert solo piano in a more 'indirect' way, can be seen in Fig. 11.49: two cardioid microphones (Schoeps MK4) are placed to pick up the reflection of sound from the piano lid, which resembles much more the sonic impression which also the audience in the floor area gets from the instrument. Therefore, the sound tends to be less direct and aggressive, as may be the case when

Fig. 11.48 Lang/Harnoncourt recording: detail view of piano microphones, small AB, mid-size AB and 'Orchestra R' microphone [from documentary video (Mission Mozart 2014); see listing of 'A/V-references']

Fig. 11.49 Piano solo with orchestra—indirect miking by two directional microphones via reflections off the piano lid (photograph taken at Salzburg Festival hall)

the microphones are placed in such a way, that the microphones have a direct 'look' on the strings of the instrument (Fig. 11.49).

11.6 Solo-Piano Microphone Techniques

There seem to be almost as many techniques for solo-piano miking, as main microphone techniques for recording a whole orchestra. Figure 11.50 tries to show a few of them: From (Huber and Runstein 1989):

> 'Position 1—indicates a mono pickup, which is attached to a partially or fully open lid
> Position 2—Two pickups are placed in a spaced stereo-arrangement at a working distance of 15–30 cm (0.5–1 foot) with one pickup centrally positioned over the low strings and the other placed over the high strings
> Position 3—a mono or stereo 'coincident' pair of microphones placed just inside the piano, between the sound board and its partially or fully opened lid
> Position 4—a mono or stereo 'coincident' pair of microphones placed outside the piano, facing into the open lid

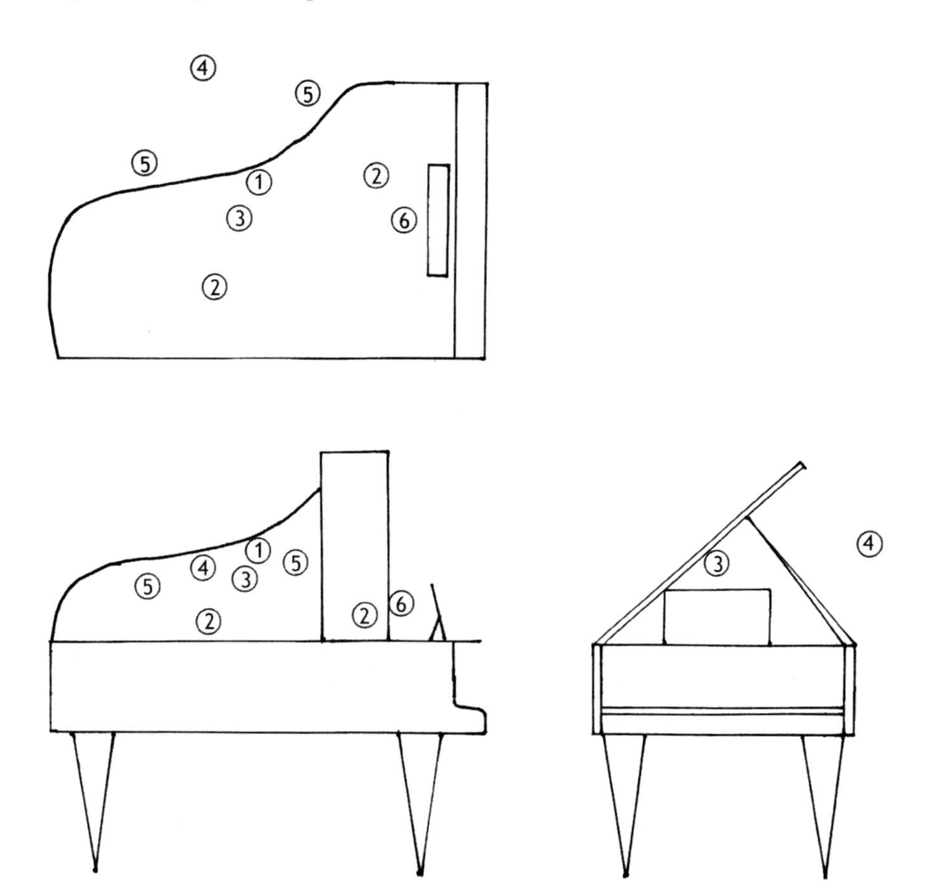

Fig. 11.50 Schematic view of the various possibilities for miking a piano (after Fig. 4-34 from Huber and Runstein 1989, pg. 88)

Position 5—placing a spaced stereo pair (usually 'omni microphones') outside the lid, facing into the instrument, using either a 'small AB' approach of 51 cm spacing, or applying the so-called 3:1 spacing rule
Position 6—for a more direct, 'pop-music' like sound one can place a monoaural or stereo coincident (XY) pair just over the piano hammers at a working distance of 10 to 20 cm '

Concerning position 5 in Fig. 11.50, some additional explanation is due. In sound-engineering, there is a rule-of-thumb called the '3:1 spacing rule' which essentially means that if you need to individually mike two instruments next to each other, or a larger instrument or sound source in stereo, in order to get sufficient separation (in the case of two instruments) or signal de-correlation (in the case of one larger sound source) the distance between the capsules should be three times the distance of the capsule to the instrument. Figure 11.51 shows, that in case of miking a solo piano, this will lead to a mid-size AB spacing of capsules inside or in front of the instrument.

Of course an alternative to the mid-size AB microphone pair (usually omnis) would be a small AB pair with a capsule spacing in the order of approx. 50 cm, as already proposed for 'position 5' (see Fig. 11.50). The photography in Fig. 11.52 from a piano recording at 'Tonzauber' recording studio at Konzerthaus Vienna shows two possible positions for such a small AB miking for the piano: The regular one in the 'bow' of the piano, as well as the so-called tail-end position for the microphones. (Rem.: as an alternative, a coincident Nevton BPT microphone (see Chaps. 3 and 4 on stereo and surround microphone techniques) has been placed, similar to position 4 in Fig. 11.50, with an absorptive panel behind in order to achieve higher separation to a solo-singer, who stood in front of the instrument).

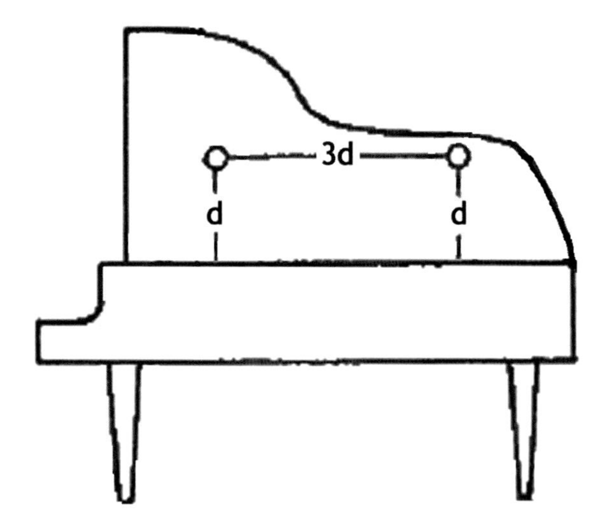

Fig. 11.51 3:1 rule applied to piano miking (after Fig. 4-29 from Huber and Runstein 1989, pg. 83)

Fig. 11.52 Small AB microphone setup (with 2 Neumann M50 diffuse-field compensated omni microphones each) for piano recording in the 'normal' (bow) as well as 'tail-end' position

On the occasion of a piano solo recital at the Salzburg Grand Festival Hall, the following microphone arrangements have been applied in order to have sonic flexibility for the final mix during post-production (Fig. 11.53):

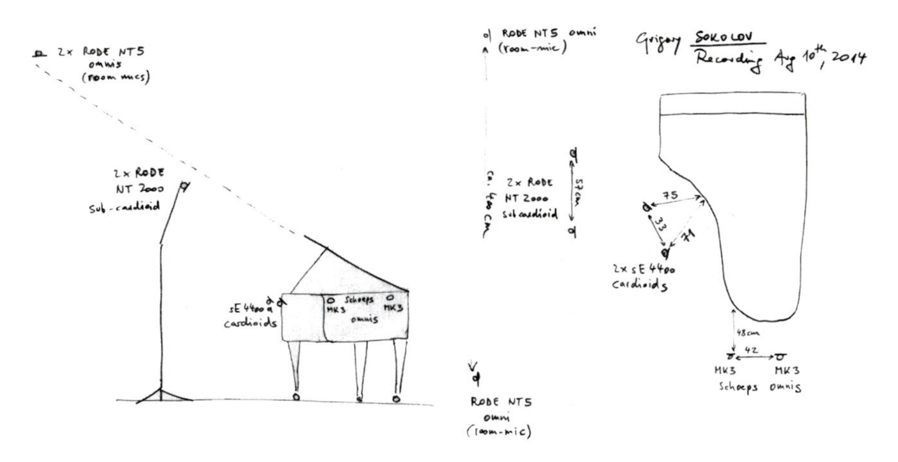

Fig. 11.53 Microphone setup for piano recording as applied for the Sokolov piano recital recording at Salzburg Festival Hall in 2014 (distances marked are in cm)

For the Sokolov-recording the following systems were used:

(a) a pair of Schoeps MK3 omnis with HF-boost at 'tail-end' position
(b) two directional microphones (sE 4400 large diaphragm condenser microphones in cardioid mode) in an 'equivalence arrangement'
(c) two hemi-cardioid capsules (Rode NT2000 large diaphragm condenser microphones with selectable mic-pattern) in a small AB configuration
(d) two suspended 'room microphones' with a capsule distance of 4 m (Rode NT5 with omni capsules)
(e) From this arrangement, the most direct sound can be achieved by use of the Rode NT2000, which have a direct look onto the strings. Sounding somewhat closer, but profiting of the first reflection off the lid of the piano, the sE4400 offer a quite different sonic impression (see also the photograph of Fig. 11.49). For a more open sound, the MK3 (below the rim of the piano) come into play and—if more 'feel of the room' is needed—the signals of the Rode NT5 omnis can be blended into the mix (Fig. 11.54).

Another example of solo-piano recording can be found in the documentation of the microphone setup used for Mauricio Pollini at Herkules-Saal in Munich, which used only two pairs of Sennheiser MKH20 microphones:

Fig. 11.54 Sokolov piano recital recording at the Salzburg Festival Hall in 2014 (note the visual angular distortion for the position of the left suspended room microphone [Rode NT5] due to the camera perspective; the right suspended room mic [Rode NT5] is not visible on the photograph)

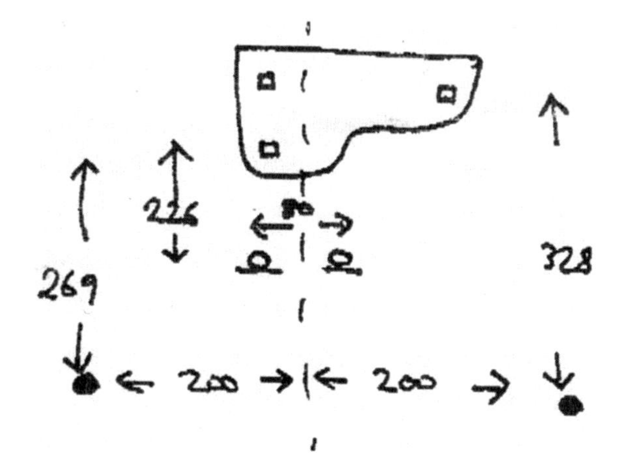

Fig. 11.55 Hand drawing of mic setup schematic for Pollini piano recording at Herkules Saal Munich (for Deutsche Grammophon)

(a) a small AB pair of omnis, spaced 80 cm apart at a distance of 226 cm from the front leg of the piano and 217 cm above stage level (quasi at the extension of the piano lid)
(b) a suspended ‘room microphone’ pair of omnis, with a capsule spacing of 400 cm at a distance of 269 cm from the front of the piano and at a height of 260 cm above stage level

Such detailed instructions where to position the microphone capsules make sense in this case, as Mauricio Pollini always travels with his own Fazioli-Steinway piano in order to achieve consistent pianistic, as well as sonic quality (Fig. 11.55).

11.7 Studio Ensemble Recordings with Acoustic Instruments

When studio recordings of orchestral works with a solo instrument take place, the sound-engineer or Tonmeister often has the possibility to re-arrange the seating order and instrument positions in the room to accommodate the microphone technique used. In the example in Fig. 11.56, the main microphone system to record the orchestra is a DECCA-tree (over the conductor’s head), with adjoining outriggers (i.e. large-AB omnis). As can be seen, the horns have been singled out to the back on the left side with a partially absorptive, partially reflective surface behind them, while the brass section has been pushed a bit toward the back, with some absorptive baffles in front of them. It is also interesting to note that the solo-piano has been singled out, well in front of the orchestra and turned around (in comparison with the normal concert orientation), most likely in order to let the orchestra musicians hear it properly. Two spot microphones have been setup for the piano in what seems to be a small AB arrangement of AKG C12 microphones or a similar model. The setup of the piano actually behind the DECCA tree is no surprise: This is done on purpose, as the

Fig. 11.56 London Mozart Players in a recording session for orchestra and piano at Abbey Road 'Studio 1'

instrument would inevitably appear too loud (and probably slightly blurred) in the main DECCA triangle if setup right underneath or in front of the tree (especially with the piano lid removed), in relation to the orchestra. Even though the recording session shown is not necessary one by the DECCA recording company, however 'singling out' the solo instruments in an orchestra-with-solo-instrument recording and miking it up separately was—apparently—very common practice in DECCA recording sessions, according to Mr. Siney, one of the last staff-engineers at DECCA (this information was obtained from a personal conversation of the author with Phil Siney on the occasion of a 'West Side Story' musical recording at the Salzburg Festival in summer 2016).

For the live recording of orchestra with solo-piano, a similar approach (of singling out the solo instrument) can be taken, as shown in Fig. 11.57: the central part of the main microphone system (in this case an ORTF or coincident (XY, Blumlein-Pair or BPT) microphone arrangement) has been moved to a position behind and above the piano. With the piano lid in place, this mic arrangement will be shielded off reasonably well from the strings of the piano, but low-frequency sound will of course bend around and also some part of the mid-frequency range will be reflected from the floor below the instrument and therefore arrive at the main microphone system. With an ORTF there will be much better signal separation than with a DECCA-triangle above the piano; as an alternative the use of a figure-eight microphone system like

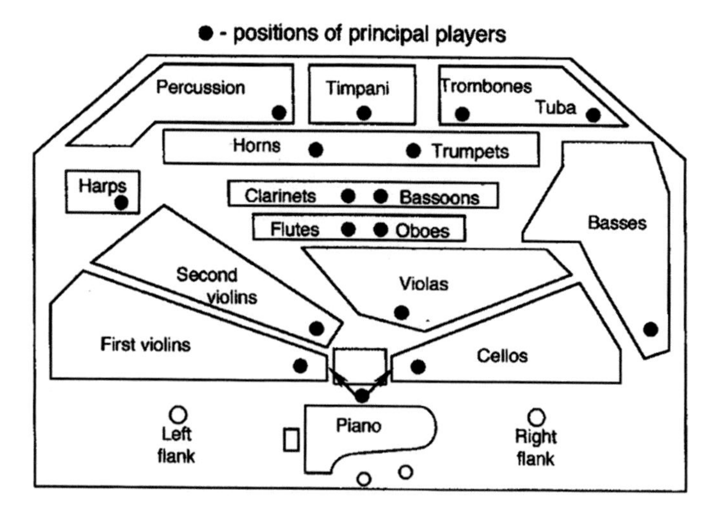

Fig. 11.57 Main microphone setup for recording orchestra with solo-piano; directional coincident main microphone with large-AB outriggers (Fig. 13-15A from Eargle 2004)

Blumlein-Pair or BPT will profit from having the axis of minimum pickup oriented toward the piano, which can be achieved quite easily.

The DECCA-triangle is also quite frequently used for the recording of a string quartet or small ensembles, as can be seen on the right side of Fig. 11.58 and the previously shown documentary photograph of the 'Wiener Oktett' in Fig. 11.23.

For live recordings, a coincident microphone system (like XY-Cardioids, Blumlein-Pair crossed fig-8s or a BPT-system) may be preferred for its compactness and for being visually less intrusive, as can also be seen in Fig. 11.59, which shows the microphone setup for a live recording of violin, cello and piano on the left side. The right-hand side shows the same instruments in a studio recording situation, being picked up by individual instrument microphones: while there is a small AB pair of omnis, which pick up both piano, as well as the violin and cello, the latter two also have their individual directional spot-microphones, which allow the sound-engineer to establish the right balance and perspective in the mixing stage.

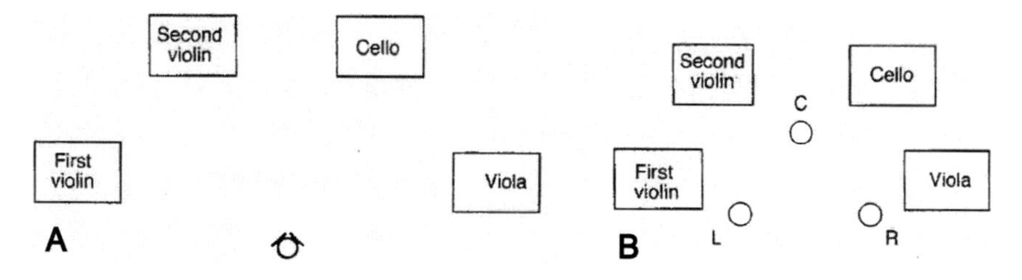

Fig. 11.58 **a** Coincident microphone system for stage setup of string quartet recording; **b** DECCA-triangle spaced microphone system for studio string quartet recording (after Eargle 2004, Fig. 13-10)

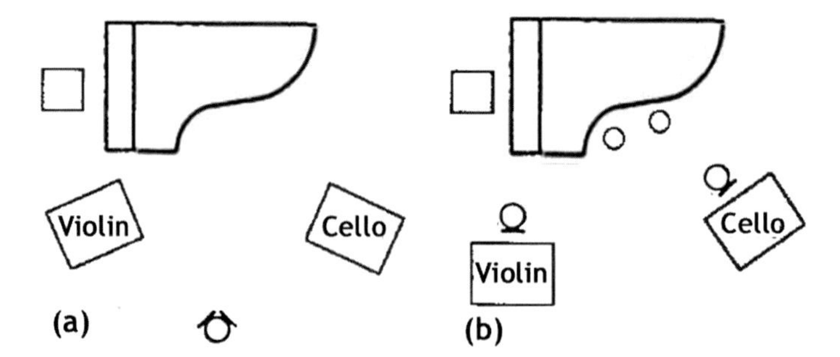

Fig. 11.59 Piano-Trio miking for **a** stage and **b** studio (after Eargle 2004, Fig. 13-8)

Another studio recording situation is pictured in Fig. 11.60: on the left-hand side, a piano quintet, of which the string players are miked with a DECCA-triangle, while the piano has its own small AB spot-microphone pair. Note that the piano is facing away with its opened lid from the other instruments for better acoustic separation.

This was also the case with a trio recording (piano, cello, clarinet), of which a session photograph is presented in Fig. 11.61. The piano was picked up with two angled cardioids (Rode NT2000) in equivalence stereophony, the cello with a large membrane condenser (Neumann U87 set to cardioid mode) and the clarinet with a directional small membrane condenser microphone (Schoeps MK4 cardioid).

On the right-hand side of Fig. 11.60 a piano recording with instrumental or vocal soloist is pictured. Again, the piano is picked up with a tail-end configuration, while the soloist has his/her own pair of small AB microphones. Assuming that both pairs will be equipped with omni-directional capsules it is very important to position them at the correct distance (direct-/diffuse sound ratio) to the respective sound-source and also in relation to each other, as there will be considerable spill from each sound-source into the microphones of the 'neighbor.'

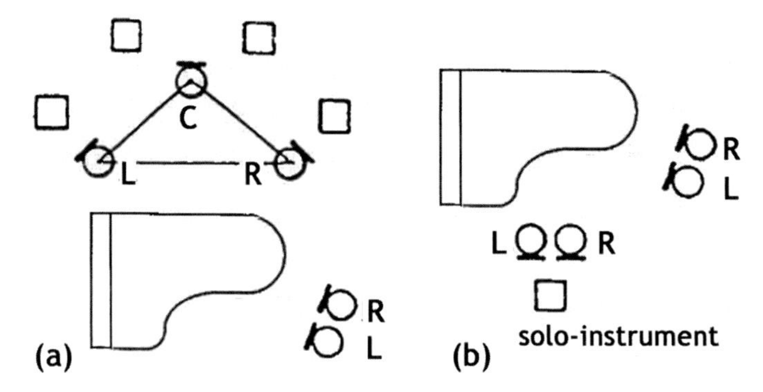

Fig. 11.60 **a** Application of DECCA-triangle for recording of piano quintet; **b** miking piano and soloist (after Sengpiel 1992)

Fig. 11.61 Microphone setup for studio recording of Brahm's Clarinet Trio with members of the Vienna Philharmonic and Mozarteum orchestras, as well as a piano soloist

Fig. 11.62 Glenn Gould with the Juilliard String Quartet, recording Schönberg's 'Ode to Napoleon Bonaparte,' op.41 at Columbia's 30th Street Studio in Feb.1965; photographer: Don Hunstein © Sony Music Entertainment; reprinted by kind permission of Sony Music Entertainment

In Fig. 11.62, we can see a photograph of the studio recording of Schönberg's 'Ode to Napoleon Bonaparte,' op.41. It seems that for the piano one (or two) microphones have been used and also for the pairings of cello/viola and first/second violin there is only one microphone in place for each, angled from above. Note that in addition to the instrumentalists there is a narrator, who—according to Schönberg—is 'equal' to the musicians and should have the same volume, not 'standing out' acoustically at any point of the piece (rem.: the narrator and his spot-microphone can be seen in the vocal booth on the far left side).

To conclude this section of case studies, we will take a look at a larger production, which is situated in the genre of religious folk musical, the 'Salzburger Adventsingen', which takes place annually at the Grand Hall of the Festspielhaus Salzburg. The ensemble is comprised of a musical ensemble of about twenty musicians, located in the central orchestra pit (with the brass section occasionally playing in remote positions outside the hall, or on stage), about a dozen vocal soloists as main singers and actors, and about twenty shepherd boys and girls. In addition, there is a large mixed choir, which is split in two groups on steps left and right of the central stage area (Fig. 11.63).

The instrumental ensemble (strings, brass, woodwinds, guitar, harp, cimbalom, portable organ) is picked up by a large-AB pair of omni microphones (Schoeps MK3, diffuse-field compensated omnis; see mics A and B), with a capsule spacing of approximately 4–5 m, with the capsules suspended about 2 m above the musicians. Of course this microphone spacing is not chosen following the 'laws of psychoacoustics,' which would require something like a 50 cm spacing for small AB, but much more with the intention to go for a medium-/large-AB or 'Wavefront' approach [see the experiments at Bell Laboratories in the 1930s in (Keller 1981)] for a highly spatial

Fig. 11.63 Documentary photograph of the cast of 'Salzburger Adventsingen' (2019): choir, instrumental ensemble, soloists—(©Salzburger Heimatwerk/Salzburger Adventsingen/Edwin Pfanzagl-Cardone)

sonic impression and low signal correlation between the capsules. Also, a bit unusual is the decision to mike both halves of the choir with only one cardioid microphone (Schoeps MK4) each (mics E and F), again in a large-AB fashion (see Figs. 11.64 and 11.65). Despite this seemingly odd approach (large AB instead of small AB for the instrumental ensemble and only 'mono' (spot-) microphones for the choir) in practice, this setup works very well and provides a nice stereophonic sound-stage and spatial impression as well. The truth is that—for parts of the production in which the instrumental ensemble and the choir are performing at the same time—for the final mix a 'combined microphone techniques' approach (see also Chap. 3, Sect. 3.4) is chosen: the large AB omnis, suspended over the orchestra pit (slightly panned-in,

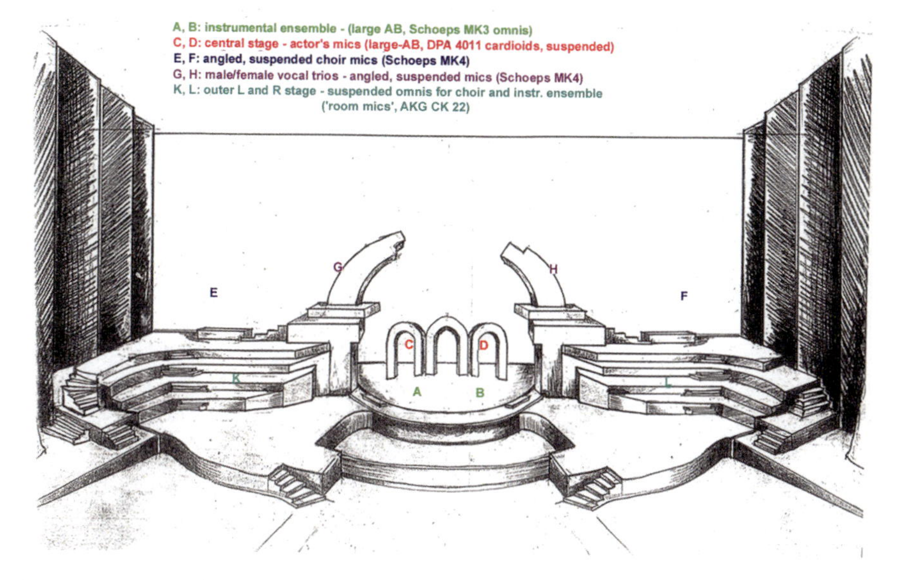

Fig. 11.64 Schematic view of microphone arrangement for 'Salzburger Adventsingen' stage setup of the 2003 performance (reproduced by kind permission of Salzburger Heimatwerk/Salzburger Adventsingen/Dietmar Solt)

Fig. 11.65 Schematic view of angled, suspended choir mics (Schoeps MK4, cardioids)

for a reduced stereophonic width), are combined with the large-AB cardioids in front of the split choir-parts (which are panned almost completely L and R).

The cardioid choir mics (angled toward the choir at a height of about 4 m above stage level; see Fig. 11.65) of course need to be time aligned (i.e. delayed) properly, according to their distance to the A and B instrumental main microphone pair in order to create a wavefront as coherent as possible for the mix of the final recording. Actors performing on the inner stage area, right behind the orchestra pit, are picked up by a pair of suspended large-AB, spaced (4 m capsule distance) cardioid microphones, pointing straight down to the stage floor (height approx. 2.5 m above stage level).

Also pictured are two additional directional microphones (mics G and H), which are meant to pick up the sound of two small vocal ensembles, a male and a female trio or quartet, which are positioned in front of the arches. There is also another pair of large-AB omnis (see mics K and L), which are suspended above the outer L and R stage area, which serve for voice pickup of actors performing on the outer parts of the stage, as well as room mics for the choir and instrumental ensemble (spatial impression).

While the drama parts of the production are preferably recorded via the suspended mics in the central area of the stage, for the singer's voices the signals of the radio mics (i.e. DPA 4088 cardioid headsets, which are primarily used for the purpose of sound reinforcement) are also mixed in with the main microphone signals in order to achieve a proper direct-/diffuse sound balance for the voices in the mix of the stereo recording.

11.8 Work Practice Recommendations

I hope the above case studies of legacy, as well as more up-to-date recordings have been of interest to the readers and I would like to conclude this chapter with the following section of advice, which is the humble result of experience gained through more than thirty years of professional engagement as recording and mixing engineer, mainly in the field of classical music.

1. Try to achieve a 'natural perspective' in the recording; i.e. capture the rear parts of an orchestra in a manner, which lets them appear to be further away from the listener than the ones in front;
2. Use as *many* microphones as necessary, but as *few* as possible;
3. Capture the acoustics of the room and the instruments with a main microphone technique that works as 'linearly' as possible (in respect to localization and depth) over the principal audio frequency range (20 Hz–20 kHz). Main microphone signals that are highly de-correlated over the whole frequency range are mandatory to achieve a convincing spatial impression and perceived naturalness in the recording
4. The 'sonic picture' created through the recording should emanate from one plane ('zero delay plane'), therefore the use of spot microphones should be restricted

as much as possible (of course with the exception of vocal- and instrumental soloists)
[Rem.: If your main microphone system consist only of a single coincident stereo microphone it is correct to measure the distance of each spot-microphone to this one-point main mic, in order to obtain the correct value for the delay time which has to be applied for the respective spot microphone signal. If you are using a main mic with outriggers, this already establishes a 'line' (or 'plane,' if different heights are used for the main mic and outriggers) of reference, to which the distance of each spot-microphone has to be measured (at an included angle of 90°)]

5. Microphones are preferably pointing from the direction of the audience toward the orchestra ('audience perspective'), in order to achieve more natural tone colors. (Rem.: possible exceptions are spot microphones for horns, for example, which can be placed *behind* the instrument to favor direct sound pickup. Also, for the sake of better signal separation, the microphones for the instrumental groups that are situated more toward the sides of the orchestra should be pointing from the center of the orchestra to the outside)
6. The use of artificial reverb or room simulation should be kept to a minimum

At first glance, these guidelines seem to be pretty much common knowledge of practicing sound-engineers; however—following them strictly, may lead to the exclusion of some well-established main microphone techniques. Choosing the main mic technique according to 'commandment 3' - for example—will rule out 'small AB' (among others), as it does not provide sufficient channel separation of low-frequency signals in order to enable a convincing spatial impression.

Above all, there is one 'final rule,' which always needs to be applied, when taking decisions: 'Trust your ears ...!' (... at least as long you have a faithful monitoring system at your disposal ....) (Fig. 11.66).

Fig. 11.66 Author at Musikverein Wien, Goldener Saal

References

Bech S (1998) The influence of stereophonic width on the perceived quality of an audiovisual presentation using a multichannel sound system. J Audio Eng Soc 46(4):314–322

Beranek L (2004) Concert halls and opera houses: music, acoustics and architecture, 2nd edn. Springer, New York

Berg J, Rumsey F (2001) Verification and correlation of attributes used for describing the spatial quality of reproduced sound. Paper presented at the Audio Eng Soc 19th int conference

Blumlein AD (1931) Improvements in and relating to Sound-transmission, Sound-recording and Sound-reproducing Systems. British Patent 394,325, 14 Dec 1931 (reprinted in: Anthology of Stereophonic Techniques. Audio Eng Soc, 1986, pp 32–40)

Cremer L, Müller HA (1978) Die wissenschaftlichen Grundlagen der Raumakustik. (Band 1), Hierzel-Verlag, Stuttgart

Decca (2003) CD booklet to 'DECCA Legends' re-release of previous Vinyl-LPs on CD

Decca (2013) DECCA SOUND—The analogue years. booklet to a 50 piece CD-Collection, 2013

Eargle J (ed) (1986) Stereophonic techniques—an anthology of reprinted articles on stereophonic techniques. Audio Eng Soc

Eargle J (2004) The microphone book, 2nd edn. Focal Press

Gottlob D (1973) Vergleich objektiver akustischer Parameter mit Ergebnissen subjektiver Untersuchungen an Konzertsälen. Dissertation, Universität Göttingen, 1973

Lester M (ed) (2002) Glenn Gould—a life in pictures. Nikolai-Verlag, Doubleday Canada (Random House of Canada Ltd) German Edition

Gray M (1986) From the golden age: the birth of Decca/London Stereo. The Absolute Sound 11(42)

Griesinger D (1987) New perspectives on coincident and semi-coincident microphone arrays. Paper 2464 presented at the 82nd Audio Eng Soc Convention, May 1987

Huber DM, Runstein RE (1989) Modern recording techniques, 3rd edn. SAMS Audio Library (Pearson Education Inc., New York)

Keet de WV (1968) The influence of early lateral reflections on spatial impression. 6th Int Congress on Acoustics, Tokyo

Keller AC (1981) Early Hi-Fi and stereo recording at bell laboratories (1931-1932). J Audio Eng Soc 29:274–280

Muraoka T, Miura T, Ifukuba T (2007) Ambience sound recording utilizing dual MS (Mid-Side) microphone systems based upon Frequency dependent Spatial Cross Correlation (FSCC). Paper 6997 to the 122nd Audio Eng Soc Convention, Vienna, 2007

Olson HF, Belar H (1960) Acoustics of sound reproduction in the home. J Audio Eng Soc 8(1):7–11

Pfanzagl-Cardone E (2012) 'Naturalness' and Related Aspects in the Perception of Reproduced Music. In: Proceedings to the 27. Tonmeistertagung des VTD, Köln, Nov 2012

Pfeiffer J (1992) An interview with Jack Pfeiffer. Aud Mag. Nov 1992:47

Rumsey F (2001) Spatial audio. Focal Press (Elsevier)

Sengpiel E (1992) Grundlagen der Hauptmikrophon-Aufnahmetechnik – Skripten zur Vorlesung (Musikübertragung). Hochschule der Künste, Berlin. http://www.sengpielaudio.de. Accessed 5 May 2004

Southall B, Vince P, Rouse A (1997) Abbey road. Omnibus Press, UK

Valin J (1994) The RCA bible—a compendium of opinion on RCA living stereo records, 2nd edn. The Music Lovers Press, Cincinatti, Ohio

A/V-Reference

Mission Mozart (2014) Mission Mozart—Lang Lang & Nikolaus Harnoncourt. C Major Entertainment. (BFMI co-production with ZDF/Arte in cooperation with Deutsche Welle TV and Sony Classical). Permalink http://dw.com/p/1E8dZ. Accessed 10 Oct 2017

Appendix A

See Tables A.1, A.2 and A.3.

Table A.1 ORCH-5.1Surround Listening test; 1-way ANOVA concerning the influence of the variable 'microphone-array' on the rating of 15 attributes (25 listeners)

'Dependent variable'	Sum of squares	*df*	Mean square	*F*	*p*	Stat. significant differences between mean-values of the surr.-mic ratings
Preference	38.5897	3	12.8632	16.7695	0	1&3, 2&3, 4&3
Naturalness	20.7831	3	6.9277	8.4185	0	1&3, 2&3, 4&3
Sound-colour	14.5719	3	4.8573	7.5008	0.0001	1&3, 4&3
High/low balance	9.7675	3	3.2558	6.3025	0.0006	1&4, 2&4
Localisation	14.9475	3	4.9825	6.3767	0.0005	1&4, 2&4, 3&4
Balance	21.6322	3	7.2107	9.4913	0	1&3, 2&3, 4&3, 1& 4
Width	79.2342	3	26.4114	30.7490	0	1&2, 1&3, 1&4, 2&3, 4&3
Distance	35.5233	3	11.8411	16.8673	0	1&2, 2&3, 2&4
Stage depth	17.6275	3	5.8758	5.9987	0.0009	1&3, 2&3
Stability	8.4431	3	2.8144	2.8690	0.0405	3&4
Differentiation	7.6853	3	2.5618	2.3699	**0.0754**	None
Spatial impression	43.7697	3	14.5899	13.4332	0	1&3, 2&3, 4&3

(continued)

E. Pfanzagl-Cardone, *The Art and Science of Surround and Stereo Recording*,
https://doi.org/10.1007/978-3-7091-4891-4

Table A.1 (continued)

'Dependent variable'	Sum of squares	df	Mean square	F	p	Stat. significant differences between mean-values of the surr.-mic ratings
Wet/dry balance	43.0233	3	14.3411	22.2303	0	1&3, 2&3, 4&3
Rear non-/intrusv.	4.0756	3	1.3585	0.7353	**0.5335**	None
Rear soft/loud	7.2100	3	2.4033	2.7409	0.0475	None

Rem.: 1 = OCT, 2 = DECCA, 3 = KFM, 4 = ABPC
Bold represents the results for attributes with a p-value $\geq$ 0.05 which have to be considered statistically non-significant

Table A.2 ORCH–Binaurallistening test; 1-way ANOVA concerning the influence of the variable 'microphone-array' on the rating of 14 attributes (10 listeners)

'Dependent variable'	Sum of squares	df	Mean square	F	p	Stat. significant differences between mean-values of the surr.-mic ratings
Difference to REF	31.3394	2	10.4465	5.6872	0.0027	1&3, 2&3, 4&3
Preference	31.4602	3	10.4867	5.5144	0.0032	3&4
Naturalness	35.4	3	11.8	10.5526	4.02E−05	1&3, 2&3, 4&3
Sound-colour	37.8214	3	12.6071	14.0776	3.14E−06	1&3, 2&3, 4&3
High/low balance	9.4583	3	3.1528	2.8363	**0.0517**	None
Localisation	23.7339	3	7.9113	3.5448	0.0239	2&4, 3&4
Balance	26.4394	3	8.8131	7.1988	6.61E−04	3&4
Width	29.2201	3	9.74	8.2485	2.63E−04	1&3, 2&3, 4&3
Distance	10.1352	3	3.3784	2.0932	**0.1183**	None
Stage depth	14.5589	3	4.853	3.251	0.0329	None
Stability	0.4451	3	0.1484	0.0775	**0.9718**	None
Differentiation	7.6208	3	2.5403	1.4245	**0.2516**	None
Spatial impression	29.0688	3	9.6896	11.5494	1.89E−05	1&3, 2&3, 4&3
Wet/dry balance	15.7089	3	5.2363	6.9765	8.08E−04	1&3, 2&3

Rem.: 1 = OCT, 2 = DECCA, 3 = KFM, 4 = ABPC
Bold represents the results for attributes with a p-value $\geq$ 0.05 which have to be considered statistically non-significant

Table A.3 'ORCH 5.1' listener ratings—Correlation-matrix for 15 attributes (Pearson product moment corr. coeff.)

	Preference	Naturalness	Colour	Hi/low Bal.	Localisation	Balance (volume)	Width	Distance
Preference	1.0000	0.8383	0.6712	0.0238	0.5875	0.6879	0.6289	−0.0964
Naturalness	0.8383	1.0000	0.7539	−0.0377	0.5967	0.5964	0.5010	−0.0530
Sound-colour	0.6712	0.7539	1.0000	0.0935	0.5113	0.5178	0.4978	0.0392
High/low bal	0.0238	−0.0377	0.0935	1.0000	0.2035	−0.064 5	−0.0054	0.2603
Localisation	0.5875	0.5967	0.5113	0.2035	1.0000	0.5292	0.4410	0.1248
Balance	0.6879	0.5964	0.5178	−0.0645	0.5292	1.0000	0.4748	−0.0219
Width	0.6289	0.5010	0.4978	−0.0054	0.4410	0.4748	1.0000	−0.2470
Distance	−0. 0964	−0.0530	0.0392	0.2603	0.1248	−0.0219	−0.2470	1.0000
Depth	0.4858	0.3996	0.4308	0.1067	0.3497	0.3813	0.4637	−0.2565
Stability	0.5407	0.6025	0.4845	−0.0402	0.4676	0.5312	0.3021	0.0675
Differentiation	0.5011	0.5017	0.4772	0.1603	0.5234	0.5742	0.2438	0.1955
Spatial Impression	0.6829	0.6398	0.5724	0.0083	0.4336	0.4867	0.6522	−0.0901
Wet/dry bal	0.4288	0.2912	0.2808	0.0112	0.1330	0.2790	0.6483	−0.4005
RearCh. non/intrusive	0.2988	0.3512	0.3836	0.1406	0.2875	0.2021	0.2987	0.0471
RearCh. soft/loud	−0.0069	0.0827	0.1738	0.0777	0.0718	0.1135	0.0220	0.0777

	Depth	Image Stability	Differentiation	Spatial Impression	Wet/dry bal	Rear non/intrusive	Rear soft/loud
Preference	0.4858	0.5407	0.5011	0.6829	0.4288	0.2988	−0.0069
Naturalness	0.3996	0.6025	0.5017	0.6398	0.2912	0.3512	0.0827
Sound-colour	0.4308	0.4845	0.4772	0.5724	0.2808	0.3836	0.1738
High/low bal	0.1067	−0.0402	0.1603	0.0083	0.0112	0.1406	0.0777
Localisation	0.3497	0.4676	0.5234	0.4336	0.1330	0.2875	0.0718

(continued)

Table A.3 (continued)

	Depth	Image Stability	Differentiation	Spatial Impression	Wet/dry bal	Rear non/intrusive	Rear soft/loud
Balance	0.3813	0.5312	0.5742	0.4867	0.2790	0.2021	0.1135
Width	0.4637	0.3021	0.2438	0.6522	0.6483	0.2987	0.0220
Distance	−0.2565	0.0675	0.1955	−0.0901	−0.4005	0.0471	0.0777
Depth	1.0000	0.2660	0.3283	0.4892	0.5055	0.2367	−0.0427
Stability	0.2660	1.0000	0.6547	0.2910	0.0415	0.2213	0.0217
Differentiation	0.3283	0.6547	1.0000	0.2661	0.0052	0.2588	0.1391
Spatial Impression	0.4892	0.2910	0.2661	1.0000	0.5099	0.3826	0.0632
Wet/dry bal	0.5055	0.0415	0.0052	0.5099	1.0000	0.1328	0.0108
RearCh. non/intrusive	0.2367	0.2213	0.2588	0.3826	0.1328	1.0000	0.5092
RearCh. soft/loud	−0.0427	0.0217	0.1391	0.0632	0.0108	0.5092	1.0000

Appendix B
Reference-Email No. 1 Concerning OCT-Surround (by Cornelius van der Gragt)

Von: sursound-bounces@music.vt.edu im Auftrag von ThomasChen@aol.com
Gesendet: Donnerstag, 07. Februar 2008 15:53
An: sursound@music.vt.edu
Betreff: Re: [Sursound] Multichannel rec techniques

I have used the OCT array however the hypercardoids are about 1 m apart and instead of a cardoid in the center I use M/S Schoeps. I find it gives a better impression of size in the recording.

Also there is less of the shift between speakers when listening

Thomas Chen

In a message dated 2/6/2008 3:32:57 A.M. Pacific Standard Time, cvdgragt@koncon.nl writes:

Hello List,

Indeed, from the recordings that my students and I made the last few years, we consider OCT to be one of the valuable tools for 5.0 recordings of classical music and larger musical formations (in our case symphonic, and wind orchestras, organs, choirs, etc.). However, when using it we should bear in mind that:

1. The supercardioids involved, which are responsible for the very good separation between the L and R signals, must have excellent off-axis frequency responses, normally not the case with directional mics. For this reason and specially for the OCT configuration, Schoeps developed the capsule MK41V.
2. the system only works well when the low frequency roll-off of supercardioids (60% pressure gradient driven) is compensated by 2 omnis that must be low-passed at, say, 100 Hz.
3. Never use the proposed backwards-directed 2 cardioids at about 1 m from the OCT support. The crosstalk from the direct signal ruins the result.
   We had good results with a Hamasaki Square at considerable distance or simply a backwards directed 'traditional' X/Y(90) pair. Again, positioned at considerable distance, the latter with its more than 90° coverage angle, very smoothly captures the ambient signals at the back of a hall

E. Pfanzagl-Cardone, *The Art and Science of Surround and Stereo Recording*,
https://doi.org/10.1007/978-3-7091-4891-4

cheers
Cornelis H. Van der Gragt
Consultant for Acoustics, Audio and Music Recording
Jan Steenlaan 8
NL-1213 EL Hilversum
phone: int 31 35 624 61 64 (home)
mobile: int 31 62 551 77 23
E-mail: Cornelis van der Gragt <cvdgragt@hotmail.com>
or:
Royal Conservatoire
Music Recording Dept./Art-of-Sound Program
Nl-2595 CA The Hague
phone: int 31 35 624 61 64 (home)
mobile: int 31 62 551 77 23
E-mail: Cornelis van der Gragt <cvdgragt@koncon.nl>

Reference-Email No. 2 Concerning OCT-Surround (by Jan Korte)

Von: sursound-bounces@music.vt.edu im Auftrag von Jan Ola Korte [mail@jan-korte.de]
Gesendet: Mittwoch, 06. Februar 2008 20:04
An: Surround Sound discussion group
Betreff: Re: [Sursound] Multichannel rec techniques

Hello,
Am 06.02.2008 um 12:31 schrieb Cornelis van der Gragt:

> 3. Never use the proposed backwards-directed 2 cardioids at about 1
> meter
> from the OCT support. The crosstalk from the direct signal ruins
> the result.

I used this setup recording a symphonic orchestra with choir in a big shoebox-hall and can absolutely confirm that. As the recording was part of a student paper about recording techniques for 3/2-Surround I had some different setups to compare. Setups that delivered best performance for the rear speakers were definitely the ones placed in the diffuse field a couple of meters from the reverberation radius.

The one I considered best were two fig-of-eights facing the side walls with a base of 2 m and about 6 m from the reverberation radius (kind of a half Hamasaki-Square).

Best,
Jan
Sursound mailing list
Sursound@music.vt.edu https://mail.music.vt.edu/mailman/listinfo/sursound

Appendix C
The MAAT "2BC multiCORR" Cross-Correlation Meter Plug-In

This DAW plug-In was released in April 2019 and essentially delivers a realtime-display of what the MATLAB function "mcorrfrequ48k" (see next section below) does offline (for any sample of appropriate length): to calculate the frequency dependent cross-correlation coefficient for the two channels of a stereo-signal.

It is intended mainly as a monitoring / visual-display tool for mastering purposes and features 31 bands of correlation with carefully chosen center frequencies, an exclusive 'trueLinear™' scale for higher precision with visuals, which—as the programmers claim—better match psychoacoustic perception than the traditional 'cosine'-scale, as used with old analogue meters, and in addition it also offers a stereo-balance meter (Fig. A.1).

E. Pfanzagl-Cardone, *The Art and Science of Surround and Stereo Recording*,
https://doi.org/10.1007/978-3-7091-4891-4

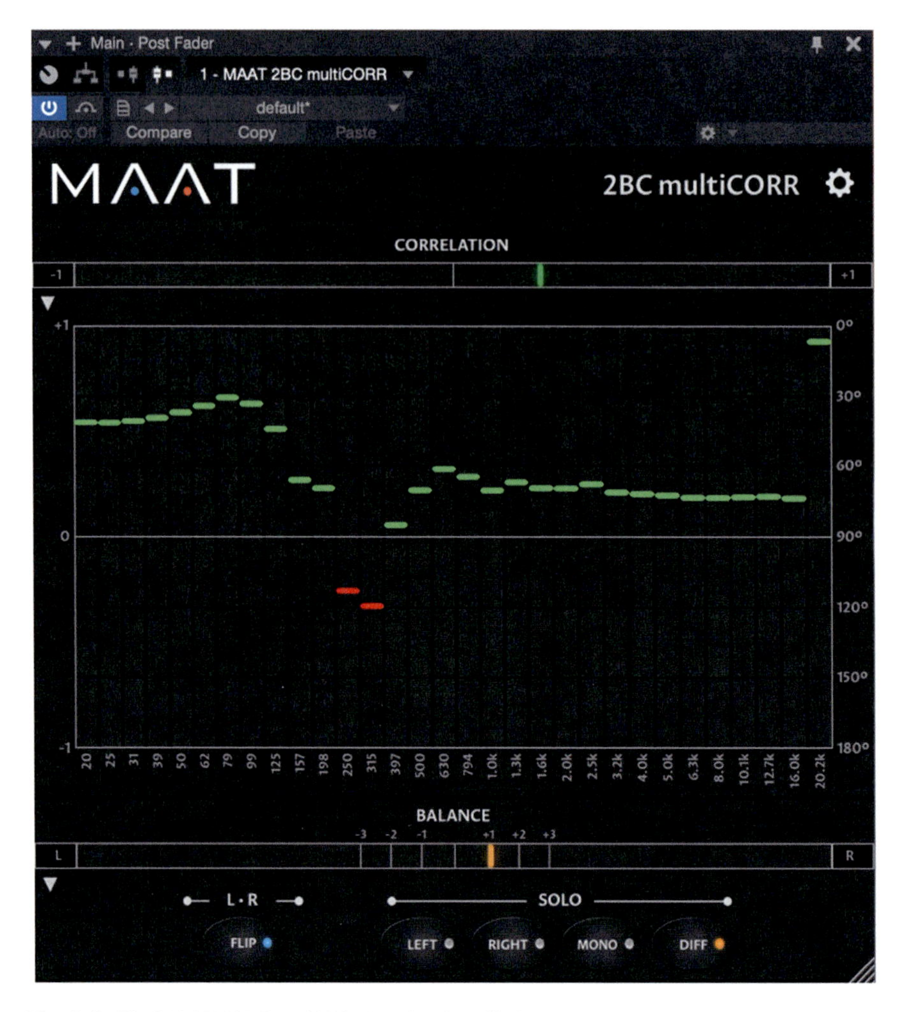

Fig. A.1 The MAAT "2BC multiCORR" plug-in, offering Top-to-bottom: (1) a correlation meter (overall-signal 20Hz-20kHz), (2) frequency dependent cross-correlation readout of the stereo signal split into 31 frequency bands, (3) a "Stereo-Balance" Meter (graphic from:https://www.maat.digital/2bcmulticorr/ accessed May 24th, 2019)

MATLAB- programming code:

The following is the MATLAB-code of the function "mcorrfrequ48k" which calculates the cross-correlation coefficient over frequency for two samples with 48 kHz sample rate. It does so by first splitting the L and R channel signals into 31 filter bands (tuned to ISO center-frequencies) and then applying a cross-correlation function.

Toward the end of the code, also the correlation coefficient for a low frequency band up to 400 Hz is calculated, as well as the correlation coefficient for the

high frequency band from 1200 Hz upwards, as well as the correlation value of the full-bandwidth signals of the L and R channel. In addition to the numeric read-out of the cross-correlation values (rho), also a figure of correlation over frequency is plotted.

This is the code from which many of the figures in Chaps.7 and 8 have been generated. (code written and © by Edwin Pfanzagl-Cardone)

```
function mcorrfreq48k(wave1, wave2)

% Bandfilter Band #1:
[b,a]=cheby1(2,1,[19/24000 22/24000]);
Hd=dfilt.df2t(b,a);
filtwaveL=filter(Hd,wave1);
filtwaveR=filter(Hd,wave2);
rho1=corr(filtwaveL,filtwaveR);

% Bandfilter Band #2:
[b,a]=cheby1(2,1,[24/240000 27/24000]);
Hd=dfilt.df2t(b,a);
filtwaveL=filter(Hd,wave1);
filtwaveR=filter(Hd,wave2);
rho2=corr(filtwaveL,filtwaveR);

% Bandfilter Band #3:
[b,a]=cheby1(2,1,[30/24000 34/24000]);
Hd=dfilt.df2t(b,a);
filtwaveL=filter(Hd,wave1);
filtwaveR=filter(Hd,wave2);
rho3= corr(filtwaveL,filtwaveR);

% Bandfilter Band #4:
[b,a]=cheby1(2,1,[37/24000 43/24000]);
Hd=dfilt.df2t(b,a);
filtwaveL=filter(Hd,wave1);
filtwaveR=filter(Hd,wave2);
rho4= corr(filtwaveL,filtwaveR);

% Bandfilter Band #5:
[b,a]=cheby1(2,1,[47/24000 54/24000]);
Hd=dfilt.df2t(b,a);
filtwaveL=filter(Hd,wave1);
filtwaveR=filter(Hd,wave2);
rho5= corr(filtwaveL,filtwaveR);

% Bandfilter Band #6:
[b,a]=cheby1(2,1,[59/24000 68/24000]);
Hd=dfilt.df2t(b,a);
filtwaveL=filter(Hd,wave1);
```

```
filtwaveR=filter(Hd,wave2);
rho6= corr(filtwaveL,filtwaveR);

% Bandfilter Band #7:
[b,a]=cheby1(3,1,[75/24000 86/24000]);
Hd=dfilt.df2t(b,a);
filtwaveL=filter(Hd,wave1);
filtwaveR=filter(Hd,wave2);
rho7= corr(filtwaveL,filtwaveR);

% Bandfilter Band #8:
[b,a]=cheby1(3,1,[94/24000 107/24000]);
Hd=dfilt.df2t(b,a);
filtwaveL=filter(Hd,wave1);
filtwaveR=filter(Hd,wave2);
rho8= corr(filtwaveL,filtwaveR);

% Bandfilter Band #9:
[b,a]=cheby1(3,1,[118/24000 136/24000]);
Hd=dfilt.df2t(b,a);
filtwaveL=filter(Hd,wave1);
filtwaveR=filter(Hd,wave2);
rho9= corr(filtwaveL,filtwaveR);

% Bandfilter Band #10:
[b,a]=cheby1(3,1,[150/24000 172/24000]);
Hd=dfilt.df2t(b,a);
filtwaveL=filter(Hd,wave1);
filtwaveR=filter(Hd,wave2);
rho10= corr(filtwaveL,filtwaveR);

% Bandfilter Band #11:
[b,a]=cheby1(3,1,[188/24000 215/24000]);
Hd=dfilt.df2t(b,a);
filtwaveL=filter(Hd,wave1);
filtwaveR=filter(Hd,wave2);
rho11= corr(filtwaveL,filtwaveR);

% Bandfilter Band #12:
[b,a]=cheby1(3,1,[235/24000 270/24000]);
Hd=dfilt.df2t(b,a);
filtwaveL=filter(Hd,wave1);
filtwaveR=filter(Hd,wave2);
rho12= corr(filtwaveL,filtwaveR);

% Bandfilter Band #13:
[b,a]=cheby1(3,1,[296/24000 340/24000]);
```

```
Hd=dfilt.df2t(b,a);
filtwaveL=filter(Hd,wave1);
filtwaveR=filter(Hd,wave2);
rho13= corr(filtwaveL,filtwaveR);

% Bandfilter Band #14:
[b,a]=cheby1(3,1,[374/24000 430/24000]);
Hd=dfilt.df2t(b,a);
filtwaveL=filter(Hd,wave1);
filtwaveR=filter(Hd,wave2);
rho14= corr(filtwaveL,filtwaveR);

% Bandfilter Band #15:
[b,a]=cheby1(3,1,[470/24000 539/24000]);
Hd=dfilt.df2t(b,a);
filtwaveL=filter(Hd,wave1);
filtwaveR=filter(Hd,wave2);
rho15= corr(filtwaveL,filtwaveR);

% Bandfilter Band #16:
[b,a]=cheby1(3,1,[591/24000 681/24000]);
Hd=dfilt.df2t(b,a);
filtwaveL=filter(Hd,wave1);
filtwaveR=filter(Hd,wave2);
rho16= corr(filtwaveL,filtwaveR);

% Bandfilter Band #17:
[b,a]=cheby1(3,1,[749/24000 860/24000]);
Hd=dfilt.df2t(b,a);
filtwaveL=filter(Hd,wave1);
filtwaveR=filter(Hd,wave2);
rho17= corr(filtwaveL,filtwaveR);

% Bandfilter Band #18:
[b,a]=cheby1(3,1,[940/24000 1075/24000]);
Hd=dfilt.df2t(b,a);
filtwaveL=filter(Hd,wave1);
filtwaveR=filter(Hd,wave2);
rho18= corr(filtwaveL,filtwaveR);

% Bandfilter Band #19:
[b,a]=cheby1(3,1,[1175/24000 1355/24000]);
Hd=dfilt.df2t(b,a);
filtwaveL=filter(Hd,wave1);
filtwaveR=filter(Hd,wave2);
rho19= corr(filtwaveL,filtwaveR);
```

```
% Bandfilter Band #20:
[b,a]=cheby1(3,1,[1495/24000 1720/24000]);
Hd=dfilt.df2t(b,a);
filtwaveL=filter(Hd,wave1);
filtwaveR=filter(Hd,wave2);
rho20= corr(filtwaveL,filtwaveR);

% Bandfilter Band #21:
[b,a]=cheby1(3,1,[1880/24000 2150/24000]);
Hd=dfilt.df2t(b,a);
filtwaveL=filter(Hd,wave1);
filtwaveR=filter(Hd,wave2);
rho21= corr(filtwaveL,filtwaveR);

% Bandfilter Band #22:
[b,a]=cheby1(3,1,[2350/24000 2695/24000]);
Hd=dfilt.df2t(b,a);
filtwaveL=filter(Hd,wave1);
filtwaveR=filter(Hd,wave2);
rho22= corr(filtwaveL,filtwaveR);

% Bandfilter Band #23:
[b,a]=cheby1(3,1,[2955/24000 3405/24000]);
Hd=dfilt.df2t(b,a);
filtwaveL=filter(Hd,wave1);
filtwaveR=filter(Hd,wave2);
rho23= corr(filtwaveL,filtwaveR);

% Bandfilter Band #24:
[b,a]=cheby1(3,1,[3745/24000 4300/24000]);
Hd=dfilt.df2t(b,a);
filtwaveL=filter(Hd,wave1);
filtwaveR=filter(Hd,wave2);
rho24= corr(filtwaveL,filtwaveR);

% Bandfilter Band #25:
[b,a]=cheby1(3,1,[4700/24000 5390/24000]);
Hd=dfilt.df2t(b,a);
filtwaveL=filter(Hd,wave1);
filtwaveR=filter(Hd,wave2);
rho25= corr(filtwaveL,filtwaveR);

% Bandfilter Band #26:
[b,a]=cheby1(3,1,[5910/24000 6810/24000]);
Hd=dfilt.df2t(b,a);
filtwaveL=filter(Hd,wave1);
filtwaveR=filter(Hd,wave2);
```

```
rho26= corr(filtwaveL,filtwaveR);

% Bandfilter Band #27:
[b,a]=cheby1(3,1,[7490/24000 8600/24000]);
Hd=dfilt.df2t(b,a);
filtwaveL=filter(Hd,wave1);
filtwaveR=filter(Hd,wave2);
rho27= corr(filtwaveL,filtwaveR);

% Bandfilter Band #28:
[b,a]=cheby1(3,1,[9400/24000 10750/24000]);
Hd=dfilt.df2t(b,a);
filtwaveL=filter(Hd,wave1);
filtwaveR=filter(Hd,wave2);
rho28= corr(filtwaveL,filtwaveR);

% Bandfilter Band #29:
[b,a]=cheby1(3,1,[11750/24000 13550/24000]);
Hd=dfilt.df2t(b,a);
filtwaveL=filter(Hd,wave1);
filtwaveR=filter(Hd,wave2);
rho29= corr(filtwaveL,filtwaveR);

% Bandfilter Band #30:
[b,a]=cheby1(3,1,[14950/24000 17200/24000]);
Hd=dfilt.df2t(b,a);
filtwaveL=filter(Hd,wave1);
filtwaveR=filter(Hd,wave2);
rho30= corr(filtwaveL,filtwaveR);

% Bandfilter Band #31:
[b,a]=cheby1(3,1,[18800/24000 21500/24000]);
Hd=dfilt.df2t(b,a);
filtwaveL=filter(Hd,wave1);
filtwaveR=filter(Hd,wave2);
rho31= corr(filtwaveL,filtwaveR);

% Low-Pass Filter Band #LF-Corr:
[b,a]=cheby1(6,1,400/24000,'low');
Hd=dfilt.df2t(b,a);
filtwaveL=filter(Hd,wave1);
filtwaveR=filter(Hd,wave2);
rhoLF = corr(filtwaveL, filtwaveR);

% High-Pass Filter Band #HF-Corr:
[b,a]=cheby1(6,1,1200/24000,'high');
Hd=dfilt.df2t(b,a);
```

```
filtwaveL=filter(Hd,wave1);
filtwaveR=filter(Hd,wave2);
rhoHF= corr(filtwaveL,filtwaveR);

% Correlation of full-bandwidth signal:
rho= corr(wave1,wave2);

% Vector of correlation values of the ISO frequency-bands:
RHO=[rho1 rho2 rho3 rho4 rho5 rho6 rho7 rho8 rho9 rho10 rho11 rho12 rho13
rho14 rho15 rho16 rho17 rho18 rho19 rho20 rho21 rho22 rho23 rho24 rho25
rho26 rho27 rho28 rho29 rho30 rho31]

% display rho-values:
fprintf('rho-value for LF (fc=400Hz) and HF (fc=1200Hz), as well as correlation-
value of the full-bandwidth signal')
rhoLF=[rhoLF]
rhoHF=[rhoHF]
rho=[rho]

ISO31=[0.020 0.025 0.032 0.040 0.050 0.063 0.080 0.100 0.125 0.160 0.200
0.250 0.315 0.400 0.500 0.630 0.800 1.000 1.250 1.600 2.000 2.500 3.150 4.000
5.000 6.300 8.000 10.000 12.500 16.000 20.000];

axis([0.020 20.000 -1 1]);
plot(ISO31, RHO)
xlabel ('Frequency [kHz]')
ylabel ('Correlation')
```

Following is the MATLAB-codeof the function "mbinauralqualindx" which calculates the cross-correlation coefficient over frequency for two samples with 44.1 kHz sample rate. It does so by first splitting the L and R channel signals into 3 octave filter-bands (with centre frequencies at 500, 1000 and 2000 Hz) and then applying a cross-correlation function.

The BQI is calculated as: BQI=(1-IACC_e3) [e3 being the mean value of the cross-correlation values found for the frequency bands of 500, 1000 and 2000Hz]

At the end, the numeric values of the cross-correlation (rho) for the octave bands is read-out, as well as the value for BQI.

(code written and © by Edwin Pfanzagl-Cardone)

```
function mbinauralqualindx(wave1, wave2)

% Filter-band #1:
[b,a]=cheby1(3,1,[353/22050 707/22050]);
Hd=dfilt.df2t(b,a);
filtwaveL=filter(Hd,wave1);
filtwaveR=filter(Hd,wave2);
rho1=corr(filtwaveL,filtwaveR);
```

```
% Filter-Band #2:
[b,a]=cheby1(3,1,[707/22050 1414/22050]);
Hd=dfilt.df2t(b,a);
filtwaveL=filter(Hd,wave1);
filtwaveR=filter(Hd,wave2);
rho2=corr(filtwaveL,filtwaveR);

% Filter-Band #3:
[b,a]=cheby1(3,1,[1414/22050 2828/22050]);
Hd=dfilt.df2t(b,a);
filtwaveL=filter(Hd,wave1);
filtwaveR=filter(Hd,wave2);
rho3= corr(filtwaveL,filtwaveR);

% Binaural Quality Index BQI=(1-IACC_e3):
bqi= (1-(rho1+rho2+rho3)/3);

% Vector of the single values of the 3 frequency bands:
RHO=[rho1 rho2 rho3]

% show rho-values:
fprintf('rho-values for the octave-bands 500, 1000, 2000Hz and BQI')
rho500=[rho1]
rho1000=[rho2]
rho2000=[rho3]
BQI=[bqi]
```

Index

E. Pfanzagl-Cardone, *The Art and Science of Surround and Stereo Recording*,
https://doi.org/10.1007/978-3-7091-4891-4

N

O

P

Printed in the United States
By Bookmasters